The BODY

The BODY

Anthony Smith

VIKING

VIKING
Viking Penguin Inc.
40 West 23rd Street,
New York, New York 10010, U.S.A.

First American Edition
Published in 1986

LIBRARY OF CONGRESS CATALOGING IN PUBLICATION DATA
Smith, Anthony, 1926–
   The body.
   Includes index.
   1. Body, Human.   2. Human physiology.   3. Human
reproduction.   I. Title.
QP38.S67   1986   612   85-40556
ISBN 0-670-80846-6

Printed in the United States of America by
R. R. Donnelley & Sons Company, Harrisonburg, Virginia
Set in Times Roman

For Anne

Contents

The Body

The Body

The BODY

Foreword

Strict scientific writing requires wholesale qualification, and no generalisations are acceptable; every statement is therefore beset with qualifying appendages. General scientific writing demands generalisations, and most of the qualifying precision has to go by the board. It cannot be otherwise. Without such a seeming disregard for the truth a contentious nightmare soon results: human beings produce babies; female human beings produce babies; female human beings aged between 15 and 50 produce most babies, most female human beings aged between 15 and 50 can and do produce most babies, assuming fertility, prowess and desire from their mates; most female human beings aged between 15 and 50, although both younger and older ages have been plentifully recorded in the literature, can be fertilised and most of these women will successfully produce babies, although a number of the offspring will be stillborn, this number varying according to race, area, age. ... The nightmare is readily accessible, and has to be avoided without falling foul of gross over-simplification.

Science has precise meaning for most of its words, even such ordinary words as average, mean, median, normal and mode. To follow such precision into the more casual everyday world, which happily blurs their respective meanings, can be disastrous. Strictly, and on average, human beings possess less than two ears. The actual number of ears per person, bearing in mind those born without ears and those who have earlessness thrust upon them, must be a figure like 1·9999. Similarly, science may contend that an average family has 2·4 children. Plainly, absolute accuracy cannot be the sole guide.

Common meanings can be extremely troublesome. Man is a mammal. Mammals are animals. Man is not an animal. The word man is often awkward. Man stands upright on two legs. Man cooks food and hunts. Man has babies. Even human temperature is no one thing. 'Normal' is usually 98·4 in Britain, 98·6 in the United States, and 37 on the continent of Europe. When does an embryo become a foetus? Some say at eight weeks, but others hate the words and talk only of the products of conception. A mother uses no such phrase and refers to her baby even before it has been born. Early attempts to respect current usage and common understanding have led, on occasion after occasion, to my own prejudices riding determinedly and rough-shod over the opinions of others.

Perinatal mortality, infant mortality and neonatal mortality all mean dead babies, but a neonate, for example, is a baby less than four weeks old whereas an infant is less than one year old. Most people do not talk of infants and certainly not of neonates; instead they talk only of babies, and let them become children at some indefinite age. However, the words baby and child are almost as meaningless in their imprecision to some people as the word neonate is as meaningless to others. Both sides have to be respected, and babies have to be qualified.

Even the word England demands extreme caution. It ought to be feasible to use the word, and to refer to the English, but facts solely from the whole of England are very rare. More commonly they come from England and Wales, or from England, Wales, Scotland and Northern Ireland. Very rarely do facts relate to the entire British Isles. The Americans can also be confusing (although they never match the British in such inanity) by having to specify whether they mean the contiguous states or the United States.

Science is written (usually) in grams, cubic centimetres, litres and degrees Centigrade. English-speaking people, despite pressures from decimal and metric enthusiasts, still speak of ounces, cubic inches, pints and Fahrenheit. To add to the problem, English-speaking people from Britain have a different pint from the American one, and are often happy to give their weight in stones, a procedure which confounds most Americans. I have a further conviction that most people are wonderfully vague about nearly all weights and measures. An 8-week embryo weighs about one-fifth of an ounce, or some 6 grams. Does either weight really indicate anything other than extreme lightness? At eight weeks a mother's uterus contains one ounce of amniotic fluid. What is that? A thimbleful or an egg-cupful or a tea-cupful? Therefore, whenever I have felt the precise units of measurement to be particularly meaningless, I have tried other methods. An 8-week embryo, for example, has to be 600 times heavier to be an average baby seven months later. There is a danger that such circumlocution can annoy (and I was once extremely irritated to be told that a missile's nose cone flying through the air was as hot as Regulo 7 on my oven) but the intention is to clarify, however unsatisfactory the result may be.

If any of these personal prejudices, mannerisms and preferences fail to achieve their object of making more sense out of the subject matter I apologise here and now for the irritation which they will surely cause.

2

A final point is that this book has had to be extensively revised since it was first written in the mid-1960's. It might be thought – indeed it was fondly thought by the author – that a subject like the human body would never need revision. This has proved not to be the case and amendments, adjustments and even wholesale re-writings have had to be inserted into this brand new edition. Certain changes were to be expected, such as neonatal mortality, maternal mortality, cot-deaths, population statistics. Others were known to have occurred, such as the altered attitude to abortion, contraceptive preferences, and hospital birth versus birth at home. But alterations were necessary seemingly at every level. 'The last war' is now not automatically assumed to be World War Two. Negroes have become blacks – at their instigation. Women have entered the scene, and the common practice twenty years ago was to assume female parents and therefore male offspring. (The child-care Dr Spock also used this simple rule – until forced by a changing world to admit that parents are of two sexes and so are babies.) The original version took eighteen months to write. This edition demanded a further six months merely in adjusting the 1960's to the 1980's.

1. Introduction

It is highly dishonourable for a reasonable soul to live in so divinely built a mansion as the body she resides in altogether unacquainted with the exquisite structure of it.

ROBERT BOYLE

Before I came here I was confused about this subject. Having listened to your lecture I am still confused. But on a higher level.

ENRICO FERMI

'These, gentlemen, are the opinions on which I base my facts.' No one is certain who originally wrote this text but, without doubt, it should be written again and again, and especially at the outset of any work which professes to be objective. Facts do indeed speak through an interpreter, and are subject to his wishes. The facts in this book, culled from a bewilderment of sources, have been influenced by three major contentions.

The first is that mankind is no isolated being. Admittedly it is well over a hundred years since the modern theory of evolution was first propounded, and few people today accept the simplicity of Adam and Eve's creation, but the human species is still considered most distinct. There is mankind on one side and animal life on the other; there is *Homo sapiens*, and there are all the beasts of the field and birds of the air. Given any opportunity I have been happy to break down this barrier. Man is just another bit of biology, and therefore much of the animal kingdom is entirely relevant to his body, to his mechanism of sperm transfer, to his sex ratio, to his brain, to his sense of smell. He is not an isolated piece of creation, like a meteorite which has flown in from somewhere else. He is a part of terrestrial evolution, and facts about animals can make sense of facts about him. To my mind the menstrual cycle (which affects women and apes) has to be compared with the oestrous cycle (which affects most other mammals). Even so, this is not a book about animals. It is a book about the human body, and animals are used to lend emphasis, to clarify and to illuminate whenever they can.

My second main contention concerns the collection of facts. I have used an extremely wide net in the belief – and hope – that most of the subjects will benefit. It is possible to deal with

4

haemophilia in a strictly biochemical fashion, and discuss the missing clumping factor most academically, but it is preferable – according to this second contention – to describe Queen Victoria's knowledge of this affliction as well, and to show its ramification throughout her family. There is the straightforward human stomach, and there is also the remarkable tale of the French-Canadian whose stomach was permanently open to view. There are twins, and there are also cases of twins where each of the pair is the offspring of a different father. There are forceps deliveries, and there is the roguish and unhippocratic story of their development. Behind every aspect of the body there are relevant but more indirect facts which can be most enlightening. What about inbreeding, and the famous men who have been products of cousin–cousin marriages? Or about speaking, and those who – like Einstein – wait until four before saying anything? The stories are best told, according to this contention, by mixing the direct with the indirect.

Thirdly, because medical men are primarily interested in medicine, in the curing of ailments and the preservation of health, most medical books concentrate on disease and on breakdown. This book does not totally disregard such failure, but regards a pumping heart as more interesting than a heart which is failing to pump. Sometimes the manner of a breakdown can be exceptionally revealing on its own, just as haemophilia has much to say about inheritance; but, in general, function is given greater space than malfunction. At the same time the classical precision of medical jargon, quite another language to most of us, is given a miss if there is a convenient way of circumnavigating it.

These, then, are the three principal contentions. Each one of them is relevant to every chapter of the book, but on page 39 is a further introduction called 'An Apologia for Reproduction' to explain why some two-thirds of this book is concerned with procreation. In between these two introductions are three chapters. Dealing with 'Evolution', 'Race' and 'Population', they are a rapid attempt to set the scene. There are some 4,700,000,000 people living today of the species *Homo sapiens* whose body this book is about (and there were 3,300,000,000 when this book was first published in 1968). They are all products of evolution, as is every other living creature, and the first chapter (the most hasty of all) runs through the biological events of the geological epochs until, in this Quaternary Age, mankind dominates the scene. The second chapter discusses our intra-specific differences for, most blatantly, we are not born equal; and the third chapter enlarges upon our actual numbers, what they used to be and what they will so soon become. Anyone alarmed by the magnitude of this book, and wishing to skip a section, could well miss out these first three chapters. Then, much like the sudden presence in that garden of Adam and Eve, the chapters on the male and the female will be abruptly encountered.

When the gamut of reproduction has finally been played, leading from fertility through growth and eventually to senility and death, the subsequent chapters (starting on page 336) are concerned with those bodily systems that cannot be brought under the all-embracing title of reproduction. Finally (on page 513) is a chapter on radiation. There is a factor 'L' used by certain scientists to indicate the length of time after a civilisation has acquired the means of self-annihilation before it uses those means. Happily, mankind is still living, but the subject of radiation is highly germane to our well-being, whether the particles come in small or large doses, and is therefore included. A postscript and an index conclude the book.

2. Evolution

The origin of life/The fossil record/Mammals/Primates/Man

'If we could first know where we are and whither we are tending, we could better judge what to do and how to do it.'

ABRAHAM LINCOLN

When Charles Darwin, chivvied into print by similar thinking from Alfred Russel Wallace, and championed afterwards so forcibly by Thomas Henry Huxley, finally caused the world to think about evolution, the world was part bewildered, part horrified by the idea. Within a few decades the horror diminished, mainly because there was reconciliation with the contradictory story in Genesis; but the bewilderment still thrived. However many facts of evolution are absorbed, however many fossils are seen, strata examined, books read, and whatever evidence is accumulated, the immensity and scale of the evolutionary epochs remain as lengthy and vast as ever. The blindness of evolution and its lack of purpose, coupled with the constant change and the eternal urgency of life, all add to the incomprehension. It has gone on for an immensity of time, and it is happening now as it was in the beginning. It is small wonder that Darwin shook the world.

The word 'million' is partly to blame, for all the epochs, eras and geological periods are numbered in millions of years. The figure is difficult to absorb. A day, for instance, rushes by and is rapidly replaced by another. We live many days. We peer into history and feel a great sense of time, of passed days, passed years. The Elizabethans lived in the remote past, and remoter still were men like the Normans, or King Alfred, or Charlemagne, or the Emperor Hadrian, or Christ. Yet Christ, who is so deeply embedded in the past, lived only some 700,000 days ago. To go back a million days means going back to the eighth century BC, before even the Greeks and the Romans had started on their glory and their grandeur. A million days is an eternity of time. The evolutionary era is a matter of thousands of millions of *years*.

The fossil record only starts being clear in the Cambrian rocks. These extremely ancient deposits were being laid down at the beginning of the Palaeozoic era, some 600,000,000 years ago. Yet the living creatures of that time, incorporated within those rocks

7

as fossils, include many forms of crustacean, of mollusc, of worms and of echinoderms – like the starfish.

Such animals, although primitive from our standpoint, are highly advanced and specialised if viewed from the time when life began as some distant protoplasmic form. The earliest recorded fossils are about five times more ancient. Some have been observed in a kind of Australian quartz, a substance thought to be 2,700,000,000 years old. Others, notably bacteria and allegedly even older, have been found in South African sediments called black chert; their age is reputed to be 3,000,000,000 years. Assuming accuracy in dating these sediments, and bearing in mind that these bacteria were found associated with some threads of organic material, the start of life was presumably distinctly earlier. And the start of this planet, whose turbulent birth had to settle down before life of any kind was possible, was thought to have occurred perhaps five, perhaps 6,000,000,000 years ago. Life began sometime after the Earth itself had been formed, and sometime before those most antique of fossils had been laid down. In short, there has been life – and evolution – for much more than half of this planet's existence. Such a quantity of continuing biological change is both hard to visualise and to accept.

A refusal to accept the idea of evolution's blindness added to the turmoil surrounding Darwin. Genesis is clear. Its story of creation is simple. Besides, it was all over in a week. There was a purpose, the making of night and day, the separation of sea and land, the creation of fish, flesh, fowl, and finally Man. Evolution has no purpose. The predator seeking out the prey has no greater objective than the grass growing, the rain falling. If there seems to be a purpose in evolution, said Sir Julian Huxley, it is only an apparent purpose. 'It is just as much a product of blind forces as is the falling of a stone to earth or the ebb and flow of the tides. It is we who have read purpose into evolution, as earlier men projected will and evolution into inorganic phenomena like storm or earthquake.'

He wrote that in 1942. Such positive and didactic clarity was not so easy in 1859, when Darwin's book *The Origin of Species* was published, or even in 1871 when he published *The Descent of Man*. Man's evolutionary origin had only been implied in the first book. It was asserted cogently in the second; hence the furore. Since then, religious beliefs and scientific opinion have managed to coexist more peacefully in most minds. Nevertheless this lack of evolutionary purpose, coupled with the eternity of its purposelessness, is a difficult pill for many to swallow. The most disquieting conclusion is that mankind, gifted and skilled, is the product of those innumerable millennia of random biological activity. Somehow Shakespeare arose from the primaeval ooze of this planet, from its mud, from the ebbing and flowing, from all those aeons of unrecorded time.

Dr G. H. Haggis, who has written and speculated about the origin of life, points out that the sudden appearance of a man in a world of one-celled organisms would be highly improbable, but his gradual evolution entirely conceivable. Similarly the sudden appearance of a cell in a world of small molecules would be improbable, but the evolution of small molecules to larger molecules, and for these to aggregate is conceivable.

Some experimentation along these lines is possible. If a mixture of the gases thought to have been present in the Earth's early atmosphere (hydrogen, methane, ammonia, water, carbon monoxide, hydrogen sulphide, etc.) are suitably encased, and if a spark – as with lightning – is made to pass through them for a time, a variety of large molecules will be formed. Some of these large molecules will even be amino-acids, the raw materials from which proteins are made. Even so, the presence of raw materials does not immediately lead to the existence of the finished product. Other conditions, and other mixtures, coupled with more energy from electrical storms, or from the sun, would have been necessary. Larger and larger molecules would then have been formed, and somehow the cell would have resulted from all this random activity. Quite how this happened naturally is a matter for speculation because too little is known about the cell, and how amino-acids and proteins were co-ordinated to form a living organism. Only when this mechanism is better understood will comprehension be possible on the origin of the living cell in the molecular bewilderment of the planet's early days. William Shakespeare did arise from the primaeval ooze – somehow.

A quaint paradox about those very early years concerns oxygen. This vital commodity has plainly been in existence at a fairly fixed level for many hundreds of millions of years. Life, in the form of most animals and plants, could not have been generated without a substantial quantity of oxygen in the atmosphere. Yet the origin of life would have been impossible, or so current theory has it, had oxygen been abundant. Had it existed the chemicals needed for incorporation in the infant organisms would have been promptly oxidised. No other planet in our solar system has our abundance of oxygen. Instead, as revealed in space exploration, some of them possess gas mixtures that may be highly relevant to the early days of the planet Earth. The other planets do not have a form of life upon them; but may have a form of chemistry germane to that existing here, say, 5,000,000,000 years ago.

The fossil record   Evolution starts becoming interesting after the Cambrian rocks had been laid down, a mere 600,000,000 years ago. The pre-Cambrian picture is extremely sketchy, and remarkably unclear. With the start of the Palaeozoic era and the Cambrian period, palaeontology becomes possible. (Geologists have divided time into three eras. Each of these is divided into periods, and each period then divided into epochs.) The Cambrian period lasted about 100,000,000 years. It was a time solely for invertebrates. Nothing existed which had a backbone, but practically every major invertebrate grouping, creatures with hard external skeletons to them such as the arthropods, or creatures with no skeletons at all, existed at this evolutionary dawn.

The vertebrates first came into being during the subsequent period, the Ordovician (roughly 60,000,000 years). These were, in general, part fish-like, part tank-like, and their armour plating looks highly impressive in their fossil forms. The true archaic fishes began in the next period, the Silurian (of 40,000,000 years), and at the same time the plants and the arthropods (the jointed-limbs – insects, spiders) invaded the land. Next came the Devonian period (of 50,000,000 years), the age of the fishes. These were abundant mainly in fresh water but also in salt water, and they lived with the first trees and the earliest amphibia. The first vertebrate invasion of the land was then occurring. The Carboniferous (80,000,000 years) saw a flowering of those trees into the coal forests, a great expansion of those amphibia, and also the first reptiles. The Permian (45,000,000 years) saw the reptiles take over from the amphibia. At this point in time geologists draw a line and call it the end of the Palaeozoic era. It totals 375,000,000 years, it lasts for over half of the clear fossil history of this planet, and it covers the change from the invertebrates to the reptiles. At the end of it, poised like some lizard all ready to go, the great reptilian empire is about to flourish.

The Mesozoic era lasted for the next 155,000,000 years. Its three periods are the Triassic (45,000,000 years), the Jurassic (45,000,000 years) and the Cretaceous (65,000,000 years). The whole era can be called the age of the reptiles. They flew in the air, swam in the sea, and either ran about on land or wallowed in its swamps. In the Jurassic age they were dominant. Great forms like *Diplodocus*, eighty feet from head to tail, gorged their way through the vegetation. At sea the icthyosaurs were powerful carnivores of the time, and in the air pterodactyls flew. Yet it was in the Jurassic age, with so much reptilian activity on every side, that the first mammals evolved about 180,000,000 years ago. Certainly not then the dominant species, their triumph did not begin until the reptiles were on the wane at the end of the Mesozoic era. Quite why the all-powerful reptilian world collapsed so dramatically is a matter of conjecture. Perhaps they were too specialised at a time of too rapid change. Perhaps the new cunning little mammals, warm blooded, furry, and producing living young, were able to supplant them from their comfortable niches in the ecological world.

The word dinosaur now implies an implacable and monstrous inability to change, and some defence of the reptiles is therefore necessary. Reptiles did flourish and dominate for 155,000,000 years. They still thrive as crocodiles, lizards, turtles, tortoises and snakes some 300,000,000 years after the first true reptiles made their appearance. The mammals have so far dominated for only 70,000,000 years, and man himself is far, far younger. He has lived as the genus *Homo* for perhaps 4,000,000 years, and has dominated as *Homo sapiens* for a hundred thousand. The reptile age of the Mesozoic was vast by comparison.

The third great geological era is the Cenozoic, the age of mammals. Unequally it is divided into the Tertiary period (of 70,000,000 years) and the Quaternary (of 1,000,000 or so years). The main subdivisions or epochs of this time are, in order of antiquity, Palaeocene, Eocene, Oligocene, Miocene, Pliocene, Pleistocene and Recent. The first five are major quantities of time, the latter two extremely short, with Recent being shortest of all. It is only in the Recent epoch of the Quaternary period of the Cenozoic era of geological time that modern man has made his mark.

Mammals There is a prime and basic difficulty about our classification of the mammal group. Their characteristics are distinct and well known – warm blood, hair (as against scale), lactation, care for the young and viviparity (or live births as against the production and hatching of eggs). But such characteristics are not those which lend themselves to fossilisation, and to easy recognition in the fragmentary and rocky bas-relief of the fossil form. Instead the palaeontologists have been forced to take more interest in the mammal's skeletal condition, such as the presence of several bones in the lower jaw (a reptilian characteristic) or only one (as in the mammals).

Nevertheless the picture is clear. The first mammals were small creatures, and mainly flesh-eaters that preyed only upon such modest meat as grubs, insects and worms. Much like modern shrews they lived privately for millions of years before the near extinction of the reptiles permitted their freedom and their ramification. Then, like the reptiles before them, they spread over the land (elephants, mice, lions, squirrels), into the sea (seals, whales, sea-cows), and into the air (bats). Out of this stock were to evolve the primates, yet another stage en route to man.

The primate group has its roots in the first days of the Cenozoic, a time when the last of the dinosaurs were dying out. Undoubtedly some of the very early shrew-like mammals living even earlier in the reptile age had taken to the trees. It is highly probable that these tree-living creatures gave rise to the primates, for such a life is entirely compatible with the characteristics which distinguish the primates from the bulk of the mammalian stock – the improvement of visual acuity, the deterioration of smell, the development of limbs for grasping, the fore-limbs for investigation, and an expansion of the brain for the agility and co-ordinated skills necessary for an active tree-top existence. Coupled with these characteristic features, and seemingly paradoxically, is a great lack of specialisation. On the ground the mammals tended to concentrate on running, on digging, on hopping, on one particular activity and on eating only a small range of foods – like the anteaters' ants or even the lions' raw meat. The early primates had simple teeth. They were capable of dealing with a wide selection of foods, and there was little specialisation in other parts of their body. They could survive in a wide variety of circumstances. Many mammals cannot do so, and perish when their precise habitat is suddenly destroyed. To adjust to a changing world is extremely important.

Broadly speaking (and here I am indebted to Professor Le Gros Clark's summary) the primates are distinguished by:

1 A generalised limb pattern with a primitive five-fingered and five-toed arrangement (contrast the hoofed mammals on two, or the horses on only a single digit per limb).
2 Free mobility of the digits. (Most primates have feet almost as skilled as their hands, although man does not.)
3 The replacement of claws by nails.
4 A shortening of the nose (with the loss of interest in smell, although primates which have taken to the ground once more tend to have developed long noses again, e.g. the baboons).
5 Good eyesight, with varying degrees of binocular vision (only possible when both eyes have moved to the front).

6   Teeth with a wide variety of tearing, biting and chewing possibilities. (The basic insectivore stock which gave rise to the primates had sharp teeth without the blunt cusps of the normal primate molars.)

7   A steady expansion and elaboration of the brain (mainly the cerebral cortex).

8   A small number of offspring – either twins or singletons, a long gestation, and a long period of nursing.

9   A habitat either within or very near the tropics. (Man is the great exception.)

Man   Unfortunately, the evolution of mankind is not a clear-cut course of events. The horse, for example, or the elephant tell a definite evolutionary story, well supported by a good fossil record; not so mankind. Instead about 1,000 fragments of fossil bone have been found that form part, somewhere or other, of the patchwork ancestry of the genus *Homo*. There is certainly not unanimous agreement about their piecing together. Inevitably much of the trouble stems from mankind's introspective devotion to his own past. Had the Piltdown fake, where the lower jaw (of an orang-utan) and a skull (of a man) were forcibly linked with each other, been of some other creature, the interest in the discovery would have been marginal, and the subsequent revelation of the fake would have been of equally slight concern. Also, wherever facts are few and evidence is lacking, theories multiply. Whole treatises have been prepared on a fragmentary find. Peking man, and a wealth of argument, was once based upon nothing more than a single tooth. (Most of the argument was entirely vindicated when more of this man came to light, the remains of fifteen skeletons being found between 1926 and 1941.)

It is very easy to distinguish man from his nearest living relatives, the anthropoid apes like the gorilla and the chimpanzee. It is far less easy to distinguish man's ancestors from the ancestors of those anthropoid apes. Briefly, and sweeping aside many contrary views, a sort of man emerged about 4,000,000 years ago, and modern man (*Homo sapiens*) appeared about 100,000 years ago. Nothing is firm about these assessments, and they change almost annually. Even so, by dinosaur standards, man's life on Earth is extremely short. There are only 200,000 generations at most between the New Yorker and those earliest *Homo* individuals. There are only 5,000 generations between today's men and the first men who could be called modern, the first representatives of the species *Homo sapiens*.

There is even less time, and not more than 2,000 generations, between the first cave-painters and us.

Coming down from the trees, and becoming terrestrial, was a hazardous undertaking. To move upright and rather slowly in a world of fast-moving herbivores and agile carnivores exposed all the weaknesses of the unspecialised ape, our ancestor; but the descent to ground level also taxed his latent ingenuities. Made possible by a versatility of hand and arm, and by many of the characteristics developed for the tree-tops, the new skills were to save him. He could invent things. He could manipulate objects to his advantage. He could fashion sticks and stones. He could provide himself with something more useful than a claw. The new terrestrial life meant a change to a more carnivorous existence, to hunting, to a greater need for tools and weapons. Finally, and with its ancestry back in an arboreal world, there existed the social organisation of mankind, the community, the co-operation of activity, the communication of information, the lengthy period of time available for learning by the young.

No one knows where man's ancestors lived throughout this vital time. Initially, with inevitable myopia, European scholars assumed Europe to have been the location. The discoveries of Peking and Java man then directed attention eastwards. Nowadays, supported by hundreds of exciting finds, the continent of Africa is thought to be the most likely spot. Southern and Eastern Africa in particular have been favoured with some exceptionally diligent and determined investigators, amongst them the Leakey family.

It had been known for twenty years before Louis Leakey went there that the Olduvai Gorge was a good fossil area. It was another twenty-eight years before the first hominid remains were discovered in that hot, arid, lonely and fascinating valley, a scar through the Serengeti plains of Tanzania etched out to reveal the remains of 150 extinct mammals – and then man. Leakey insisted that the two score teeth, the bits of four skulls and three jaws, plus a few other bones and bone fragments from seven individuals, to which the name *Homo habilis* was given, were parts of the earliest known representatives of mankind. Standing erect, some four feet high, with an advanced foot, a less well developed hand, and a brain about half the modern size, this individual became the centre of controversy.

The principal attraction of *Homo habilis* is that he possessed a good few of the features which bridged the gap between the ancient apes (who possessed some human features) and the far less ancient men (who possessed some ape features). Whether

14

Homo habilis is called an ape-man or a man-ape is obviously excellent material for perpetual argument. Evolution, it should be remembered, is the gradual merging of one type into another type or types, and taxonomy is the ruthless classification of the resulting types. The student of any evolutionary line is therefore torn both ways, towards rigid classification and towards a breaking down of that classification. Is it man-ape or ape-man, primitive Australopithecine or *Homo habilis*, an aberrant offshoot or direct descendant, a missing link or just a fragment from the past?

The *Homo habilis* supporters justify his importance by saying he had:

1 A brain capacity between the Australopithecines and *Homo erectus* (of whom more soon).
2 A skull rounded more like man's than the Australopithecine skull.
3 Man-type teeth.
4 A capable hand, although not a modern one.
5 A modern collar bone.
6 An advanced (or human) form of foot.
7 Many stone tools associated with his own fossil remains.
8 Animal bones in association with his own which indicated both that he hunted and cracked them open for their marrow.

Moving on in time means moving towards both greater agreement and a much more advanced type of individual. *Homo erectus* is a blanket term for many fossils found in strata sometimes a million years old, sometimes much less old. Java man, Peking man, and others tend to get lumped together into this group. Walking upright, with a thick skull, a heavy jaw, a cranial capacity of 1,000 cc. or more, with an ability to hunt and use fire, these creatures were undoubtedly human. (Unfortunately only modest portions of these ancient people are ever found. The most complete – so far – *Homo erectus* discovered surfaced in northern Kenya during 1984. Virtually everything was found of this twelve-year-old boy except the left arm and hand, the right arm from the elbow down, and most of both feet. The hands would have been particularly interesting, as mankind's evolution was partnered by increasing dexterity, but no *erectus* hands have yet been found.)

Later on, and moving from the Middle Pleistocene to the Upper Pleistocene, *Homo erectus* is supplanted by modern man. By no means are all or any of these modern men our ancestors. Neanderthal man, although big brained and burying his dead (occasionally and with flowers), and although active in much of Europe, Asia and Africa only 70,000 years ago or less, was definitely not our ancestor.

Quite suddenly, and entirely replacing the Neanderthal remains, appears Cro-magnon man, with a modern skull and a fondness for art. (Neanderthal man never drew anything.) Cro-magnon appeared on the scene in Europe, having apparently come from the East about 40,000 years ago, right in the middle of the last great Ice Age. Mankind had been making things for over 2,000,000 years but Cro-magnon was the first artist. From 40,000 to 10,000 BC he painted cave walls, he carved bones and was even making limestone statuettes (all buttocks and breasts) 30,000 years ago. Artistic man had therefore arrived.

The line *habilis* to *erectus* to *sapiens* had, somehow, been drawn. The blindness of evolution had finally created a species capable of dominating the entire planet, capable even of leaving the planet, of travelling to other worlds. The few amino-acids mingling in the primaeval ooze had finally led to man, the species with the wit and skill to organise the resources all around him. Several million species, including all the invertebrates, exist today. Each one is the culmination of an evolutionary line, but there must have been hundreds of evolutionary lines that led nowhere, for each one existing today, except to extinction. Hundreds of millions of evolutionary lines, therefore, and then the creation of man. Quite suddenly, out of 4,000,000,000 years of random biological activity, a species has appeared with the means to control its fate.

3. Race

Racism/Racial classifications/Jews/Negroes/Brain size/
Superiority/Purity/A biological necessity?/The scientific
conclusions

'The human race, to which so many of my readers belong...'
                                    G. K. CHESTERTON

Racism 'In today's world anybody's weakest chapter', wrote
Paul Bohannan in the *Scientific American*, 'is the chapter "On
the Concept of Race".' At this stage in this chapter I am happy to
agree. He gives as reasons that no reader can face up to it
dispassionately, even if the writer can present it so. The word is
hideously loaded. As jokes about Jews are almost forbidden by
society to non-Jews, so the word 'race' almost implies racism if it
is not dismissed almost as soon as it is brought forward.

Benito Mussolini said, 'Race is a feeling not a reality'. Lancelot
Hogben wrote that geneticists believe anthropologists know what
a race is, ethnologists assume their racial classifications are
backed up by genetics, and politicians believe that their
prejudices have the sanction both of genetics and anthropology.
Actually, none of them, he adds, has any grounds for such beliefs.
We all use the word race to suit our prejudices and our feelings.
'But', wrote John Stuart Mill back in 1848, 'of all the vulgar
modes of escaping from the consideration of the effect of social
and moral influences on the human mind, the most vulgar is that
of attributing the diversities of contact and character to inherent
natural differences.' Had he been able to foresee some of the
vulgarity attained in the subsequent century his *Principles of
Political Economy* would have boomed even louder.

To some degree racism only began its course in the nineteenth
century. The first Negro arrived in the New World a mere two
years after Christopher Columbus, but he was not a slave.
Countless Negroes were never slaves, but the anti-slavery
agitation assisted in the creation of an irremediable linkage
between 'slaves' and 'Negroes'. In large parentheses it should be
added that many a scientific wrangle over race has nothing to do
with racism. Professor Carleton S. Coon, for example, was not
only a professor of anthropology but wrote a famous book called
The Origin of Races (in 1962), having earlier revised W. Z.

17

Ripley's *The Races of Europe* (first published in 1899). John R. Baker wrote *Race* in 1974, and earned an accolade from Sir Peter Medawar: 'The idea of race or raciality has been systematically depreciated for political or genuinely humanitarian reasons, and it was high time that someone wrote about race as Baker does . . .'

Other scientists have found themselves unable to use the word. Sir Julian Huxley repudiated it back in 1936. Professor Ernst Hanhart wrote that there are 'no true races' in man anyway. Professor L. S. Penrose, of London, could see no virtue in the apologetic retention of the term. 'The biological view of race has been more misused and distorted than perhaps any other scientific conception', wrote Thomas F. Pettigrew of Harvard in his *Profile of the Negro American*. In general, scientists prefer the term 'ethnic group' because (as summarised by Professor Ashley Montagu) it emphasises distinguishability, it does not indicate status, it suggests cultural influence and it eliminates 'obfuscating emotional implications'. However, the word race still exists. So does racism.

Foundations of the Nineteenth Century by Houston Stewart Chamberlain sounds innocuous enough, but this book, published in 1910, was one of the first racist tracts. Kaiser Wilhelm II of Germany called it his favourite book, and bought many copies. John Oakesmith purloined a phrase (from Sydney Smith) to name it 'the crapulous eructations of a drunken cobbler'. Anyway it followed an *Essay on the Inequality of the Human Races* (in four volumes) by Count Joseph Arthur de Gobineau, a French anti-revolutionary of the mid-nineteenth century who held that the Revolution's 'egalitarian philosophy was the hopelessly confused expression of a degraded rabble', and it was followed in its turn by more recent efforts. Among these is *The Biology of the Race Problem* by Dr Wesley Critz George, the writing of which was commissioned by the Governor of Alabama in 1962. It admits that some Negroes are intelligent, but likens them to the Trojan Horse. Let in some, and you let in the lot, with an aftermath like that of Troy. Dr George, drawing on facts from Latin America, has decided that miscegenation is the root of evil there for its people are an assortment of Whites, Indians, Mestizos (White–Indian crosses), Negroes and Mulattos (White–Negro crosses). He cites Brazil as an example for, despite her lavish natural resources, she is unable even to feed herself. 'There is no advanced civilisation in any area where there has been a high degree of absorption of Negro genes.' 'Nowhere in the world have the Negroes demonstrated that they have the creative capacity to make a civilisation'. Etc., etc. – with a few occasional comments about some disastrous crossings in the dog world.

Rough things have been written about the Negroes. Rough things have been written about all groups. 'It has been said against the African Negroes that they never produced a scientist; but what kind of scientist would he be who had no weights and measures, no clock or calendar, and no means of recording his observations and experiments? And if it be asked why the Negroes did not invent these things, the answer is that neither did any European, and for the same reason – namely, that the rare and perhaps unique conditions which made their invention possible were absent.' That was a quotation from Lord Raglan. This is one from Cicero: 'Do not obtain your slaves from Britain, because they are so stupid and so utterly incapable of being taught that they are not fit to form part of the household of Athens.' The Emperor Claudius was just as trenchant in his assertions that the inhabitants of Britain were no good.

Nevertheless, despite all this, despite the rejection of any belief in innate racial superiority, and despite attempts at an objective assessment of superiority, it can never be argued that men are born equal. It is also undeniable that groups of people do exist. There are Negroes and Eskimos and Pygmies. There are still Bushmen, and there were Tasmanians until 1876. Mankind is of many kinds, but quite how many is virtually anybody's guess.

Racial classifications   Scientists, normally all in favour of useful classifications, are quite unable to agree about the species *Homo sapiens*, of the subfamily Hominae, of the family Hominidae, of the superfamily Hominoidea, of the suborder Anthropoidae, of the order Primates. Theodosius Dobzhansky, then Professor of Zoology at Columbia University, concluded from this disagreement that there must be some basic fault in the methods of description and classification. The more carefully human observations are studied the less clear cut the races become.

Taxonomy really began, in 1740, with Carolus Linnaeus, the Swedish prince of classifiers. Since then over 2,000,000 species of animals and plants have been described, most of which are insects and one of which is man. Linnaeus contented himself with four variants: European, American Indian, Asiatic and African. Thirty-five years later J. F. Blumenbach, an early anthropologist, divided man into a spectrum of five: White (Caucasian), Yellow (Mongolian), Black (Ethiopian), Red (American) and Brown (Malayan). Back in the eighteenth century classification was simpler partly because fewer groups of men had been discovered. By 1900 the age of exploration was over, everything was more complex and Joseph Deniker distinguished six groups with twenty-nine smaller classifications:

Woolly hair, broad nose
Yellow skin	— Bushmen
Dark skin	— Negrito, Negro, Melanesian

Curly or wavy hair
Dark skin	— Ethiopian, Australian, Dravidian
Tawny white skin	— Assyroid

Wavy brown or black hair, dark eyes
Clear brown skin	— Indo-Afghan
Tawny white skin	— Arab or Semite, Berber, Littoral European, Ibero-Insular
Dull white skin	— Western European, Adriatic

Fair, wavy or straight hair
Reddish white skin	— Northern European, Eastern European

Straight or wavy hair, dark eyes
Light brown skin	— Ainu
Yellow skin	— Polynesian, Indonesian, South American

Straight hair
Warm yellow skin	— North American, Central American, Patagonian
Brownish yellow skin	— Eskimo
Yellowish white skin	— Lapp, Turkish or Turco-Tartar
Pale yellow skin	— Mongol

In 1934 Egon von Eickstedt reduced everything to three major stocks: white, black and yellow (Shem, Ham and Japheth?). He called them Europiform or Leucoderm, Negriform or Melanoderm, and Mongoliform or Xanthoderm. Not surprisingly, these names have never caught on, and preference is given to Caucasoid, Negroid and Mongoloid for the big three. What was surprising was that von Eickstedt's sub-divisions (eight, nine and twelve respectively) added up to twenty-nine, the same as Deniker's total, but their sub-divisions were quite different.

Recently there has been a retrenchment. (This often happens with classifications. First there is one type, then far too many, and finally a reasonable few.) A more modern American anthropologist, S. M. Garn, has concluded that nine are sufficient. These are:

1   American Indian
2   Melanesian–Papuan (New Guinea to Fiji)
3   Micronesian (most Pacific Islands north of the equator)
4   Polynesians (the rest of the Pacific, including New Zealand,
    Hawaii and Easter Island)
5   Australian
6   Asiatic (a huge group, including the Filipinos, Japanese,
    Eskimos, Mongolians and Tibetans)
7   Indian (all of that sub-continent)
8   European (including North Africa and the Near East)
9   African (south of the Sahara)

But even Garn, having trimmed the races to this extent, admits these groups are 'neither fully discrete nor internally uniform'. He indicates that six of his nine groups could easily be split into thirty-two, making thirty-five in all.

Garn's division was published in 1961. Five years earlier W. C. Boyd, another American, and a professor of immunochemistry, had made much use of blood groups and reached a total of thirteen groups, namely:

1   Early Europeans
2   Lapps
3   North West Europeans
4   East and Central Europeans
5   Mediterraneans
6   Africans
7   Asians
8   Indo-Dravidians
9   American Indians
10 Indonesians
11 Melanesians
12 Polynesians
13 Australian Aborigines

At first glance this may seem a different grouping; but if all Boyd's Europeans (1–5) are banded together to form Garn's number eight, there is remarkable similarity between these two recent assessments of major groups. In other words, there are about nine human types/sub-races/infraspecies/subdivisions, give or take a group or two.

Jews   A word is necessary here about the Jews. They do not appear (or should not) in anyone's grouping of the human races. There is no such thing, wrote Professor Montagu, as a Jewish

21

physical type, and there is not, nor was there ever, anything remotely resembling a Jewish race or ethnic group. The conventional picture of a Jew, short of stature, with dark eyes, a Shylock nose, greasy skin and hair, thick lips and fat women is neither universal nor confined to Jews (think of Armenians). Chinese Jews look Chinese; 45% of Polish Jews have light eyes; 18% of Hungarian Jews have fair hair, 51% of Jews in Romania have fair hair, and many from all countries have red hair. Most certainly Jews are not a race. George Sacks, who wrote *The Intelligent Man's Guide to Jew-Baiting* in 1934, said in that book that the Jews are a race only because others consider them a race. In fact, they are a socio-religious (Huxley and Haddon) or a quasi-national (Montagu) group. 'Anti-semitism is so profoundly irrational', wrote Anthony Barnett, a zoologist then at Glasgow University, 'that it can hardly be made the subject of a straight-forward biological discussion.'

Negroes Negroes (or Africans or Congoids) appear, or should appear, in everybody's grouping of mankind, but the definition of a Negro is not easy. Melanesians also have woolly hair, dark skin and a broad nose. Many included within the Negroid camp, such as Bushmen, have much lighter skins than many Indians and Australians. Definitions can become tortuous, but comparisons are easier. A lot of work has been done on this subject in the United States of America. The characteristics of the whites (who mostly come from Western Europe) and the blacks (who are mostly from the Gulf of Guinea) have been compared and the following conclusions reached. The American Negro has: a slightly narrower, longer and less high head; a smaller brain-case (by about 50 cubic centimetres – but more on this later); a wider gap between his eyes (pupil to pupil); a lower hair line on the forehead; a broader nose with a lower bridge; a more jutting (prognathous) jaw; thicker lips; a shorter external ear (pinna); a shorter torso (by 3.1 cms., 1¼ ins.); a shallower chest; a smaller, narrower pelvis (female Negroes have hips 21 millimetres broader than Negro males as against a 10 mm. difference in the white sexes); a longer leg (by 5 cms. or 2 ins.); a similar stature but greater weight; and of course a darker skin. (Pigmentation increases until puberty – reaching its maximum by the age of fifteen – and then decreases, less rapidly, after the age of thirty-five.) Typical Negro hair is more frizzy and less profuse, having both fewer follicles and less growth. There are more sweat glands (reasonably enough) and Negroes have slightly larger teeth.

There are also differences which do not exist between black and white Americans. Professor Montagu has listed them. There is no difference in intelligence (despite an abundance of funds set up to prove a difference); the brain looks the same; suture closure, or cessation of skull growth, is similar; nose cartilage is similar although the Negro's has often been alleged to be more primitive; the Negro forehead does not slope more (it just looks that way due to the jutting jaw); the Negro hand is not longer, only the fingers are; there are no disharmonies between pelvic outlet and baby's head in mulatto births (a strong racist assertion); the Negro eye is not larger, but often looks so due to its less angular orbit, and finally the Negro penis is not larger. This particular story is 200 years old, dating from that pioneer anthropologist, Blumenbach, who saw 'an Aetheopian with a remarkable genitory apparatus'. Whereas he, the scientist, only wondered whether this prerogative was 'constant and peculiar to the nation', many others for reasons of their own have assumed it to be the case. Professor Montagu is adamant. It is not.

Brain size   A difference in brain size has naturally been used in anti-Negro literature. Major R. W. Shufeldt, M.D., wrote that 'The Negro brain is undersized with a cranial capacity of 35 fl. ozs. as against 45 fl. ozs. for whites. Also the cranial bones are dense and unusually thick, converting his head into a virtual battering ram. . . .' Few agree with his figures. Also, brain size and intelligence are not related. The average male capacity is between 1300 ccs. and 1500 ccs., but 1050 ccs. to 1800 ccs. is still considered normal. Two particular writers, Anatole France and Ivan Turgenev, are frequently quoted because the Russian's brain was exactly twice the size of the Frenchman's – 2000ccs. to 1000 ccs. (For more about brain size, and why the capacities of this pair of writers are so frequently mentioned, see *The Mind*, companion volume to this book.)

The brain of the average European male is 1450 ccs. (The female brain is about 10% smaller in most populations, and therefore around 1300 ccs. in Europe.) This European average dwarfs the anthropoids, such as the chimpanzee (400 ccs.), the orang-utan (420 ccs) and the gorilla (500 ccs.). It is also bigger than many fossil men, such as *Australopithecus africanus* (600 ccs.), *Telanthropus* (850 ccs.) and *Sinanthropus* (1043 ccs.). And it is bigger than some modern men: Australian aborigines (1246 ccs.), Veddas – cave-living people of Sri Lanka (1285 ccs.) and Andaman Islanders (1264 ccs.). In other words, there does seem a certain logic in the order thus far. But Eskimos, some southern

Africans, Mongols, Japanese and American Indians have brains of larger capacity than the Europeans. (I have here left out the actual capacities, for these vary between different authors, although the trends are similar.) Bigger still are the brains of Neanderthal man, the thick-set people who buried their dead but knew no art and fashioned only simple stone implements. Their capacity was about 1550 ccs. Yet 100 ccs. bigger than their brains (and therefore 200 ccs. bigger than modern Europeans) were the brains of the cultured Cro-magnon, who made the cave paintings of Europe, and who killed off, absorbed or survived better than the Neanderthal men. Again there exists, in part, a certain logic, for Cro-magnon was undoubtedly brighter than Neanderthal, but was Neanderthal brighter than us today?

It is assumed instead that, because brain size and intelligence do not correlate precisely, they did not always do so in the past. In short, to borrow a statement from N. A. Barnicott, of London University's Anthropology Department, 'although increase of brain size is a notable feature in earlier phases of Hominid evolution, the significance of brain volume in relation to mental functioning is by no means clear'.

Superiority and Purity Whatever anybody says in any piece of polemics about race, there are always two dominant questions, whether spoken or unspoken, whether answered or not. The first asks about racial superiority, and the second about mixed breeding.

Racial superiority tends to be uppermost, with intellectual prowess at the top. The fact that Negroes are more likely to get sickle-cell anaemia, and less leukaemia, and more diabetes, and fewer 'blue' babies and far less haemophilia and less colour-blindness does not really concern people (although these points intrigue scientists). It is intelligence that is disturbing. Adolf Hitler and some Germans may have been horrified by the success of the American Negroes (Jesse Owens in particular) at the 1936 Olympic Games in Berlin, but the Americans were jubilant. Today American Negroes are three times more successful in athletics than their population size would suggest, and no one complains. But there was nation-wide interest in the results of a huge and famous investigation carried out by the US Army into its recruits. The so-called 'Alpha and Beta' tests were given to 1,726,000 men, and Negroes scored unmistakably worse than whites. Only much later was it pointed out that, while Southern Negroes scored badly, Northern Negroes, on average, scored more than Southern Whites. As Otto Klineberg has said: 'It is apparently not "race" but environment which is the crucial variable.' Philip Vernon, of London University's Institute of Education, has said '. . . it is unprofitable to investigate racial difference in intelligence . . . because intelligence . . . is always the interaction between genetic potentialities and environmental pressures, and because intelligence is no one thing, but rather a name for a group of overlapping skills. . . .' J. B. S. Haldane said at a Ciba symposium: 'I do not believe in racial equality, though of course there is plenty of overlap; but I have no idea who surpasses who in what.'

This overlapping is a vital concept. Just as American Negroes have a shorter stature than American Whites – by 2 mms. – so may there be an intelligence difference between the two by the intellectual equivalent of those 2 mms. Stature is, of course, easy to measure. Intelligence is not. 'But even', wrote Dr Barnett, 'if future investigation does show a correlation between intelligence and physical characteristics, every large human group will overlap every other in the distribution of intelligence.' Every lot will have people both high and low, idiots and geniuses, mental dwarfs and mental giants. Hence intelligence tests will never justify discrimination against any group. 'But so far,' wrote Professor Montagu, 'if there are any inborn mental differences associated . . . with the different mental groups, then science has been unable to discover them.'

Mixed breeding is the second emotional firecracker. In one sense the question has no right to exist. Mankind is a heterogeneous assortment and, were it being viewed unknowingly by a biologist, he would consider mixed breeding impossible. The physical differences between, say, a Hottentot (short stature, yellowish and loose skin, big steatopygous rump, peppercorn hair) and a Nordic European, are far greater than many animal species which look more alike, yet cannot breed. As it is the 'Reheboth bastaards' of South Africa, the results of Hottentot/ White matings, are taller and more vigorous than both parental stocks. Groups of normal creatures, unlike man, are kept apart genetically, even if occupying the same territory, partly by having a set of signals with which they communicate with their own kinds. 'There is little desire', said Professor Coon, 'for them to inter-breed.' 'But man', he added, 'communicates by speech.' We learn each other's language. We do interbreed. Wives have been traded or captured, and men have frequently travelled alone without any compunction about mating only back home. So, he concluded, despite racial differences, that is how we are still one species.

Is further mixed breeding beneficial? Even though the word

'pure' can hardly be applied to the mixed ancestry of everyone of us ('man has always been a mongrel lot', said Professor Dobzhansky), ought efforts to be made to prevent further impurity? The largest mixed racial groups are in South Africa, in the Southern United States, in South America and in certain Polynesian islands, notably Hawaii. Strangely the first two areas have and have had laws against further mixing, the second two have not. There were fifteen states in the USA which had laws prohibiting marriage between whites and Negroes: Alabama, Arkansas, Delaware, Florida, Kentucky, Louisiana, Mississippi, Missouri, N. Carolina, Oklahoma, S. Carolina, Tennessee, Texas, Virginia and W. Virginia. Punishments varied between thirty days and ten years imprisonment, but in June 1967 the US Supreme Court declared that miscegenation laws were unconstitutional. Evidence for breakdowns in fitness (equivalent to some of the disastrous crossbreeding of dogs, where they have been fashioned with legs of insufficient length to keep their stomachs off the ground) has not been found. Heterosis (or hybrid vigour – well known in the plant and animal breeding worlds) has even been claimed, as for the off-spring of the Polynesian women who lived with the *Bounty* mutineers.

Unfortunately, as G. A. Harrison stressed in the book *Human Biology*, 'no really systematic studies have been undertaken of human hybridisation'. Its consequences are, to all intents and purposes, unknown. 'Confronted with this lack of evidence', said Professor Stern of Berkeley, California, 'the conservative will counsel abstention, since the possible ill-effects of the break-up of races will not be reversible.' 'The less conservative', he added, 'will not raise their voice against racial mingling since, from a long range point of view, it is probably bound to occur anyway.' There is also the argument that in today's world the half-caste is frequently an outcast. Most mixing is in defiance of social convention. Therefore the half-breed may suffer, and may in consequence attain less than his racial peers. 'After all,' said the American geneticist, W. E. Castle, 'attainments imply opportunities as well as abilities.' In other words, it is society, not genetics, which can kill that hybrid vigour.

To summarise: no biologically harmful effects have been discovered following mixed marriages, but scarcely any real (objective) work has been done to find them. However, so long as prejudice exists, social harm will always be at work, hindering, thwarting and stopping.

A biological necessity? Particularly in the past it was considered that racial strife of whatever form was merely a manifestation of man's vital and necessary aggression, of his competitive struggle to live. It caused his castigation of other groups, his dislike of aliens, foreigners, savages, wogs. This resentment still found its supporters when it bubbled over into war. In 1912 General Friedrich von Bernhardi, who had been a young Uhlan officer at the Siege of Paris, wrote: 'War is a biological necessity; it is as necessary as the struggle of the elements in nature; it gives a biologically just decision, since its decisions rest on the very nature of things.' Two years later the war began which led to 8,000,000 dead on the biological battle field. The Nazi war, started a mere twenty-one years later, which was to kill 50,000,000 in its six years, had a racist creed at the back of it. 'A nation', wrote Alfred Rosenberg, the official spokesman, 'is constituted by the predominance of a definite character formed by its blood, also by language, geographical environment, and the sense of a united political destiny. These last constituents are not, however, definitive; the decisive element in a nation is its blood.'

Perhaps the final say should be given to Voltaire: 'As long as people believe in absurdities, they will continue to commit atrocities.'

Here are four facts which touch more lightly on the subject. First, there are even racial bed-bugs: *Cimex lectularius* is the white man's bed-bug, and *Cimex rotundatus* is the black man's (the two species even have different chromosome numbers). Secondly, as Charles Darwin first pointed out, lice that decide to move from Hawaiians on to English sailors die before the week is out. Thirdly, in a book entitled *The Natural Superiority of Women*, it has been pointed out that everything ever said of any 'inferior race' has also been said of women. And fourthly, in deference to the prevalent black wish (ever since the early 1970's) to be known as blacks rather than Negroes, this edition will usually refer to blacks as blacks. It may be muddling, as Indians, Pakistanis and others are often included under this heading, and some blacks are extremely light in their shade of black, but if blacks wish to be known as blacks, and if the pinks like to be called whites, so be it.

The scientific conclusions There have been two major and definitive statements on race by UNESCO, each drafted by an international team of scientists. Both are some 1,500 words long, and the second was published in September 1962. Among its fourteen authors were J. B. S. Haldane, Theodosius Dobzhansky, Sir Solly Zuckerman, Sir Julian Huxley, Professor Harry

Schapiro and Professor M. F. Ashley Montagu, and they proclaimed that:

1. Scientists are generally agreed that all men living today belong to a single species, *Homo sapiens*, and are devised from a common stock. . . .

2. Some of the physical differences between human groups are due to differences in hereditary constitution, and some to differences in the environments in which they have been brought up. In most cases, both influences have been at work. . . .

3. National, religious, geographical, linguistic and cultural groups do not necessarily coincide with racial groups, and the cultural traits of such groups have no demonstrated connection with racial traits. . . . The use of the term 'race' in speaking of such groups may be a serious error, but is one which is habitually committed.

4. Human races can be, and have been, classified in different ways by different anthropologists. Most of them agree in classifying the greater part of existing mankind into at least three large units. . . . So far as it has been possible to analyse them, the differences in physical structure which distinguish one major group from another give no support to popular notions of any general 'superiority' or 'inferiority' which are sometimes applied in referring to these groups. . . .

5. Most anthropologists do not include mental characteristics in their classification of human races. . . . It often happens that a national group may appear to be characterised by particular psychological attributes. The superficial view would be that this is due to race. Scientifically, however, we realise that any common psychological attribute is more likely to be due to a common historical and social background, and such attributes may obscure the fact that, within different populations consisting of many human types, one will find approximately the same range of temperament and intelligence.

6. The scientific material available to us at present does not justify the conclusion that inherited genetic differences are a major factor in producing the differences between the cultures and the cultural achievements of different peoples or groups. . . .

7. There is no evidence for the existence of so-called 'pure' races. . . . No biological justification exists for prohibiting intermarriage between persons of different races.

8. . . .We wish to emphasise that equality of opportunity and equality in law in no way depend, as ethical principles, upon the assertion that human beings are in fact equal in endowment.

9. We have thought it worth while to set out in a formal manner what is at present scientifically established concerning individual and group differences:

(*a*) In matters of race, the only characteristics . . . (effective) as a basis for classification are physical.

(*b*) . . . no basis for believing that the groups of mankind differ in their innate capacity for intellectual and emotional development.

(*c*) Some biological differences between human beings within a single race may be as great or greater than the same biological differences between races.

(*d*) Vast social changes have occurred that have not been connected in any way with changes in racial type. . . .

(*e*) There is no evidence that race mixture produces disadvantageous results from a biological point of view. The social results of race mixture, whether for good or ill, can generally be traced to social factors.

4. Population

Past/Present/Future/Urbanisation/Immigration/
Prediction

'Be fruitful, and multiply, and replenish the earth, and subdue it.'
GENESIS

'Only a scientific people can survive in a scientific future.'
THOMAS HUXLEY

*'The time has already come when each country needs a considered
national policy about what size of population, whether larger or
smaller than at present, or the same is more excellent. And having
settled this policy, we must take steps to carry it into operation.'*
JOHN MAYNARD KEYNES

*'I would suggest that it is time to consider a fifth freedom – freedom
from the tyranny of excessive fertility.'*
SIR DUGALD BAIRD

In Mauritius in the year 1900, 37 babies were born for every 1,000
people on the island. Sixty years later the figure was still 37 babies
per 1,000 people. Yet during those decades Mauritius suffered a
powerful population explosion, the like of which the island had
never known before. No more babies were being born, but far
fewer babies were dying. An intensive post-war D.D.T. campaign
on the island against malaria was so successful that infant
mortality fell from 150 per 1,000 to 50 per 1,000 in just ten years.
It is now 35 per 1,000. Hence the phenomenal expansion, the
sudden transformation from stability to rapid growth, the
'explosion'.

Any student of bacteria, or mice, or rabbits, or practically any
creature, is amazed at the constant repetition of the word
explosion in this context. Given good conditions bacteria will
double their numbers in twenty minutes, quadruple them in
forty, and octuple them in a single hour. Humanity raised the
alarm when it reached a state whereby its numbers, as viewed
from the 1960's, would double in the subsequent thirty years.
Mice, needing only three weeks to produce a litter of ten or so,
take about as many days as man now takes years for a doubling to

occur. Even large animals like horses, or cattle, or the huge blue whale are sexually mature in two or three years, and can double their numbers far faster than man. The word explosion, used to describe a duplication over three decades, is hardly accurate. A growth rate of 2% in many countries, or 4% in the most rapidly growing, is precisely equivalent, arithmetically, to the growth rate of money invested in unrewarding companies. It is even lower than the annual increment of money invested governmentally, and nobody calls that explosive. Nevertheless the word demonstrates mankind's concern at his swelling numbers, a growth which resembles the inexorable expansion of rice grains in water rather than a great big bang.

'We have been god-like', said the historian Arnold Toynbee, 'in the planned breeding of our domesticated plants, but rabbit-like in the unplanned breeding of ourselves.' Until advances in death control had advanced sufficiently to cause a need for birth control we were entitled to be as rabbit-like as the rabbits. No one knows, but men on earth at the time of Christ are thought to have numbered about 250,000,000 with most of them living in Asia. How many men there were before then is naturally even more obscure. They probably numbered only about 30,000,000 to 50,000,000 before the discovery of agriculture, but there was then a slow, slow climb over the millennia to reach that AD figure of 250,000,000. The start of towns, of farming, of animal husbandry must all have been associated with expansions of population in various areas at various times; but disease and wars and poor food and the lean years were powerful obstacles to any major flowering of the human race. Between them, despite modest technological and medical progress, they prevented any further doubling of the human population for the next 1,600 years. It was such a delicate advance that one wonders why it did not retreat, and why the birth-rate was *just* a match for the death-rate.

The delicacy stopped in the middle of this current millennium. There were about 500,000,000 people on earth in the year 1600. The British figure then was about 5,000,000 (according to G. M. Trevelyan). The North American figure was about one million. Suddenly the situation altered rapidly. There were 1,000,000,000 people in the world by 1850, 1,500,000,000 by the turn of the century, 3,000,000,000 by the early 1960s, and there will almost inevitably be 6,000,000,000 by the end of this current century. The four doublings since Christ will have taken some 1600 years, 250 years, 90 years and 30 years, respectively, to bring the human race from a scattering of 250,000,000 people to over 4,000,000,000 people by 1980. All finesse has gone from the balance of birth versus death. In some of the Asian, Latin American and African countries the life expectancy for a newly born child has been rising recently by one year per year. A mediaeval birth-rate has been combined with a twentieth-century death-rate. There has never been a time quite like the present in the whole history of the earth. In the 1960's, when the population issue appeared particularly critical, it was considered by many that the huge figure of 6,000,000,000 by AD 2000 might be an under-estimate. If the fertility rate of those 1960's had continued at the same pace, and if the mortality rate had continued to decrease in similar fashion, the year 2000 might have been welcomed in by 7,410,000,000. This figure was calculated by the United Nations' demographers, but most estimates now hover around 6,100,000,000 for the turn of the century. Nevertheless three times as many people will welcome the twenty-first century as welcomed the twentieth century back in 1900. Current projections are that 7,000,000,000 will not be reached before a decade of the twenty-first century has passed, and there will be 7,810,000,000 people by 2020.

To try to stir the conscience of the world, the propagandists who are in favour of a less warren-like approach to breeding have hammered home their message in varied fashion. They tell us there are 84,000,000 more mouths to feed every year, that this annual enlargement is more than the population of Denmark, Finland, Iceland, Ireland, Norway, Sweden, and the United Kingdom, and that it represents the population of Greater London appearing every five weeks. The present annual increase means 230,000 more mouths to feed every day, 9,600 more every hour, and 160 more every minute. Two people die every second, but four babies are born, and these two extra mouths are the remorseless tick-tock addition to the world's bulging population.

If put another way round the unyielding growth also means that more people are in danger of starving today than at any time before. (Even if everyone at the time of Christ had been starving, they would only represent 7% of today's population, and no one believes that 93% of today's people are adequately fed.) Despite improvements in education there are more illiterates in Asia than there were ten years ago. Nehru once said that every problem in India 'has to be multiplied by 400,000,000'. Each year that number goes up – by 10,000,000 or so, and had he been living in 1983 he should have said 'by 730,000,000'. Even Alice was told in her fairy-tale world that it was necessary to run twice as fast actually to get anywhere. A further difficulty is that a rapidly expanding population is swamped with children, who eat well but do not – in general – support families. In countries with a 3% growth rate, half the population is under fifteen. In Britain less than a quarter is under fifteen.

In 1950 Sir Julian Huxley, who was never one to pull his demographic punches, warned that the world population was probably increasing by 0.8% a year, perhaps by 1%. By the end of the century, he said, the global population would be 3,000,000,000. Within a dozen years that total had been reached. Revised estimates now put the world increase at 1.8% a year. Asia's growth rate is 1.9%, and there will be 3,564,000,000 Asians in AD 2000 as against today's 2,730,000,000. Even this huge increase is proportionately overshadowed by the Latin American growth rate. In 1800 two men out of three on this planet were Asians. By the year 2000 only 58% will be Asian, mainly because the Americas as a whole will then possess 14% of the world's people, whereas in 1800 they possessed 2.8%. The overall growth rate today in Latin America is 2.3%, with some countries reaching 3.5%. Such an expansion means a doubling in twenty years. It hardly emulates the bacterial twenty minutes, but it is revolutionary for mankind. Every three years man is now adding as many people to his planet as were existing in all five continents when Christ was born. Putting the facts yet another and final way round, about as many people will have reached adulthood in the twentieth century as reached adulthood during the whole of the first millennium AD.

Coupled with the explosion of numbers there has been an implosion towards the towns. Instead of pushing outwards, and making new frontiers, like the pioneers of North America, mankind has been concentrating on the crowded areas. He is doing this at an even faster rate than he is breeding. The world growth rate is less than 2%. The town and city growth rate is 4%, and it is even 8% in some rapidly developing countries. In 1960 one-quarter of the world's people were living in towns of 20,000 or more inhabitants, and they are continuing to crowd themselves together. In a place like Monaco there are 40,000 people per square mile. Each person therefore has on average only 75 square yards, or a plot of ground slightly less than 9 yards long and 9 yards wide. In London and its home counties there are 2,733 people per square mile, but in a place like French Guiana there is one person per square mile, Such emptiness is beaten by Antarctica, with 2,300,000 square miles and not a soul as resident population.

All the signs are that the world's cities will grow larger and larger, the world's population will increasingly forsake its age-old role of agriculture, and the world's empty places will remain empty – unless tremendous governmental pressures are exerted in their favour. In 1950 only a fifth of the world's population was living in towns. By 1960 one-quarter was living in a town. Already in Latin America half the people are living in towns, and in the United States the proportion is two-thirds. The World Health Organisation reckons that only 10% of us will actually be working on the farm by AD 2000, and the lure of the towns will continue. São Paulo in Brazil leapt from 1,320,000 people in 1940 to 5,000,000 in 1980. Santiago, in Chile, went from 952,000 to 4,000,000 in the same time. It is small wonder that town planners often make a poor job of good accommodation for such invasions – 1,800 more people a week in São Paulo *every week* for forty years.

Immigration has always been thought of as the great pressure relief valve for crowded areas. Certainly the Atlantic steamers took millions to the New World, notably in the nineteenth century. The people of Ireland, made desperate by starvation, emigrated in such numbers as to leave the mother country not just with a reduced pressure but with a small fraction of her former population. Even so, despite the exodus of human traffic, and the beckoning fingers of a whole new continent, the number of people leaving Europe for the New World, since 1850, has been estimated at only 35,000,000. And the number of new people being added currently to the world's population every year is over double that figure. So much for immigration providing a solution. Were some magic to happen, and were Antarctica to blossom like the warm prairie fields of the mid-West, the new lands would only be of marginal assistance to the over-crowding problem. Before very long the frozen sixth continent would have a population hazard of its own, and the rest of the world would have benefited from no detectable easing of its own pressures.

Besides, unlike the Irish, most people do not move. Not long ago in England many people had not been outside their county. In the United States many have never seen the sea (I was amazed to discover the truth of this when staying in Nebraska). The Isle of Man, located not far away from busy Lancashire, and often thought of as a spot primarily for low-tax retirement, keeps most of its citizens to itself. On a recent census night some 67% of the people then on the island had actually been born there.

The New World is still the great recipient of the foot-loose, but there is no longer an open door. For the past eighty years potential immigrants to the United States have been either barred or subjected to restrictive quotas (based on race and nationality). Great Britain used to have a quota of over 65,000 annually, and recently had only been making use of about half its allocation. Italy had a quota of 5,666 and Greece of 308, and both such quotas were easily consumed. In a law which became effective in July 1968, such national prejudices were curtailed. The Western hemisphere can now send 120,000 annually, with 20,000 being the maximum for any country. The rest of the world can send 170,000 annually. This immigrant total of 290,000 is a sizeable number of people if thought of in terms of boat-loads. It is minuscule when set beside the world's problem, because the world adds that number of mouths to its population every 30 hours. Besides, the United States itself is making its own contribution to all those extra mouths. The 1967 population of 200,000,000 reached 268,000,000 by 1983, is expected to reach 302,000,000 by the turn of the century and 333,000,000 by 2020. If all quotas are taken up there will have been almost 10,000,000 new immigrants between 1967 and 2000, but about ten times that number of native-born Americans to swamp them. (The US Department of Commerce has a census clock in Washington. Every 17 seconds one American is added to its total.)

The fact that 84,000,000 people are being added to the planet every year, and that the average population increase in the less developed countries is well over 2% a year (with Kenya, for example, being over 4%), does not seem to call for complacency, but there has been a distinct shift in attitude in recent years (and since this book was first written). At the United Nations' conference on population, held in Mexico City during August 1984, the tone was quite different from the previous such conference (in Bucharest in 1974) when the talk had been of a planet unable to feed its growing numbers. At Mexico many countries even welcomed the idea of adding to their numbers (West Germany, Hungary, Kuwait, Iraq, Zaire, Somalia, Sudan, for example) and the general feeling was that food supply was more than sufficient, if only it could be distributed properly. It was thought likely that the world population would stabilise at about 10,000,000,000 people towards the end of the next century, namely slightly more than double the current total.

Nevertheless demographic projections can be considered blood brothers to weather forecasts, and it may just be that the prevailing optimism is quite unjustified. As a futuristic digression various calculations have been made about the whole planet's ultimate potentialities. One set has been worked out by Dr J. H.

35

Fremlin, of Birmingham University. He has divided progress into stages. Briefly they are:

Stage 1. Proper agriculture, using roofs over cities plus the oceans as growing areas, would permit seven doublings of the existing population. Therefore 330,000,000,000 people 260 years from now.

Stage 2. The plants and animals eaten in the twentieth century would be replaced by more efficient converters, e.g. plankton are better than fish. Ten more doublings therefore permitted, meaning 3,300,000,000,000 people by the year 2330. There would then be $\frac{1}{30}$ of an acre per person.

Stage 3. Satellite reflectors, bringing energy to the poles, abolishing night, and bringing equatorial conditions everywhere. 15,000,000,000,000 people by AD 2410.

Stage 4. Direct synthesisation of all foods. Use of waste-products. 1,000,000,000,000,000 by AD 2640.

Stage 5. With two people to every square metre there would be problems in keeping the planet cool. There could be ways and means of losing heat . . . 12,000,000,000,000,000 people by AD 2760 – or, instead, the 10 billion that are currently forecast.

Whether futuristic whimsy of this sort proves to be wide of the mark or not, it is highly relevant that such predictions do not have to probe very far into the future before our current and customary methods of living become hopelessly upset. Dr Fremlin's forecast does not peer at some Jurassic age of future time, but keeps itself within the third millennium. The year AD 2760 is only as far in the future as the year AD 1210 is in the past; only as many years as lie between King John (of Magna Carta) and Elizabeth II.

Such phenomenal doublings of the population are inevitable unless either the death-rate goes up or the birth-rate goes down. As no one advocates an increase in the death-rate, and as the prospects of huge population increases are alarming in most countries, attention should consequently be focused on birth control. Most countries hope that contraception will be readily available for all, and that their citizens will be reasonably sensible in planning their families. The technique is much like fostering education solely by printing cheap textbooks without any schooling being compulsory. Even countries with determined family planning programmes do not find themselves in control of the situation. At the 1965 United Nations conference on world population in Belgrade it was stated that 'not one of the major fertility declines in the Western World was the result of an organised programme of birth control by a governmental agency. It resulted from individual decisions. . . .'

Not too much is known about these individual decisions. Similarly no one seems to know why in the late eighteenth century the British population, for example, suddenly began to increase so sharply. Of course population increase is quite different from birth-rate increase, and in Britain in 1870 the birth-rate started to go down even though the population was still leaping up. No one is too sure about that birth-rate decline, particularly as the widespread use of contraceptives in Britain developed almost entirely after 1910. The availability of good contraceptive methods is plainly not the answer. The slump in the birth-rate which paralleled the lean years of the early 1930's was not marked by any change in contraceptive usage. In France, where contraceptives were used less, the birth-rate fall between the wars started even earlier. (The French have ever since had a particular population problem. M. Jacques Chirac, Gaullist politician, said in 1984 there should be a tightening of the abortion law and special grants for families with three or more children in an attempt to stop the 'terrifying' depopulation of France: 'In demographic terms, all of Europe is in the process of disappearing. In about twenty years' time our countries will be empty. And, whatever our technological power, we shall be incapable of exploiting it.'

So why is the birth-rate falling? Why did the United States reach a peak in 1957 (with 125 births per 1,000 women of child-bearing age) and why did it then fall steadily (to less than 100 births per 1,000 women in 1966) so that the rate was almost identical to the lowest years of the depression? Why did Britain continue to have more and more babies until 1964 (its highest birth-rate for 17 years) and only then begin to slacken its puerperal pace?

Theories abound for both rises and falls, indicting social prestige, economics of the family, women's status, education, the likelihood of a child surviving childhood, and available birth control devices. Interaction among these and other factors helps to distort the picture still further; higher education may mean having more children – but later, causing a greater number of children per family but a national fall in the population growth rate. If average family size is four but average parental age is thirty the population increase is almost identical to that caused by parents aged twenty who have only three children per family. This calculation is extremely relevant to India where the average maternal age of a first birth is $16\frac{1}{2}$. India would be far better off if this could be raised to the European average which is about 22 years.

Thomas Robert Malthus, born on February 17, 1766, a cleric, a denouncer of Rousseau, and a gloomy prophet of inevitable over-population, is no longer having his predictions proved so woefully accurate. He contended that population numbers would increase geometrically while food production would only rise arithmetically. Consequently he saw little hope for mankind, only failure. 'Any attempt', he wrote in 1798, 'to check the superior power of population will produce misery or vice.' Today population is not growing faster than the food to support it. Malthus did not distinguish between mankind and animals in the power of both to breed excessively and inexorably. 'I can see no way in which man can escape from the weight of this law which pervades all animate nature.' Until now, mankind's willingness to breed without much concern for the consequences does put it in the animate or rabbit-like class. The Communists, for example, do not regard Malthus as a progressive and forward-looking thinker: they bluntly call him a reactionary. However, it is up to mankind to call him and prove him merely a false prophet, and to prove that the first species with the power to arrange its own destiny actually does so.

Dr I. J. Good, the Oxford scientist, who once prefaced a book by saying that its intention was to raise more questions than it answered, has expressed his own particular and extremely final view of the present growth of people. 'Should the population continue to expand at the current rate it will only be 3,500 years before we shall have converted into people all the matter that can be reached during that period.'

5. An Apologia for Reproduction

My main assertion in this book is that reproduction deserves most emphasis. Mother Nature, the eponymous heroine of all natural history, certainly insists that everything is secondary to survival, not of the individual but of the species. Without the ability to reproduce there can be no biological survival. Without the ability to reproduce in a manner which induces change there can be no such survival in a world of constant change. Reproduction is uppermost, and all internal systems of excretion, respiration and so on are ancillary to it. 'What are people for?' asked Sir Julian Huxley in another context. They are, like every other living thing, to survive, to change, to reproduce.

Reproduction has a wider scope than mere replication. Unlike some berserk factory which cannot alter its own output, which pours out identical scrubbing brushes or crankshafts in a relentless stream, biological reproduction is more than the mere creation of offspring. It is the creation of induced change. It is also development. It is maturity. And then, first for individuals and then for species, it is senescence, it is death. The old must be prepared to die to make way for the new. If there is not death there is not change; instead there is stability, and that way lies extinction. 'Qu'est ce que la vie?' asked Claude Bernard, 'La vie, c'est la mort.'

Therefore reproduction is all-pervasive. It is the egg. It is the growth of that egg. It is puberty. It is the manufacture of another ovum. A chicken, said Samuel Butler, is nothing more than an egg's way of making another egg. Rather, a chicken is an egg's way of making, from time to time, a slightly different egg. Reproduction has to change its chickens in order to have the ability to change its eggs, and it must get rid of earlier examples. Therefore reproduction has to include post-maturity, old age, death and decay; immortality has no place in a changing world.

Just as fundamentally, reproduction is also the creation of sex. Human beings may revel in sex, or currently revel in talking about it; but only rarely do they ask themselves why there should be two sexes. After all, put at its crudest, the existence of two sexes does complicate creation. Why should there not be one? Why should we not all be mother figures who bud off infants, much as any hydra makes two grow where one grew before?

The fact that there are two human kinds can, of course, never be forgotten. Apart from the primary distinctions between male and female, there are the secondary characteristics, the type of hair growth, of stature, of shape, of fatness. And there are what can be called tertiary characteristics. Women have a different suicide pattern to men. They live longer. They have a different proportion of red blood cells. Their ulcer tendencies are not like men's. Their brains are lighter. Fewer of them are conceived; fewer of them are stillborn.

The differences between man and woman are by no means confined to the chapters distinguishing the male and female reproductive systems. So, both reproduction and the existence of two sexes spread a wide net. All in all some three-fifths of this book is encompassed by reproduction in the broad sense of that word. Its breadth, to my mind, includes not just the basic entities of sperm and egg, but of conception and infertility, of implantation and contraception, of pregnancy and delivery, of twins and malformations, of maternal and infant mortality, of lactation and puerperal fever, of growth, development and size, of circumcision and castration, of puberty and adolescence, of sex crimes and marriage, of mating, of the proof of paternity, of inheritance, inbreeding and incest, of family size and population, of the menopause and old age, of suicide and death.

To kill oneself, or not to conceive or to have the infertility of castration thrust upon one, is quite as crucial to reproduction as stillbirths and abortion. To be genetically unfit means a failure to reproduce – from whatever cause.

The word sex does not indicate reproduction; it indicates male and female. The system of sexual reproduction arose extremely early in evolution – no one has any idea when – but it has been of paramount importance because it has such commendable advantages. At first sight it might seem that other systems, and there are many, have more to be said for them. With sexual reproduction only one sex can produce offspring, and it cannot produce those offspring without the intervention of its nonproductive sexual partner, the male. With asexual reproduction the offspring can be manufactured as soon as the adult is ready to make them; every adult is therefore a producer, and no delay is necessary while waiting for, finding, cajoling, stimulating, and mating with a partner. Asexually, one makes two, or two hundred as the case may be, and the two or two hundred become four or forty thousand as soon as they are ready to do so. Many lower animals and plants breed in this fashion, and do indeed multiply most rapidly; but mere multiplication is not entirely desirable. Individual insects or bacteria or protozoa can produce a population explosion entirely on their own individual accounts, but their teeming descendants have nothing but numerical superiority on their side. They are all carbon copies of their solitary ancestors; they have strength in numbers but not in diversity.

Sexual reproduction gives that diversity. Instead of receiving all genetic material from a single parent, as in asexual reproduction, sexual offspring receive half from each parent. In other words, they cannot be carbon copies of either parent, but must be different from both of them. This may or may not be a good thing in immediate terms, but the reproduction of dissimilarity is extremely advantageous in the long term. Instead of producing those satisfactory and teeming copies, it leads both to superior and inferior offspring. Mother Nature has most casual maternal characteristics for she considers the ninety and nine inferior offspring who die quickly are a fair price to pay for one exceptional infant. What does it matter if the herd is cut to pieces by a predator provided that one escapes due to an exceptional turn of speed?

Assuming that the successful escaper finds a mate, the offspring are likely to possess something of that swiftness even if the mate is normal; but – and this is a second major advantage of sexual reproduction – there is a chance the mate will also be exceptional to some degree. Any offspring from these two will then be the product of a particularly advantageous combination. This can never happen with asexual reproduction; there can be no such thing as a combination, advantageous or otherwise.

Human beings can only reproduce sexually. There have been claims for virgin birth (see page 74) but the occurrence, whether believed or disbelieved, is certainly irregular. Not so with many animals. From the perpetually asexual (as with many simple forms) to the perpetually sexual (as with man) there is a wide, ambivalent range. Earthworms, for example, are hermaphrodite in that they always possess both male and female organs; but they still have to mate with other worms. During these encounters the two worms exchange sperm. Later on each worm will use the stored foreign sperm (certainly not its own, as that would defeat the object) to fertilise its own eggs. Some creatures, such as the oyster, are not truly hermaphrodite, but regularly alternate from Hermes to Aphrodite, from male to female. Others become female

41

only in adversity, when short-term numbers are more advantageous than long-term adaptability. Still others, such as the slipper limpet, change their sex according to the first individual of their species that comes their way; if the newcomer is male the original limpet becomes a female. Some segmented worms are male at one end of their body, female at the other end. And some males of other species actually live inside their females, or are attached equally parasitically to the outside – the ultimate gigolos, to use a phrase from the zoologist Martin Wells.

Having such diminutive masculine forms also seems the ultimate compromise: the male provides the genetical advantages of sexual reproduction and yet does not consume half the available food supply. In fact, such dwarf males are rare. Odder still, considering that most males in most species are capable of fertilising many females, the sex ratio between male and female is customarily 1:1. Even in species which do favour harems, and which have single bulls serving a score of females, such as seals, there is still parity; do not forget all those other bachelor seals in the back-room who are frustratedly just as male as the active bulls. A beehive contains only one mature female but dozens of mature males.

Whenever the 1:1 ratio of male to female is significantly different in the animal kingdom, an overwhelming preponderance of males is a more probable cause of the disunity. There must be advantages in making sure that every female is found by a mate even at the risk of having an unnecessary number of unproductive males occupying the same ecological niche, consuming the same food, and competing in a sense with the females they never serve. Those bachelor seals eat fish just as determinedly as the females who will at least be producing more pups for the community, even if half those pups will be just another crop of males.

Reproduction by means of two distinct sexes, the paramount system, has certainly dominated evolution for hundreds of millions of years. Whereas lower forms make use of it, but can breed asexually, the higher forms can only use it. Nevertheless, despite ubiquity, despite its history and indispensability, the sexual distinction is incomplete. No male is wholly male; no female is wholly female. Not only do human males, for example, possess breasts and frequently experience transient, small-scale changes to them during adolescence, but both females and males possess hormones to some degree which, given the opportunity, can develop both forms of sexual system. Sex-reversal is common among many animals; something similar can even happen with man, with woman. This is not homosexuality, nor transvestism, nor intersexuality; it is a change of sex. The two sides, so crucial to reproduction, so vital to our society, so integral a part of us, are nothing like so far apart as we generally imagine. The Gender Identity Clinic, of Baltimore, for example, proclaims a general dictum that 'if the mind cannot be changed to fit the body, we should consider changing the body to fit the mind'. There are a total of six such clinics in the United States, all set up for the benefit of trans-sexual patients. The world total for such operations is now over one thousand. Any Martian inspecting this planet, having comprehended our bisexual arrangement, would be astounded to learn of the slender distinction between the two sides of the sexual fence.

Perhaps it is to that Martian that I bequeath this book. He would, one hopes, be primarily interested in the species man. He would not, I hope, be initially as intrigued by our diseases as by the organism itself, how it multiplies and grows, how it feeds and achieves its energy. He would be told of evolution, and would then know that mankind is really another animal, however much mankind wished to distinguish itself from such other forms. Therefore, and this I fervently hope, he would like the approach to be biological, and to see man as quite the most intriguing and exciting of all the animals here on Earth; but, first, he would have to know how it all began. Only then, having examined evolution and having received a broad picture of humanity, would he start to learn about its reproductive system. As with Genesis, he would start with male and female.

6. The Male

Means of sperm transfer/The penis/Potency/Aphrodisiacs/
Spermatozoa/Testicles/Ejaculation/Sperm prints/
Nomenclature/Mating time/Sexual offences/Homosexuality/
Castration/Circumcision

The penis is the accessory organ used for the transfer of sperm. Low down in the animal kingdom and in water-living forms there is no need for such an organ. Both sperm and eggs are liberated into the water, and fertilisation occurs in that water. The new forms are truly independent right from the start. As soon as either shelled eggs or viviparity are involved, with the mother producing either a shelled form or a free living form, both of which have already been fertilised, there is need for sperm transfer. Somehow or other she must receive the sperm fairly near the start of the operation, and the father must be capable of giving them to her. The average human couple may consider this interchange an extremely elementary procedure, but occasionally they should spare a thought for sharks, for whales, for porcupines, for seven-ton elephants, for giraffes, and for tortoises; the mechanics of sperm transfer does have its problems.

Male sharks and rays have a pair of claspers, extensions of their pelvic fins, which are stiffened by the skeleton. Each pair has tube-like folds of skin along which the sperm travel, and the claspers can be inserted into the female cloaca. Various other vertebrates are much more haphazard about the process, merely presenting cloaca to cloaca, with the proximity helping the sperm to get across. (A cloaca is a combined vent for faeces, urine and reproductive products possessed by many of the lower vertebrates. It is Latin for sewer.) Most birds – ducks and geese are exceptions – have no specific organs of copulation, although it is thought that their ancestral forms were not so lacking in this respect. Many reptiles, the group from which birds evolved, do have a special organ. Snakes and lizards have hemipenes; at the right moment these are turned inside out, extruded and then inserted into the female. Various other reptiles, such as crocodiles and turtles, do have an organ which seems to be the true forerunner of the mammalian penis, and therefore merits the same name. It is in the mammals that the penis is most highly developed. Nevertheless the mere possession of a copulatory organ is inadequate for copulation. It has to be made capable for insertion into the female.

Such insertion is made possible both by increasing length and rigidity. Neither of these is necessary at times other than copulation, and the penis is normally a more flaccid and reduced possession. There are very many variations within the mammals concerning this reduction. The penis may disappear from view completely, it may be made to fold, or retreat within a special sheath, or merely remain completely external but shrunken. A cold external temperature can cause this shrinkage to be even more marked. Erection is achieved when the organ is engorged with blood. Suitable nervous stimulation, whether directly physical or merely mental, starts this process. The penis's dorsal vein is compressed, thereby restricting the blood's outlet. At the same time the muscular coating around each arteriole (or miniature artery) is relaxed, and so is the spongy tissue which forms much of the penis. Hence, with a greater area to occupy, the blood distends the penis. Similarly, due to the continued pressure from the heart and a smaller venous outlet, the pressure within the penis is built up and rigidity follows. (An analogy is a hose pipe. Provided the end is open, water will flow smoothly along the pipe. When it is sealed, the water pressure will distend and stiffen the pipe. The hydraulics of both pipe and penis are the same.) Full engorgement of the human penis can be most rapid. It may occur within three to five seconds from the start of stimulation. The average final length is $6\frac{1}{4}$ ins. (16 cms.), with its diameter at the base being $1\frac{1}{2}$ ins. (4 cms.). According to R. L. Dickinson, author of *Human Sex Anatomy*, there is relatively little difference in penis size as compared with body height and weight differences.

Naturally with animals, and with the great size differences that are involved, there is considerable variation in the process of erection and retraction. Usually the penis is concealed when at rest within a fold of skin in the lower abdomen known as the preputial sac. The penis, for example, of a bull, a stallion or an elephant is customarily invisible. Ability to sex humans at a glance when they are naked is easy. Such simplicity does not exist with many animals, despite their nakedness; the genital organs can be totally invisible. For a long time the sex of London Zoo's giant panda, Chi-Chi, was unknown. Only after the animal had been several years at the London Zoo was Chi-Chi definitely pronounced to be a female, and then sent fruitlessly to Moscow to mate with the male An-An. It was frequently held in the past that the hyena must be hermaphrodite because all hyenas appear externally similar.

Sometimes, as in most monkeys and apes, the preputial sac is less effective, and the penis is only partly hidden. Or, as in man, the tree shrews and certain bats, such a sac is absent, and the penis is then said to be truly pendulous. Occasionally as with bulls and guinea-pigs, the normal penis is longer than the sac into which it retreats. In these cases the retracted organ folds conveniently into an S-shape. With very large animals the penis naturally has to be very large to be effective in sperm transfer. In the boar it is 18 ins. (46 cms.), in stallions 30 ins. (76 cms.), in bulls 3 feet (91 cms.), in elephants 5 feet (152 cms.), and in the blue whale – the largest animal there has ever been – it is 7 to 8 feet (213–244 cms.) long.

Many animals, but not man, possess a bone in their penis. Known as the os penis, or baculum, it can be of bone or cartilage or a bit of both. Presumably it helps in the acquisition of rigidity. Presumably, too, such a bone was possessed by the mammalian ancestors because its present distribution is so widespread. All carnivores (dogs, cats, seals, bears), insectivores, bats, rodents, and primates (except man) have one. Man, the ungulates (hoofed mammals), most marsupials, and whales do not have one. The shape of the bone among those that do possess one, is itself highly variable, far more so than any other bone. A femur, whatever the animal that walks on it, always tends to look like other femurs. Not so the os penis. Professor A. S. Romer of Harvard University, author of exemplary zoological textbooks, has said that it may be 'a plate or a rod of varied form – straight, bent or doubly curved; round, triangular, square or flat in section; simple, pronged, spoon-shaped or perforated; long or short'. As an examination query for a student, the bone has a host of confusing possibilities.

The tip of the penis, cone-shaped in man, also varies greatly. It may be in two halves, as in some marsupials, or single, as in all other mammals. It may be extended into a long, thread-like part, as in the ram. It may be covered with spiky protuberances, as in rodents, carnivores and insectivores. It may, as in some primates, be very small. In man it occupies about one quarter of the total penis length, it is the most sensitive part, and normally (without circumcision) it is covered by a fold of skin called the prepuce.

The ability of the human male to achieve erections, and to produce sperm, continues in time long after the human female has met her menopause. One American survey showed that over 90% of men still had 'erectal potency' at the age of fifty, over 80% at the age of sixty, over 70% at seventy, and even 25% at eighty. These proportions do not necessarily indicate the ability to achieve orgasm. The survey, carried out by A. C. Kinsey and his colleagues, also indicated that the American male of fifty was still having about two orgasms on average a week, and of seventy about one a week. As for sperm production this probably continues at a diminishing rate until the time of death. The mechanism of erection and ejaculation can, like presumably every other bodily function, go wrong. Ejaculation, for instance, can become retrograde. The sperm, instead of being ejected correctly, can be shot backwards into the bladder at the moment of orgasm. Erection can also become permanent. Known as priapism, the condition has a multitude of causes, some sinister, some merely inconvenient. For a long time it has been observed that erections can be caused by severe injuries to the spinal cord. Fatal injury to the cord caused by judicial hanging can have the same effect.

Aphrodisiacs Many societies, including our own, have great faith in the ability of certain substances to arouse sexual appetite and performance. The beliefs of other societies are pooh-poohed, such as the belief in rhinoceros horn; but there is a general feeling that aphrodisiacs do exist. The beetle *Cantharides*, which is used to make 'Spanish Fly', is said to be one of them. Certainly this concoction, also used by animal breeders, does act as an irritant, and some irritation is caused in the genital area, perhaps stepping up the flow of blood to this area; but such action by no means implies improvement of prowess or desire. Similarly there is yohimbine from the yohimbine tree of Africa, and phosphorus, and caffeine, and strychnine. Great faith has been attached to all of them, yet little satisfactory scientific evidence exists. According to Ford and Beach's book *Sexual Behaviour*, there is no real evidence to indicate 'that any pharmacological agent is capable of directly increasing the individual's susceptibility to sexual arousal and capacity for sexual performance'. There may well be such an agent, they admit, but there is not one as yet.

The greatest faith of all is attached to alcohol. 'Lechery, sir, it provokes and it unprovokes'; said Shakespeare's Porter, 'it provokes the desire, but it takes away the performance.' Scientists only admit that it unprovokes social inhibitions, thereby releasing desire from its customary cage. Otherwise alcohol is a sexual depressant. It interferes with the capacity both for erection and for orgasm. In large amounts neither is possible. Alcoholics may be totally impotent. Drugs such as morphine and heroin reduce sexual excitement and activity, but abrupt withdrawal of these drugs from addicts can lead to frequent erections and ejaculations in males, and orgasms in women.

The most bizarre addition to the list of reported aphrodisiacs, bearing in mind its current status of warm gentle goodness, is

47

cocoa. The Aztecs of Montezuma said their drink chocolatl was a sexual reviver, and the Spanish priests believed them. For over a century cocoa had resplendent fame as a violent inflamer of the passions, and then it quickly subsided into its present day role. Sir Hans Sloane holds the credit for the final indignity accorded to the Mexican brew when he invented milk chocolate. A drink fit for the halls of Montezuma became a cosy night-cap for the old world.

Spermatozoa   The first man to see bacteria and protozoa, the first to see red-blood corpuscles, and the first to describe the insect's compound eye was also the first man to see spermatozoa. Living a simple life as a rich merchant of Delft, Antony van Leeuwenhoek, spent much of his ninety-one years constructing his own microscopes, grinding his own lenses and then looking through them at the far more miniature world beyond them. His microscopes looked more like a draughtsman's compasses than optical instruments, but they could magnify by some 160 times. Therefore they could unlock much of the invisible world beyond the normal reach of the human eye.

It was in 1677 that he and his friend L. Hamm looked at human semen, and saw minute objects swimming in it. The Royal Society of London was then very new, having been formed the moment political stability and Charles II returned to England, but it was the leading scientific society in the world, and the Dutch microscopist sent it a letter detailing his findings. Simultaneously, bearing in mind the fact that human semen might be considered an indelicate subject for discussion, he begged the Society not to publish his findings should it consider them either obscene or immoral. The Society found them fascinating, and published without delay. The existence of human spermatozoa has therefore been known for 300 years.

The existence of such a human seed had already been assumed in many societies for countless centuries. In fact some considered the male seed to be all that was necessary, bar a little nourishment from the mother, for the production of the next generation. A contrary view held, for example, by many Australian tribes is that intercourse has no relation to pregnancy: they say it is a matter of spirits, not semen. The Ancient Greeks held that the mother of a child was no real parent to it, merely a nurse to the young life that had been sown in her. The Egyptians had also held that the father alone is the author of generation. Aristotle, who lived in the fourth century BC, could not quite accept these earlier views, and gave rather more credit to the mother. The embryo was attached, after all, for nine months to the mother's uterine wall, and therefore could be receiving not only nourishment but other influences.

It was up to the greatest polymath of all time, Leonardo da Vinci, to give the mother full credit. 'The seed of the female', he wrote, 'is as potent as that of the male in generation.' And he came to this conclusion after examining the offspring of a white Italian woman and a black 'Ethiopian' man. The children were neither black nor white, for both male and female inheritance had decided the mixed outcome. The remarkable part of this story, as in so many others, is not so much that one man saw the light, but that so many had failed to see it. It must have been common knowledge, if not in mediaeval Italy at least in any Egyptian century, that blacks crossed with whites produce a broad spectrum of browns. Leonardo was employing no microscope to see what could so conspicuously be seen by everyone else.

Nevertheless, spermatozoa needed an optical aid to become visible. They are minute. A micron is a thousandth part of a millimetre, and there are twenty-five millimetres to the inch. The sperm head is 2·6 microns in diameter, and 4·6 microns long. The tail is the longest part, being twelve times longer than the head, or fifty microns, or one-twentieth of a millimetre, or about one five-hundredth of an inch. Its thinness means that the fatter but smaller head is always easier to detect. Within this head, despite its minute volume, is all the tightly packed nucleoprotein of the chromosomes that are the male factors of inheritance. Within it is all that the father contributes to his offspring. When joined with the maternal chromosomes (and they are no bigger) all the information exists, all the nucleoprotein coding, for the next generation. The fact that the offspring will be a vertebrate, will be a mammal, will be a human, will be a type of human, and will have both paternal and maternal characteristics, is all ordained by the miniature bundle of chromosomes. And the masculine contribution to that bundle, precisely half – give or take a fragment of chromosome number 23 – is all contained in an active, motile, invisible, flagellating sperm with a head measuring 2·6 by 4·6 microns. One wonders why the Royal Society did not despatch Leeuwenhoek's letter about the microscopic human seed back to Delft on the grounds not of indecency but of incredibility.

Each human sperm is formed in about forty-six days. It is fashioned within one of the two testicles, and from a fragment of the tubular convolution – some say a mile long – of sperm-making cells within each testicle. It is a progressive manufacture. Upon the tubular membrane are cells called spermatogonia. These divide to form spermatocytes. These then divide, and each division marks a change as well as duplication, to form spermatids, and they then divide to form spermatozoa.

During this process the normal cell complement of 46 chromosomes has been changed to 23, following the so-called reduction or meiotic division when spermatocyte becomes spermatid. At fertilisation, when 23 fuse with 23, the normal complement for a human cell will again have been restored. Whereas the male production line is producing perhaps a couple of hundred million sperm a day, the female is maturing just one egg a month. Similarly, while a male may produce some infinite number around a million million in his lifetime, the female ovary actually sheds only some strictly finite number of eggs like 400, and no more, in her reproductive lifetime. Both egg and sperm contain the same amount of genetic material. Therefore the male's production swamps the female's, although both male and female contribute a precisely equal amount actually used to form the next generation.

Human testicles are strangely placed. No one designing an organism, having packed every internal organ within the confines of the body cavities, would be content with such an extra-mural position for the vital testes. The whole future generation of the species hangs not by a thread but is certainly less secure than with some internal system. Most animals do have such an internal system. Only in the mammals do the testes (or testicles) migrate from their original position up near the kidneys. The movement happens in early life, and long before birth. However, there is great variation within the mammal class about where, when and how this happens.

In some species it does not happen at all, and the testes remain in their original place somewhere near the small of the back. In other species they move down to the bottom end of the abdominal cavity, but still remain within the body, and in some (as in man) they move into a pair of pouches known as the scrotal sacs or scrotum. A male human baby almost always has his diminutive testes, which will not produce sperm until puberty, within his scrotum by the time he is born. (There is a widespread but false belief that the testes do not descend until puberty.) In humans the testes stay in the scrotum, or should stay there, for life; but this weakness in the defences of the abdominal wall means that the testes can move inwards, and parts of the intestine, for example, can move outwards as in inguinal hernia.

In no species is such a hernia either an acceptable or regular occurrence, but in many species the testes do migrate with the seasons. When unwanted they return to the abdomen. During the breeding period, they move again into the scrotum. This annual movement hammers home the point that, for many creatures, the testes must be external to be effective. Certainly the outside scrotum is at a lower temperature than the internal abdomen, and certainly temperature does seem to be involved; but the reason and mechanisms are much less clear. Rats have been made infertile by wrapping their scrotal sacs in cotton wool, and there is even a fertility theory that the breeziness of the kilt is better for man than the restrictive confines of pants and trousers.

It is difficult to see why sperm grow best at such different temperatures for such similar species. Why do some marsupials have internal testes when some do not? Why do the whale and the armadillo, for example, have them internally when their lives have so little comparison, and why are some scrotums firmly – and warmly – attached while others (as in most carnivores, and horses, and cattle, and many monkeys and man) are loose and pendulous?

There is even variation in the position of the scrotum among those animals possessing one. In man it lies behind the penis. In marsupials it may be in front, and in gibbons it may lie alongside the penis. In bats the scrotum may even be behind the anus. Some birds carry that matter of temperature still further by lowering their whole body temperature when sperm are being produced. For example, the house sparrow is normally at 110°F (43°C). Its night temperature, when most sperm are being produced, is 104°F (40°C).

Dimensions are also highly variable. In man each testis is about 2 ins. (5 cms.) long, 1 in. (2·5 cms.) wide, and less than 1 in. (2·5 cms.) thick. They stay that size throughout the seasons. With hibernating animals they shrink and then regrow. So too with animals having a definite season of sexual excitement. As man has no sexual respect for the seasons, sperm production is continuous. When formed, each spermatozoon is complete with its head, neckpiece, middle piece and tail – but this tail is without movement. Nevertheless each sperm has a long way to go – over a foot – before it even enters the female. Initially it is jostled along by the production of subsequent sperm until it reaches the epididymis. (Almost all biological names make classical sense despite their appearance. This one is from two Greek words meaning 'on' and 'testicles'.) Within this small lump of tissue, which does indeed sit on the testicle, the sperm mature, they become motile, and they wait – for this is the main sperm store-house. Both fertility and motility of sperm can last for several weeks, but if they are not ejaculated during this lifetime, the sperm degenerate and eventually liquefy.

Ejaculation If they are to be emitted, the long journey is accomplished very quickly. To begin with they travel up the thin vas deferens, the tube that is surgically cut during male vasectomy. This initial journey is longer than strictly necessary because of the original migration of the testis. Instead of leading straight to the penis, the tube leads up and over another tube leading from the kidney to the bladder, and can only then descend in the general direction of the penis. (Like a cat trailing a length of wool caught around a paw, the wool will indicate any back-tracking in its path. Similarly, the apparently random detour of the vas deferens behind the bladder positively indicates the original location of the testis.)

Having made this loop the tube then meets up with the seminal vesicles. These produce a yellow fluid which forms much of the semen, and which mixes with the sperm on its journey. The mixture then passes through the prostate gland, and this adds its complement of secretion, a thin fluid which gives semen its characteristic smell. The next addition to the mixture is from the small bulbo-urethral glands. And that is the lot. The resultant assortment of liquids and sperm, totalling between 2 and 7 ccs., and containing 90·3% of water, is then ejaculated down the urethra along the penis and into the female vagina. Swimming energetically at half an inch or so a minute the spermatozoa, perhaps two hundred million of them, perhaps more, set about achieving their goal. From this huge armada of an assault only one individual sperm will be successful.

Various niceties of co-ordination and timing have to be achieved during the headlong flight of orgasm. The urethra is the tube used by both urine and sperm. Therefore a sphincter has to be closed to prevent either sperm being shot into the bladder or urine being mixed disastrously (for urine is spermicidal) with the sperm. Removal of the prostate gland, a fairly frequent operation in the aged, usually destroys this sphincter. Therefore retrograde ejaculation – into the bladder – will result at least partially. The normal ejaculation process is a refinement of ordered co-ordination. The muscle coatings of that original epididymis, of the vas deferens, of those seminal vesicles, and of the prostate all contract, all shrinking the available volume, all forcing the sperm and semen towards the final urethra. And when they reach it a wave of contraction from the muscles around it, the bulbo-ischio-cavernosus, shoots those few vital ccs. straight out of the penis.

The haste of orgasm is succeeded by the haste to reach the egg. Human sperm have a short life. If still in the vagina after an hour they cease to move. If within the uterus or cervix they live longer, perhaps twenty-five, perhaps forty hours. In any case forty-eight hours is generally reckoned to be the limit. The sperm have to swim upstream because the normal uterine flow is towards the vagina; but, if the woman has an orgasm as well as the man, the uterus, or so many scientists believe, suddenly contracts and then relaxes, thus helping to suck up the semen towards the Fallopian tubes where fertilisation will take place. Female orgasm is not necessary for fertilisation. It is believed that without its aid the sperm will reach the tubes within an hour, although possibly within minutes if orgasm helps.

There is no attraction by the egg, if egg there be at the time, for the sperm. They cannot detect its proximity, but can only start the process of fertilisation if an actual encounter is made. This then involves digesting away the egg's outer layer. Many sperm help in the task by liberating the necessary enzyme (called hyaluronidase merely because much of that outer layer contains hyaluronic acid). Once digested, the egg is exposed, and ready for fertilisation by a single spermatozoon.

There is not such continuous haste in all species. Fertilisation can often take place months after copulation. Female bats can store sperm, and so can various reptiles. The female rattlesnake is often mated in the autumn, although ovulation will not occur until the spring. The record lack of urgency that I could find, and presumably the record length of life for viable sperm as well, was of a snake that produced fertile eggs in captivity five years after the last possible copulation. The only way of improving upon this record is to freeze the sperm artificially. Semen banks, used in the artificial insemination departments of animal husbandry, store sperm for long periods at −110°F (−79°C), or colder. Thus a bull's semen can be used long after the bull himself is dead. For some obscure reason different semen from different bulls responds either well or badly to the technique. And human semen responds (so far) even less well to lengthy periods in the deep freeze, although techniques for conserving it are improving. (It is not so much the length of time at the low temperature that is important, but the freezing procedure in acquiring these temperatures.)

Sperm character The difference in character between sperm from different individuals is supported by what have been called

53

sperm prints. Finger prints are well known; sperm prints less so. Nevertheless they exist, for all individuals tend to produce sperm with varying degrees of abnormality. And they tend to produce these abnormalities in fixed ratios. There may be twisted heads, or double tails, or immobile flagella, or any one of a wide variety of deformities, and they are produced by cells which, presumably, cannot make them otherwise. Hence the constancy of their mistakes; hence the ability of a scientist examining sperm through a microscope to match one collection of deformed sperm with a similar collection from the same man. The number of items which can be left with impunity at the scene of the crime is steadily decreasing.

Another major difference is in the fertilising power between individuals. The various possible disabilities of the human sperm are mentioned in the chapter on fertility, but all males tend to become infertile if their ejaculations are too frequent. Remember that it takes forty-six days to make a sperm; the available stock can be quickly consumed. Determined virility can lead to unintended infertility if production is not permitted to keep pace with demand. At the other end of the virility scale is impotence. This too can mean infertility, despite an abundance of sperm. Generally it is manifested in one of three main ways. The first is an inability to achieve an erection, either at all or for sufficient time – often associated with alcoholism. The second is an excessive ability to ejaculate sperm, possibly the moment erection is achieved, and before penetration is possible. The third and rarest is an inability to emit sperm: erection is achieved, but ejaculation does not follow.

Nomenclature   Once again there is a problem of names. A quite remarkable aspect of our society is that there is no word both universally acceptable and accepted for the sexual act. There is no equivalent of 'to eat', or 'to drink' for the equally basic function of sperm transfer. Instead our plethora of blunt or evasive terms, including both scientific, euphemistic and barrack-room phraseology, has to be used according to circumstance. Coitus, copulation, intercourse, intromission, make love, sleep with, go to bed with, fornicate, stuff, couple, fuck, prod, poke, mate, mount, lay, knock up – all (and presumably others) have their uses – but in fairly closely confined contexts. Animal husbandry has words of its own. To serve is the general description, but stallions cover, dogs and cats mate, bulls bull, and rams tup. The Bible, the law, the newspapers, the ante-natal clinics, science, literature, mess decks and the general populace have their individual preferences. Consequently there is much scope either for misunderstanding or for total inarticulation when different cultures meet, and wish to discuss this subject. The same is true both for defaecation and excretion. I well remember a fellow school-boy's insistence to a nurse that, so far as he knew, he had never opened his bowels in his life. 'What on earth are they?' he protested.

Mating time Whatever word is used, mankind is also odd in restraining most of its sexual activity to the time customarily reserved for rest and least physical activity. Man is a diurnal (day-time) creature, and yet most sexual play is in most human societies at night. Contrarily, the Chenchu of India believe nocturnal intercourse to be dangerous, producing blindness in a child. The anthropoid apes, just as diurnal as man, keep the night for sleeping; all their sex activity is in daylight. Nocturnal animals keep the day for sleeping, and have their sex activity at night. Day-time birds (most birds are diurnal) mate during the day. Night-time birds, such as the owls, mate at night. A. C. Kinsey and others suggested that human males prefer light, whether at night-time or not, whereas human females like the dark. (Kinsey, incidentally, so much an exponent on human sex was originally an entomologist, primarily concerned with wasps and with the intricacies of their sex.)

Mankind is also one of the few species where mating continues throughout the year, and with even fewer restrictions than most. The female can be receptive to the male virtually always and not solely – as with so many animals – during the short periods when in oestrus (on heat).

There are various taboos about when and where human sexual intercourse is forbidden, many allied to the belief that it weakens the male. For the rest of the time, according to C. S. Ford and F. A. Beach, 'in most of the societies on which information is available, every adult normally engages in heterosexual intercourse once daily or nightly'. For American couples Kinsey reported a slightly lower frequency, four times a week when aged fifteen to twenty, three times a week by the age of thirty, twice a week at forty, and once a week at sixty. Naturally such average figures contain considerable individual differences. In every age-group from fifteen to sixty there were men who reached orgasm at least ten times a week.

In the animal world there is less regularity of behaviour. There is usually a mating season, possibly with all the masculine activity being confined to only a few of the males, as in seals. Or there may be several oestrous cycles for the female in one year with each oestrous female being mated by several males before the period passes. Females in oestrus seem, in general, to be insatiable. Males are likely to find that the sexual capacity of the available females exceeds their own.

With animals the sexual act is often markedly brief. In the bull and ram it lasts a few seconds. In the stallion slightly longer. In chimpanzees it is less than ten seconds on average, rarely more than fifteen. In the elephant, aided by an independently mobile penis, it is less than thirty seconds; in the boar perhaps several minutes. Felines, such as lions and cats, may be seconds; but canines, such as dogs and wolves, may be far longer. They can become locked, as the penis only swells to its maximum size after insertion; premature attempts at unlocking can cause actual injury. Mice can continue for twenty minutes or so, producing perhaps 200 thrusts during that time.

In humans the time can vary widely. Generally it takes longer for the female to achieve orgasm than for the male, and the time of coitus may be controlled by the masculine ability to restrain his climax in the hope of coinciding with hers. An American survey (by R. L. Dickinson and L. Bean) concluded that 9% of couples took over thirty minutes from intromission to ejaculation, that 17% took fifteen to twenty minutes, 34% five to ten minutes, and 40% less than five minutes. Mae West, in a more personal survey, recounted a possible record in her autobiography – a man 'called Ted' made love to her for fifteen hours. He later said 'he was both astounded and pleased at his own abilities'.

The Balinese, although unkindly forbidding sexual intercourse to the sick and malformed, conveniently believe that a hurried love-making will result in a malformed baby. Female orgasm, although unnecessary for conception, is important in a well adjusted sexual partnership, but many women never experience it. It is unbelievable that the characteristic of a feminine climax is confined to the human species, but female animals do not in general appear to reach a similarly blatant pinnacle during coitus. The male's mounting climax and fulfilment is usually only too easy to observe. Equivalent behaviour in the female is markedly absent. Ford and Beach, from whose *Patterns of Sexual Behaviour* so many of these facts have been culled, suggest that the clitoris may be a clue. This organ, highly sensitive and directly equivalent (or homologous) to the male penis, usually causes great sexual excitement when stimulated. It so happens that if the penis is inserted into the vagina from the rear, with both male and female facing the same direction, the clitoris is less likely to be stimulated by the penis. Hence, according to the theory, a lack of sensation. Hence, possibly, an inadequate stimulus for orgasm. All mammals except man favour rear entry. Mankind, although wildly experimental in positioning, generally favours a face to face encounter. Such a situation, with such a method of entry, does generally lead to stimulation of the clitoris, and most human females experience orgasm.

The clitoris hypothesis is admitted by its authors to be 'highly speculative'. Perhaps female animals experience orgasm, but less demonstratively than their frantic partners. Or perhaps the human female is curiously unique in her climactic ability.

Sexual offences come most properly under the general heading of 'male reproduction' because it is so much simpler for a male to offend the law in western society. Sexual acts between male homosexuals were punishable in Britain (until the July 1967 legislation); those between female homosexuals were not. Similarly, penalties can be severe for any male convicted of rape. For a female accused of raping a male the act is not a specific crime; she may be convicted only of an indecent assault. The Institute for Sex Research in the United States put the situation bluntly by saying (in the book *Sex Offenders*): 'If a man walking past an apartment stops to watch a woman undressing before the window, the man is arrested as a peeper. If a woman walking past an apartment stops to watch a man undressing before the window, the man is arrested as an exhibitionist.'

The main law in Britain is the Sexual Offences Act of 1956. C. J. Polson, in *The Essentials of Forensic Medicine*, summarised the various crimes coming under its aegis. They can be grouped as:

(*a*) Intercourse by force, intimidation, false pretences, or the administration of drugs to obtain or facilitate intercourse;
(*b*) intercourse with girls under sixteen;
(*c*) intercourse with defectives (now called 'severely sub-normal persons');
(*d*) incest;
(*e*) unnatural offences;
(*f*) indecent assaults.

Male homosexuality came under group (*e*). Rape, which does not have to include the emission of sperm, although this can help to provide proof of the crime, comes under group (*a*). Under group (*c*) – intercourse with defectives – there is scope for quite unintentional guilt. An usherette in Glasgow once took her mongol daughter to the cinema, and showed her to a seat. During the performance the daughter had intercourse with a man who afterwards denied all knowledge of her deficiencies, particularly as the cinema had remained dark throughout the encounter. He was later convicted, but later still the conviction was quashed. Under group (*b*), particularly with increasingly early puberty and the increasing ability of young girls to appear older than their juvenile years, many males must be victims of their own assumptions and the girls' appearances. As a kind of proof, babies are born every year to many mothers aged thirteen, sometimes aged twelve, sometimes eleven. As nine months have to be subtracted to reach the actual date of the offence, making their ages probably twelve, eleven and ten, the girls are jumping the gun to a considerable degree. Nevertheless it is their male consorts who are found guilty. The Act of 1956 does not recognise 'consent' by any girl until she is sixteen, although it does recognise the difference between girls who have reached their teens and those who have not. Intercourse with a girl over thirteen, but under sixteen is only an offence. Sexual intercourse with a girl under thirteen is a felony, a crime subject to severe punishment. Only boys over fourteen are considered capable of rape – in England. In Scotland there is no such watershed; theoretically all boys are capable of rape.

The 1956 Act was supplemented by the Indecency with Children Act of 1960. This makes a criminal of any person who, without committing an assault, 'invites a child to handle him or otherwise behaves indecently towards a child under fourteen'. If the girl is under thirteen and the man tries sexual intercourse he can get seven years' imprisonment. If his assault is merely 'indecent' he can get five years. Incest is defined as sexual intercourse between a man and his granddaughter, daughter, sister or mother, or between a woman and her grandfather, son, brother, or father. The practice must be commoner than the few facts indicate, and information about most incestuous indiscretions is likely to remain within the family circle. The cases which reach the courts usually involve a father–daughter relationship or one between brother and sister. Sex crimes, in general, have increased so far as the courts are concerned. In 1938 there were about 5,000 (in Britain). In 1945 the total was little short of 10,000. In 1960 it was 20,000, a four-fold increase in twenty-two years, and roughly there the figure has stayed since then.

Newspapers give considerable coverage to any association

between the murder of a child and sexual motives. In fact a child has only to be reported missing for the assumption to arise that its naked and assaulted body will be found a few days later; but in 1964, for example, thirty children were murdered during the year. Of this number two were unsolved, twenty-seven were murdered by relatives (almost always a parent) and only one was murdered by someone who was not a relative. Offences by strangers make the most headlines, but most detected sexual acts with children are still committed by relatives or friends. Children should therefore beware of strangers, but statistically they are far more likely to be harmed either by friends and relations, particularly their parents, or by accidents. Sexual murders of children are outnumbered by ordinary child murders, and both are quite outnumbered by road accidents involving children and by accidents the children bring upon themselves.

Homosexuality Facts on homosexuality tend to be obscured by society's attitude to it. Estimates on the number of male homosexuals vary between 3% and 5%. A round figure frequently quoted for Britain is 500,000. Under particular conditions, such as prison camps, the proportion rises. A. C. Kinsey and others, in their survey of sexual behaviour, reckon that 40% of adult American men and 13% of women had experienced homosexual contact to the point of orgasm. Greater proportions had experienced homosexuality to some degree. Male homosexuality is generally commoner than lesbianism, both with man and animals. Bryan Magee, in his book *One in Twenty*, is referring in the title to the number of men considered to be homosexual in Britain. Estimates on lesbian numbers are usually smaller and more vague.

On average, homosexuals (and this means male from now on) have more brothers. The normal average of families is 105 boys to 100 girls, while the homosexual average is 119 boys to 100 girls. Lesbians have more sisters than normal. Mothers of homosexuals are older than average, and homosexuals also tend to appear late among the family births. (These two last facts may seem to be a rewording of the same phenomenon, but both are quite distinct.) Homosexuals have no detectable chromosome abnormality, and are therefore nothing like the intersexes. Despite much current opinion, homosexuals do not have an excess of feminine features. They are as physically capable of fathering children as hetero-sexuals. There is no necessarily hard and fast line between the two. Many men can be both, with preferences verging one way or the other.

Most homosexuals are not, as is commonly supposed, interested in young and pre-pubertal boys. The proportion of formerly illegal homosexual acts to known offences was phenomenal. Estimates vary from 2,500:1 to 30,000:1. Societies still have and have had totally different attitudes towards homosexuality, ranging from complete acceptance to complete intolerance. The Greeks put it on a high plane. The Romans regarded it in a more basic fashion. The higher primates frequently engage in it, whether in zoos or not. No one knows whether it has a genetic cause or not. E. Maurice Backett, to whom I am indebted for many of these facts, has written, 'Most biologists would put heavy bets upon the environment as the main contributor in the production of homosexuality'.

Theories about this contribution, although similarly obscured by society's attitude to homosexuality, are plentiful. The men's mothers have been called demanding, over-protective, domineering, seductive and inhibiting. Their fathers have been described as weak, indifferent, absent, hostile, abusive or rejecting. In general, more positive characteristics are ascribed to the mother, weaker ones to the father. The family as a whole is often said to be rigidly puritanical. Of course, there are exceptions even to this broad spectrum of suggested possibilities. Countless mothers must be protective and fathers indifferent without the son becoming homosexual. And homosexuals must arise from the most well-balanced families. Such exceptions appear with monotonous regularity in every biological generalisation, and of course they cannot be forgotten. Nevertheless, the general picture is of an ill-balanced parental situation, with the mother highly emotional. It seems likely, says Backett, that a son subjected to such a charged relationship becomes disturbed by, or horrified at, the 'implicit sexuality of the relationship, and tends thereafter to renounce the female as an object of his sexuality'.

Can homosexuals be treated? Can they be helped to revert to normal heterosexuality, or adjust themselves satisfactorily to their problem? Backett says attempts have not been 'strikingly successful'. A *British Medical Journal* leader in 1965 said treatment for the established homosexual was 'frequently ineffective', although the journal was then jumped on by one doctor who said he had converted 16 out of 32 patients to heterosexuality. Plainly the task is not easy and must depend on the degree of homosexuality, the willingness to co-operate, and the desire to revert to heterosexuality. In Britain the clause which made criminals of homosexuals was part of an Act, and some say it was included inadvertently, designed to protect women from assault. However inadvertent, the Criminal Law Amendments Act of 1885 could put a man in prison for a private homosexual 'offence' with another man – until 1967. Soliciting, pimping, homosexual brothels, and homosexual acts with minors are still punishable in Britain. It is only acts performed in private by consenting males over twenty-one that have been freed. There is no law against lesbianism, but there is a story that Queen Victoria had a royal hand in this omission. The original wording of the law embraced both male and female homosexuals. She refused, allegedly, to countenance the idea of lesbians, and so struck them out before signing.

In theory the word homosexuality should suggest either male–male or female–female relationships; but, save in some very correct circles, it is reserved for the men who already possess a far wider vocabulary for their kind. *Time* magazine once compiled a list: fairy, nola, pix, flit, fag, faggot, queer, homo, agfay, fruit, nance, pansy, queen, she-male, mary, but no doubt there are many others. The word gay arrived in the 1960's, and entered general usage in the 1970's, having been selected by the homosexuals themselves rather than the other way about.

Finally homosexuality is not intersexuality. In the chapter on inheritance the various types of intersex are mentioned, such as the male hermaphrodites, the pseudo-males, and the female pseudo-hermaphrodites. These physical intersexes, or between-sexes, may be due to abnormal chromosomes or abnormal glandular development; but there is no simple relationship between them and abnormalities of human sexual behaviour. A homosexual cannot be seen to be a homosexual merely by looking at his chromosomes. Physically he or she is usually a typical specimen of his or her sex: it is only the behaviour that is abnormal.

Castration   The word eunuch comes from two Greek words meaning to guard the bed. It was to frustrate any sexual intentions on their part that so many men suffered this indignity. Nevertheless such intentions may not be halted if the operation is done too late. One suspects that many a eunuch, castrated after puberty, had quite a merry time in the harem because neither desire nor performance need be diminished by the total lack of sperm. If the testicles are cut off before puberty the boy is never affected by all the changes of puberty. His voice does not break. No hair grows in the masculine places, such as the face and the chest. His pubic hair grows in a feminine fashion. His body acquires fatty deposits like a girl's, and his muscles are weak. He

61

may grow very tall, although this seems to depend upon the growth spurt and the precise time of castration. His skin will be pallid. Despite all this, despite the absence of the hormone testosterone which is produced in the testicles, claims have been made that some early castrates have been able to produce erections, some even to copulate. They are the exceptions. Most prepubertal castrations are not followed by any form of sexual competence.

The story is different with adult castrates. Their voices have broken. They are already the shape of a man. They know about sexual desire. Their penis, assuming this part is retained, is of a normal size. And many of them, with a total lack of testicles and of sperm, and with a withering of various sexual accessories like the prostate and the seminal vesicles, will maintain normal desire and potency for decades after the operation. Their sexual activity will be satisfactory, and ejaculations of a sort will be possible. Other men, influenced perhaps by an assumption that castration is followed inevitably by a cessation of previous sexual activity, do find that desire and capacity both wane fairly rapidly. A third group claim desire but an inability to satisfy it. Castration can, reasonably enough, be a psychological trauma, producing various consequences in its wake, but its direct physical manifestations are very slight, save the over-riding effect upon reproduction.

The Chinese, and various other Eastern groups, were not generally content to give their eunuchs any chance of sexual potency. They performed a total removal of the genitals with a hallowed operation involving one deft stroke of a sickle-shaped knife. The urethra, the tube leading from the bladder, was then plugged. The wound was bandaged, the victim ate nothing, and the pewter plug was removed after three days. A city called Ho-chien-fu, near Tientsin, had the honour of producing most of the eunuchs for the Imperial Court, and the practice has not been confined to days of old. Even in this century a European doctor in Hong Kong was introduced to 'an irascible Chinese gentleman from the north'. He proved to have no external genitalia, and his urethra was found with difficulty in a great mass of scar tissue. He admitted the operation had been at the age of twelve, and his home had been 'near Tientsin'.

Eunuchs, one might have thought, would have been ideal not just as bed-guards, but in those professions which demand the absence of sexual activity, such as the Roman Church. In fact they are banned from ordination. The Law of Moses stated that eunuchs could not enter into the sanctuary, and Leviticus even considered that castrated animals should not be permitted as sacrifices. The Greek Church (and I am indebted to Dr Charles W. Lloyd's book on human reproduction for this account) did not accept this discrimination against eunuchs, but the Roman Church – according to legend – felt that anyone elected Pope should prove that his genitalia were intact. A special chair was fashioned, one of which is in the Louvre, that had a horseshoe-shaped seat, much like the old birth-stool, upon which the Pope would allegedly sit. The cardinals would pass by, checking the papal possession and proclaiming *'Testiculos habet et bene pendentes'*. It is not known – for sure – whether the tale is strictly accurate. If it is true, how on earth did Pope Joan get elected? Or is she totally apocryphal? Or, as suggested by Friedrich Gontard (in his book *The Popes*), the various wild stories about a woman sitting on the papal throne around the year 855 may have led to the examination. The custom was apparently discontinued after the papacy of Leo X in the sixteenth century.

Circumcision Any mother wishing her child to retain his prepuce 'would be well advised to maintain permanent guard over it until such time as they both leave the hospital'. So said Dr W. K. C. Morgan. This British doctor practising in the United States was expressing concern at the 'rape of the phallus', the habit of circumcising new-born males with routine indifference. He has a hard task ahead of him if he intends altering the present attitude towards the operation of circumcision. It is as old as the hills (Egyptian male mummies were all circumcised). It is supported by religions (with the Bible and the Koran constantly haranguing the uncircumcised). It is practised in innumerable societies (where there are puberty rituals, or merely traditional mutilation). It is advocated by large segments of the medical profession (who refer to greater hygiene, less cancer of the penis for circumcised men and less cancer of the cervix for their wives). It is promoted actively and doggedly by many women, who have no valid reasons for their prejudice, save conformity, tradition and, say some, a desire to influence masculine characteristics. It is also a remunerative procedure for many of those who carry out the operation. Any crusader fighting such a formidable battery of opinion, prejudice and custom is going to have a very long crusade.

He will also find the enemy very determined. Considering the insignificant size of the new-born foreskin, and considering its relative unimportance, the arguments for and against its retention have generated great heat. In Britain the operation is now being done less frequently than it was between the wars, but it is hard to find out the actual incidence. In the United States it is still being carried out fervently, and often two to three days after birth. In Britain, when it is done, it is usually done a little later. With Jews it is virtually as important as it has always been. In Britain many parents can find their request for a baby's circumcision bluntly refused by hospitals. In the United States, says Dr Morgan, most American hospitals have an insatiable urge to circumcise.

In Britain it is often held to be an archaic procedure, and Sir James Spence in 1950 proclaimed this view in a delicious letter to a colleague. 'If you can show good reason why a ritual designed to ease the penalties of concupiscence amongst the sand and flies of the Syrian deserts should be continued in this England of clean bed linen and lesser opportunity, I shall listen to your argument; but if you base your argument upon anatomical faults, then I must refute it. The anatomists have never studied the form and evolution of the preputial orifice. They do not understand that Nature does not intend it to be stretched and retracted in the Temples of the Welfare Centres, or ritually removed. . . .' On the other hand, and on the other side of the Atlantic, to quote from a representative textbook, it has been 'almost universally recognised that circumcision is a valuable procedure'. Someone must be right. Or at least more right than the others.

There are facts; but there is also disagreement about their interpretation, and difficulty about their acquisition. One major problem associated with questionnaires is that a wife is often confused about her husband's penis, and the husband too is often at fault. In a survey, in Buffalo, New York, one-third of the men claiming circumcision were wrong, and one-third disclaiming it were equally wrong. The women were even more inaccurate. A fact generally agreed is that cancer of the penis is virtually eliminated by circumcision. This cancer is common in many primitive countries but not in the United States or Britain. England and Wales have 300 to 400 cases a year. To counter this long-term reason for circumcision are various other reasons, mainly short-term, such as the pain it presumably causes the child, the ulcers more likely to occur on the circumcised, the increased risk of infection – some hospitals have banned the operation on this account alone, and the possibility of death. Dr Morgan quotes fifteen deaths in Britain a year as attributable to circumcision (although this was based upon a statement in 1949).

Another fact is that nuns (never) and Jewesses (hardly ever) get cancer of the cervix. Admittedly nuns are different from the normal community in many ways – no intercourse, no contraceptives, no children – but Jewesses are generally similar save that they are all married to circumcised men. Another difference with Jews is that they are supposed to follow the Niddah ritual, or no intercourse for a week or more after menstruation begins, but some Jewish groups are very lax about this. Also the Jewish freedom from cervical cancer varies geographically; North African Jews are not so free.

Does the foreskin, and in particular the smegma, the glandular secretion which lies under it, help to cause either cancer of the penis or cancer of the female cervix? No one knows, but many want to know, particularly as cancer of the cervix kills about 2,000 women in England and Wales every year. Why should most Jewish women, notably those from central and eastern Europe, evade it so successfully? Is it that Jewish standards of hygiene are higher? There are bacteria which live on smegma. Perhaps they, less rife in cleaner people, carry the carcinogen which does the harm? Existing evidence suggests that a clean foreskin is no more harmful than no foreskin. At present, according to a leading article in the *British Medical Journal*, 'there is no case for recommending circumcision rather than hygienic measures to adults as a means of preventing cervical cancer'. The same leader was more in favour of circumcision for babies 'when advice on hygiene is unlikely to be followed'.

A further argument frequently cited – one suspects mainly by women – for removing the foreskin is that the tip of the penis lying beneath it, the highly sensitive part called the glans penis, becomes less sensitive when exposed to the daily abrasion of clothes. Hence, and there is some evidence for this argument, the circumcised male can take longer to reach his orgasm, thereby being more likely to coincide with his slower partner (although *Human Sexual Response* by William H. Masters and Virginia E. Johnson reports no difference in glans sensitivity between circumcised and uncircumcised males). Sometimes, but rarely, there is a straightforward call for circumcision: phimosis, for example, is a condition which can affect infants and is caused by the foreskin blocking the flow of urine. Balanitis, an inflammation beneath the foreskin, which was caused occasionally by the universal sand of World War Two's North African campaign, can occur at any time. Nevertheless, when compared with the daily slaughter of foreskins, such complaints are extremely rare.

7. The Female

Anatomy The male's reproductive function is to manufacture sperm, and then transfer them to the female. The female's function is to receive the sperm, produce the egg, provide protection and nourishment for the developing form, and then expel it from her body. There is some economical duplication of function: the vagina acts both as a receptacle for the sperm, then as the birth canal for the offspring 266 days later. Uneconomically, there are two ovaries and two Fallopian tubes, when one of each would do. The human female has an undeniably more complex reproductive role to play than the male, but her basic reproductive anatomy is no more complex than his.

Equivalent to the male testes are the female ovaries. They are slightly smaller than the testes, when both are fully grown, and they lie internally roughly three inches on either side of the midway point between vagina and navel. This human similarity in the size of the testis and ovary is odd, bearing in mind their totally different approach to the identical task of producing germ cells for the next generation. The male testis does not really begin its manufacturing abilities until puberty; thenceforth sperm are produced in their millions. The female ovary has all its eggs at birth – a dozen or so years before puberty. Moreover, by the time puberty is reached there are far fewer than there were at birth. Women live on their capital, while men start with nothing and then produce in abundance. It is generally reckoned that a newborn female has 200,000 to 400,000 ova in her ovaries, but, by puberty, the number has shrunk to about 10,000. Of that smaller figure only about 400 will actually be shed as mature eggs from the ovaries. Only one in a thousand of the original supply (or fewer if the woman has a lot of babies) will have a chance of being fertilised. A normal man is manufacturing far more sperm every second than a woman is producing mature eggs in her lifetime.

Basically the female reproductive anatomy is in the form of a Y. At the extremities of the two top arms are the ovaries. Leading from them, and forming the two arms of the Y, are the Fallopian tubes. Sometimes called the oviducts, or uterine tubes, or ovarian

66

tubes, they are very thin. They are able to be so thin because the egg which will pass along them is a speck one-tenth of a millimetre across and just about visible to the human eye. Where these two tubes join at the centre of the Y is the uterus. Traditionally called pear-shaped, and roughly pear-sized, this muscular lump of tissue will house, nourish, swell enormously, and then expel any future baby. Unlike a pear it has the stalk pointing downwards. This thinner end of the pear is the cervix, the neck of the womb. It virtually seals off the diminutive cavity of the uterus, but it changes its shape dramatically and effectively in the last few hours before birth. After the cervix, and forming the bottom leg of the Y, is the muscular tube of the vagina, the connecting link with the outside world, the receiver of sperm, the exit passage for the baby.

The symmetrical arrangement of two ovaries leading to a united or single vagina is the rule, but just occasionally there have been cases of total symmetry, i.e. two ovaries leading through two uteruses (or uteri) to two vaginas (or vaginae). With such a system two quite different pregnancies have been known to occur, each with different conception times and different delivery dates. The basic simplicity of the female anatomy is paralleled by that of the male but, unlike the male system, which is consistent throughout its post-pubertal life, the female system has its cycles. Quite apart from the cyclical programme of a baby's gestation, there is the cycle of egg production, the cycle of menstruation, and the finality of the menopause.

Oestrus As processes the cycles need independent elaboration. Before doing so, and in order to emphasise some of their peculiarities, a look at the rival systems used by animals is germane. After all, to criticise it bluntly, the human system means that: successful matings are only possible for a couple of days a month; there is no co-ordinated mechanism whereby copulations are more frequent at this time; there is no general indication of ovulation; there is all the unpleasantness of the time both before and during menstruation; and the whole system packs up long before the woman herself is likely to die.

Most mammals, unlike humans and various other primates, follow the ritual of the oestrous cycle. (Oestrus is the noun, oestrous the adjective, but estrus in the US.) It is characterised by a time when the female is on heat, when she most attracts the male, and when she will mate with him. It is also characteristically co-ordinated with ovulation – the production of an egg or eggs. Sometimes the egg appears spontaneously, as in rats, mice and whales, but sometimes the very act of mating, as in rabbits, cats, mink and ferrets, causes ovulation. The cat needs three matings for ovulation to occur, and the egg is produced twenty-six hours later. Spontaneous ovulation, not dependent on coitus, is the commoner process. The sexual attraction that goes with it and which assists in the coordinated arrival of sperm and egg, is usually followed by a period of repulsion, when the female will have nothing to do with any male, and the males are not attracted by the non-oestrous females.

With the majority of animals sexually ruled by the behaviour of the oestrous cycle, it is usually the female that calls the tune; either she welcomes mating, or she does not and her consort just takes the opportunity when it comes. In a few species the males also have a cycle of their own, known as the rutting season; the males are not sexually inclined until they are 'in rut'. Camels, elephants and deer are examples.

Some creatures carry such a sexual season still further. In the mole and spotted hyena, for example, the vagina closes up at the end of each cycle. Hence the old countryman's idea that female moles can only be found in the spring, and the African notion that hyenas are hermaphrodite. Neat co-ordination of ovulation with mating can go wrong; rabbits, cats, cows and horses, for instance, can come on heat during pregnancy. Subsequent matings are, of course, useless as the uterus is already occupied , but in the hare it can lead to what is called superfoetation. With this animal it is a regular occurrence. Matings can successfully occur three days before delivery of the gestating litter. This means, for the wild hare, that litters can be born every thirty-nine days even though the actual gestation period is forty-two days. Rats and mice, although producing litters very rapidly, cannot quite emulate the hare's ability to jump the gun. However, they do achieve oestrus within twenty-four hours of the birth and, if mating then follows, it cannot be said that much time has been wasted.

For a long period it was thought that no difference existed between a woman's menstruation and the ovulation of animals in oestrus. After all, a bitch, say, loses blood when in oestrus, and so does a menstruating woman every month. In fact the two things are quite distinct because ovulation occurs *between* menstruations and not at the time of menstruation. This particular manifestation, which is so different from the traditional oestrous cycle, occurs in apes, monkeys, and a few shrews, as well as in humans. By the time human or monkey menstruation occurs, and when the uterus is shedding its prepared surface, the ovulated egg has long since died and withered away. Oestrus and ovulation go together: menstruation and ovulation do not. As menstruation occurs in the higher mammals, and is certainly a subsequent development to the oestrous pattern, the method presumably has advantages, even though they are not strikingly apparent.

The menopause, or complete cessation of female reproductive ability, appears to be confined solely to womankind. It is true that animals in the wild have little chance to live to a ripe and infertile old age, and animals in zoos are not truly representative of their kind, but study of both shows that there is no stage in the animal world comparable to the human menopause. Animals do definitely breed less as they get older, but there is no such climacteric as that of the human female. Rather, as with men, their powers gently lessen as they grow more senile.

As a system, the human reproductive method could plainly be improved. A compensation is that, as intercourse is so casually geared to fertilisation, there is much more desire and opportunity for it than with animals whose oestrous cycles forbid, prevent or discourage any sexual behaviour for large portions of every year.

Ovulation   The production of one egg occurs in one ovary at a time. Recent work, principally from France, suggests that the two ovaries tend to ovulate alternately, possibly because the hormone progesterone exists at a higher level in recently ovulated ovaries and this inhibits egg production. The female egg is produced in a much more complex fashion than any sperm. For one thing it has lain dormant since before birth, and therefore is probably twenty, possibly forty years old before its sudden awakening. Although one egg is shed regularly some two weeks after the start of menstruation, the mechanism of its shedding has started a few weeks before that particular menstruation. During this procedure the chosen egg is surrounded by a swelling bundle of cells. Known as a Graafian follicle, this bundle slowly enlarges within the ovary. Containing just one egg the follicle is almost half an inch in diameter fourteen days after the start of its particular menstrual period.

Usually just one follicle from one ovary is ready then, but two may have developed. When this has happened here is a possibility of two-egg – or dissimilar – twins being initiated. Their two eggs may have developed. When this has happened there is a possibility random procedure. Anyway, whether two follicles are ripe, or merely one, they move to the edge of the ovary. Then, because of the increasing pressures within them, they burst and the eggs are liberated. This is ovulation. It takes place usually thirteen to seventeen days after the start of its particular menstrual period. It can create slight pain, called mittel-schmerz, and this pain has been the cause of many an appendix operation when doctors, not realising the monthly regularity of its appearance, have confused it with appendicitis. Ovulation is associated with a change in temperature, and the time of ovulation can thus be assessed, but most women remain quite unaware that one follicle ten millimetres in diameter has ruptured and has freed its solitary egg.

The egg itself, having waited so long within the ovary for its crucial hour, suddenly has very few hours to live, probably no more than forty-eight, possibly no more than twenty-four. The preceding period of inactivity could well have been twenty-four *years* or even forty-eight *years*, but suddenly the length of time when it is ripe for fertilisation is extremely brief. If an encounter with sperm does not occur the egg will become infertile, and deteriorate. By then it will have passed along the few inches of Fallopian tube, and will have reached the small cavity within the uterus. From there, at the next menstrual flow, the unused and aged egg will be flushed out of the system.

The follicle that produced the egg has a greater role to play than the mere production of that egg. It secretes hormones, initially oestrogen, then progesterone. The chemistry is complex, but the tasks of the hormones are basically straightforward. Oestrogen, which comes from the ovary itself, gives rise to oestrous behaviour, and prepares the vagina for mating. Progesterone supports the changes going on in the uterus – all the hopefully receptive preparations for the embedding of a fertilised egg. Each disrupted follicle is thus continuing to care for its solitary egg after that egg has escaped.

The care persists – assuming no fertilisation and no pregnancy – for twelve to fourteen days. Known by then as the corpus luteum the follicle has grown further in size and has poured out its hormones; then it stops doing so. The uterus, suddenly deprived of the vital chemicals for its preparatory changes, suffers menstruation. And the follicle, without any further need for its activities, suffers degeneration. However, should there be fertilisation, and should there be a successful implantation of the developing egg, the follicle will then continue to produce hormones. The uterus and its prepared receptive layer have to be suitably maintained, and progesterone has to be manufactured to maintain that suitability.

So the follicle grows, and the large yellow object it has become remains vital to the continuance of the pregnancy. Eventually, when the placenta has grown, and is itself producing hormones for its own maintenance, including progesterone, the old follicle will degenerate. This happens about half way through pregnancy. By then the carefully nurtured egg, cared for, cast out, and then more distantly supported, is a foetus of sizeable proportions.

Menstruation   These days, in societies where pregnancy is an exceptional event rather than the inevitable rule, most follicles do not have pregnancies to care for. More frequently they degenerate after those twelve to fourteen days, and trigger off menstruation. They do not cause it so much as they suddenly fail to prevent it. Whether this is a tautology or not, menstruation is undoubtedly a bizarre phenomenon. Some have called it a paradox and a puzzle, a seemingly useless procedure, inexplicable, wasteful and disturbing.

The menstrual cycle begins at puberty. The outward manifestation of blood flow takes a good many girls completely by surprise. Some surveys indicate that about a third of the women in Britain have received no pre-menstrual counselling, and the appearance of the blood is the first they learn about the menstrual cycle.

The flow is symptomatic of the end of the cycle, and not the beginning. The menstrual cycle had started on its monthly course several days earlier. The innermost layer of the uterus is called the endometrium, and initially it is very thin – about one millimetre. Then it thickens. Within its substance various glands and arteries also lengthen and enlarge, and by the time of ovulation, normally some eight days after the earlier menstrual bleeding has stopped, the endometrium is about 3 millimetres thick. For girls having their first menstruation this preliminary process is likely to take longer than normal.

Anyway, whether for the first or any subsequent menstruation, the glands within the endometrium, which secrete a mucus, become straight and tubular. During the next fourteen days these glands become more convoluted, more like concertinas than tubes. The endometrium containing them continues to thicken, and soon reaches 5 millimetres (half a centimetre, one-fifth of an inch). Roughly mid-way through these fourteen days any fertilised egg will have been implanted. It takes about a week after ovulation before implantation occurs. The egg will both have implanted itself and will also have been implanted by the uterus, for the process is mutual; both developing egg and welcoming uterus assist in the embedding. Should no implantation occur there will be menstruation. The bloody disruption is the product of a frustrated womb, one which has not experienced implantation in that particular month. The menses are the outward sign of the internal changes.

With the sudden stopping of the flow of progesterone, cut off by a finally disillusioned corpus luteum, the uterine preparations abruptly end. The top section of the endometrium is shed; the 4 millimetres built up in the preceding three weeks are discarded. Not all of this layer is shed at once, but patches here and there. The shedding is due to changes in the various hormone levels, but it is still not known precisely which does what to what, as they are all inter-related, and experimentation is difficult. In any case, four to five days after the start of the shedding all the carefully prepared layer has been affected. Together with blood and a lot of mucus the discarded endometrium is exuded first from the uterus, then from the vagina. The average blood loss in each period is 40 millilitres, or less than one-tenth of a pint, but it may be five times as much, or a fifth as much. The range of variation in this and other aspects of menstruation is considerable.

No woman is absolutely regular. The 'monthly' period may have a normal cycle of between nineteen days and thirty-seven, although twenty-eight days is an average interval. Usually, the more average the length of the period, the more fertile the woman. The irregular lengths of time are generally between ovulation and the preceding menstruation; there is far greater constancy between ovulation and the succeeding menstrual bleeding. When the cycle is finally over, the endometrium then starts to grow again. And once again, assuming no pregnancy, it will be shed only to start to grow all over again. Dr George Corner, of Princeton, has called menstruation 'an unexplained turmoil in the otherwise serenely co-ordinated process of uterine function, a puzzling paradox whereby a normal function regularly displays itself by the destruction of tissues'. For the average woman the paradox will irritatingly happen about 440 times in her lifetime, if she has no children, and about a dozen times fewer for each child. Its regularity will end finally with menopause. But before that happens she will have learned that there is more to menstruation than those merely physical changes regularly taking place within her uterus.

The pre-menstrual syndrome, cyclical syndrome, and pre-menstrual tension are names given to the host of complaints experienced by many women (some say 75%) in the few days before menstruation begins. These include headaches, irritability, tiredness, nausea, and aching joints. A survey at Holloway Prison completed shortly after World War Two showed that '93% of female crime was committed during the pre-menstrual phase'. Other surveys have shown that well over half of the accidents in the home occur in the pre-menstrual week, and that women driving cars at this time are more likely to be involved in road accidents. Millions of working days are lost from the pre-menstrual cause.

Physiologically the main difference at this time is water retention, often by as much as 6–7 lbs. (or two-thirds of a gallon). This is caused by progesterone. The oedema, or liquid swelling, may be too diffuse to be noticed, or it may manifest itself in the face, the ankles, the breasts, or the abdomen. Perhaps the oedema directly affects the brain, which would help to explain many of the symptoms. Oddly, women who suffer pre-menstrual tension seem to have little or no pain during the menstrual flow and, conversely, women who suffer a painful flow – dysmenorrhoea – rarely suffer pre-menstrual tension. There is a rough justice in this arrangement.

The retention of water is taken many stages further by those animals which possess sexual skin. Although not a feature of human beings, such skin is part of many non-human primates, of which the mandrill, with its blatant colouring, is a good example. Some monkeys have their sexual skin around the vulva, some around the anus, some on the face. It starts to swell with the start of each cycle, and reaches a peak of size and coloration at ovulation. In fact, the sexual skin may, by then, be so swollen that it incorporates one-sixth of the animal's total body weight. After ovulation the balloon collapses, the water is suddenly lost via the urine, and the colour recedes. The cycle will then begin all over again. As a digression one can speculate upon the possible changes in human behaviour if the human female had been endowed with such a blatant demonstration of her fertile hour.

The other important change during the menstrual cycle is temperature. Its prime attribute is its guidance to the timing of ovulation. In the days preceding ovulation, when the uterus has once again started on the task of preparing the endometrial layer, a woman's temperature remains reasonably constant, but tends to get slightly lower as ovulation day approaches. Then, roughly fourteen days after the preceding menstrual bleeding began, there is an abrupt but brief drop of 0·2°F to 0·4°F (0·1°C to 0·2°C) below the normal temperature. This is considered to indicate the crucial hour, the bursting of the follicle and the liberation of the long-imprisoned egg. To make slightly more of a landmark in the temperature chart, this small and brief depression is followed by a larger and longer rise, the so-called thermal shift. Temperatures vary from woman to woman, but a typical temperature pattern may start off at 98°F (36·7°C), fall slowly to 97·8°F (36·6°C) in the early days of the cycle, fall abruptly to 97·6°F (36·4°C) at

73

ovulation time, rise to 98·5°F (36·9°C) and stay there until a day or so before the next menstrual flow. It will then fall to 98°F (36·7°C) again. Should it not fall just before the next flow this can indicate pregnancy, although the subsequent lack of flow will help to confirm this.

Apart from indicating ovulation, and when couples urgently desiring a child should best have intercourse, a woman's temperature chart can also indicate either normality or abnormality. Faulty ovulation, faulty progesterone quantities and other defects can be implied by a faulty temperature rhythm. There should, on average, be a difference of 0·9°F to 1·3°F (0·5°C to 0.7°C) between the highest and the lowest temperatures of the cycle. The best day to have intercourse for those couples who want a baby is the day of the thermal shift; some 50% of conceptions are thought to occur on that day. The remaining 50% occur either slightly before or slightly after that time.

Sex determination and virgin birth Brief reference has to be made to two of the oldest 'old wives' tales' of all time.

There are two ovaries, and offspring are of two sexes. Hence it was almost inevitable that a theory should arise connecting the two. Whether the right ovary produces boys, and the left girls, or whether the system is the other way about, varies from place to place, but neither tale has a whisper of validity about it. The male sperm are the sole arbiters of the offspring's sex. The female eggs, whether from left or right ovary, are identical. The only way a determined feminist could advocate a mother's influence over the sex of her offspring would be to produce a new theory. As only sperm can carry the essential maleness for boys, the new theory would have to suggest that the mother's egg or Fallopian tubes were less eager, on occasion, to receive either the male-determining sperm or the female-determining sperm. Such a theory would find few backers. On the other hand no one knows why more boys are conceived. Is it because more male-type sperm are produced? Or because male sperm are swifter? Or because the female-type sperm are tolerated less happily by the women they are invading? Despite the sperm being the arbiters, women may have some influence in the matter after all.

The importance of sperm is also crucial to the possibilities of parthenogenesis, or virgin birth. The ability of an egg to divide on its own, and to produce a normal adult of either sex, without the initial assistance of the fertilising sperm, is a common occurrence in many of the lower animals, but it is highly improbable, to say the least, that virgin birth should occur in the higher and highest animals. However, several claims have been made by woman-kind.

The onus upon such women is to prove the non-intervention of any male. Parthenogenetic eggs, which have started to develop by themselves, have been found in human ovaries, and it would be hard for anyone, male or female, to assert categorically that virgin birth is totally impossible; but what becomes particularly difficult to suggest, assuming that such a freak of nature as a virgin birth did happen, is that the offspring might be a male. Females possess two so-called X chromosomes as their sex chromosomes. Males possess one X and one Y. This minute difference controls the sex of offspring. All normal women have 44 chromosomes in each cell plus two X chromosomes. All normal men have 44 chromosomes in each cell plus one X and one Y. The difference between a chromosome arrangement containing two X's and one containing one X and one Y is clearly visible under the microscope.

Each male is bound to receive his X from his mother, for she has nothing else to contribute, and his Y from his father. Had the father contributed an X instead, the offspring would have been XX – therefore a girl, and not a boy. As females are XX any sudden and unfertilised egg development must lead to an XX. The vital male chromosome cannot be conjured up from nowhere. Hence the inevitability of virgin births leading, should they occur, solely to females. The most famous virgin birth of all time, Jesus, was awkwardly a male, but in this instance, bearing in mind the complex role of the Holy Ghost, it is plainly out of step for biologists to continue to assert the importance of such a thing as a Y chromosome. The manner of Christ's birth must be a matter of belief, not of microscopy.

In November 1955 the *Sunday Pictorial* asked women who had reason to believe there had been no father to their child to come forward. It was stressed that the baby would have to be a girl who bore a striking resemblance to her mother in looks and gestures. Nineteen parthenogenetic claims were made to the newspaper, and these were sifted and investigated. Eleven of them were negated even in the preliminary interview because the mothers were under the impression that an intact hymen inevitably indicated a virgin birth. After further sifting other reasons reduced the original nineteen to one. Mrs E. Jones and her eleven year old daughter, Monica, were then subjected to a still more critical examination.

Their blood, saliva and tasting powers were all examined, and attempts were made to graft skin from each to the other. The results were that blood, saliva and tasting powers were almost identical, but the grafts did not take. The panel of doctors who helped with the investigation made special mention of the fact that Mrs Jones made her claim before she could have known about the tests to which she and her daughter were to be subjected, and before she could have known of the results and extreme similarity of the blood groupings, etc. In June 1956 the *Sunday Pictorial* stated that, after six months of investigation, the results were consistent with a case of virgin birth. Profesor J. B. S. Haldane disagreed. He said the evidence led to the opposite conclusion – that the child in fact had a father. His wife, Dr Helen Spurway, wrote – in the *Daily Worker* – that the mother's claim was disproved. Nevertheless it was, I feel, a brave and interesting attempt by a mass circulation newspaper to use its enormous readership in an attempt to solve a scientific riddle.

Menopause At one end of a woman's reproductive life is puberty; at the other lies the menopause. One marks the beginning, and one the conclusion. Known also as the climacteric, after the Greek word for the rung of a ladder, it is an inevitable step in every woman's life, provided she reaches the requisite age. Just as the age of puberty is relentlessly becoming younger these days, the age for menopause is growing older. Those who have examined the ancient literature on the subject believe that reproduction used to end at forty. By the middle of the last century it was ending, on average, at forty-five. Now, certainly in most of Europe, the median age of menopause is fifty. One recent report even gave an average of fifty-two.

Therefore, in the last hundred years, with the age of puberty falling from fifteen and a half to thirteen, and with the time of the menopause rising from forty-five to fifty, the reproductive span has stretched from three decades to nearly four. No one knows for sure why either end is stretching, but better food and better conditions are probably relevant. Although fifty years is the current average, this change of life can occur at thirty-eight or thirty-nine at the lower end of the scale and extend as far as fifty-five, though rarely higher. There is no relationship between the age at which puberty is reached and the age at which the menopause occurs; if a girl reaches puberty early, there is no greater likelihood of her reaching menopause early.

Although it emphatically changes a woman's reproductive ability, the menopause does not necessarily change either sexual ability or sexual appetite. For some the impossibility of an accidental conception steps up enthusiasm. Similarly, for others, the change induces a last-fling feeling and an increased determination to enjoy sex while they can, and while their mates are ready and willing to gratify their desires. Sometimes desire simply increases as a direct physical consequence of the menopausal changes. More often there is a loss of libido, of lust, and the change is frequently an excuse to stop sexual relations that were none too happy beforehand. However, there is no reason, given satisfactory sexual compatibility, why the cessation of egg production should be accompanied by a cessation of coitus.

No woman can tell which period is going to be her last. Should she keep diligent record of her changing cycles, a gynaecologist ought to be able to give her a good guess; but for most women the realisation that the last menstrual bleeding of their lives has occurred will only come when no successor has followed it. Whether or not each last twitch from a dying reproductive system is the final effort is of less concern to the woman experiencing this change than is the cessation of the symptoms accompanying it.

Many of these are extremely unpleasant. They include hot flushes, headache, giddiness, obesity, nervous instability, itching and insomnia. The flushes can range from a temporary blush-like manifestation to frequent and long-lasting effects often partnered by excessive sweating. Whether drenched by day or soaked by night, the woman can be left cold, exhausted and thoroughly fed up with the whole business. Coupled with the other symptoms, and with various emotional feelings of inadequacy, unattractiveness and depression, the menopause can be a dismal experience. Some people have harangued the researchers for doing so little to ease this inevitability, to make the change something less cataclysmic. There is much to their complaint. Common unromantic ailments do not make glamorous research.

Despite there being so much on the debit side, there is also contrary evidence that the change is not generally severe. To counter those reports, which state that '75% of all women suffer distressing symptoms at the climacteric', there are others from equally reputable sources which state that '70% to 90% of women experience no symptoms materially interfering with general health or with domestic or social activities'. Both quotes are taken from the *British Medical Journal*, and both may be accurate, despite the general implication being otherwise. It all depends, as C. E. M. Joad used to say, on what you mean by the words being used. Dr M. E. Landau, who wrote *Women of Forty*, stated that 'for most women menopausal discomforts are too trivial to demand the attention of a doctor, and can be dealt with by the patients'. She also writes that if sexual response was good before the menopause, it will be good after it, or better. There is general agreement that those least susceptible to menopausal troubles are the emotionally well-balanced women who are healthy, whose marriages are good, and whose families are entirely satisfactory.

The detectable physical changes may be sudden or gradual. Menstruation may stop without warning. Or it may continue regularly, but produce less and less. Or it may be irregular, but with an even flow. Or it may be regular, but with an uneven flow. In other words, the change rings all possible changes. Those women, usually plump, who go to hospital with a stomach-ache, and come out a week later with a brand new, totally unsuspected, full-term baby have often been assuming that the nine-month long cessation of menstruation, coupled with the marked weight increase, has merely been another manifestation of the 'change'. The error can be made by women who have had babies before. A more trivial error is to blame upon the menopause events that are taking place independently of it, and which are more a feature of increasing years than decreasing ovulation.

The internal physical changes are primarily a gradual atrophy of the reproductive organs. First on the list are the ovaries, each of which becomes smaller. Their Graafian follicles, which earlier were the protective containers of solitary eggs, disappear and the unused eggs disappear along with them. In their place fibrous tissue is formed. No longer are corpora lutea produced, and no longer do the ovaries manufacture their internal secretions. The hormones, created elsewhere, which used to galvanise the ovaries into their regular cycles of activity, are still being created by those other glands, but the ovaries no longer respond. Their day is over. Similarly the Fallopian tubes, no longer having eggs to transport from ovary to uterus, become smaller and shorter and they lose their surface layer of epithelial cells. The uterus itself, with no more child bearing ahead of it, also atrophies under the influence of the menopausal changes. It hardens and shrinks, possibly to one-quarter of its former size, and loses its original rounded shape. The vagina also shortens slightly, narrows, and loses some of its elasticity. The breasts lose some of their glandular contents and therefore wither, but such atrophy may be accompanied by a compensatory deposit of fat. The urine becomes markedly different in its hormonal content.

Such internal changes are unlikely to be as vexing as outward and visible signs. Certain male characteristics may manifest themselves, especially beard growth, and particularly at the mouth's corners. Many of the troubles can be alleviated by the administration of hormones, and the finality of menopause can be delayed and delayed if need be. The old rhythm of menstruation can be artificially perpetuated, and many thousands of American women are doing so, but hormones are more frequently used in attempts to lessen some of the unpleasantness of the change. Tranquillisers and sedatives are generally thought to be preferable to hormone treatment. Although headaches are common at this time, one benefit is that migraines are nearly always cured by the menopause. An evil to offset this good is that menopausal women are more prone to become diabetics. That such an association exists is proved when a dose of oestrogen is added to the necessary insulin, for it promptly reduces the insulin requirements. Oestrogen is the hormone manufactured by the developing follicle. Without an active ovary these egg-bearing follicles do not develop, and neither does their oestrogen. The actual chemical relationship between oestrogen and diabetes is both complex and unclear.

Some women get fatter at menopause. Once again the causes are complicated, and interrelated. Sometimes a woman just allows herself to get fatter by caring less about diet, but sometimes the decreasing oestrogen, which is associated with a decrease in energy output, is not coupled with a decrease in food intake; the woman just eats what she has always eaten but gets fatter. The reduction in oestrogen is also linked with a thinning of the skin, and with osteoporosis. The latter can occasionally be serious, with pain and general demineralisation of the bones, but such long term changes, although initiated by the menopause, are not what is normally thought of as part of the menopausal syndrome. It is the short term manifestations, the flushing and so on, that are most typical.

Unfortunately, although short term, the menopause is no sudden matter. Women can have troublesome symptoms for several years, or just for months, or not at all. Those who are unlucky have longer time to reflect upon this further blatant inadequacy of the human system. It seems so unnecessary, and it appears to be peculiar to the human species. It causes an upheaval and it certainly has no equivalent in those animals living sufficiently long for their reproductive systems to wane.

And, just to confuse the issue still further, there are cases on record of women who have conceived after an apparent menopause. Unlike the women who have confused a new pregnancy with the menopause these women suffer an absence of menstruation and *then* become pregnant. No one knows precisely when ovulation ends for any woman. A generalisation is that it has finally stopped within a year or two of the end of menstruation.

Circumcision Many of the emotive words so closely allied with the reproductive system, such as castration, sterilisation, rape, homosexuality and circumcision, are immediately allied in most minds with the male system. Yet females can be castrated, by removal of their ovaries, and they can be sterilised. The usual operation, known as salpingectomy, cuts the Fallopian tubes. Females can rape, although in England this is not a specific crime, merely an indecent assault. Females can be homosexual, although more rarely than men. Finally, females can be circumcised. The extent of the operation varies from place to place. In Australia, for example, one Aborigine group removes both the clitoris and the labia of its young girls at puberty; all the men subsequently have intercourse with the newly circumcised.

In Northern Africa various other forms of mutilation have been, and presumably still are on occasion, carried out. Apart from the excision of either clitoris or both labia, there is also infibulation. Basically this is the sewing up of the vulva in very young girls, leaving only a small hole for the passing of urine and for the effluent of menstruation. When the girl is married the hole is enlarged, and is totally opened up shortly before the birth of any child. After delivery the infibulation is again performed. Apart from being sewn together the vulva can also be held together with special clasps or with thorns. Evidence that such operations are still performed was shown when a Sudanese woman was examined at a Sheffield hospital recently. She was then in the middle of a pregnancy, although her vagina was exceptionally small, totally preventing normal intercourse. The infibulation had been done in 1951 when she was eleven, and somehow she became pregnant after her marriage eleven years later. The surgeons at Sheffield restored her vagina to normal and the subsequent delivery of a boy was entirely satisfactory. The girl later admitted that, although the procedure was illegal, female mutilation was still the rule in certain high class Moslem families.

8. Fertility

The infertile couple/Artificial Reproduction/The case of the 'strange males'/Age, ardour, mood

'Give me children or else I die.' – Rachel

GENESIS XXX. I

Only half of humanity is responsible for the next generation. The other half plays no part in reproduction, and therefore fails to pass on any of its genetic material. Members of this second, unreproductive half fail either because they die too soon, or have chosen not to have children, or because they have not been able to have children. In evolutionary terms this half can be called genetically unfit. To have 50% of humanity classified as 'unfit' is considerable, but Professor L. S. Penrose, of London, has added up the various classes of unfitness to reach this total. Some 15% of humanity dies even before birth, and therefore can play no part in reproduction. These miscarriages etc. have to be included for they are failed human beings just as much as a baby or child who has died. Some 3% of humanity is stillborn, another 2% dies shortly after birth, and another 3% dies before maturity. Of those that do reach maturity 20% do not marry, and 10% marry but remain childless. Reproductively speaking, 50% drops out of the race in each generation. Only 25% of our grandparents' generation were responsible for the current generation and only 12% of our greatgrandparents' contemporaries. When Cain slew Abel and put him out of the running, he was also inadvertently establishing the pattern of 50%.

On the other hand, nothing like 50% in many other species is able to reproduce, to produce the next generation. One million fish eggs, laid and fertilised, may all start development but only a couple of mature specimens may survive. Most bachelor seals, living in a harem-ridden world, will die before having the strength and cunning to possess a single wife, let alone a harem. And a 2% neonatal mortality rate – the human figure – must be uniquely low in the animal kingdom.

A further anomaly of mankind is the small number of offspring produced by each fertile male. A man is considered highly fertile if he produces ten babies in his lifetime. A male seal will produce treble that number in a year. A good bull, in these days of artificial insemination, can be made to serve forty to fifty cows from each ejaculation with a 60% success rate being expected. Rams can serve thirty to forty ewes from one ejaculation, stallions eight to twelve mares, and boars two to four sows. A good human male could doubtless be made to serve comparable numbers of women, and even without artificial insemination he could, if society and custom permitted, be enormously more potent than at present. After all, to be an average father it is only necessary to fertilise an egg on two or three occasions, and in a lifetime of hundreds – or more probably thousands – of ejaculations.

The infertile couple   About one in six of all marriages are sterile. Despite a strong male belief to the contrary it is just as frequently a deficiency in the husband causing the sterility as in the wife. An inability to conceive after, say, two years of marriage does not necessarily indicate infertility. There is much that can be done for the infertile couple, and it has been reckoned that only 35% of such couples are hopelessly sterile. Incapacities like azoospermia (or complete absence of sperm) make for such sterility. Other male defects, which are all more frequently encountered, are: oligospermia (when there are too few sperm in a given volume of the semen); asthenospermia (when too few are capable of movement); and spermatic dysplasia (when too high a proportion of sperm is abnormal). By no means can males sit back and assume that the female, plus the greater role she has to play, must be at fault.

Nonetheless there are very many ways in which the female can fail. She may not have eggs to produce. Or she may not be able to produce them because of a cyst on her ovaries. Or her uterine tubes may be blocked. Or her uterus may be diseased or absent. Her hymen may be intact. She may be rejecting the semen. Or she may be incapable of receiving her husband's penis. This last defect is really a combined one, as is premature ejaculation by the male, which wastes the semen, or a failure to practise intercourse at the right time, or an over-frequent indulgence. Finally, by no means is every case of infertility understood. Some women have an incompatibility with their husbands, and can be fertilised easily by others. In many cases, there should – by all the signs – be fertility, but there is not. Conversely, when couples have been considered to be infertile, the wife has suddenly become inexplicably pregnant, despite a defect and despite fidelity.

Much can be done for the infertile couple both by advice and by surgery. The advice can be mainly concerned with body temperature, and the way in which this suddenly rises at the time of ovulation, because conception can of course only occur when an egg exists to be fertilised. This time has to be known if it is going to be aimed for, whether naturally in normal intercourse or unnaturally in artificial insemination (or A.I., A.I.D., or A.I.H., which stand for artificial insemination, either by a donor or by the husband).

Artificial Reproduction　It never rains, as the old saying would have us believe, but it pours. It certainly seems to have been doing something of the sort with what could be called science-assisted babies, and much – if not everything in this area – has changed since this book's first publication. The rightness and wrongness of abortion was one complex issue that engulfed Western society, notably in the 1960's (although the debate still continues), but the argument was clear-cut relative to those that arose following scientific work in connection with *in vitro* fertilisation, the growth of those eggs into embryos, the implantation of those living bundles of cells into suitable wombs, the possibility of surrogate motherhood (with the baby being handed over to a different mother as soon as it is born), and the much more remote possibility of an embryo becoming a foetus and then a baby without any female involvement beyond the original donation of an egg. This would be the ultimate in what is generally termed test-tube science, however poor is this term in its accuracy but all-embracing in its usage. Ectogenesis, the full-term procedure, is still for the future to unravel, as a technique, as an issue; but the others are all with us today, as techniques, as issues. There is, to put it mildly, a general unpreparedness for their arrival. Not only ordinary individuals but official groupings (such as the churches) and the scientists themselves are extremely divided over the various issues. There seem, at times, to be as many opinions as people who have examined the subject. The most recent British committee to review this topic, suggest legislation, etc. (presided over by Mary Warnock) surfaced after eighteen months with its conclusions in mid-1984, acknowledged being split on key issues and said it would be producing 'minority' reports to accompany the main recommendations.

　Strictly speaking, artificial insemination is in a different category from the new *in vitro* fertilisation techniques reviewed by the Warnock committee. Nevertheless, from some points of view, the issues raised are similar in kind if not precisely. At least A.I., as it is generally known, has been with us for some time, and therefore opinions have less immediacy about them. A.I.H. was first recorded in the 1770's when John Hunter, quite the most exciting medical man of his day, assisted a fellow Londoner to inseminate his wife via a syringe, normal intercourse being impossible as the man suffered from hypospadias (a defect of the urethra). Today the more common procedure is A.I.D., as defective sperm are commoner than defective means of insemination. Over 2,000 babies are born annually in Britain with the aid of A.I.

The results of A.I.D. are good, with 30% of women conceiving in the first month, 18% in the second month, and 18% in the third, namely 66% in three months. This may increase to 70% or 80% after six months. Insemination has to be done at the right time, paying due respect to the temperature shift. Dr Walter Williams, a famous authority on this subject, was among the first to publicise guide-lines that should be followed. For example, both husband and wife should give evidence of a harmonious marriage, they should be in complete agreement about the sperm donor's suitability, and the husband should actually assist in the operation. The man who is chosen from a description of ten to fifteen donors by the couple, and who gives the sperm, should do so shortly before the operation. He should always remain anonymous, even to the inseminating couple. He should look like the husband, be of the same race, be of unquestioned integrity, and aged between twenty and thirty-five. He should neither be sexually promiscuous, nor indulge in excesses of any kind. His moral qualifications should be of the highest order, and he should be either a university student or a graduate. Medical students are frequent donors. Afterwards, it is best if the woman is treated by a doctor who is unaware of the cause of her pregnancy, and assumes the husband to be the true father. The Warnock committee recommended that no single donor should father more than ten children, as it was necessary to reduce the chances of half-brothers and sisters meeting and mating in later life. It also urged that A.I.D offspring are legally legitimate. Currently, and surprisingly, the 2,000 or so A.I.D. babies born annually in Britain are presumed illegitimate and have to be formally adopted.

Infertility is not normally associated with multiple births; but it became so. Quins, sextuplets and septuplets are blue-moon affairs, all very rare and pounced upon by the world's press; but on Tuesday, July 27, 1965 a woman gave birth to quins in New Zealand. The press was delighted. Two days later quins were born in Sweden; the press was amazed. Earlier Munich had been amazed when quins and sextuplets had been born within three weeks of each other. Sweden had had an earlier encounter when septuplets had been born both in August 1964 and February 1965. Behind this spate of multiple births was a new technique for bringing fertility to certain infertile women. It was also plain that the technique had not been completely mastered.

This story began in 1958 when Professor Carl-Axel Gemzell and his team reported that they had brought ovulation to several young women who had not been menstruating. The women's production of the correct hormones had been at fault. They were given extra F.S.H., the follicle stimulating hormone produced by the pituitary gland, which ripens some of the cells in the ovary and stimulates them to produce an egg or eggs. In 1966 Gemzell summarised the situation. Half of the one hundred infertile women chosen for the new procedure had become pregnant, and of those pregnancies half had been multiple – either twins or more. Other researchers were trying to make the dosage more accurate, both in its quantity and its timing, because the dividing line between no pregnancy and a multiple pregnancy was apparently a slim one. The quantity of pituitary extract had to be correct for each woman.

An added difficulty was in the supply of pituitary extract, it being recovered either from mortuaries or the urine of older, and post-menopausal women. Either the whole pituitary gland was taken from a dead woman, or the F.S.H. substance was taken from the urine of women whose pituitaries were still producing it, and who were past child-bearing age. Nuns were particularly helpful. Either ten whole glands from the dead, or a month's supply of urine from the living, was necessary for one course of ovulation treatment.

However the pituitary supply problem became less critical when scientific advance moved in a different direction. Instead of stimulating the ovaries to produce eggs that might or might not lead to multiple births, the new plan was to remove these eggs from the woman, fertilise them outside her body, and then replace the developing embryos within her uterus. There would, in this fashion, be less hit-and-miss about the operation and rather more control as a known number of developing embryos could be inserted.

A crucial variable is timing. The eggs have to be fertilised and then partly developed so that they are at a correct stage for implantation, and the uterus itself has to be at the correct stage for receiving this potential implant. Hence it was inevitable that the thought, and then the act, should arise of freezing embryos so that the thawing and the insertion can be achieved when the mother's uterus is at its most suitable.

The first such baby, fertilised outside her mother, frozen, and then inserted, was born normally in Australia in March, 1984. By the end of that year Australia – at the forefront of this research – had witnessed the birth of 130 so-called test-tube babies, some frozen as embryos, some not, and a further 170 had been born elsewhere in the world. Sometimes the eggs have been from the mother herself and some have been provided by other women. Sometimes the sperm have been from the mother's husband, and some have also been donated. Legislation has, in general terms, not yet caught up with what is already occurring, and has to clarify the legal status of children, parents and donors.

It also has to clarify the position for the scientists and doctors involved, not just in the kind of work already achieved but in future research and practice of this kind. The Warnock committee did not recommend a time limit for embryos to be kept frozen – this can be influenced by the success of implantations and the spacing of subsequent children – but it did consider fourteen days post-conception to be the maximum for embryo research (which is the very end of the period when embryos implant in the uterus). The Royal College of Obstetricians and Gynaecologists would like the research period to be seventeen days, which is when the neural tube starts to become the brain and spinal cord (and is therefore of particular interest for the investigation of neural tube defects, such as spina bifida).

The committee was adamant that surrogate motherhood should never be permitted for profit, either by agencies or the surrogate mothers. With such births, artificially conceived by one means or another, the incubating mother hands over the result of her nine months' labour to the parents who supplied either the egg, or sperm, or both (or even neither), and who paid for the surrogate pregnancy. Fourteen of the sixteen Warnock committee members were against all forms of surrogate motherhood on principle, partly because this mother might then wish to retain the baby (with extremely complex consequences for all concerned); but two members argued that such motherhood is already here and there may be occasions when it is the only feasible recourse. There are 'womb-leasing' agencies operating within the United States, and by mid-1984 there were already two British women acting as surrogate mothers for one of these agencies. The babies were to be handed over a few days after their births, and £13,000 would also change hands.

As for the future there is ectogenesis. The embryo would grow to full-term entirely within the precincts of a laboratory/intensive care/premature ward. The possibility is feasible but highly unlikely, mainly for the practical reason that it would be extremely costly. Nature's way, however interfered with at the start, is very much cheaper and always will be, however brave the new world we are about to enter.

'Strange males' About the strangest story of all in recent research into infertility is the case of the 'strange males', as it was called. It has nothing to do with man, so far as is known. It has to do with mice, but there is just a possibility that it has a moral for humans. Dr Hilda Bruce was working during the 1960's at the National Institute for Medical Research, London, when, by chance, she discovered the 'strange male' phenomenon. Some normal female mice, which had been normally mated, were moved to a new cage. Such females, already pregnant, were expected to follow the traditional pattern of rejecting all further male advances. Generally, further acceptance only comes after the litter has been born.

However, the traditional apple cart was upset. Within the new cages were new and strange males. Promptly the females all mated again. Their original pregnancies vanished, and their final offspring were emphatically sired by the strange new males, and not by their earlier mates. Dr Bruce had therefore opened up an extra dimension to conceptions, and had showed that smell was a crucial factor. The mice were recognising each other not only by smell, but were able to distinguish between one type of mouse and another. The more curious the smell of the newcomer the less likely was the female mouse to maintain her original pregnancy.

The moral for humans? Well, so little is known about those first chancy days of pregnancy, and what causes a developing embryo to implant itself firmly, that women may be mimicking those female mice. No grass grows where many people walk, say the Scandinavians. Mate too soon afterwards with a strange male, and the original short-lived pregnancy may quietly wither away. The hypothesis is highly unlikely, but the case of the 'strange males' was also totally unexpected.

Age, ardour and mood Even though males have no menopause or its equivalent physically, and they retain their powers of fertility into their advanced years, an older man is not statistically an ideal father. Not only do foetuses from him die more frequently in the uterus, but more babies die at birth if conceived by ageing fathers. It is widely believed, notably among males, that prowess and fertility go hand-in-hand. But excellence in the former can lead to nothing in the latter, as the requisite number of sperm may not exist in each skilled but depleted ejaculation. With old people, whose production is tending to fall off anyway, a mimicking remembrance of youthful ardour can also decrease fertility quite unnecessarily, and below the level for a successful conception. Such over-taxing of the system may be an explanation of the prompt manner in which many women conceive shortly after adopting a baby. The adoption may follow years of failure, followed abruptly by the arrival of two babies; one adopted, one conceived. There may previously have been a fervent amount of sexual activity, which was then relaxed and which therefore permitted greater fertility.

On the other side of the fence, when pregnancies are equally fervently undesired, and when rape has taken place, they still happen. Emotion does not seem to play a powerful part. Nor, strangely, does chronic underfeeding. In those parts of the world of perpetual food shortage, conceptions are never in short supply. (Perhaps even noise is important. Professor Bernhard Zondek, of the famous Aschheim–Zondek pregnancy test, reported from Israel in 1966 that rats and rabbits copulated just as frequently in noisy surroundings but their fertility was much impaired – until the noise ceased.)

9. Contraception

Historical ideas/The birth controllers/Religious opinion/The modern situation/The pill/I.U.D.'s/Sterilisation

---

There was a young woman who lived in a shoe.
She had no children: she knew what to do.
> Quoted in THE MALTHUSIAN, September, 1906

Mary Mother, I believe, without sin thou didst conceive.
Mary Mother, still believing, let me sin without conceiving.
> Attr. to LORD BYRON

Mankind has had some idea what to do about birth control for millennia. Yet, in its modern sense, it is less than a century old. The even more modern notion of having clinics to dispense information is less than a lifetime old, and the Church of England did not give any blessing to the limitation of parenthood until 1930. The story of the contraceptive pill did not begin until the 1950's, nor did the idea gain much acceptance until the 1960's. Therefore, despite a history stretching back to the world's first medical tracts, most of the progress in attitudes and techniques is recent.

Both the Petri or Kahun papyrus of 1850 BC (use crocodile's dung as a pessary, place honey in the vagina) and the Ebers papyrus of 1550 BC (sit astride a burner producing fumes of wax and charcoal) gave recommendations. Earlier the Chinese had suggested the consumption of minerals, such as mercury and lead, to offset conception. Both ideas, the taking of pills with a metallic base to them, and the use of materials with a multitude of effects (including spermicide and womb-contraction) were to last for thousands of years. No one seemed short of ideas. Catch a wolf, kill it, boil its penis and add some hair, wrote St Albert the Great (1193–1280) in his *Book of Admirable Secrets*. Use pessaries made of cabbage, colocynth pulp, pomegranate skin, ear wax, elephant's dung and whitewash, wrote Rhazes of Baghdad in 882. (Plainly spermatozoa encountering such a chemical battery of conflicting properties cannot have benefited from their presence.) Use rock salt, said Ali Ibn Abbas in the tenth century. Use alum, said Avicenna, a few years later.

The world's herbalists plucked apparently at random from the plant kingdom. On the list are rosemary, pea, yarrow, asparagus, barrenwort, ivy, marjoram, white poplar (all used somewhere in Europe), thistle (North America), green coconut (Pacific Islands – there is a seeming inevitability about this particular choice), and pineapple (Malaya). According to B. E. Finch and Hugh Green in *Contraception Through the Ages*, it is extraordinary that no natural contraceptive was found in so many centuries of testing and tasting. The American Indians came nearest to success with the plant *Lithospermum ruderale* (the chemistry of the substance involved has not yet been worked out).

The world's folklore was naturally steeped in suggestions. Pliny the Elder (first century AD) cited instructions concerning the dissection of a spider, the mixing of part of it with a deer's skin, and the attachment of all this to a woman before sunrise. The complexities of such recipes presumably added to one's trust in them. In Russia they collected menses (from menstruating women), cooked them up and then listened for the sound of weeping children to indicate sterility. In modern Egypt, less gruesome but more courageous, women have lain between the railway lines, face upwards to cure sterility, downwards to prevent a pregnancy. The passing train is the effector.

Sperm have also met with physical barriers for centuries. Sponges, suitably soaked with honey or wine, were put into the vagina. Aetius of Amida recommended half a pomegranate. The Arabs of AD 1200 suggested using the right testicle of a wolf, partly because it would help to stop sperm if wrapped in oil-soaked wool and partly because such an insertion would lessen feminine desire. Distraction of this kind, particularly at the moment of coitus, was often held to be effective. The Chinese in 1100 BC recommended Kong Fou – for the girl. At the exact moment she should draw a deep breath and think of other things. Soranus said the woman should hold her breath, then get up quickly, squat and sneeze. Avicenna felt that ejaculation should be followed by the woman leaping backwards seven times, sneezing all the while, if conception was not desired. Sheaths also have an ancient history, although much of the spur for their use was the fear of venereal disease rather than conception. Some Romans used animal bladders for this purpose, but the real inventor seems to have been Gabriello Fallopio, the sixteenth-century Italian whose name was given to the Fallopian tube or oviduct. His sheath was a modest thing of linen, fashioned to fit over only the tip of the penis.

The dominant mystery of the whole sheath saga is the origin of the name condom. There is no mention of this word in Britain in the seventeenth century; then it suddenly blossomed in the first twenty years of the eighteenth century. Spelt first as condum, then condon, then conton and condom, the word has no known derivation. Suggestions are that it comes from the Persian kemdu (a kind of haggis of grain), or from the Gascony town of Condom, or from an English doctor or colonel. Neither John Evelyn nor Samuel Pepys mentions such a man; neither do the medical nor military lists. Perhaps, having made the invention and given his name to it, the man could no longer tolerate such a name with such a fame.

Giovanni Jacopo Casanova de Seingalt called them 'Redingotes d'Angleterre', or English riding-coats. Some time later the English retaliated by calling them French letters. Casanova also favoured a gold ball, two-thirds of an inch in diameter. Inserted into the vagina, it theoretically acted as a sperm barrier and was undoubtedly cheap after the initial capital outlay. Nevertheless he approved 'of the overcoat that puts one's mind at rest'. So did James Boswell who makes frequent reference to his 'armour'. And armour many considered it to be, effective but insensitive. The arrival of rubber in the nineteenth century meant that equal effectiveness could be achieved with far greater sensitivity. The current European giant in this fertile field is the London Rubber Company. It started in 1916 in one room at the back of a city tobacconist. By 1960 the company was producing over two and a half million sheaths a week and, coupled with its other steady line of toy balloons, the company is still doing very nicely. Some 3 million British couples make use of the sheath.

Important names of pioneers in other areas of contraception include Walter John Rendell, Dr Friedrich Adolph Wilde, Dr Wilhelm Mensinga and Dr Ernst Gräfenberg. Rendell in 1880 opened a pharmacy in Clerkenwell for pessaries based on quinine. They were extremely successful. Dr Wilde, of Germany, invented the rubber cap in the 1830's; but Dr Mensinga, also of Germany, popularised it. The Mensinga diaphragm reached England via Holland, and en route became known as the Dutch cap. Dr Gräfenberg was experimenting with 'rings' in Berlin in the 1920's, and was thus bringing up to date various ancient ideas. These intra-uterine devices, known today as I.U.D.'s, have had a controversial history, but are now part of the scene. (For a fascinating account of the story of birth control propaganda, its promoters and detractors, read *The Birth Controllers* by Peter Fryer. The battle really began a century before Marie Stopes with Francis Place (1771–1854) being the first propagandist and Richard Carlile (1790–1843) fighting for the freedom to discuss and publicise all forms of opinion. Dr Marie Stopes's famous birth-control clinic, the 'first in the British Empire', was opened at 61 Marlborough Road, Holloway on March 17, 1921. Among its patrons were Arnold Bennett and Dame Clara Butt.)

Quite apart from all physical aids to contraception there is also coitus interruptus and the safe period. The former, mentioned in the story of Onan (Genesis 38), has undoubtedly been used by millions for millennia, despite its obvious dissatisfactions and its failure as a method among those whose desire, timing and self-control do not match. It was only in 1930 that scientists in Austria and Japan (B. Kraus and D. Ogino) discovered almost simultaneously that a maternal egg was regularly released from an ovary twelve to sixteen days before the start of menstruation. Ordinary mortals had known ages beforehand about the safe period, about rhythm, about the likeliness of contraception on certain days, but they had not known the precise reason for such timing. Soranus in the second century AD said coitus should not take place at certain times if children were undesired. The ancient Hebrews said that a woman was 'unclean' for fourteen days after the start of menstruation. Hence the clean time came at a time of maximum chance for conception.

The present Roman Catholic attitude, which still approves only of the 'safe period' method of contraception, really stems from 1931. In 1930 Pius XI's Encyclical on Marriage stated that any frustration of the natural power to generate life was against God's law. The following year another encyclical added that married people were not acting against the order of nature if 'they make use of their rights ... even though no new life can thence arise on account of circumstances of time ...'. The advisory council on Catholic marriage gives the following tongue-twister of a clarification about these circumstances of time: 'First fertile day is found by subtracting nineteen from the number of days in the shortest recorded cycle, and the last fertile day is found by subtracting ten days from the longest cycle the woman has ever recorded. This formula gives at least ten possible fertile days in the middle of the menstrual cycle. And the greater the difference between the shortest and longest cycles, the greater the number of possible fertile days.' Two main troubles with such a method are that many women (about one in four) are wildly erratic with their periods, and all women tend to be erratic for the first few periods after the birth of a baby.

Pope Paul VI's famous encyclical, reaffirming the Roman Catholic Church's faith in the safe period technique of contraception, was issued on July 27, 1968. It aroused considerable resentment, and 2,000 scientists gathered at Dallas called it 'repugnant to mankind'. Despite the Catholic attitude it is not necessarily the Catholic countries which have high birth-rates but the under-developed countries. The eleven Catholic countries of Europe have an average birth-rate of 14 per 1,000 people, which is slightly lower than that prevailing in the fifteen non-Catholic European states. In Catholic countries with high birth-rates (and high abortion rates), such as Latin America, poverty and illiteracy are blamed more than religion, even though religion is not helping to bring down these high birth-rates.

In 1984 Pope John Paul II went even further than Paul's encyclical by stating that Roman Catholic couples should on occasion not even use the rhythm method of control. 'The use of infertile periods in married life can become the source of abuses, if the couples seek in such a way to avoid, without just reasons, procreation lowering procreation below the morally correct level of births for their family.' Others might argue that the planet as a whole is 'just reason' if its swelling population is to be curtailed.

Other religions have other ideas. The Jews customarily forbid methods of contraception for males, although extreme orthodoxy forbids any form of contraception except abstinence, and extreme unorthodoxy tolerates any contraceptive behaviour. The Moslems, following a fatwa issued by the Grand Mufti in 1937, are allowed to take any measures by mutual consent to prevent conception. In India, according to Jawaharlal Nehru in 1960, there is no major organised group opposed to family planning. Orthodox Hinduism demands abstention on certain days. The Church of England was against the birth control movement throughout its most difficult years, and was supported by, for example, the *Sunday Express* – 'John Wesley was an eighteenth child. Birth control would have deprived the world of Wesley.' Many others with a similar view to the *Sunday Express* drew up long lists of the famous who had been born well down the family order and whose births would have been frustrated by family planning.

At the Lambeth Conference in 1930 the voting was 193 to 67 for a resolution which approved of birth control 'when there is a clearly felt moral obligation' to limit parenthood, and a morally sound reason for avoiding complete abstinence. Contraception was condemned if the motives were of selfishness, luxury or mere convenience. In that same year the Ministry of Health permitted local authorities, for the first time, to give birth control advice to married women who needed it.

The modern situation In the mid 1960's, shortly before science was about to change the old ways (in revolutionary fashion, as it was thought at the time), the Consumers' Association surveyed contraception in Britain. It interviewed 1,500 people married between the year of the favourable Lambeth Conference and 1960. Main methods used were condom reported by 49%, withdrawal, i.e. coitus interruptus 44%, safe period 16%, cap 11%, suppositories 10%, douche 3%. By 1980 a comparable survey elicited: pill 53%, I.U.D. 20%, sheath 9%, cap 8%, none 5%, other systems including sterilisation and rhythm 3%. In the previous five years the only major change had been a reduction (13%) in those using the pill and a gain (6%) in those using I.U.D.'s. Apart from withdrawal or the safe period, which are free, the cost of a year's contraception varied widely, but was within the range spent unthinkingly by those who have as against those who have not. Cost of contraception is a different matter in, for example, rural India. It was Dr S. Chandrasekhar, then Professor of Sociology, University of California, who wrote in 1961: 'Shocking as it may seem, in many rural areas the cost of having a baby would be cheaper than the price of birth control equipment.' This single statement hammers home the problem with one quick and resounding stroke.

The Consumers' Association concluded that no brand which had been tested was, if used alone, a certain way of preventing conception. Earlier Sir Alan Parkes, then President of the British Institute of Biology, had said, 'Contraceptive methods are so crude as to disgrace science in this age of spectacular achievement' and the 1960's have become known as the age when contraception no longer disgraced it. Three lines of attack made remarkable strides in the first half of the decade. They were the pills, the loops (or other shapes) and surgical sterility.

Nevertheless, in those 1960's, some of the old reaction against birth control still festered in various guises. In 1965 only one British medical school compelled students, during their six-year course, to attend a Family Planning Association clinic for instruction. In France the old (1920) law forbidding birth control propaganda was not revoked until 1967 (and yet it was estimated that 50% of all French pregnancies ended in abortion). The birth control pill was not permitted to be sold in France, even with a medical prescription, until 1969. In 1961 a birth control clinic in New Haven, Connecticut, was forced to close down after only nine days and it took almost four years for the closure order to be rescinded. In the United States tobacconists and garages still sold more contraceptives than chemists did, and in 1965 twenty-eight states still had laws on their books limiting contraceptive advertisements or sales outlets or the spread of information. The unmarried girl in Britain still found it extremely difficult to get contraceptive advice. The first Italian birth control clinic was opened in 1969 in defiance of both the Church and a law left over from Mussolini's campaign for larger families. (Italy then had over a million abortions a year.) The old battle for more knowledge about contraception, first called birth control in 1914 (by Margaret Sanger, of the United States), later called family planning, is not yet over. A few more skirmishes have yet to come, but the pill, the loop, and surgery have all made extremely significant and independent advances. Science is at last leaping ahead from the régime of English overcoats, American tips and Dutch caps – and about time.

The pill The story of the pill has been extremely rapid. It was in 1955 that Dr Gregory Pincus, an American research director, spoke at the Fifth International Conference on Planned Parenthood in Tokyo. His subject was the inhibition of ovulation in women who had taken progesterone or norethynodrel. Ten years later his few female volunteers had been replaced by ten million women. When Pincus died, in 1967, he knew that his work had revolutionised the old ways of contraception.

An important step in making the transition from the few to the ten million was the intensive trial undertaken in Puerto Rico in 1957. First trials in Britain began in 1960, in Birmingham, Slough and London. (Thirteen of the thirty-eight Birmingham women had conceived by June 1961 as the dosage had been much too weak.) By 1962, according to the Family Planning Association's clinics, 3,536 women were taking the pill in Britain. By 1963, it was 13,760. By 1964, 44,000. By 1968, the estimates were 1,000,000 in Britain, over 7,000,000 in the United States, and 2,000,000 in the rest of the world. By the 1980's, according to the United Nations, 23% of all those who used any form of contraception in the world outside China were on the pill, with the Chinese proportion being 9%. (Male graffiti referred to this new advance as 'even better than sliced bread because it provided unlimited crumpet'.)

The pill, which should strictly be called a tablet, quickly became big business. By 1968 there were twenty different types on the British market. In the United States, with an even higher proportion of women taking this oral contraceptive, the advertisers were immediately busy pushing their new products. 'Simple setting, built-in "memory" recording, push-button ease, crush-proof tablet protection, safety from small children, plus the look and feel of a fashionable compact . . . this is what she will like about it.' Long, delicate, feminine hands, plainly ready for bed, stroke the latest offer. 'It costs no more . . . offers much more.' In the June 18, 1965 issue of *Medical World News* there were 70 pages of advertisement. Thirteen of them were for oral contraceptives only. The remaining 57 dealt with everything else. Considering the number of other pills and products on the medical market, such concentration upon the new idea was phenomenal.

There were doubts about the pill even in those heady days. Dr John Rock, the American Roman Catholic whose name is linked with that of Dr Pincus as co-developer, was scornful of these wrongful notions. The right pills taken in the right way, he stressed, *do* prevent ovulation (it had been said that they did not always do so). They in *no way*, he said, act as an abortifacient. (It had been said that they cause an abortion if started when an unknown pregnancy already existed.) They in *no way* act as a condom by preventing sperm entering the womb (some had argued that they were sperm killers, not ova preventers). There is *no* relationship, he continued, except perhaps a favourable one, between the pills and cancer of the female organs. (This fear of a cancerous side-effect had initially been strong.) *No* association has been found with thrombosis (a few women taking the pill had died of thrombosis, but not in greater numbers than is normal for their age group). The pills *can* bring out symptoms and side-effects similar to those encountered in pregnancy. The pills are *not* dangerous. They do *not* postpone the menopause. Dr Rock was most emphatic. So too was the special committee set up by the US Food and Drug Administration. In August 1966 it reported finding 'no adequate scientific data, at this time, proving these compounds unsafe for human use'. Like similar reports from Britain and the World Health Organisation it warned there was no proof they were safe even though there was no proof they were unsafe.

In 1967 the *British Medical Journal* published results of work which showed that hospital admissions for blood clotting in the leg veins (deep vein thrombosis) or for blood clots in the lung arteries (pulmonary embolism) was about ten times greater for pill users. There was no relationship between the pill and heart attacks, but some evidence for an association with strokes. These added up to the conclusion that there were one to two deaths per year in every 100,000 previously healthy pill-takers aged between twenty and thirty-four, and three to four deaths a year in the age group thirty-five to forty-four. The pill was therefore not totally free from guilt. However pregnancy itself, the state prevented by the pill, was even less free in those two age groups: the comparable figures for deaths caused by all the risks of pregnancy, including abortions, being 23 and 57. Also the pill may even be of benefit with some diseases, and there is a theory that it actually helps with multiple sclerosis.

In the United States, where many magazines campaigned against the pill, some 3 million people abandoned it before the end of the 1960's. At a conference in 1969 Dr Anne Lawrence, of Chicago University, said she would not prescribe it for someone with a family history of breast or cervical cancer, or circulatory problems, or diabetes, and 'I am a concerned physician pleading that the drug be used with a certain circumspection. But I wouldn't even try to deny that the pill has been a boon to millions of women.' In fact 95% of American gynaecologists were then prescribing the pill, with 4% refraining on religious grounds, and the final 1% on medical grounds.

During a BBC broadcast at that time Dr Gordon Wolstenholme, then director of the Ciba Foundation in London, said, 'I consider that Dr Pincus's pill was one of the most risky and foolhardy measures ever to be put into general use, particularly against an underdeveloped people such as the Puerto Ricans. The fact that it has come off is quite remarkable.' On a later occasion he amplified this point. 'It was a bigger risk among the Puerto Ricans, not so much because they could not appreciate the risks as because if anything went wrong the emotional racial reaction would have been overwhelmingly strong. Nevertheless, Pincus accepted the responsibility, he introduced all possible safeguards and controls, and I admire him immensely for doing it. In the history of this century I have a feeling that the introduction of the pill will turn out to be one of the most responsible things done by any one individual.'

It has already been one of the most rapid advances. From a few women to nearly 10,000,000 in ten years was remarkable. So too the leap to 23% of the world's contraceptive users in the subsequent decade and a half. The pace of growth has now slackened but, as two-thirds of all couples (of reproductive age) in the developed world and one-quarter of the rest of the world use some form of birth control, and as there are those 9% in China, this means that some 200 million couples are currently making use of the pill. It is just a quarter of a century since the Puerto Rican trials. The pill is still not without risk, and new 'scare' associations arise in some quarters which are then, almost inevitably, attacked in other quarters for not taking all factors into account (such as age at first intercourse and number of sexual partners, both known to be associated with cervical cancer), but plainly there is also tremendous benefit to set against the modest risk. However the pill itself, or rather the many pills on the market, can always be improved, and those women at greatest risk from them must be warned. As a recent report suggested: 'If the oral contraceptive is to be used to its fullest potential during the remainder of this century we must move towards the situation in which a low-risk pill is taken by low-risk women.'

The I.U.D.'s Scarcely had the pill started on its remarkable career when it had a rival. In 1959, only two years after the Puerto Rican trials had begun, two important papers were published, one by an Israeli and one by a Japanese. Both reviewed recent progress and thinking concerning intra-uterine devices. For centuries humanity had been inserting metallic objects into the vagina, notably of animals, in the hope of preventing conception. (Camel pregnancies were particularly unwelcome during any long and dangerous journey, and their prevention was extremely important.)

Then, after World War One, Dr Ernst Gräfenberg put the procedure on a more scientific basis. Working in Berlin he inserted coils of silver or gold wire into the wombs of hundreds of women who did not wish to conceive. His results were good, but others trying his methods were less successful. The women bled, or did conceive, or aborted their foetuses; and the method fell quickly out of favour. The two 1959 papers drew attention again, and favourably, to the idea of 'Gräfenberg rings', to intra-uterine devices (or I.U.D.'s) and work began in many countries. All kinds of shapes were devised, inserted, and the results noted. By 1964 a large conference in New York summed up the situation. The five shapes then available, plus their polyglot names, were the Zipper nylon ring, the Hall-Stone ring (stainless steel wire in a coil), the Lippes loop (a plastic snake-like coil), the Birnberg bow (a plastic figure of eight), and the Marguilies spiral (a plastic Catherine-wheel shape, with a long tail). The last three were all used in the early British trials. The Lippes loop was the one recommended by the Planned Parenthood Federation of America, and 3 million I.U.D.'s of one sort or another had been distributed in the United States by the end of the 1960's.

At the time it was thought they would rival the pill in popularity and usage, mainly because – at a penny or two – they were so cheap to make. In the intervening years enthusiasm for

I.U.D.'s has waxed and waned in the various countries, often for no readily detectable reason (and perhaps the dictates of fashion are not irrelevant even in the world of contraception). In the rich world 7% of those couples of reproductive age using some form of birth control have had an I.U.D. inserted, which means over 20 million people. In the less developed world, excluding China, the equivalent total is similar. However China has taken this form of control much more seriously, it being used by 49% of its women with need of contraception (and there is great need in that country as severe financial penalties are imposed on families with more than one child, so much so that considerable infanticide is suspected in many areas, the latest census having revealed a startling rise in the proportion of baby boys). Only 9% of Chinese women are on the pill, as against 23% in the remainder of the world.

All I.U.D.'s permit fertility to return soon after they have been removed. All discourage conception when correctly in place, although none is 100% effective. And all are inserted right up in the uterus. (The procedure is much easier if the woman has already had a baby.) A minority of women find that their loop/ring/spiral is ejected. Sometimes another type will stay; sometimes it will not. Also a minority of women are made to bleed by these devices. And a very small minority conceive, with the device being born at the time of birth or, very rarely, remaining within the womb even throughout all the turmoil of parturition. Someone else has to do the original insertion, and that someone – according to one report – can place 60 to 75 in position a day. An important point, particularly for under-developed countries encountering a device suitable for their threadbare pockets, is that a person without medical qualifications can be taught to do the inserting.

It is not known, for sure, how I.U.D.'s work. Somehow the presence of this shape within the uterus either prevents sperm and ovum from meeting, or prevents the bundle of cells from implanting itself within the uterus or – not the same thing – prevents the uterus from being receptive to the cells. In which case, is the device therefore causing a very early (one week old) abortion or preventing the new eggs from ever being fertilised? Dr Eleanor Mears, a British family planning expert, considered this to be a philosophical rather than a medical matter. At what point does life begin? A working party set up by the British Council of Churches' Advisory Group on Sex, Marriage and the Family came to the convenient conclusion that biological life became human life when the embryo was implanted and attached to its mother. Thus the devices can only destroy biological life. The Group could see no objection to their use.

It has even been thought that the ring might solve the problem of the sacred cow, one of the most formidable aspects of religious tolerance. There are 176,000,000 such cattle on the loose in India. Their castration is about as unthinkable to the Hindus as their slaughter. So an American research scientist, long stationed in India, fitted eighteen cows with plastic spirals, and was pleased with the result. Whether cows or people or both make use of such contraception will have the same result; there will be more food for the remaining millions. India's over-population and starvation problems must be aggravated by the 176,000,000 hungry, useless, ownerless, and wretched sacred cows. However India has grown disenchanted with I.U.D.'s even for its human population in recent years, and fewer are being fitted despite the annual increment of 15 million people. The sub-continent is currently more attracted by the more dramatic alternative, cutting off either the supply of eggs or of their fertilising sperm.

Sterilisation   At the International Conference of Endocrinology held in London in 1964, a figure of 100,000 American males a year was quoted. *Time* magazine reckoned in January 1965 that one and a half million Americans had by then been voluntarily sterilised. Known as vasectomy the operation takes about fifteen minutes, and can be done with a local anaesthetic. Each of the two tubes called the vas deferens is cut, and the ends are tied off. As these tubes carry the sperm from each testis up into the body, the sperm can no longer travel. Hence there is ejaculation, normal libido, normal capacity for intercourse – but no sperm. Should a man change his mind, and wish for more children, he can ask for the tubes to be rejoined. As yet this operation (reversible vasectomy or vasovasotomy) is not always successful, but it can be – on occasion. In Britain the operation was then less common. For a long time there was argument about its legality, as an Act of 1861 seemed to forbid it, but its lawfulness is now more securely established. Female castration, or salpingectomy, is not quite so simple, but was fairly common as an operation, even in those early days.

At all events sterilisation, either of males or females, is no longer a rare event anywhere, and in the developed world is practised by about twice as many couples as use I.U.D.'s (which means it is about two-thirds as popular as the pill). In the less developed world, despite I.U.D.'s and the pill being used – by those who use anything – as frequently as in the developed world, this sort of equality no longer applies. Sterilisation of one or other party is the commonest contraceptive method employed, accounting for 45% of the methods used. Its advantage, of course, is its effectiveness and cheapness, once the original outlay has been spent.

In the earliest days vasectomy was much more frequently performed than the equivalent female operation. By mid-1967 there had been 2½ million sterilisations in India, 95% of them on men. Even ten years ago more men than women had been sterilised in, for instance, India, Bangladesh and South Korea; today the opposite is true. Female sterilisation, which is arguably the most dangerous form of contraception to initiate, is now the most popular form in South America. Doctors are not notably in favour of it, but South American men are famous for their belief that nothing must imperil their masculinity, even if it might do so only in their minds, as there is no relationship between vasectomy and performance: everything is the same, save for the lack of sperm. It has also been suggested that the lack of male sterilisation may be due only in part to male conservatism. It may also be due to the family planning programmes giving most of their attention, if not all in certain areas, to women. And how much easier it must be for those women to agree to their own sterilisation rather than suggest that the operators make their first move towards the men.

Breast-feeding A final point is that breast-feeding still prevents more pregnancies in developing nations than any other form of contraception. The longer that such feeding can be maintained the greater the delay before the next conception. As only 25% *on average* use modern forms of contraception in such countries this means that the proportion is much lower in the poorest countries, such as Bangladesh, where the figure is 10%. Papua New Guinea has actually demanded that all feeding bottles and powdered milk should be available only on medical prescription, and other countries poor in everything save a burgeoning population could follow this example. It is an unhappy contradiction that the world's population only took off in numbers this century when contraceptives first became available, and leaped ahead in even greater strides after World War Two when science made contraception easier than ever before.

Carlile, Place, Sanger, Stopes and all the rest would be amazed by the suddenness of the world's changed attitude to contraception. Will the actuality be equally sudden? Will the good news percolate down to the deepest layers where many people live? I am thinking of a report about a desperate Mexican girl, already with nine children, who entered a clinic for advice. If another child came, she said, she would kill herself. The girl, a Catholic, was informed about the rhythm method of control. Her floods of tears suddenly flowed even faster. 'How can I try that? I haven't had a menstrual period since I got married.'

10. Marriage

'All men are equal – but the women aren't.'

'All women should be married, and no men.'

'The dread of loneliness being keener than the fear of bondage, we get married. For the one person who fears being thus tied there are four who dread being set free.'

The Unquiet Grave by PALINURUS

Within the animal kingdom there is a great assortment of forms of association and relationship, but generally a consistency of behaviour within each species. Some species find a mate for life, some for the season, some just in passing. Some find no mate, such as those fish which liberate sperm and ova into the sea at random. Some have many mates, such as the bull seals. Some have their mates fastened to them, as in the case of the dwarf males who resemble appendages rather than partners. And some eat their mates after intercourse. There are myriad systems; but normally – at least – there is intraspecific consistency.

With the single human species there is far less conformity. There is both polyandry and polygyny, the two forms of polygamy. There is marriage with or without choice by the partners involved. There are marriages which are totally indissoluble, and there are marriages with every stage of solvency from such totality down to the most casual and temporary pairings. Some societies never expect the pairings to last. Even in some which do the pairings can be brief – the current record number of divorces in the United States is sixteen, held by both a man and a woman. Some widows have been traditionally expected to follow their husbands even through the barrier of death; some, as in India today, are not expected to marry again; some, such as 7% of America's widows, manage to find another mate and re-marry. Due to the different interpretations by different groups of these human partnerships it is hard to equate one country's ideas of marriage with another's, and to compare statistics. Even in modern societies various forms of association are blurring the picture, but the State does not recognise any marriage which its officials and representatives have not themselves authorised. (For facts concerning suitable partners in any marriage as defined by the State see the chapter on inbreeding.)

In Britain marriage, even as the State understands it, is booming. Despite a widespread unwillingness to conform, despite the tax drawbacks, despite the removal of so much stigma from illegitimacy, marriage is a far from outmoded custom. In England and Wales, despite all the other changes that have been going on, with cohabitation, pre-marital sex and all manner of mutual arrangements on the increase, there has still been a steady affection for the old ritual. Since the early 1950's, even though the population of youngsters has been variable, there have never been fewer than 350,000 marriages in any year (and never more than 420,000). The 1980 figure, for example, was 370,000 and there is no sign of any major trend, either upwards or – more to be expected – downwards. The spring is still a good time, with March being top month; September is customarily second, October third, and January is usually least favoured. Since 1935 the average age of bachelor bridegrooms has been falling. In 1967 their average age was twenty-five and of spinster brides was twenty-two years, six months. In that year both bride and groom were under twenty in 27,000 (6·5%) of the year's official pairings; ten years earlier the number of such teenage marriages had been one-third of 1967's level. Since then it has been more constant. Teenage girls give birth to 8% of the country's children, and a quarter of these are illegitimate. Early marriage is frequently hastened by a premature pregnancy, and two-thirds of all girls who marry when less than twenty years old have produced a baby within twelve months. In the past the women either worked or got married; at least that was the custom. Today one-third of British wives go out to work, and they form over half of the female labour force.

Along with marriage, divorces are also booming. Over 120,000 new petitions were filed in 1980, as against 51,000 in 1967 which, at the time, was a record year. Currently each new year witnesses a new high in the divorce rate. Also rising inexorably are the marriages in which either one member of the couple or both members have been previously divorced. To some extent the divorce increase explains the consistently high marriage total because there is a tendency for divorced people to re-marry (and therefore swell the marriage total). The number of people marrying for the first time has been steadily falling, with the figure being 300,000 a year in the early 1960's (or 84% of the total) as against 240,000 in the late 1970's (or 65% of the total). Nevertheless, despite this divorce boom, most British marriages do endure, and 50% of those which reach the courts have lasted ten years. In 1967 the median duration was nearer 12 years, and people seem to be less dilatory these days in deciding that their marriage has failed. The peak years for divorces are three and four after marriage, partly because it is still (although suggested changes are afoot) difficult acquiring a divorce after less than three years of marriage. Two-thirds of divorced people re-marry and a third of all divorced couples have no children. Less than 10% of children born in England and Wales are born to parents not married at the time. Nevertheless, even though the picture is not wholly rosy, it is important to emphasise the positive side of marriage. As was expressed at a recent conference of the British Society for Population Studies: 'Most people in Britain still marry, the majority of couples still produce at least one child, the majority of children are still reared by their natural parents, and the majority of marriages are still terminated by death and this is likely to remain so for the foreseeable future.'

With women living so much longer than men, and with the gap between the lengths of male and female lives growing steadily longer in the Western world, the problem of widows is growing. In the 1890's the American wife outlived the husband in 56% of families; now she does so in 70% of families. What had been a patriarchal society, and what is now often said to be a matriarchal society, is becoming a widowed society. Men leave their money to their wives, and their wives live on – wealthily. Women collect 65% of all legacies in the United States. They are thought to own 60% of all stocks and bonds, 65% of all savings and 40% of all houses. Taking everything else into consideration the country's women, many of them elderly, are thought to own 70% of the country's assets.

By no means are all widows rich, or even the majority of them; it is a time also of great financial hardship made harder by society's attitude to the old. The increasing number of widows, who are now widowed on average aged fifty-nine, means a growing number not just of rich or poor old women but of lonely old women. Re-marriage is difficult, partly because of the modest number of available males. Dr Victor Kassel, who practises in Salt Lake City, thinks polygyny should be permitted for all men over sixty-five provided that their additional spouses are also over sixty-five. (Polygamy means more than one mate; polygyny means more than one wife.) At least loneliness might be reduced were such vintage cohabitation to be encouraged. Even without such a kind-hearted law the marriage of old people is blossoming. The wisdom of Dr Kassel's suggestion therefore appears yet wiser as there are four times as many widows over sixty-five as widowers in the United States. In England and Wales, with a perennial housing shortage, 12% of its houses are occupied by just one inhabitant, and a very large percentage of these are the homes of old and lonely widows; in years to come more and more of our population will be widows. Currently in Britain slightly fewer widows than widowers marry in any year, with the merry widower apparently having more appeal (and certainly a wider choice) than his counterpart.

In India the widows are an even greater sociological problem. Most widowers re-marry, but society frowns on the re-marriage of widows. Hence, as Indian women often die young and do not in general have the longevity of the American female, there is a great shortage of girls. Men seeking marriage or re-marriage have to take younger and younger brides in a country where the ceremony is a quasi-religious duty. 'I am not old enough to be a virgin' is a remark attributed to an English girl, but many young Indians could say the same on their wedding day. Child marriages, i.e. for males under eighteen and girls under fifteen, are forbidden and punishable under Indian law, but the 1951 census revealed innumerable law-breakers. Although government officials collected the facts for the census, and although one might have expected some reticence at declaring illegal marital states, the final figures revealed that there were (at least) 2,833,000 married boys, 6,118,000 married girls, 66,000 widowers and 134,000 widows all between the ages of five and fourteen. In that same year only 6% of Indian females over fifteen were unmarried; equivalent percentages for unmarried females over fifteen in the United Kingdom and the United States were about four times as high. The Indian government has even considered the imposition of a ban upon all female marriages under twenty-one. One wonders for the success of any such scheme in a country with 134,000 widows not yet fourteen, and where – as already reported –the current *average* age of first childbirth is sixteen and a half.

Although English people are not allowed to vote until they are eighteen and although they cannot join the Army until seventeen, English boys and girls are allowed to marry (and care for any subsequent children) at sixteen – provided their parents have given consent to the marriage. Certainly the age of puberty is decreasing in Western society, and therefore argument exists that the sixteen-year-old of today is precisely the same as the eighteen-year-old of, say, two generations ago. The yardsticks of mental and physical development are more relevant than the mere passing of a certain number of days and years. In time the requisite age for marriage, voting, driving, etc. will be lowered to take account of today's speedier maturity.

There is a theory that past puberty and growth was speedier still, and great play is made of Shakespeare's Juliet being thirteen when she longed for, lusted after and even went to bed with (once) her beloved Romeo. 'Younger than she are happy mothers made,' said Paris, making the same point even more strongly. 'And too soon marr'd are those so early made,' added Capulet. In truth, although many a modern marriage still founders if started too early, it was not the custom in that first Elizabethan age to marry anything like so young, and Shakespeare was not affirming the custom of his day. Instead, as was his habit, he was re-telling a traditional tale, and paying steadfast attention even to its details, such as the extremely early age of the lovesick Juliet. (Professor Jim Tanner put me right on this important score.)

11. Pregnancy

The signs/The tests/Embryology/Fertilisation/Ectopic
pregnancies/Implantation/Ectoderm, mesoderm, endoderm/
The calendar of a pregnancy/The puzzles of pregnancy/Its
normality/Pica/The placenta/The umbilical cord/Evolutionary
parallels/The foetus and the law/Abortion

No woman knows the precise time at which one of her eggs has
been fertilised. At present she has no means of knowing, no signs,
no symptoms, nothing. She may be stepping on to a bus, for all
she knows, when sperm actually meets ovum, and a new
life has been begun. And she may be stepping on to a bus again,
for all she or anyone else knows, when that fertilised egg takes the
first positive and vital step of dividing into two. Finally, she is
equally unaware when, some six days after fertilisation, the
round bundle of divided and dividing cells settles down from
its apparently chancy and mobile existence and implants itself
securely within the receptive wall of the uterus.

A week later she misses the normal start of menstruation, a
firm pointer towards the activity of which she has been ignorant –
although some women claim an ability to be aware from a
markedly early date. By then, fourteen days after it began, the
minute embryo is already on its way, many hundreds of cells in
size, with its sex decided at the moment of fertilisation, and with a
phenomenal rate of growth and development both behind and
before it. A week later, when menstruation has been missed for a
week, and when maternal doubts are growing into certainties (if
they have not already done this before), the embryo within has
a lump of heart, the rudiments of brain and eyes and
limbs budding from its surface. The new life is then one-tenth of
an inch (2·5 mms.) long.

Perhaps a week later still, and four weeks from the silent
undetectable initiation of the explosive spurt of growth stemming
from that single cell, the mother will be given official
confirmation of her state. Within her the small streak of
humanity, curved, a quarter of an inch in length, and with traces
of all organs being developed, is still a minute and delicate
thing. Yet, 238 days from that date, and 3,000 times heavier, it is
due to be born as a viable, loud, independent and single-minded
human being.

Signs of pregnancy Most women suspect or know because:

1. Their expected menstrual flow does not happen (amenorrhoea), although it can fail to happen for a host of other reasons (such as menopause), and can even seem to happen in a modified fashion during the first month or two of pregnancy.
2. They (some two-thirds of them) have 'morning sickness', either the feeling or the actuality, which is not necessarily confined to the morning. This can start soon after the first missed menstruation, and is usually over a couple of months later.
3. They feel abnormally tired, and can and do sleep longer than usual. Sleepiness can and does occur in early pregnancy when the actual physical burden is as nothing to what it becomes in late pregnancy.
4. Their breasts start to tingle or itch, and swell (by the second or third month). The thin fluid of colostrum can be expressed from the breasts even at this stage. The areola darkens, and its glands become more noticeable.
5. They have to empty their bladder more frequently.
6. They can feel the foetus moving (the quickening) by the eighteenth or twentieth week. At the same time the foetal heart (ticking like a watch at 140 beats per minute) gives emphatic proof.
7. They, particularly brunettes, acquire a dark line running down from their navel. Their forehead and cheeks can also darken – the 'mask of pregnancy' – and their vulva becomes darker and bluish.
8. They get bigger. The enlarging uterus is out of the pelvis at the end of the third month.
9. They get heavier.

It is one or more of the first five that usually sends a woman along for medical confirmation. Certain of the signs can be caused by occurrences other than pregnancy (such as tumours). A whole group of the signs can be nothing more than false pregnancy (or phantom pregnancy, or pseudocyesis); this is usually confirmation not just of a glandular derangement but of a frantic desire to have children. It often leads not only to all the subjective symptoms of pregnancy but weight gain, abdominal enlargement and breast changes. To some degree it can even happen with men.

Tests for pregnancy Medical examination means partly taking note of all the symptoms, partly checking the obvious signs (they are what someone else sees; symptoms are what a patient has), and partly doing tests which, in the not-so-long-ago days, smacked more of the apothecary's den than a modern surgery.

The signs include inspection of the breasts for any changes, of the skin and genitals for any discoloration, and a manual examination of the abdomen for any clues about the uterus within. The various laboratory tests first made use of rabbits, mice, rats, and toads. The time these tests took and their reliability depended upon which test was being used, but they exploited the fact that the hormone called chorionic gonadotropin (known as HCG) abounds in the blood and, more conveniently, in the urine of pregnant women. It is produced by the placenta, and the amount produced steadily increases with pregnancy for some 90 to 100 days. Thereafter its level decreases, but the later detection of pregnancy, with swelling abdomens and increasingly active foetuses, requires a less subtle procedure. The old laboratory examinations underline the fact that human hormones are not rare and unique chemicals in some rigid class by themselves, but can have wide-ranging and powerful effects on many members of the animal kingdom. After all the toad is not even a mammal – by a long way.

The first hormone test to be used, which formed the basis for all the others, was the Aschheim–Zondek. It employed immature female mice. Urine from the woman under test was injected into the mice, and 100 hours later they were slaughtered and their ovaries examined. If these had been made mature, the woman was pregnant. In the Friedman test virgin doe rabbits are used. Once again human urine is injected, and within forty-eight hours the rabbit ovaries are examined. Again activity there means pregnancy. Yet another test uses rats in similar fashion. The test most strange of all makes use of the South African clawed toad (*Xenopus laevis*). Called the Hogben test, it necessitates the injection of concentrated urine into the female amphibian. Within fifteen hours the toad will spawn 50 to 200 eggs.

All these tests have a high degree of accuracy, 95% or thereabouts. Although they are intriguing, particularly the last one, they did take time. This delay, and the complexities, were their undoing. In the 1960's a rash of less time-consuming alternatives was proclaimed. Urine and the same hormone are still involved in most of them, but not the living animals. To quote from three claims made in 1965: the first 'takes two minutes, gives no false positives'; the second 'requires only three minutes to be performed and interpreted'; the third shows up a pregnancy 'within ninety-six hours following a missed menstrual period'.

In other words, miss a period, become suspicious, visit the surgery, give some urine, flick through a magazine, and get the answer while you wait, all within, say three weeks of the successful mating. It is quick. Even so, three weeks is a long period in the development of an embryo. A litter of mice would have been born in that time. And a human, although with 245 intra-uterine days to go, is in quite an advanced state at three weeks, not in size but in differentiation. In fact pregnancy testing these days can be made even before the woman has missed her period and become pregnant, as a week post-conception is ample time for modern methods of immunoassay to check the HCG levels. The days of toads are over, and also the days of waiting to know if the longed for/dreaded happening has come to pass.

Embryology, the study of an embryo's development, is a hideously complicated subject. It starts off simply enough with that solitary, scarcely visible, and recently fertilised egg. It then steadily embroils itself in greater and greater knots. Every organ, and every part of every organ, and every nerve and blood vessel leading to every part of every organ, has to be accounted for. It is small wonder that books on embryology are so generally decried.

Imagine, for an analogy, writing the history of some Norman accomplice of William I at Hastings. First he takes a wife, and they beget children. Their children's activities and characteristics must be described and so must those of the grandchildren, whose activities and characteristics cannot be the same. And each succeeding generation, with more descendants, greater differen-tiation, greater dissimilarity, greater complexity – all leading one to the other, all interwoven, all intermingled, must all be described. The Norman pair become a people, a population, a country. Like some grotesque Russian novel every character is playing a part at every stage. To my mind embryological books have a perfect right to be as bad as they like, for the human baby, stuffed full of organs, brimming over with reflexes and instincts, more chemically complex than any laboratory, is no simple egg. Its development is a bewilderment quite of its own.

To begin with, like the Norman pair, events are concise and straightforward. The female ovum (previously called the oocyte) is only serviceable for twelve to twenty-four hours after leaving the ovary, and the male sperm (or spermatozoon) is only effective for forty-eight hours. Therefore fertilisation is only possible when the arrival of sperm coincides reasonably accurately with

111

ovulation, with the entry of an egg into the oviduct. There is usually only one egg (dissimilar twins disprove this rule) and there are initially 400,000,000 sperm in the 3·9 ccs. of the average male ejaculation. Many are necessary, as weak sperm counts are rarely fertile, but only one can fertilise the egg. (An exciting exception to this is dispermy when two sperm both – somehow – fertilise one egg.) Many of the one sperm's companions have previously helped to detach the ovum's outer layer, and the solitary sperm then moves forward into the ovum, losing its tail in the process and swelling its head. Twenty-three chromosomes have met 23 chromosomes to form the central part of a single egg. The long chain of events, hardly any of which is understood, that leads to the birth of a baby has begun. It is not even known why that union of two half cells to form a single cell is the trigger for such abundant cellular multiplication in the months to come.

Only very recently has the early life of a human being been seen at all. Sperm were first observed in 1677 by Antony van Leeuwenhoek, but the first human egg cell was not seen until 1930. Experimental union of human sperm and egg was not seen until 1944. And one of the biggest mysteries of all, the appearance of a human being in its first few days of life, when nothing more than a blackberry jumble of cells, remained a mystery for a long time. Only in 1952 was the first free blastocyst seen, the first human being ever to be observed before implantation in the uterine wall. That individual was less than one week old.

A triumph, announced in September 1965 from the Vanderbilt University School of Medicine, was a photograph of the nuclei of both ovum and sperm, lying side by side, with the sperm's discarded tail lying nearby. This was found after washing out the contents of a woman's oviduct that had been surgically removed for other reasons. It was similar surgical operations that were responsible for providing initially the means whereby some understanding was achieved of mankind's earliest hours and days. Now that *in vitro* fertilisation is possible, as outlined in the chapter on fertility, the actual process whereby one cell becomes two, two become four, and so on, can be examined far more easily. The reproductive biologists assisted the *in vitro* fertilisers in their work, and these teams have been able to reward the reproductive biologists by showing them just what does happen in those earliest, pre-implantation days.

The first step of all, the passing of the egg from ovary to oviduct, is one of the least well understood. Normally there is a gap, almost as if the egg were expected to jump from the ovary to the funnel-ended tube. It is thought that the funnel becomes erect, thus lessening the gap, but no one knows. It is known however that the egg can fail to make it and so pass by error into the abdominal area, and even grow there having been fertilised. Assuming instead that the egg does reach the oviduct, and the sperm's arrival is synchronised, some evidence shows that it may take the sperm only an hour to reach the egg. (As they live for longer this does not mean that fertilisation must occur in an hour or so.) Fertilisation is the union of the two nuclei, of the two sets of genes, of the father's sex-determining genes for the offspring (it is either male or female from that moment), of the father's dominant characteristics and of the mother's, and of the interaction of both their genetic influences. So much of the human-to-be has been decided already that it is mainly a matter of development right from the start of the fertilisation stage.

Having united, the next step is to divide. Although sperm and ovum are so different, the two cells from the first cell division are virtually the same. (An analogy is mixing a glass of water with a glass of ink, and then dividing the mixture.) It is thought that the unification and then the division takes thirty hours. The human form is then still pinpoint sized, measuring roughly a tenth of a millimetre across. Its two cells, to judge from studies in the laboratory, do not divide again simultaneously. There is first a three-celled and then a four-celled stage, possibly reached fifty hours after the start. (Two days gone; 264 still to go.) At the end of the third day the form is composed of twelve cells and has probably reached the diminutive (at this time) cavity of the uterus, having left the long thin oviduct (or Fallopian tube, or uterine tube). How the developing group of cells moves along the oviduct is another obscurity, complicated by such facts that the time taken to do so is equal in the pig and the mouse even though the embryo pig's journey is forty times longer. Multiplication is now proceeding apace, and 58 and 107 cells have been counted for two human blastocysts which were thought to be four and four and a half days old respectively.

The next really crucial stage for this free-floating object, which apparently has every chance of just floating straight out of the uterus – and must do so on occasion – is to attach itself to the walls of the uterus. Called implantation, or embedding, or nidation, no one knows how this happens, whether the embryo digests its way into the uterus or whether a niche is prepared for it by the uterus, or whether it is a bit of both. (A clue is that the embryo can become attached elsewhere, as in ectopic pregnancies which happen away from the uterus and the normal sites.) Anyway, the attachment probably happens between five and a half and seven days after the start. This seems late for such a necessary and positive step, but scarcely later than the mouse (four days) which is born a little over two weeks afterwards, and the guinea pig. Later still are many other creatures, such as the macaque monkey (tenth day), the cat (fourteenth day) and most ungulates (even later).

The cavity of the uterus, now small, and minute relative to its final pregnant size, is filled with salty and sugary fluid. Presumably the multiplying embryo takes some form of nourishment from it. Once implantation has begun, and the embryo sinks into the specially receptive layer of uterine wall, the process for the acquisition of future nourishment via the placenta can begin. It is the loss each month of this receptive layer, the weeping of the womb, that is the basis of menstruation, the monthly 'curse'. (Incidentally, although I could find no kind of date for the changeover, the curse used to be called the blessing. Another four weeks' grace – of *not* being pregnant – had been granted.)

Ectopic pregnancies Before carrying on with those embryos happily implanted in the correct uterine fashion, I would like to digress to those far less happily implanted in a bewildering assortment of other spots, the so-called ectopic pregnancies. They can be within the Fallopian tube (the commonest place), or down at the bottom end of the uterus (very rare), or on the ovary itself, or even within the abdomen just attached to some organ like the large intestine.

Not surprisingly nearly all these pregnancies fail, and they kill about twenty-five women in Britain a year. One difficulty is that they can be outstandingly hard to detect. (The obstetrician Elliot E. Philipp says no wrong diagnosis is made more often and no condition is more often missed than ectopic pregnancy.) A general practitioner will meet – according to *The Lancet* – about five or six in his lifetime, for they occur once in roughly 300 pregnancies (although more frequently with blacks and older mothers). Sometimes they can have a secondary implantation, by bursting through the narrow confines of the oviduct and settling down somewhere else.

For most women severe pain indicates early on that something is wrong, and for nearly every woman a laparotomy (opening up of the abdominal cavity) will put things right by showing the surgeon the cause of the trouble and extent of the damage. For very few women such an operation will show the surgeon not a six-week embryo (a frequent time of trouble with tube ectopics) but a well-developed foetus capable of survival. One such baby, weighing 4 lbs 6 ozs. (2·0 kgs.), was born in this fashion at Chesterfield in 1965. The mother had felt no pain for 7½ months, and she then had the operation two weeks later. Even normal weight can be achieved outside the uterus. In India, also in 1965, a 7lb. (3·2 kgs.) baby was found after laparotomy to be lying free in the peritoneal cavity. In both these cases, the mothers (and babies) were well after the strange delivery and even stranger pregnancy. Yet one misplaced foetus within a woman is as nothing compared with some pregnancy errors: a three-month old boy in Hong Kong had three foetuses removed from his abdominal cavity after his mother had grown suspicious at his swelling condition. Of the three foetuses one was two inches long and well formed.

Implantation Now to normality, and normal implantation within the pear-sized uterus. There are several general points which need to be borne in mind during any discussion on development. The first is the speed of differentiation. Most mothers tend to regard nine months as long, while the weeks go wearily by, as leaves come and leaves fall. It is indeed long, and for the bulk of the time the foetus has just been growing; but actual differentiation, or the original distinguishing of its various parts, is phenomenally rapid. The foetal heart is usually beating by the twenty-fifth day. And by the end of the eighth week the creature looks human, it has all the major external features (eyes, ears, mouth, limbs, toes) and the main internal organ systems. But, and this is the second general point, the human object is still minute. When the heart starts beating, it does so within a pea-sized creature less than a quarter of an inch long. By the eighth week, when liver, pancreas, kidneys and so forth are all distinguishable, the new life is not even one and a half inches long (or less than two joints of an adult finger). It still weighs next to nothing. Even by the end of the third month, when differentiation is virtually complete, the young human being weighs less than an ounce.

The third general point is that cell division and growth rate have slowed down by birth time to a fraction (2%) of the fastest rate. The first cell division takes place after thirty hours, the next twenty hours later. Thenceforth growth and division proceed apace but, although the new-born baby has some twenty thousand million cells, there have been only thirty-four cell generations between fertilisation and birth. The second generation is two cells, the third is four cells, the fourth is eight cells, the fifth is sixteen cells, the tenth is 500 cells, the twentieth is

1,000,000 cells, and thirty-four will give the baby's formidable total. (It is the same mathematical enigma as was used by the legendary individual who asked, as a reward for some legendary deed, to have a grain of wheat for the first chess-board square, two for the second, four for the third, etc. His host, delighted by the apparent humility of the request, agreed greedily, and ordered the requisite grains of corn. As a chess board has sixty-four squares, the request soon proved itself inordinately demanding.)

So, too, with human replication. With two divisions taking place in fifty hours, and only thirty-two more being necessary, the initial rate does not have to be maintained. The three general points therefore are:

1. Differentiation occurs very rapidly.
2. It occurs when the embryo is very small.
3. The rate of growth after birth is as nothing compared with growth before it.

To get from one translucent cell to 20,000,000,000 in nine months is fantastically rapid when set against a baby's growth from 7 lbs (3·2 kgs.) at birth to 28 lbs (12·7 kgs.) some eighteen months to two years later, or to 140 lbs (63·5 kgs.) eighteen years later.

As I have said, embryology is a hideously complicated subject. It tends to get bogged down in its own convolutions right from the start. Therefore, instead of mimicking customary descriptions of development, I shall give the briefest possible general picture, and then outline specific events week by week in calendar form.

The egg divides to become two cells. Those two divide, and soon there is a bunch of cells (a morula). The bunch then develops a hole in the middle (the blastocyst stage) with more cells at one end, and the whole thing becomes implanted in the uterus. So much for the first week – and for simplicity. The situation now becomes more fraught. Not all the blastocyst cells become the future human being. Some, by growing into the uterus, become placenta. Some become yolk sac, a relatively large (in early days) but yolkless and vestigial structure that can be forgotten. Some become amnion, the protective and liquid-filled balloon that surrounds the developing form. And only some become the human being.

From the lump of cells at one side of the blastocyst three layers begin to develop. To begin with they are vague, but they are crucial in all development (whether human, or right down the animal tree). When they first appear they are flat, something like the layers of sponge cake. They are called ectoderm, mesoderm and endoderm (the New World prefers entoderm for the third layer). Initially the three of them lie on top of one another, but the flat cake has to become a circular cake, with ectoderm moving right round to meet ectoderm, with endoderm being central, and with mesoderm being the jam in between the inside and the outside. Right from the beginning one end is the head, and the other is the tail.

The first complication of this three-layer tube is that the outer layer – the ectoderm – forms an extra tube, a swelling that runs from front to back, from head end to tail end. From this swelling will grow all nervous tissue, the brain, the spinal cord, and all nerves leading from them. From the rest of the outer ectodermal layer will grow, reasonably enough, all outer skin. From the inner endodermal layer will grow the innermost organs, such as the lining of the digestive tube, and the organs linked to that tube. From the middle layer will grow the middle things, such as muscle and bone. A human is really no more than a complicated worm, with an outer skin, with a gut running through the middle, and with various extras like muscles and bone and blood vessels in between. But even during its early days – two weeks after conception, the human is plainly not looking like any worm. It has already started to distinguish itself from the basic formula.

With the three layers firmly in place, with a head and a tail, and with the nerve tube running from one end to the other, the various parts of the body then begin to manifest themselves. Buds appear, little local swellings become apparent in certain areas of tissue, and they grow. They become limbs, or organs like the liver, or glands. If endodermal, like the liver, they grow into the area of the mesoderm but they do not grow through the outer and skin layer of ectoderm. Some of the skin layer, such as the nervous tissue, grows deep down from the skin, and ramifies to every part of the body. Cut yourself anywhere, and you will cut ectoderm and then mesoderm, but only if you stab yourself centrally will you meet tissue from the original internal layer of endoderm. The tasks of these three primaeval and distinguishable layers, all formed so early in development, are quite finite in the formation of a human being. Briefly, here are some of their end products to indicate the basic pattern of the body's outer, inner and intermediate layers:

Ectoderm (the outer layer) – outer skin, nails, hair, sweat glands, eye lens, mouth lining, tooth enamel, lining of nose, salivary glands, and all nervous tissue.

Mesoderm (the middle layer) – muscle, bone, cartilage, inner skin, blood vessels, kidneys, and connective tissue.

Endoderm (the inner layer) – lining of oesophagus, stomach and intestine, liver, pancreas, bladder, lining of lungs, glands – such as thyroid and thymus.

A customary belief about human development is that the embryo passes through all the evolutionary stages of fish, amphibian, reptile and primitive mammal before emerging as a proper human being. This is not so, despite the fact that every human being is a direct descendant of creatures from these zoological groups. But it is true that there are many similarities in the early development of each group. Plainly each egg, then each two-cell stage, or the later morula, or blastocyst, looks much like another, and later similarities also exist.

During the fourth or fifth week a series of grooves, for example, appear within the human embryo (pharyngeal pouches) on the rudimentary throat walls, Were this a fish embryo these structures would become gills. As it is a human embryo they do not (hence the statement that man does not pass through a fish stage) and they develop instead into other organs. The first pouch becomes part of the middle-ear cavity. The second forms the basis of the palatine tonsil. The third becomes parathyroid gland tissue, and the basis of the thymus. The fourth becomes the superior parathyroid gland, and the fifth pouch becomes the ultimo-branchial body, a vague and little understood object, but emphatically not a gill.

Similarly, although the human embryo has a short stub of tail for a while and this is precisely similar to the short stubs that become tails in many other species, the human stub only forms the basis of the human coccyx. Mankind does not travel up the trunk of the animal tree with each new embryo; it just makes use of a similar cellular differentiation which in the past has led to all the branches of that tree.

The calendar of pregnancy   So much for the general story. Now to the calendar. It starts with fertilisation. The practice of referring to pregnancy as a period of 280 days or ten lunar months considers the pregnancy to have begun at the start of the last menstruation. This calendar assumes instead that it begins two weeks later – with fertilisation. It also assumes, as a further indicator of time passing, a fertilisation date of January 1. Therefore the start of the last menstruation was approximately two weeks earlier, i.e. December 17.

First week	Fertilisation. First cell division (after thirty
(1–7 days)	hours). Whether dissimilar twins or not

(1–7 Jan.)

now established. Cell mass moves along the 4 ins. (10 cms.) of oviduct, and becomes a hollow blastocyst en route. Enters uterus (third to fifth day). Implantation occurs (sixth to seventh day). Mother probably unaware of the situation, but pregnancy can already be confirmed by modern methods of immunoassay. Customary menstruation not due for another week. Uterus still weighs about 2 ozs. (57 gms.).

Second week
(8–14 days)
(8–14 Jan.)

Bleeding may very rarely occur (on the thirteenth day) as a result of increased blood flow to implantation site. May be confused with menstruation which is due at the end of this week. Embryo is now plate-shaped with hundreds of cells, some of which will form the true embryo and some structures like the yolk sac, umbilical cord, placenta, and amnion. The yolk sac and amnion are the first to be formed. Whether or not identical twins are to be born is usually established at about this time.

Third week
(15–21 days)
(15–21 Jan.)

Embryo one-tenth of an inch long (2·5 mms.) and pear-shaped. Neural tube formed. Menstruation now overdue. Morning sickness, and nausea can start. So can breast tenderness. Primordial germ cells, which will eventually produce the cells that produce the sperm and eggs, already present but the embryo's sex cannot as yet be determined. Both eyes and ears start development by the eighteenth day.

Fourth week
(22–28 days)
(22–28 Jan.)

Embryo a quarter of an inch (6·3 mms.) long. Heart (one tenth of an inch – although more of a tube than a heart) starts beating on twenty-fifth day, and a circulatory system of a sort exists. Tongue has started to form. Limb buds appear on the twenty-sixth day. The yolk sac, often said to be useless as its function is not known, is now as big as the embryo (but will grow little more, although it will never quite disappear). Umbilical cord still very rudimentary. Mother often now asks for and receives medical confirmation of her state. Given a date 238 days ahead.

Relative size increase is never again so great as in this first month. The embryo is now ten thousand times larger than the egg. Also the extent of physical change never again to be equalled.

Fifth week
(29–35 days)
(29 Jan.–4 Feb.)

Heart now pumping frequently. External ears start taking shape. Arm buds differentiate into hand, arm and shoulders (on the thirty-first day). Finger outlines appear (on the thirty-third day). Foot still a flat and budding protuberance. Nose and upper jaw start to form. So does stomach. Embryo half an inch (13 mms.) long.

Sixth week
(36–42 days)
(5–11 Feb.)

Tip of nose visible (thirty-seventh day). Eyelids begin to form. Five separate fingers. Toe outlines appear. Marked skeletal growth, but still of cartilage. Stomach, intestines, reproductive organs, kidneys, bladder, liver, lungs, brain, nerves, circulatory system are all being actively developed. Embryo three-quarters of an inch (19 mms.) long.

Seventh week
(43–49 days)
(12–18 Feb.)

One inch (25 mms.) long. Stomach already producing some digestive juices, and liver and kidneys have started functioning. Muscular reflexes can work (as can be shown experimentally, but nothing yet felt by the mother). Outer and hearing part of the ear almost complete. Upper and lower jaw very clear. Mouth has lips, something of a tongue, and first teeth showing up as buds. Thumb markedly different from fingers. Heel and toes definite. First true

bone cells appear. Circulatory system now
effectively operational.

Eighth week
(50–56 days)
(19–25 Feb.)

A sort of neck now visible. The head of the
embryo is very large compared to the rest
of the body. The uterus is now about 4
inches (10 cms.) long, and wide, and deep.
The corpus luteum which produced the
egg is now about the size of the original
ovary. The ovary itself has not grown. The
corpus luteum will continue to grow until
it becomes almost hen's egg-sized. It will
then regress, having fulfilled its hormone-
producing role.

Ninth week
(57–63 days)
(26 Feb.–4 Mar.)

Embryo often called foetus after this time.
(Some call it foetus – or fetus – after fifth
week, with embryo existing beforehand.)
Sex of individual can be detected
externally. (Until now organs have
appeared similar even to expert scrutiny,
although chromosomal assessment of sex
possible under the microscope.) Head is
three-quarters of an inch (19 mms.) and is
half crown to rump length (i.e. total length
of foetus minus the legs). Footprints and
palmprints now indelibly engraved for life.
Spontaneous movements occur. Also eyelids and palms sensitive to touch (as shown
by reflex squinting and gripping attempts).
Nails start to grow. Dissimilar twins start
looking dissimilar. Eyelids close over the
eyes for the first time, and eyes still look
outwards rather than forwards. Amount of
hormone HCG (used in pregnancy tests)
now reaches its maximum level. Eighth
and ninth weeeks generally considered
best time for abortions, for 'terminating
pregnancies vaginally'.

Tenth week
(64–70 days)
(5–11 Mar.)

Quarter stage reached on sixty-sixth day
(but foetus will have to multiply its weight
over six hundred times in remaining three-
quarters of pregnancy). Mother's average

weight gain is now 1 lb. (0·45 kgs.). (But this *average* figure can be misleading, as many women can lose weight in first three months and many start gaining weight from the start.) Weight of placenta is less than an ounce (28 gms.) but may be four times heavier than the foetus which is about a fifth of an ounce (6 gms.). Uterus weighs 7 ozs. (200 gms.) and contains 1 oz. (28 gms.) of amniotic fluid, perhaps 2 ozs. (57 gms.), perhaps 3 ozs. (85 gms.). Breasts have increased in size, some say by about 2 ozs. (57 gms.). ('No one', said one gynaecologist, 'has been enthusiastic enough to cut a breast off, and weigh it. What is certain is that women have either let their brassières out to the full by now, or bought themselves a larger size.') Commonest time for miscarriages is at tenth week, i.e. time of third missed period. The second most common time is at the sixth week, i.e. time of the second missed period. It is much rarer for miscarriages to occur at time of fourth missed period.

Eleventh to fourteenth weeks (71–98 days) (12 Mar.–8 Apr.)

Many more reflexes possible, such as frowning. Thumb can be moved to the fingers. Swallowing starts. If foot is tickled (experimentally) the whole leg will be withdrawn. Two halves of palate fuse together. Vocal chords completed. Urination has begun (and urine is removed with regular renewal of the amniotic fluid). Swallowed amniotic fluid can be digested. Sperm cells or egg cells exist. Uterus moves up out of pelvis, and can be felt, especially in thin people. At twelfth week foetus's crown to rump length is $2\frac{1}{4}$ ins. (5·7 cms.), and crown to heel is $2\frac{3}{4}$ ins. (7·0 cms.). Its weight is $\frac{3}{4}$ oz. (21 gms.).

End of Development Period, even though the developed form weighs so little. Start of Growth Period. Growth has of course been considerable in first three months, but in subsequent six months foetal growth is from less than 1 oz. (28 gms.) to 7½ lbs. (3·4 kgs.).

Foetus and placenta roughly equal in size. Heart pumps 50 pints (28 litres) a day. Sex of individual now physically definite, as organs clearly distinguishable. Uterus half way between pubic bone and navel. By this month the placenta is producing the hormone progesterone in sufficient amount to maintain pregnancy. This job was formerly done by the ovary's corpus luteum.

Fifteenth to eighteenth weeks (99–126 days) (9 Apr.–6 May)

Growth of head hair starts. Eyelashes and eyebrows also begin. Nipples appear. Nails become hard. Heart beat can be heard externally by listening to the mother's abdomen. The separate heart beats of twins should be detectable with a stethoscope. Mother starts feeling foetal movements (the quickening) although certain movements started less obtrusively six weeks before. Thin women may notice movements earlier than normal or fat women. So do women not having their first child. Foetal hiccups also occur. By now amniotic fluid (which has been steadily increasing) is over half a pint (0·3 litre). Placenta weighs 6 ozs. (170 gms.) at eighteen weeks, amniotic fluid 9 ozs. (255 gms.), and uterus 20 ozs. (567 gms.). Half-way time.

Foetus's crown to rump length is 6 ins. (15 cms.), and crown to heel length is 9 ins. (23 cms.) (approximately half that of the new born baby). Its head is now only one-third of total body length. The foetus's weight is 11 ozs. (312 gms.) (approximately one-tenth of a new born baby). Mother's weight gain (at eighteen weeks) is 9 lbs. (4·1 kgs. or twelve times that of foetus). Her breasts may have gained ½ lb. (0·23 kg.) in weight

123

by now, and colostrum may be expressed from them. Usually breasts do not increase much in size between the twelfth and twentieth weeks. The customary picture is of breast growth at the beginning and end of pregnancy, rather than the middle. The mother is now putting on 1 lb. (0·45 kg.) a week, and will do so for the next two months; thereafter the weekly gain will fall slightly. If foetus is aborted at this time it may possibly survive but, if it does, probably only for a few minutes.

Nineteenth to twenty-second weeks (127–154 days) (7 May–3 June)

Eyelids can and do open. Premature life now possible, although infant only size of a large man's palm. It can now grip firmly with its hands (although some degree of the grip reflex was seen twelve weeks earlier). Lanugo, the hairy growth on arms, legs, and back appears. This is generally seen on premature babies, but has normally gone by birth. The foetus's face is red and wrinkled, although fat is now being deposited. Uterus at end of month is up to the navel, but height varies according to the mother's posture. Foetus crown to rump length is 8 ins. (20 cms.), and crown to heel length is 12 ins. (30 cms.). (Leg length is gradually becoming relatively more important.) Weight is 1 lb. 6 ozs. (0·62 kg.).

Twenty-third to twenty-sixth weeks (155–182 days) (4 June –1 July)

Many prematures (strictly called immatures until twenty-eighth week) of this age able to live. Volume of amniotic fluid perhaps 1½ pints (0·85 litres), but after thirtieth week may either not increase or even decrease to allow for foetal growth. Head hair may grow long (although many babies are born bald). Most of lanugo goes. Thumb sucking a frequent habit. Umbilical cord has reached a maximum length. At end of month uterus 2 ins. to 3 ins. (5cms. to 7·5 cms.) above the navel. Foetus's crown to rump length is 9 ins.

(22·5 cms.), crown to heel length is 14 ins. (36 cms.). Weight 2 lbs. 11 ozs. (1·22 kgs.).

Twenty-seventh to thirtieth weeks (183–210 days) (2–29 July)

Three-quarter stage reached during the twenty-ninth week (and existing weight needs to be multiplied two and a half times in remaining quarter of pregnancy). Chance of survival if born now better. Finger nails reach finger tips (and may actually need cutting at birth). Foetus (generally) settles down into head-down position. Earlier acrobatics cease. Fat is being deposited, and smooths out the skin. Mother's average weight gain is (at twenty-eight weeks) about 19 lbs. (9 kgs.) (or nearly six times that of baby). Her breasts have increased by about 14 ozs. (0·4 kg.). Placenta weighs 15 ozs. (0·43 kg.) (at twenty-eight weeks). Foetus now four times heavier than placenta. Uterus is half way between navel and lower end of breast bone and weighs 2 lbs. (0·9 kg.). Its height above the navel may fluctuate as the baby's head does or does not fit into the mother's pelvis. This fitting in (or engaging) usually comes later. Efforts to rearrange breech presentations often made now, sometimes as late as thirty-second week. Foetus's crown to rump length is 10½ ins. (27 cms.), crown to heel length is 16 in. (41 cms.). Weight, 3 lbs. 12 ozs. (1·7 kgs.), is half final weight (this is very much an average figure. The biggest babies at this time may be twice the weight of the smallest).

Thirty-first to thirty-fourth weeks (211–238 days) (30 July–26 Aug.)

Premature babies of this age reasonably handsome (less like pink old men) as body is rounded and skin is smooth. Crown to rump length is 12 ins. (30 cms.), crown to heel length is 17½ ins. (43 cms.). Weight 5 lbs. 1 oz. (2·3 kgs.). Most airlines do not like women to fly after the thirty-third week (with five weeks to go) for long distance flights, or after thirty-fourth week (four weeks to go) for short trips.

Thirty-fifth to thirty-eighth weeks (239–266 days) (27 Aug.–23 Sept.)

Heart pumping 600 pints (340 litres) a day (although blood content only slightly over half a pint). Growth usually stops for each individual shortly before its birth, but by then weight of that original fertilised egg has been increased 5,000,000,000 times. Now, and in the next twenty years, weight only has to be increased another twenty times. Mother's average weight gain is now 27·5 lbs. (12·5 kgs.) (but may be nil or 60 lbs (27·2 kgs.). Breasts have gained about 1 lb. (0·5 kg.). Maternal surface area has increased by 1½ sq. ft. (0·14 sq. metres).
Uterus, now 14 ins. (36 cms.) long, reaches highest point a couple of weeks before birth. It weighs 2¼ lbs. (1·0 kg.) or some twenty times original weight. Placenta weighs 1½ lbs. (0·7 kg.) at term, and is 7–9 ins. (18–23 cms.) in diameter. Baby's crown to rump length is 13 ins. (33 cms.), crown to heel length is 20 ins. (51 cms.). Head is now a quarter of total body length. Weight at thirty-eight weeks is highly variable, but an average for white communities is 7 lbs 4 ozs. (3·3 kgs.).

The puzzles of pregnancy   There are several mysteries in all this development. They concern, firstly, the ability of the human body to contain such explosive growth within one section of it without the growth urge disastrously affecting the mother's body as a whole. The second is the maternal ability to produce and tolerate a 'foreign' being; an immunological puzzle of extreme proportions. A third mystery, which I shall describe first, is the mother's weight increase; she puts on more pounds than can readily be accounted for.

It is easier to take an average woman, and her weight increase, rather than those others who swell up by 60 lbs. (27 kgs.), or those who have babies without increasing their weight at all. An average woman, who is not dieting overtly during pregnancy, puts on 27·5 lbs. (12·5 kgs.). And the same average woman expels a baby of 7 lbs. 4 ozs. (3·3 kgs.), a placenta of 1 lb. 7 ozs. (0·65 kg.), and amniotic liquid of 1 lb. 12 ozs. (0·8 kg.). Her uterus has gained over 2 lbs. (0·91 kg.). Her breasts have increased by 14 ozs. (0·4 kg.). She has 2 lbs. 12 ozs. (1·25 kgs.) more blood, and 2 lbs. 10 ozs. (1·2 kgs.) more extra-cellular water within her. All this adds up to less than 19 lbs. (8·6 kgs.). It therefore leaves almost 9 lbs. (4·7 kgs.) unaccounted for. Plainly the extra weight must be somewhere, whether in fat or maternal stores of other kinds, but such deposits are difficult to measure even though they weigh more than the baby that caused them. For instance, some of the fatness of many pregnant women disappears fairly rapidly after birth, but it is not known whether this lost fatness was fat, or just extra blood in the skin, or just extra water. The subject of weight gain is by no means completely understood.

The immunological puzzle is presented by the fact that both mother and child are two unique people, even though attached. As with all people (except identical twins) they are not able to exchange grafts with each other because of vigorous resistance from both sides, and yet a foetus is grafted on to a mother by the placenta for the best part of a year. Sir Peter Medawar, then director of the National Institute for Medical Research, London, suggested that the foetus's privileged position (in not being destroyed by the mother, and sloughed off like some skin graft) is due to one or all of three things: lack of antigenicity (or the ability to reject 'foreign' tissue) of the placenta; anatomical isolation of the foetus from the mother; immunological inertia (or general acceptance of foreign tissue) of the pregnant woman. Having posed these three attractive possibilities, Sir Peter then added that: the placenta *is* antigenic; there is *no* question of the placenta plus its foetus being anatomically isolated; there is *ample* evidence that the pregnant mammal is not immunologically inert. In other words, a valid conclusion is that pregnancies are impossible – or that some additional research needs to be done urgently to restore one's faith in the system.

As for growth, it is the mother's own endocrine system that is responsible for successfully harbouring the enlarging embryo. Growth stimulants pour from her pituitary glands, and yet she does not grow. Her own structure does not change. The only parts to enlarge are those which need to enlarge. When the effects of pregnancy are finally over, say a year after it all began, the mother is certainly the same height, virtually the same weight, and only marginally altered by the experience.

Its normality Is pregnancy a more normal state for a woman than non-pregnancy? Just as the curse of menstruation was called

127

a blessing, was it a rare event in former times with repeating pregnancies preventing its regular occurrence? Presumably the answer to both questions is yes. Girls would become pregnant before they were even full grown, as is so frequent in India today, and regular pregnancies would then relentlessly follow. As soon as the repressive effects of each pregnancy and the subsequent breast-feeding began to wear off, and as soon as ovulation and menstruation started again, the time for a new pregnancy had arrived.

Perhaps the interval between each, between one birth and the next fertilisation, was six months. Children would then have been born every fifteen months. Within twenty years sixteen offspring would have been born. For twelve out of those twenty years, therefore, the mother would have been pregnant, and she may not have experienced a single menstruation during the whole time. The production of, say, thirteen children in twenty years is still not an uncommon event in many parts of the world, and used to be common in Britain. Even this rate, whether the babies grow up or not, means that the mother is pregnant for half the time, and possibly breast-feeding a child throughout the other half. The average modern mother of today produces two to three children and cannot compare her experiences of motherhood with those women who scarcely experienced anything else.

Apart from the obvious changes which occur during the pregnant state (and which could be called a reversion to normal, bearing in mind that traditional regularity of childbirth), there are others less obvious. The heart output rises by a third. Heart beat goes up, and more blood is pumped with each beat. Pressure in certain veins, such as those in the leg, becomes high (sometimes giving rise to trouble with varicose veins and haemorrhoids). The heart increases in size, and changes its position. The flow of blood to the kidneys rises considerably, as it does to the skin, sometimes making hands 'clammy' and the days feel unaccountably hot. (A certain Madame X of France made gynaecological history by having white and insensible fingers which became normal in pregnancy; it was her first sign of another conception.) Although the bladder becomes more active in getting rid of water during pregnancy, an early sign is also increased thirst. The bladder's flow and general rate of excretion are not what might be expected. They are greatest in early pregnancy, when the embryo is smallest, and least or even below normal towards the end, when the mother may be 2 stone (12·7 kgs.) heavier.

Pregnant women may get special cravings (or pica). A BBC broadcast once solicited 991 craving stories, of which 261 were for fruit, 105 for vegetables, 187 for other special foods, such as pickles, 35 for coal, 17 for soap, 15 for disinfectant, and 14 for toothpaste. It is the oddities which are true pica. One explanation is desire for taste, as all sweet, salt, sour, and bitter things have less taste during pregnancy.

Also, during pregnancy, there is more of the clotting chemical fibrinogen in the blood, and there is a small rise in the incidence of thrombosis. Respiration is increased, by breathing more deeply, not more frequently. The sensation of shortness of breath (which is not backed up physiologically by any actual shortness) may exist more often when sitting down rather than moving about. There are more fats in the blood during pregnancy. Also the average woman can then do less work for she has more weight to shift about. Consequently she feels less efficient, but it is not known whether her actual efficiency for some given and accustomed task has been lowered. Her area has, on average, gone up from 17 sq. ft. (1·6 sq. m.) to 18½ sq. ft. (1·7 sq. m.) by the end of pregnancy. Practically everything she takes can cross the barrier between her and her child, such as all gases, alcohol, barbiturates, all antibiotics, and tranquillisers. The effect of smoking is to produce smaller babies (by up to 7 ozs. or 198 gms.). According to the Food and Agricultural Organisation a woman needs an extra 80,000 calories to produce a baby, or 300 more per day. Other experts feel this is excessive and, by changing her activities slightly, by sleeping a little more a day, a woman can easily halve the amount. Greater changes will decrease her extra requirement to nil. Generally, the reaction of the fact-finders is one of astonishment at the minute amount of food a woman needs to produce a healthy child.

Whereas pregnancy is now considered to be an abnormal state of affairs in contrast with former days, the very normality of today's pregnancies is questionable. Countless women resent the idea of that average 27·5 lbs. (12·5 kgs.) increase, and wish for an average baby without an average increase. So they diet. Such behaviour was notable in the 1920's with increases of not more than 14 lbs. (6·4 kgs.) being desired. One man who aimed at this maximum with 119 mothers achieved anything from a loss of 20 lbs (9 kgs.) to a gain of 41 lbs. (18·5 kgs.).

Today's unnatural habits also include the taking of iron although, according to Hytten and Leitch's *The Physiology of Human Pregnancy*, there is 'no convincing published evidence that a normal pregnant woman is at an advantage if she takes extra iron'. Other customary extras are vitamin preparations, such as folic acid, more calcium or more diuretics (which step up urine excretion), and countless pills, stimulants, sedatives, hypnotics, antibiotics, and alcohols, virtually all of which cross the placental barrier to affect baby as well as mother. The modernity of today's world impinges upon the modern child long before it is born. And, finally, no child is born anywhere without some man-made radioactivity, such as strontium-90, already residing within its bones. It has encountered and entered the nuclear age long before its birth.

The placenta has been mentioned frequently already. It has the apparently contradictory role of keeping mother and foetus apart, and yet being permeable to everything the foetus needs from the mother. It does not let her blood through, and the two blood systems never meet; but oxygen, salts and nutrients (plus all the mentioned medicaments) pass through the barrier. Initially there was curiosity about the method whereby the oxygen in the mother's blood is transferred to the foetal blood. Now it is known that foetal blood is different and has a greater affinity for oxygen. Hence the ready diffusion of that gas from one to the other. Conversely, the foetus gets rid of its waste products by having a lesser affinity for them; so they pass through rapidly to the mother's side.

The word placenta means a flat cake, and flat-cake-shaped it is, being 1 in. (2·5 cms.) thick and 7 ins. (18 cms.) in diameter for most of the pregnancy. The side facing the foetus is smooth, with the umbilical cord (usually) coming from a spot between its centre and its edge. The side of the placenta towards the mother, together with the uterine wall facing it, is a thick pile of protuberances. This increases the interchange area between foetal and maternal circulations. Although the placental cake has a surface area on the smooth side only of some 40 sq. ins. (258 sq. cms.), the contortions of its rougher side increase this area to 140 sq. ft. (13 sq. m.), the floor area of a fair-sized room. Gases, foods and drugs permeate across from one system to the other throughout this huge area. So can antibodies, which will help the new offspring for a time to ward off certain diseases already experienced by the mother. So can some bacteria cross the barrier. So, on occasion, can cancer cells. And then, quite suddenly, when pregnancy ends, the whole delicate arrangement comes to an end as the placenta is sloughed off, and ejected summarily about twenty minutes after the birth.

In the wilds of nature the placental afterbirth, rich in blood and tissue, is eagerly consumed by any scavenger. Hospitals have generally considered it as valueless waste. Now it is being collected because gamma globulin can be extracted from it. Essentially this substance contains the blood's antibodies, and can therefore help those in need of greater immunity. Placentas can also help those researchers wishing to develop drugs for use during pregnancy. They can provide an excellent solution, if speedily collected, to many of the ethical problems concerning the testing of drugs on pregnant women. In fact there is much to be said for realising that the placenta, such a miraculous organ in separating foetus from mother and yet nurturing one from the other, can be of benefit after it has completed its nine-month labour within the uterus. Animals jettison or even eat it. Humans are realising they should do rather better than throw the thing away.

The umbilical cord   The most outstanding anatomical variant in the production of a baby is the length of its umbilical cord. In fact it is about the most remarkable variant in the whole baby. Some babies are admittedly born bigger than others, some pelvic girdles are narrower, some cervixes rounder, skulls sharper, abdomens fatter, and so on; but the amount of variation is held reasonably in check, presumably by the power of evolution and natural selection. The umbilical cord, the one and only link between mother and foetus, is – to put it facetiously – as long as a piece of string. From 7 inches (18 cms.) to 4 feet (1·2 m.) is considered normal. Much shorter cords have been known and much longer ones; but to have the long normal cord seven times longer than the short normal cord is a highly abnormal anatomical situation.

All of this cord growth is completed before the twenty-eighth week of pregnancy. It then stops. Every cord has a helical, or screw-like, form which is fully established by eight weeks. About three-quarters of cords have a left-handed twist, and the number of twists (as few as four and as many as twenty-nine have been counted) bears little relation to the length of the cord. Boys, on average, have cords a couple of inches longer than girls. Very short cords are less than one-third of the length of the foetus; a very long cord may be three times the length. There is no correlation of cord length with mother's weight, with her height, with the length or weight of the baby, with the duration of pregnancy, with the number of offspring born from the same womb, with age, with the placenta's weight, or with anything, apart from the foetus's sex. Long cords get themselves curled round the foetus's neck fairly easily, and may even put two, three, or four coils around it. Percentage occurrences of such loopings have been recorded as 22%, 3·2%, 1·4% and 0·1% for one, two, three and four loops. Strangulation is only likely when the cord has coiled itself around the neck and is short as well. Fortunately those cords long enough to coil are rarely short enough to strangle. Even though the cord is firmly attached at both the foetal and placental ends, and has no knots to start with, it can have one or more by the end of pregnancy. Presumably a coil has slipped right over the baby to make the knot. Fortunately, again, the knots do not give trouble unless pulled tight, notably during birth, and only the long cords get knots in them.

A particularly dreaded form of cord trouble is known as prolapse. Part of the cord slips down from the uterus before birth, and may even appear externally from the vagina before anything else. It occurs once in perhaps 300 pregnancies, although more frequently with twins, but is highly dangerous because it can cause asphyxia to the baby within. Part of the cord gets caught, or compressed, or wedged in the manoeuvres of birth, and its blood supply then slows down or stops. There are two arteries and one vein in each cord, and normally about a seventh of a pint a minute runs each way. Any blockage cutting off all that maternal oxygen for long is likely to be disastrous. Some 35% of babies who formerly suffered cord prolapse used to die as a result. In the 1950's it was believed that 1,750 babies a year in England and Wales were killed by this one cause. Even today many stillbirths are still caused by prolapse, and many livebirths suffer permanent damage as a result of the temporary lack of oxygen. Some spastics can blame the transistory complication for their life-long handicaps. The current situation is that some thirty-three deaths a year (in England and Wales) are directly attributable to prolapsed cord, but that another 330 die because of other compressions of the cord. The foetus's life-line can therefore also be a death-line if the birth does not proceed correctly.

A perennial argument with the cord is when to cut it. The problem does not arise with animals for it is usually torn with herbivores and chewed by carnivores. Humans used to let it be, at least until the placenta was born (usually within twenty minutes), until the seventeeth century. The practice then began of cutting it and tying it off, leaving a short stump. Today's practice is to leave it slightly over an inch long. The lump of cord dries out quickly, and falls off naturally within a couple of weeks, probably by eight days. If the tying or clamping of the stump has been badly done, the new baby can bleed to death very quickly (which happens to about one in a thousand babies).

The argument in favour of delaying the cord cutting is to permit more of the blood left in the placenta to pass into the baby. In the old days, when sitting positions were the rule for labour (most of the early civilisations appear to have done it this way), there was more chance for the placenta's blood to flow down, assisted by gravity. With an uncut cord it could then enter the baby. With a cord hygienically clamped with stainless steel almost the moment it appears, the blood can do nothing of the sort. Dr P. Vardi, of Budapest, for instance, has recently been advocating a method of using more of the placental blood. The new baby is kept between the mother's legs until the placental afterbirth arrives. This is then placed in a plastic dish, and held a foot above the baby. The new life is therefore charged with extra blood for a few minutes. Dr Vardi reckoned that it gains 80 to 90 millilitres in this fashion. As the average foetal blood capacity at birth is otherwise 310 millilitres (less than half a pint), the proportion gained is not inconsiderable. Even without using Dr Vardi's technique the delayed clamping of the cord for, say, five minutes can well augment a baby's blood volume. Incidentally, a final use for the umbilicus is that if properly used it makes a first class and ready-made transfusion site for any new baby in need of extra blood.

Thereafter the navel remains primarily as something allegedly contemplated for countless hours by those who find such trance-like introspection rewarding. There also remains, for ever, the debate whether artists should or should not have given one to Adam. Or, for that matter, to Eve.

The foetus and the law Any offspring has a legal right to a separate existence as soon as it has reached the twenty-eighth week of gestation. This right is according to the Infant Life Preservation Act of 1929. On the other hand, many people consider any arbitrary date is totally wrong. By the twenty-eighth week the young human being, powerless but sucking its thumb, is no less human than any full-term baby. Even during the third month of gestation, when nearly all enforced abortions take place, the foetus being so summarily aborted still is and always has been a miniature human being. According to a United Nations conference on world population, some 30,000,000 of these undeveloped lives, or one in five of all pregnancies, are bluntly stopped every year. Over 80,000 foetuses a day are expelled, and dumped behind a bush, or in a hole, or down the lavatory, or in some antiseptic container in some antiseptic hospital. Abortion has been called 'the greatest epidemic of all time'.

The world's countries had very different ideas on the problem.

In Britain, and until 1803, the law permitted medical abortions, provided quickening had not begun (which is usually in the fifth month). The various nineteenth century recommendations on the subject congealed in 1861 as part of 'The Offences against the Person Act'. It was this law which was the basic ruling until 1967. Section 58, well over a hundred years old, enacted that 'every woman being with child who, with intent to procure her own miscarriage, shall unlawfully administer to herself any poison or other noxious thing, or shall unlawfully use any instrument or other means whatsoever with the like intent, and whosoever, with intent to procure the miscarriage of any woman, whether she be or be not with child, shall unlawfully administer to her or cause to be taken by her any poison or other noxious thing, or shall unlawfully use any instrument or other means whatsoever with the like intent shall be guilty of felony'. The maximum penalty for both the aborted mother and abortionist was imprisonment for life.

There was much play over the word 'unlawful' as this implied the possibility of a lawful use of instruments, a lawful administering of poison. The lawfulness question was first clarified by the 1929 Act. This Act, more on the side of the mothers than the babies, said abortion was lawful if done 'in good faith and for the purpose only of preserving the life of the mother'. No law yet concerned itself with the health of the mother. One big step forward in this direction was taken in 1938 with a famous trial. Mr Aleck Bourne, a distinguished obstetrician, performed an abortion without a fee on a girl of fourteen who had been raped. He was subsequently acquitted on the grounds that continuation of the pregnancy would have risked the girl's mental and physical health.

The clarity of the ruling in this test case became less clear the following year. Mr Norman Birkett, chairman of the two-year-old Interdepartmental Committee on Abortion, then pointed out that the judge's ruling in the Bourne case was not binding upon judges of the High Court. Juries attending similar cases might be directed by the judge 'on somewhat different lines'. Various people had tried to allay the existing fear that such acquittals were not always to be expected, even if it is clearly shown that the mother's mental and physical health had been at stake. In 1960 Professor T. N. A. Jeffcoate, of Liverpool University, wrote that no doctor need fear the consequences of the present law 'if, after proper consultation with colleagues, he terminates a pregnancy which he honestly believes is causing or is likely to cause significant harm to the physical or mental well-being of his patient'. Similarly, Dr Glanville Williams (in *The Sanctity of Life and the Criminal Law*), has written, 'Some doctors are afraid that the decision in the Bourne case will be reversed. . . . It may be safely said that this fear is unfounded.'

If so unfounded, if no fear is necessary, why not amend the law? This was a major argument, notably of the Abortion Law Reform Association. In 1964 the Home Secretary, Mr Henry Brooke, was asked about possible reform. The time, he replied, 'is not yet ripe'. It was then 103 years after the original Act, and twenty-six years after the contradictory Bourne ruling.

The situation at that time was therefore fraught with problems, with hypocrisy, with unpleasantness. It was estimated that 50,000 to 150,000 abortions were carried out annually in Britain. Either one in eighteen or one in six, therefore, of all conceptions were crudely halted. Most of the abortions were illegal, the back-street abortionist was still in business, and the highly-qualified abortionist was still content to take £150 a time. Of all doctors removed from the Medical Register 19% were expelled because of criminal abortion (the number lay second to the 26% removed for adultery and improper conduct). Several million pounds must have changed hands annually for abortions (if there were 100,000 and, say, £60 an average fee for each). Nevertheless, populations can only be controlled by infanticide, abortion or contraception. According to some, the most widely practised of these three methods in the world was certainly not infanticide, and not even contraception, but abortion.

In those days a woman could succeed in getting a lawful abortion if she admitted a label of mental unbalance. Would this label catch up with her whenever her medical history was called for in the future? This was one of the arguments put forward by the reformers. There were others. Was it right to consider only the mother's condition? Ought it not to matter what kind of child it would be, perhaps hideously deformed, perhaps totally unwanted? Ought it not to matter that the child had been subjected to German measles during its early days, making its chance of a defect perhaps as great as one in two? A doctor may earnestly believe an abortion to be necessary, and recommend it, but then come up against the scruples, religious or otherwise, of the local gynaecologist. Finally, a girl under sixteen had, by law, to have the baby she conceived but she could not, by law, get married.

The abortion controversy was most actively debated in Britain in the 1960's due to the efforts to amend the law. *The Lancet*'s correspondence columns were more active than most. 'I do not like to perform the operation, and those who assist me like it even less. I can think of no comparable operation in which normal tissue is removed, and that tissue is a potential new individual.' 'As prevention is not a disease, its prevention is not a proper object for a medical service.' 'Is the present hesitancy to recommend or carry out abortion due to the profession's deep regard for the preservation of life or to uncertainty in the profession about the present state of the Law in the United Kingdom?' 'No woman wishes to have an abortion. Either she desires a child or she wishes to avoid a pregnancy.' 'I have strong objections to emptying the womb for social reasons, and in this I am probably among 95% of my colleagues in gynaecology who practise in the National Health Service.' On therapeutic abortions: 'One of the few agreed facts on abortion is that it can never be "therapeutic" to the foetus except in the sense that euthanasia is "therapeutic".' 'Ultimately the only way to abolish illegal abortions would be to make it legally possible to terminate any pregnancy. If this is not acceptable illegal abortions will continue.'

Early in 1967 the Royal College of Obstetricians and Gynaecologists and the British Medical Association issued a joint statement on abortion. This made the point that it should be lawful 'for a registered medical practitioner to terminate pregnancy if, in good faith, he considers it to be either in the interests of the physical or mental health of the mother or because of the risk of serious abnormality of the foetus. In deciding whether termination is in the interests of the health of the mother the doctor is entitled to take into account the total environment of the mother, both actual and reasonably foreseeable.'

In the United States of those days, forty-four of the fifty states considered the sole legal ground for abortion to be that a pregnant woman would otherwise die. As elsewhere, the law in America was made to seem an ass by reality. One million induced abortions were thought to occur annually, and 1% were induced in hospitals. Even of this small percentage the majority were probably not in strict accordance with the law, according to Dr Robert E. Hall, president of the Association for the Study of Abortion. In 1965, the American Medical Association's Committee on Human Reproduction recommended that licensed physicians ought to be able to terminate pregnancy if they could reasonably establish that: there is substantial risk that the continuance of the pregnancy would gravely impair the physical or mental health of the mother; or there is substantial risk that the child would be born with grave physical or mental defect; or the pregnancy resulted from legally established statutory or forcible rape or incest.

The Abortion Law Reform Association in Britain was a major force in efforts to change the situation. Its recommendations were that the doctor's discretion should be legally permitted in four types of case: when the physical or mental health of the woman is endangered; when there is a serious risk of a defective child being born; when the pregnancy results from a sexual offence; and when the pregnant woman's capacity as a mother will be severely over-strained. Others were more radical, feeling a decision to abort should rest not with any law, but with the individual consciences of the man, the woman and the doctor.

Attention to the defective risks had been drawn by the Eugenics Society. Its honorary secretary wrote in 1965 about various severe human ailments, and the chances of their inheritance:

Classical achondroplasia, epiloia, multiple polyposis of the colon – a one in two risk of the affected parent passing it on to a child.

Classical haemophilia, Duchenne muscular dystrophy – a one in two risk for sons, and a one in two risk that daughters will be carriers.

Cystic fibrosis of the pancreas – if parents already have one affected child, there is a one in four risk for further children.

Maternal rubella (German measles) – if the mother gets it during the second half of first month of pregnancy, the risk to the child of some defect may be 50%.

The odds against these complaints, and many others, being transmitted from parent to offspring are very short indeed. Many parents may accept the odds, and welcome any form of child. Many may not.

Eventually, in Britain at least, the old laws were finally jolted out of their supremacy. In October 1967, having bounced about between the Commons and the Lords, the Abortion Bill was passed. It came into effect on April 27 1968 and, according to its conditions, a legally induced abortion must be:

(a) performed by a registered medical practitioner,
(b) performed, except in an emergency, in a National Health Service hospital or in an approved place, and
(c) certified by two registered medical practitioners as necessary on any of the grounds:
    1. the continuance of the pregnancy would involve risk to the life of the pregnant woman greater than if the pregnancy were terminated,

2. the continuance of the pregnancy would involve risk of injury to the physical or mental health of the pregnant woman greater than if the pregnancy were terminated,
3. the continuance of the pregnancy would involve risk of injury to the physical or mental health of any existing child, or children, in the family of the pregnant woman greater than if the pregnancy were terminated,
4. there is a substantial risk that if the child were born it would suffer from such physical or mental abnormalities as to be seriously handicapped,

or in emergency, certified by the operating practitioner as immediately necessary:

5. to save the life of the pregnant woman, or
6. to prevent grave permanent injury to the physical or mental health of the pregnant woman.

It was also considered that the upper limit for the age of the foetus should be twenty-eight weeks.

What was happening in other countries? Sweden has always had a reputation for liberality (much publicised when Mrs Sherri Finkbine travelled there in 1962 from the United States to be relieved of a foetus which had been subjected to thalidomide). Sweden had passed a liberalising Act in 1946, which tolerated abortions if there was a defect in the mother, a risk of severe defect in the child, a social 'weakness' in the mother, a humanitarian reason (as with rape), and an 'emergency' reason. Before the war (in 1939) there had been 439 legal abortions in Sweden, and a suspected 20,000 illegal ones. After the war and the passing of the new Act there was a steep rise in legal abortions, 5,503 in 1949, reaching a maximum of 6,328 in 1957. Since then the numbers have dropped. The law, although considered fairly liberal, was not as liberal as all that. Nearly all foreign girls then travelling hopefully to Sweden returned home still pregnant, and in 1965 many Swedish girls were discovered to be going to Poland for abortions.

In Latin America, where the situation is not helped by the attitude of the Roman Catholic Church towards contraception, abortions are illegal but common. It has been reckoned that in Chile, for example, there is one illegal abortion for every two live births. In Japan, where the accent has been on population control since the American occupation authorities said that a total of 80,000,000 people was the maximum the land could support, abortion has been very common. (Japan's population is now 120 million). Professor Shiden Inoue, of Nanzan University, has said that 80% of Japanese women with at least one child have experienced abortion, and most gynaecologists in the country were financially dependent upon abortion income. Japan's prime minister asked the health ministry in 1967 to tighten its abortion laws after it had been revealed that a Japanese doctor was offering easy abortions to US patients: 'Simply let her fly to Tokyo, and leave the rest to us.' In the early 1960's legal abortions in Japan outnumbered births. Eastern European countries also have liberal laws. East Germany recently revised its laws to permit abortions if the mother's health is endangered, if she is over forty or under sixteen, if she has had four children at an average interval of less than fifteen months, if any pregnancy starts within six months of the previous birth, if there are already five children, if the pregnancy results from a crime, or if the child is expected to have a serious abnormality. Only the first and the last had been considered legal before this latest legislation. Oddly, Romania announced the banning of abortion at the end of 1966. The first Communist country to take this step said a new law would permit interruption of pregnancy only in very rare cases. Yugoslavia, with liberal abortion laws, has been expressing concern at the falling birth-rate, and is now trying to cut down the number of legal abortions. Russian abortions have been legal since 1955, but the Russians are now being more liberal about the supply of more efficient contraceptives than those formerly available.

In Britain, and until the 1968 Act, the official abortion statistics were as nothing compared to the estimated abortion rate. In 1950 there were 103 women officially recorded in the year's register as having had abortions which resulted from criminal interference. In 1955 there were 66 women thus registered, and 62 in 1960. With 300 or so abortions a day before the Act, the official register was woefully misleading. Of the offences for procuring abortion known to the police (according to Keith Simpson's *Forensic Medicine*), which totalled about 200 a year, the number of people committed to trial was about one a week. In other words, there were 2,000 abortions a week, and one trial. At least the passing of the Act meant that more abortions became known about, whether or not there were more abortions. In the first eight months there were 22,256 legal abortions, and some experts predicted 50,000 by the end of that first liberal year. The final figure was 37,700, or five abortions per 100 live births. This compares with 6·7 per 100 for Denmark, 7·9 for Sweden, 32 for Poland, 38·5 for Japan, 44 for Czechoslovakia, and 126 for Hungary.

The British figures (now that the Act has, as it were, settled down) have been disappointing to both sides, to those in favour of the new legislation and those against it. It was to be hoped that the new methods of contraception would bring down the numbers of needed abortions, but there has been a fairly steady rise. In 1970 there were 75,000 official abortions to women normally resident in England and Wales. By 1975 this figure was 106,000, and by 1980 it was 128,000. To begin with, and when abortions first became readily available in Britain, many foreign girls arrived to take advantage of the new British ways. Since then fewer such girls have been making the journey, largely because of more liberal laws passed in their own countries. Nevertheless the numbers are still considerable. In 1970 there were in Britain 10,000 of these foreign abortions, by 1972 51,000, by 1973 56,000 (the peak year) and by 1980 there were 31,000.

The number of British girls at risk from unwanted pregnancies has varied over the years, with the post World War Two baby boom having a secondary effect when these babies were having babies of their own; but, allowing for the number of vulnerable women, there has been a remorseless rise in the abortion rate. Per 1,000 English and Welsh women aged between fifteen and forty-four there were 8 abortions in 1970, 11 in 1975, and 12 in 1980. In one survey more concerned with age the increase is most noticeable with younger women, such as the single girls less than fifteen, the teenagers in the fifteen to nineteen group, the twenty to twenty-four-year-olds, and (but less so) the twenty-five to thirty-fives. Over thirty-five the rates have been more static, at about one-third the rate (per 1,000 women) of those aged twenty-five and under. Without doubt, as many asserted would happen once the Act had been passed, many women (and girls) are using abortion as a form of contraception. Although contraceptive methods can fail even if properly used, many of today's women requiring an abortion were failing to use any contraceptive method either through neglect or in the belief that their male partner was suitably equipped or because they chanced to luck, with an embryo rather than luck coming their way instead. Although abortion is generally considered (by the majority, at least in Britain) to be a necessary alternative if other methods have failed, there is little enthusiasm anywhere for the official fact that 163,000 (in 1982) human foetuses had their development curtailed. It is about a third of the number that are permitted to be born.

Nevertheless official and legal abortion must be preferable to the unofficial variety which makes frightening reading. Unqualified 'back-street' abortionists usually make use of instruments, such as bougies or curettes, to initiate abortion. Although instruments are the most consistently successful method (according to Simpson) the method of interference is beset with dangers, such as fatal shock, air embolism, instrumental injury, and sepsis. The women themselves are more varied in their efforts to dislodge pregnancies. They try drinking gin, or gin with iron filings, or hot baths, or skipping, or riding horses, or bicycling, or lacing themselves tightly, or falling downstairs. Violence of this sort is rarely effective. A woman once deliberately fell 25 feet and, although she died twenty-four hours later, the two-and-a-half month foetus stayed put. A man once knelt on a woman's abdomen, and then trampled on her back. He later used scissors – but the baby was born at term. In 1963 a pregnant woman was struck by lightning on a Welsh mountain-side, and was temporarily knocked out. Naturally this case was no deliberate attempt at abortion, but the foetus happily survived the incident. Even women in iron lungs and a coma have delivered healthy children.

Women and their procurers are a frighteningly determined lot on occasion. C. J. Polson, in *The Essentials of Forensic Medicine* has listed some of the methods. Instruments, apart from normal medical equipment designed for other purposes, include knitting needles, hat pins, pokers, paint brushes, or just fingers. Then there is slippery elm bark, an ancient household remedy. (This is self-lubricating in that a jelly-like layer is formed on it by any available moisture.) There are also vegetable poisons, like saffron, juniper, rue, brook, and laburnum. In the last war potassium permanganate (generally used as a disinfectant) was much in vogue as it was frequently recommended by United States servicemen. Long before they arrived, and after they went, there have been preparations like 'Dr Reed's Extra Strong Female Pills' and 'Widow Welch's'. Keith Simpson says women who abort after taking drugs are more likely to be doing so because they have been made gravely ill by the toxic effects rather than by any abortifacient value in the drugs.

Without doubt, even if the number of abortions has not actually been reduced (or even raised) by the new legislation, the mediaeval procurer of abortions is no longer an essential feature of society. Preventing conceptions is better than aborting them, but aborting them legally is better than illegally. If the world estimate of 30,000,000 abortions a year is either correct or incorrectly high by ten, twenty, or thirty times, the problem is still monumental.

12. The Time Before Birth

Gestation length   There is much similarity between a weather forecast and the expected day of a baby's arrival. It is eagerly noted down and even believed in a kind of way; but rarely does it become precise reality. 'The 8th of December', says the Doctor. 'Ah the 8th of December', repeats the woman sitting before him. 'The 8th of December', she informs her mother on the phone. In fact only 5% of babies actually arrive on the forecasted date.

The average period between fertilisation and birth is generally reckoned at 266 days, or eight and three-quarter calendar months. A traditional method of arriving at the date of birth is to add 280 days to the start of the last menstruation. The time of 280 days is often said to be the length of human gestation mainly because the date of the start of the last menstruation is a commendably positive date. Another method of assessing the birth date, which also cares little about individual variation, is to add 273 days to the day on which the last menstruation ended. If irregularity in calendar months is casually forgotten, and if the last menstruation began on the first of the month, and if it is assumed that the flow ended a week later on the eighth, then the magic date is also the eighth, but nine months further ahead (or three months backwards if that is easier to count). The round figure of 280 days is also ten lunar months, or around ten times the average menstruation period of 28 days. Despite efforts to complete this happy circle by finding a correlation between the length of the period and the length of gestation, no collusion has been found.

Two hundred and eighty days, or ten lunar months from the start of the last period, or eight and three-quarter calendar months, are therefore the rules which are surrounded on both sides by all the exceptions. Apart from premature babies, which are born at least four or five weeks before expectation, there are many babies which are just born early. From 240 days onwards the babies are not actually considered premature (with an arbitrary definition of $5\frac{1}{2}$ lbs (2·5 kgs.) or less) whatever the mother, caught unawares by the event, may think. Some babies of only thirty-two weeks (224 days) weigh those $5\frac{1}{2}$ lbs, and are being born normally a full fifty-six days before the doctor's computed forecast.

After 280 days (and still with no sign of activity) no one is greatly concerned for a while; but, when that while has passed, thoughts of inducing the labour become more positive. By the time a further month has passed beyond the forecasted date, the pregnancy is generally agreed to be abnormal. One woman is on record who, twice in succession, had pregnancies of over eleven lunar months (308 days). At each traditional time of ten lunar months she felt labour pains, but these then subsided and did not manifest themselves again for another month. It would seem from this one case as if the long-forgotten menstruation period had something to do with the onset of labour.

Two hundred and eighty days is too firmly rooted as the hallowed pregnancy time in Europe for it to be treated with less respect, although the bull's eye is rarely hit, and women steadily pepper the target on either side of it. In the United States the mean for white women is 279 days for male offspring and 279·9 for females, and yet 12% of the mothers wait for over 300 days before producing their child. Blacks' gestation times are shorter, being 273·3 days for male offspring and 274·7 days for females. Black babies are smaller and so the shorter time seems reasonable, but Indian children are smaller still and they take longer to appear, 285·6 and 286·6 days respectively for boys and girls. Both birth weight and gestation length are racial characteristics in India, and not a direct cause of harsh living, because they are consistently low and long amongst the well-to-do, or Indians living abroad.

With all groups of humanity everywhere girls are carried, on average, a day longer than boys. This is just as basic for *Homo sapiens* as the fact that males die off more rapidly from conception onwards.

Legal pregnancies The longest pregnancies on record are those reported not in the annals of medicine, but of law. Men, such as husbands, have frequently wished to prove that they could not have fathered a certain child as they were elsewhere/at sea/ abroad through the vital time. The law, aware that pregnancies can exceed 280 days, has never been prepared to accept the plea of a man who is absent for 281 days before the birth.

Or, come to that, 300. But back in 1802 a man was considered guiltless who had been away for 311 days. Nowadays, pregnancies are (legally) getting longer. It is not good enough to have been at sea for only ten months, and then disown a brand new baby on stepping ashore. In 1921, with the case of Gaskill versus Gaskill, the husband was petitioning for divorce on the grounds of his wife's alleged misconduct because Mrs Gaskill had produced a baby 331 days after he 'had had access to her'. Mr Gaskill lost his case. In 1947, with the case of Wood v. Wood, Mr Wood had been away for 346 days, but he still lost. With Hadlum v. Hadlum in 1949 the period of absence, or length of presumed possible pregnancy, was 349 days – or one year bar a couple of weeks and couple of days. The court had accepted that such a lengthy pregnancy, over two months longer than the traditional forecast, was possible. Mr Hadlum was assumed to be the father. (I have read of one birth after 360 days, but there was no supporting evidence.)

A similar legal judgement has to be made in cases of doubtful legitimacy after the husband has died. Is the child the dead man's child, or has it been born too late for this possibility to be acceptable? Some countries have fixed dates. In Germany the law recognises the legitimacy of a baby born 302 days after the husband has died. In France the legitimate time is 300 days. A mere twenty days on top of the 280 is very different from the 349 days of Hadlum v. Hadlum. Presumably a difference is that if money is to be inherited only by legitimate offspring, many a young widow might hurry to become pregnant in order to secure more loot for her (illegitimate) successors.

Animal pregnancies Animals, as might be expected, vary enormously in their gestation time. At one end is the elephant, which produces its infant after twenty-three months, while the whale, larger and with larger offspring, takes about a year. At the diminutive end of the scale are the marsupials, the mammals of Australasia and, in part, of South America. Marsupials do not develop placentas, as most mammals do, and so cannot emulate their incubation system. Instead they rely upon different systems.

The opossum is typical. Its young are born when less than two weeks old. These primitive things, more like miniature slugs than young opossums, do not pass out through the vagina. Instead they travel along a rupture in the skin, the pseudo-vaginal canal. At the end of this, and when confronted by the open air, they have to squirm, wriggle, and make their way for three long, hairy inches from the canal to the pouch. The pouch is security, and also a supply of nipples. Each foetus clamps itself on to a nipple, and then relaxes until weeks and weeks of growth have passed. Finally it unclamps and becomes a more independent form. It remains in the pouch for as long as necessary, turning to the milk supply whenever thirsty, and viewing the world from its kind of external uterus. There is much to be said for the system.

AVERAGE GESTATION TIMES OF VARIOUS SPECIES

	No of days	Usual No. born
Golden Hamster	16	6–7
Mouse (laboratory)	19	8–12
Rat	22	6–8
Rabbit	31	6–9
Hedgehog	35	4–7
Polecat	40	6–8
Ferret	42	6–9
Mink	50	3–5
Fox	51	4–6
Cat	56–64	4–6
Dog	58–63	3–7
Guinea-pig	63–70	3–4
Lion	107	4
Tiger	107	4
Pig	113	12(8–16)
Chamois	140	2
Goat	150	2
Sheep	150	1–3
Monkey	150–180	1
Fallow deer	250	1–3
Homo sapiens	266	1
Cattle	280	1–2
Buffalo	300	1
Roe deer	310	2
Camel	315	1
Horse	336	1
Llama	340	1
Badger	352	2
Ass	365	1
Zebra	400	1
Elephant	700	1

Human eggs, once successfully initiated and implanted, grow throughout their gestation or development. Some other forms conceived at a convenient time have their growth delayed so that

145

they can be born also at a convenient time. Seals, for instance, only come ashore for any length of time once a year. Their pups are then born, and shortly afterwards each cow is mated by a bull for next year's pup. (Customary nomenclature frequently goes haywire in the animal kingdom.) One whole year for the development of a seal seems unreasonable when you think that large dogs, very similar creatures biologically, have gestation times of nine weeks rather than eleven months. Some bats employ similar delaying tactics, and so do roe-deer. The does are mated in the autumn, but the fawn embryo stops developing at the segmentation stage (very early on indeed) and only starts again the following spring.

Labour Surprisingly, it is not known precisely what causes labour to start. There are hormone changes at this time, but it is not known what initiates the changes or, in the chicken-and-egg manner of much of development, what other changes initiate those that alter the hormone levels. Nevertheless, there are theories, relating to the ageing placenta, increasing movement of the growing uterus, increasing size of the foetus, increasing irritation by the cramped intestines, and metabolic change (either the mother or child producing substances during metabolism which trigger off labour, quite apart from the hormones actually being liberated by the offspring in increasing quantities).

Despite ignorance about the actual trigger or combination of triggers, labour can be artificially induced if need be. Needs include: death of foetus; postmaturity – very old foetuses are not only larger, but harder to mould through the birth canal; Rh difficulties – blood transfusion may be necessary, and the sooner the better – within reason; poisoned pregnancies; premature rupture of the membranes or just anxiety. Labour is artificially induced only very rarely before the thirty-fifth week. Induction methods include oxytocin injection, rupturing of the foetal membranes, or dislodging these membranes from around the circumference of the cervix. Frequently suggested to be popular in hospitals for the convenience of hospitals (causing fewer births, for example, to occur on difficult days) the procedure of induction has undoubtedly become more common in recent years. Back in 1962 it was used in only 8% of deliveries in England and Wales. By 1974 the proportion had leaped to 39% and there, more or less, it has stayed since then. Nature, in short, has had to take a back seat, and convenience has taken its place. 'We always induce on a Wednesday,' said one London hospital. 'Everything is then in order, and the mother plus child may even have gone home, by the weekend.'

Labour, whatever its cause may be, is generally regarded as having three stages. (The following description by no means fits every case, but gives a representative picture.)

The first stage starts with regular uterine contractions. These occur initially every twenty to thirty minutes, then increase until they occur every two to three minutes with each contraction lasting perhaps up to one minute. Therefore as the intervals between contractions are reduced, the contractions themselves are lengthened. The main function of this stage, which lasts fourteen hours on average for first offspring (some say average is sixteen hours, but all agree time is less for subsequent births), is to open up the cervix, the neck of the uterus. Before labour starts, the pathway is emphatically blocked. By the end of the first stage the cervix is thinned. Its substance is moulded to each side, the former 'plug' diminishes, and a pathway appears possible. It is not known what initiates the uterine waves of contraction, or whether two waves start simultaneously on each side.

The second stage is the actual birth. The upper part of the uterus continues to contract and retract, with the lower part of the uterus more passively stretching to permit passage of the foetus. Deliberate stomach wall and diaphragm exertion helps but it is not vital; women in coma or with severed spinal cords can have normal labour. The membranes usually rupture early in this stage (or may have done so before the first stage began). The foetal head is gradually forced lower and lower until it meets the pelvic floor, the bones whose shape causes the head to rotate through 45°. The head is then born, and rotates back again to its normal position in relation to the rest of the body. The shoulders are then born, following a similar encounter with the pelvic floor and a similar rotation. The head, which is now outside, is caused to rotate again as the shoulders are born, and the rest of the body then follows. The moulding of the skull takes place during the second stage (made possible by softness of the bones and spaces between them).

The amount of pressure exerted by the uterus can be over 3 lbs per square inch, and it has been calculated that the total expulsive thrust on the foetus is about 24 lbs. (11 kgs.). (An obstetrician once had a finger bone broken by a strong uterine contraction.) The duration of the second stage can vary enormously between minutes and hours, but the average for a first child is two hours; the length of time lessens with any subsequent children.

The third stage is generally reckoned to take twenty minutes. It begins after the expulsion of the baby and ends with the expulsion of everything else – the membranes, placenta, and remaining fluid. Even during the second stage the uterus shortens by some 6 inches (15 cms.). After delivery it contracts upon the placenta, and its rhythms may cease temporarily. The placenta is then separated from the maternal tissues, a process lasting about eight minutes, but the membranes usually take longer to separate. More contractions follow, plus some exertion from the diaphragm or the abdominal wall, as the shrunken uterus rids itself of the unwanted tissues and fluid. Blood loss at this stage may be one pint, but this is still regarded as normal. Greater loss of blood is abnormal and is usually associated with the retention of the placenta, or part of it. Such excessive bleeding – postpartum haemorrhage – occurs in 4% of deliveries, and about a dozen women a year die from it.

For the 96% who do not suffer from excessive bleeding, and then for the 4% whose blood loss is stopped and replaced, labour is over. The business of birth has finished. The puerperium has begun. The most dangerous journey in the world, to borrow one man's phrase, is the four inch trip down the birth canal. Now it is over and done with.

Labour is generally considered to be prolonged if it lasts more than twenty-four hours, although signs that it is being slow or difficult can be seen long before twenty-four hours have passed. The process is generally quicker in animals, save for some monkeys who may take just as long. Cows, and many other ruminants, take two hours but sometimes up to twelve hours. Sheep take about fifteen minutes for each lamb. Sows, bitches and cats take ten to thirty minutes for each member of the litter. With the mare everything is very rapid, usually five to fifteen minutes, and any difficulty often kills the foal. There is little quite like the prolonged and drawn-out sequence of events which precedes the safe delivery of each new human being. Also, whatever general practitioners may think, or even stringently believe, most births do not take place at night. A huge Pittsburgh survey, which examined 57,000 arrivals, discovered that 55% of them had the grace to appear between 6 a.m. and 6 p.m. Even so birth is a nighttime business, for those which happen during the day probably started at night, and those which started during the day probably reached their conclusion long after the sun had set.

Finally, it is an offence in England to conceal a birth. In Scotland it is an offence to conceal a pregnancy.

Presentation Babies are normally (95%) born head first (vertex presentation) and face downwards. Very rarely (0·4% to 0·8%) the face is born first, and looking upwards at the deliverer. The most common abnormality is breech delivery (i.e. buttocks first – but may be occasionally knees first, or one foot first). Breech deliveries of all kinds add up to 3·5% of births. They are unwelcome partly because they involve three deliveries, first the breech, second the shoulder and third the head. (With normal presentations everything is relatively plain sailing after the head.) The mother is likely to fare just as well with breech and vertex births, and labour will be of similar duration although breech delivery takes a little longer on average, but the baby is likely to fare less well (the breech delivery mortality rate is three times higher than that of vertex delivery). A further presentation (0·5%) is when the shoulder attempts to be born first. Worst presentation of all (and only prematures can actually be born this way) is the transverse lie, or cross-birth. It frequently results in the death of the baby, even if the obstetrician manages to correct the foetal position during labour.

It is 'not properly understood' (this phrase is so beloved in the scientific literature) why most foetuses come to lie head downwards shortly before birth (93% are head downwards by the end of the thirty-fourth week). Certainly the foetal head within the pelvic framework makes a neat fit, and this may be the only reason. Breech deliveries are more common, understandably, in premature than mature births and also, less understandably, more common with subsequent offspring than first-born. Some women produce breech presentations every time; in fact 14% of all breech deliveries are habitual. Breech presentations can be detected in late pregnancy, and efforts are made, often successfully, to change the position. They are not attempted too early (to give the foetus time to sort things out for itself) or too late (when enforced sorting out is impossible and the breech is too firmly entrenched in the pelvis). Efforts are usually made between the thirty-fourth and thirty-sixth week. Movement is effected manually, often without any need for anaesthetic, and the foetus is coerced into the new position. Occasionally the foetus may counter all efforts by manoeuvring itself (or being manoeuvred by the uterus) back again into the original breech position. Breech deliveries almost always need a skilful attendant if the baby is not to be born dead, because the birth sticks at the head, the umbilical cord may stop pulsating, and the new life may then cease. (The Central Midwives Board does not permit a midwife to deliver any breech presentation that is also a first-born baby.)

Twinning naturally complicates the conventional downward

or vertex lie within the uterus. Compromises have to be made. Even so, both are head down (double vertex) in 30% to 45% of cases. One vertex and one breech occurs in 35% to 40% of cases, with the vertex twin usually being born first. Other arrangements such as double breech (8% to 12%) or one vertex and one transverse (4% to 12%) are much rarer. In both these last cases the horizontal twin is almost always born first. Rarest of all combinations is when both twins are transverse, and are lying acros the uterus.

Labour for twinning generally starts early (by several weeks – average shortening is twenty-five days) and generally takes longer. Twinning pregnancies put the mother in greater danger, as maternal mortality is two to three times more severe. Twins themselves are also in greater danger than conventional singletons, mainly because of prematurity and its consequences. When twins share one placenta (and therefore one blood circulation) the birth of the first-born and the cutting of its cord can cause the second twin to bleed to death unless the first-born's cord is clamped at the placental end as well as at the baby's end. Occasionally Caesarians are necessary with twins, as when both foetuses become locked together in the narrow exit. Second twins are sometimes born after a long wait (even, very rarely, after a month), but the optimum time is short.

Uterine crying A fortunately very rare occurrence in childbirth is known as vagitus uterinus, or crying in the uterus. This still imprisoned yelp must be alarming even to experienced obstetricians, let alone inexperienced mothers. Sometimes the cry is described as soft and whimpering, sometimes loud and gasping. At all times it must be the last sound anybody wants to hear. A case of it occurred recently in Inverness-shire (reported by Robert G. Blair, Lanark's Consultant Obstetrician and Gynaecologist). The mother had had difficulties in the past (two abortions, one ectopic pregnancy, and one forceps delivery) and was five days overdue when this story began. The slight lateness, plus the earlier history, plus her anxiety, caused her doctors to decide that labour should be induced. They did so at 9.20 a.m. by rupturing the membranes, a customary method. More than a pint of liquid was withdrawn with a catheter, and there then came three loud cries from the foetus. They were heard by the two doctors, the three midwives, and the mother. Everyone was startled. First thoughts were that someone had wrongly brought some other baby into the room, but these first thoughts were quickly proved wrong. It was the foetus, and labour had not even begun. On several other occasions during that day it cried again, and labour finally began at 8.30 p.m., or eleven hours after that first awesome cry. Fortunately everything else, such as the foetus itself and then the labour, was absolutely normal, and a healthy boy was born at 4.10 the following morning. He cried immediately at birth, almost nineteen hours after his first foetal call. Both mother and child were well.

The event of vagitus uterinus is certainly rare (one man added up the reported cases until 1940 to find a total of 122), but one suspects that it would have to be far more common before anyone attendant upon such a birth could hope to forget the first shattering cry, however diminutive, from a foetus still locked within a uterus. The recommended procedure, provided the foetus is well, is to do nothing but wait for it to be born.

13. Birth

The history of midwifery/The Chamberlen family and forceps/
Caesarians/Anaesthesia/Home or hospital/How long a stay/
Maternal death/Semmelweiss and puerperal fever/Infant
mortality/Infanticide/The sex ratio/The baby's weight/
Prematurity/The puerperium/Paternal labour/Ergot/Who has
babies?

The business of delivering babies started well, but the deliverers
then deteriorated and so did their methods. The skilled became
the unskilled, and a modest knowledge of anatomy became an
ignorance coupled with uncleanliness and superstition.
Midwifery and midwives reached their nadir at the end of the
Middle Ages. As the medical historian, Dr H. S. Glasscheib puts
it, the pregnant woman of those days 'was completely in the hands
of ragged old harridans, who travelled from house to house like
tinkers with their old-fashioned labour chair, and filthy hooks
hanging from their belts'. The chair, with only a horse-shoe
shaped edge on which to sit, followed the custom of sitting
deliveries.

This labour chair, however unnatural it appears to us, was a
reasonable development of the time. The habit of lying deliveries
upon a bed came much later, roughly at the start of the eighteenth
century. By no means do all cultures conform. Two Germans
described (in *Das Weib in der Natur-und Völkerkunde*, first
published in the nineteenth century) forty different positions
adopted during labour by the women of different areas. Most of
the positions were more or less upright. A later analysis of the
birth customs of 76 countries outside Europe showed that women
in 62 of them favoured upright positions, a classification which
included sitting, leaning, squatting, kneeling and standing. The
remaining 14 countries adopted positions which included lying
prone, lying supine (customary in Europe) and resting on all four
limbs. The labour chair would not be such an anomaly in many
countries today.

The midwives' hooks were in case of failure. 'Child breaking'
took place whenever a weak labour or a dead baby demanded it,
and the offspring were summarily hooked out, having been
carved up in the process as much as was necessary. One major

obstacle, not the fault of the midwives, was the prevailing prejudice against any form of exposure. Not only were men totally excluded (in 1521 a Hamburg doctor was burnt for having disguised himself as a midwife), but the attending women had to grope in the dark beneath the skirts.

The Christian Church, and its past attitudes, did not help. The new life was more important than the mother's, for without baptism it was a lost soul. Either the unsuccessful midwives were dubbed as witches 'who robbed innocent babes of the Holy Sacrament' – and such an indictment did nothing to raise the women's already miserable status – or great and diverting efforts were made to baptise the babe, come what may, birth or not. Curved syringes were fashioned which could eject holy water upon ailing foetuses still within a struggling uterus. A most unlikely reformer was Louis XIV of France who did much to revolutionise obstetrics by watching his mistresses give birth. By taking an interest the Sun King made such study respectable. Others followed suit, and gradually male midwives became fashionable, although not without considerable resentment from the well-established females.

The progress of midwifery   This can be listed as a series of 'firsts', most of which happened many, many years later than might be expected. The first book on the subject in English (*The Byrth of Mankynde*, prepared by Thomas Raynalde) was printed in 1540, a translation from a German work of 1513. The first original work in English is generally considered to have been William Harvey's *De Generatione Animalium* published in 1651. He had written it when in his seventies, shortly after the turmoil of the Civil War (he had been a fervent Royalist attached to King Charles I's court) and long after his famous work on blood circulation. It was not until 1726 that the first professor of midwifery held office in any university: Joseph Gibson was given the post at Edinburgh, when that university was just about to have its flowering as the leading medical school. (Not one of the famous European medical colleges, such as Salerno, Paris or Padua – where Harvey was a student – then concerned itself with the subject.) But even Edinburgh did not have regular lectures on obstetrics until 1756, and a maternity hospital was not established there until 1793. In the same year France was the first country to pass decrees which provided for the welfare and health of expectant mothers.

By the start of the nineteenth century there was no formal teaching of obstetrics in Britain, except at Edinburgh. The Medical Act of 1886 was the first to indicate that midwifery should be a necessary subject for medical qualification. In 1915 the first 'ante-natal clinic' was started, a particularly ill-chosen description, so dangerously close to anti-natal. The first such clinic was – once again – in Edinburgh, and by 1918 there were 120 of them throughout the country. (Marie Stopes' first anti-natal, or birth control, clinic only just missed coming first. It was established in London in 1922.) Today there are some 2,000 ante-natal clinics, and the proportion of pregnant women (in Britain) attending such clinics has been steadily rising, even in recent years.

The family Chamberlen   An intriguing story, without much consideration for the Hippocratic code but with plenty for fame and fortune, concerns the development of forceps. Edmund Chapman was the first (in 1733) to illustrate and describe the 'extractors', and before him William Giffard (in 1726) was the first to use them in a straightforward manner. But before both of them were the remarkable Chamberlens, a family of pioneers whose ethics did not match up to their inventiveness. Peter Chamberlen was born in Paris in 1560, but his Huguenot parents fled with him to England when he was nine. The boy later developed forceps, a device for clasping the baby's head and easing its passage in birth. Together with a brother-in-law, also called Peter, and the new instrument plus their own skill, the pair of them charged prodigious fees (always in advance) for attending at births. They travelled the country well armed, with the forceps locked in an elaborate chest. No one was allowed to assist them, and the delivery room was barred to every person and all eyes. Even the labouring women were blind-folded. Glasscheib says that, to confound the awe-struck and anxious relatives still further, the obstetricians rang bells, rattled chains, and banged with a hammer during the deliveries. Despite cloak, dagger and bells, they were outstandingly successful.

The original Peter died (without issue) in 1626, and his brother-in-law and partner, Peter, inherited the instrument. On his subsequent death it went to his only son who, unhelpfully, was also called Peter. This man used the instrument to good effect, having many of the Royalty among his patients, and the midwifery fortunes of the family reached a zenith. Peter number three died in 1683, but before then a mixture of unwise political dabbling plus a triumphant petitioning of Parliament by female midwives for their ancient rights, had caused the business to decline. Peter III's son, Hugh, decided to sell the valuable device. (All the others had been called Peter Chamberlen, despite one of them being a brother-in-law, and the change of name seems to have partnered the change of luck.) He tried Paris first. Louis XIV's doctor, Mauriceau, provided Hugh with an awkward birth (a thirty-eight year old dwarf with a narrow pelvis having her first child) which Hugh should sensibly have declined. Hours later both mother and child were dead, and the deal was off.

Hugh returned to England, but then had to flee the country (the reason is obscure) for Holland. There Roger van Roonhuyze bought the instrument. Hugh vanished (to die in 1728) and the Dutchman received his ill-luck along with the forceps. At his first attempt both mother and child died. A few more failures later van Roonhuyze sold the secret. Again there were more failures, and another hasty sale. More and more failures and sales followed suit. Finally the Amsterdam College of Medicine bought the forceps and showed just as little Hippocratic feeling by forbidding any doctor to practise midwifery unless he had bought the secret from the College. The price was 2,500 guilders plus an oath to reveal nothing. It was not until 1753, and twenty years after forceps had been publicly described in England, that two Dutch doctors honourably bought the famous instrument and then honourably made it public property. But what they had paid for turned out to be a crude iron lever and certainly not the Chamberlen forceps.

Who had done the switch? Who had sold this iron bar in the first place, and had kept the forceps? It is easy to suspect Hugh, the last of the Chamberlen medical line. On the other hand his personal attempts at deliveries were outstandingly unsuccessful, and had more of the touch of an iron bar about them than of nicely worked forceps. Could Peter III, so long immured in the tradition of secrecy and in conduct generally unbecoming to the medical profession, have kept the secret even from his son, and passed on to him that sinister bar? No one knows. It is still not known, even though in 1818, ninety years after Hugh's death and 135 years after Peter III's, a cache of forceps was found at Woodham Mortimer Hall, near Maldon, Essex. Peter III had once lived there.

The forceps practice did not quietly mature with the death of the inventive Chamberlens. It exploded vehemently. Perhaps the long years of secrecy, stretching from the sixteenth to the eighteenth centuries, so closely followed by the development and publication of many forms of helpful instrument, was too swift, and a reasonable use of forceps was as unlikely and improbable as reasonable behaviour by a starving man suddenly confronted with food. Anyway, instruments had their heyday. Midwives

155

revelled in them. William Hunter (elder brother of the John Hunter who experimentally inoculated himself with venereal disease) was furious. Nature, he said, had apparently abandoned her work of propagation, and left it to forceps. He himself carried a rusty pair, saying the better the midwife the thicker the rust. Eventually the craze died down, reaction set in, and nature was again left to her primaeval self.

Midwives (and many of them were now men) were suddenly content to watch, to stand and stare. Sir Richard Croft did so for fifty-two hours while Charlotte, only child of the Prince Regent (later George IV), suffered before him. She died six hours after her stillborn child had been produced by nature, and Sir Richard then ended his own life with a bullet. (The failure of this royal birth had far-reaching consequences. The Regent's three elderly bachelor brothers promptly married to beget an heir to the throne. Within a creditably short time two daughters, one daughter and one son respectively arrived for the three of them. The two daughters of the eldest brother died as children, and the mantle of inheritance fell upon the next daughter in line, Victoria.)

The forceps battle still flares from time to time. Fifty to sixty years ago medical students in Britain were bombarded with exhortations to avoid the use of them at almost any cost. Dr Grantly Dick-Read managed to write his famous *Childbirth Without Fear* without mentioning the word. Today forceps deliveries are frequent, but the personal whims of the obstetrician in charge tend to dominate. An American survey indicated that between one-quarter and one-third of US births were assisted by forceps (the figures came from an analysis of all deliveries in twenty-two Navy hospitals). Although some of the hospitals used instruments on 60% and some on 10% of births there were no differences in the death rates of the babies, either for the worse or the better. The current tendency, at least in Britain, is to use instruments more than formerly. In the 1950's the proportion of instrument-assisted deliveries was 4%. By the end of the 1970's it was nearer 13%, about twice the proportion of the next major aid to difficult deliveries – Caesarians.

Sectio Caesaria The name 'Caesarian section' indicates an antiquity far beyond that of forceps. Unfortunately, there is little evidence of the frequency of this drastic measure in the past, and no one even knows whether Julius Caesar was born in this way, as legend has it. (Probably not, as his mother lived long after his birth and the surgical simplicities two millenia ago were unlikely to have benefited both mother and child.) There is general agreement that Julius Caesar did not give his name to the operation, whether he suffered it or not. Some think the verb caedere (to cut out) is to blame, and others add that the father of Scipio Africanus called his son Caesar because he was cut out of the womb. Julius Caesar was a descendant of Scipio's and therefore, according to this theory, he acquired both the reputation of being cut out plus the name from his forefather. Yet another theory holds that the term came from the Lex Regina, later called the Lex Caesarea, which forbade the burial of a dead pregnant woman until her baby had been removed.

Whereas no one knows the origin of the name there is even less certainty about the origin of the custom. The *Susruta Samhita*, an Indian textbook of medicine written down between 800 BC and AD 400, recommended a cutting out of the baby should the mother die in labour. (The vagueness of the publication date cannot be helped, as disagreement abounds, but the textbook was a written statement of ancient oral tradition.) It is odd that, even though Dionysus and Aesculapius (the Greek god of medicine) are said – by some – to have been cut from their mothers' wombs, there is no mention of a Sectio Caesaria or anything like it in the Hippocratic writings.

There were, and still are, two sides to the Caesarian. On the one side the baby is cut from a dead mother; on the other the aim is to short-circuit a difficult birth, and save two lives. Perhaps Caesar's mother, and many others, did survive the operation, but it is reasonable to imagine few such successes. It is reported that Jacob Nufer, a Swiss butcher, operated successfully on his own wife in 1500. Not only did she live, but she bore him four more children afterwards in the normal manner. Other distraught husbands, butchers, barbers and surgeons were frequently less successful in an age without either anaesthetics or notions of antisepsis, and the practice of Caesarian section upon the living fell out of favour. It was not resuscitated until the end of the nineteenth century.

However, there was still great determination to extract babies from dead mothers, partly because the Church was fervent in its demand for souls. Councils and synods urged on the operation, but midwives were reluctant either to state positively that a wretched mother had died or even more to cut her up. Remember that the morals of the times prevented them even seeing beneath the skirts, let alone below the abdominal wall. There were occasional successes. King Robert II of Scotland was allegedly cut from his mother, Marjorie Bruce, by a swift-acting hunter after the woman had broken her neck in a riding accident. King

Edward VI, the son of Henry VIII and Jane Seymour, was said to have been cut from his dying mother, but the story is generally disbelieved.

Ashley Montagu wrote in the 1960's that 2% of all deliveries (in the United States) were by Caesarian section. A Scottish hospital reported in 1964 that over 1,000 of its 18,102 deliveries (over 5·5%) had been Caesarian, with many mothers having four, six, or more. A Lanarkshire woman has had twelve successful operations at this Glasgow hospital, and was 'young enough to have a few more yet' (she was then forty-one). A Dublin woman apparently holds the existing record with thirteen. Today's average figure (for England and Wales) is 7·3%.

Whereas operations on the living are now notably successful, operations upon the dead are more complex. Is the mother really dead? What about attempts at resuscitation? What about the next-of-kin, and is the baby still alive? A leading article in the *British Medical Journal* stated the problem. The child should be extracted as quickly as possible, and its chances are better if the mother has died suddenly rather than slowly. In any case, after five minutes the odds in favour of extracting a live infant become steadily worse and after twenty-five minutes are virtually nil. Cases of longer survival have been heard of, and one child was successfully delivered after forty-five minutes, while its dead mother was kept oxygenated by artificial respiration. The article recommended aseptic procedure just in case the mother decides to live after all.

Anaesthesia Coupled with the controversy over man or midwife interfering with the natural process of childbirth, whether by forceps or other mechanical aid, has been the storm over anaesthesia. Once again the Church was loudly vocal. The idea of women not suffering during childbirth was quite as painful to many men as the hardships normally suffered by the women. Certainly the Bible reiterates the point. It supports the theme of inevitable and proper suffering. 'I will greatly multiply thy sorrow and thy conception; in sorrow thou shalt bring forth children' – Genesis iii. 16. A similar theme of anguish is found in Galatians iv. 27; Isaiah lxvi. 7; Isaiah xvi. 8; Isaiah xxi. 3; Revelations xii. 2; and Hosea xiii. 13. The thanksgiving ceremony called 'The Churching of Women' is equally to the point: 'the snares of death compassed me round about: and the pains of hell gat hold upon me. . . . We give thee humble thanks for that thou has vouchsafed to deliver this woman thy servant from the great pain and peril of

Childbirth.' As soon as anaesthesia was developed there were assertions that no man had the right to rob God of the deep and earnest cries of women in labour.

Sir James Simpson, the introducer of chloroform, first used this anaesthetic in 1847. His action brought the Church down upon him, particularly when he used it to ease childbirth. The Scottish surgeon countered by quoting that God had caused 'a deep sleep' to fall upon Adam during the process of extracting that rib to fashion Eve, thereby condoning anaesthesia. Simpson also argued that 'In *sorrow* thou shalt bring forth children . . .' was a mistranslation for 'effort'. (Presumably there could have been counter-counter arguments in like vein, for were not all women punished with birth pain *after* Adam's deep sleep and Eve's misdemeanour?)

Anyway, the anaesthetists suddenly acquired a trump card when Queen Victoria gave it to them. She had frequently asserted that there were some displeasures in the repetitive and wearisome production of children, and on April 7, 1853, accepted chloroform. Dr John Snow gave her 'one ounce' during the birth of Prince Leopold. This courageous act did not still the critics, and they exist today, but it was a body blow to the resistance movement. (Incidentally, this early story of chloroform is steeped in cross references. Sir James Simpson, who had pioneered the use of chloroform in Britain six years beforehand, was to become a fierce antagonist of Joseph Lister, so soon to promote antisepsis. Dr Snow, here acting as anaesthetist, had already achieved immortality by removing the handle from a Soho water pump and dramatically reducing the local incidence of cholera, thereby proving it was water-borne. And poor Prince Leopold was the first sickly pointer to Queen Victoria's carrying of the haemophilia gene. It had probably arisen in her elderly father, although her child's illness was blamed by many upon the chloroform, either as a physical consequence of it, or a spiritual punishment due to its use.)

Mimicking the history of the use and neglect of forceps for child delivery, which swung wildly from one extreme to the other, is the similarly erratic pendulum concerning the use of anaesthetics in childbirth. The idea caught on. It had its promoters who used it to excess. Then came the detractors, those who said there was no pain whatsoever in childbirth, or there should not be and hence no need existed, or should exist, for pain-killing treatment. Grantly Dick-Read made his name famous as an advocate of this argument. He had written other books before 1942, but in that year *Childbirth Without Fear* was first published. In the next twenty years it sold over half a million copies. Here are some quotations. 'There is no physiological function in the body which gives rise to pain in the normal course of health.' 'The physiological perfection of the human body knows no greater paradox than pain in normal parturition(birth).' 'I am inclined to believe that babies are born entirely free from instinctive fears.' 'All fear therefore in a human being is acquired either by suggestion or by association.' 'From this it follows that faith eliminates fear.' And, finally, a statement many may find particularly hard to swallow: 'I am persuaded without a shadow of doubt that, with the exception of unforeseen accidents, the origin of every form of disease, both surgical and medical, whether hereditary or not, can be traced by careful investigation to the influence of fear upon the human mechanism.'

Many believe that too much emphasis has been placed on relaxation and the exercise of the voluntary muscles. Sir John Peel, surgeon-gynaecologist to the Queen, is one of them. At a conference in Milan in 1965 he told 400 obstetricians that Dr Dick-Read had been right for all the wrong reasons. The concept of painless labour was untrue, and the claim that women in primitive communities suffered no pain was less than a half truth. Sir John did not like any form of indoctrination which led the patient to believe that 'labour would be completely painless and all drugs harmful'. Episiotomy, for example, which is cutting the outlet of the vagina at childbirth to ease the birth, is widely practised and does not suggest a totally relaxing situation. About half of British births are facilitated by this method.

Painkilling can be dangerous at childbirth (a very small proportion of all maternal deaths are due to anaesthesia), and not only can the drugs affect the baby as well as the mother, but anaesthetising a woman in labour is difficult. Professor J. Stallworthy of Oxford has called it 'one of the most difficult tasks an anaesthetist has to perform'. At the Italian Conference Sir John summed up the current knowledge by saying pethidine was the drug of his choice once labour had started. With regard to the second stage 'what you do depends on the philosophy and the expectation of the patient, and the availability of anaesthetic and obstetric personnel'.

Home or hospital   As experts are not always readily at the beck and call of every woman in labour, even in hospitals, and more frequently at home, these facts are bandied about in yet another traditional argument concerning childbirth: should babies be delivered at home or in hospital?

In England and Wales some 99% of all deliveries are in a maternity institution, whether a hospital or something smaller. In 1927 the figure was 15%; so the revolution has been outstandingly rapid. In 1964 the proportion still having their babies at home was 29%, by 1970 13%, and by 1975 3%, leading to today's figure of about 1%. In the United States, Sweden and Australia the proportion was traditionally higher than in Britain, and the British have now brought themselves into line with these other countries. The principal argument put forward in favour of an institutional delivery is that countless forms of aid and assistance are to hand. Against this is all the familiarity of home, with the emotional comfort it possesses, and the friendly local doctor dropping in to add one more to a family already familiar to him.

Both points of view have been expressed with candour. Sir Dugald Baird, Aberdeen's Professor of Obstetrics and Gynaecology, has said bluntly that hospital delivery is safer. A birth caused too much upset in a small house. But Dr Andrew Smith, from Durham, has said hospitals pay inadequate regard to personal dignity and emotions. Maternity patients are no longer the humble poor; they are articulate women wanting the most out of the experience, and are not just 'walking wombs, and vehicles for foetuses'. Supporting him, Dr P. A. S. Lowden has said the myth of hospital superiority must be destroyed. Of hospital deliveries 72% are done by midwives, 4·2% by consultants. The argument is eternal, but some units are getting some of the best from both worlds. In them mothers are delivered by their own doctors, and then packed off home, perhaps within a hundred minutes, all within eighteen hours.

Statistics on the subject are bewildering. Some reports (such as *Perinatal Mortality*) conclude that home births should be virtually abandoned if deaths are to be much reduced. Others conclude that if hospitals cannot show themselves better than homes, then home births should continue or increase. Certainly home births cost the country less money. With over 700,000 births in Britain every year, almost all of which are delivered in institutions with a ten-day puerperium (or length of time after childbirth when a woman needs professional care) still being commonly recommended, several million maternity ward-days are spent every year. Any saving in this figure is to be welcomed, but the current policy is to maintain the high institutional attendance and reduce the length of stay. Even royalty has been encouraged to follow this custom, with Princess Diana having her babies at St Mary's Hospital, Paddington.

How long a stay   An important and frequently challenged point is this length of stay in hospital. Ideas on this subject have been altering radically. In 1955 the Central Midwives Board stated that 'the lying-in period is not less than fourteen days nor more than twenty-eight days after the end of labour'. Six years later the Board said it 'was not less than ten days nor more than twenty-eight'. In 1959 the Cranbrook Committee's report on maternity services thought that 'normal (not average) length of the puerperium should be ten days'. This committee, bearing in mind a rising birth-rate and a general bed shortage, had considerably eaten into those more leisurely two to four weeks. More nibbling was to follow. Influential work done at Bradford sent 30% of mothers home after two days, provided that birth-weight, home conditions and fitness of mother and child were all suitable. Such rapid dismissal of one-third brought down the average length of stay considerably. Since then everywhere else has tended to reduce those arbitrary ten days set by the Cranbrook Committee. (According to one surgeon this committee was responsible for ossifying the idea that there was something inherently valuable in care lasting ten days.) By 1962 the average length of stay was 9·5 days in non-teaching hospitals, 11·1 days in the London teaching hospitals, and 9·2 days in provincial teaching hospitals. By 1965 some 3–4 days, on average, had been lopped even off these times.

By 1980, and in England and Wales, the old ten-day rule had been completely abandoned, with 20% of mothers staying in hospital for less than three days, 21% for 3–4 days, 22% for 5–6 days, 20% for 7–8 days and 11% for 9–10 days, which totals (including the decimal extras) over 95% of mothers staying for less time than the previously sacrosanct ten days. Princess Diana has not only been a trend-setter in this regard, but has been more eager than most to leave hospital. With both her children she has gone home within twenty-four hours.

Length of stay includes the time before birth. Women do not, in general, produce their babies on passing through the hospital doors, but wait and labour a while. Therefore, the average length of stay, whether for the right reasons (doctor and mother both wishing it) or for the wrong reasons (the maternity departments being suddenly under pressure), is already notably less than ten days. The current feeling is for no fixed rules, and more judgement about individual cases. Philip Rhodes, Professor of Gynaecology at St Thomas's, London, has written that 'the length of the puerperium . . . should be assessed for each woman and her baby, and not preordained'.

Maternal death   A résumé of the history of obstetrics, and the rise and fall of different theories, is totally invalid without an account of the attitude towards puerperal fever. It was the scourge of childbirth, and a great dread. More specifically it was often *Streptococcus haemolyticus* invading the raw places of the new mother. These wounds were both the lacerations of the birth canal and the huge area where the placenta had been. Given opportunity the streptococci would multiply rapidly. The first symptoms of this 'child-bed fever' came on the second or third day after labour, but only when the bacteria moved into the nearby veins and lymph vessels did the symptoms become really severe, with much fever, shivering, prostration, a fast pulse and abdominal pain. General inflammation was then gaining the upper hand, as so frequently occurred. When the bacteria reached the general circulation they caused septicaemia accompanied by a very high temperature, delirium, prostration and death. Every midwife in the past knew the list of symptoms only too well. Puerperal fever was all too common. It did not always kill, but it was always feared.

Like so many infections it was rare in the days when humans lived more separate lives. It increased with the increase in town life, and it reached unparalleled peaks with the creation of large maternity hospitals in the eighteenth and nineteenth centuries. The bacteria would leap with ease, so to speak, from victim to victim, from infected woman to woman so soon to be laid equally low as the students, midwives and doctors went from bed to bed. Maternal mortality figures were terrifying. Perhaps a hospital would have a good run for a while, and then perhaps this would be broken by a death rate of one in four, or more. With such small odds, it is little wonder that mothers gave thanks in the 'Churching of Women' for their deliverance.

Quite the most desperate tale in the research for safe childbirth and puerperal fever is the life and death of Ignaz Philipp Semmelweiss. From beginning to end it has all the ingredients of melodrama, the battle with authority, his downfall, the poignant manner of his going, and his eventual vindication. Essentially Dr Semmelweiss's contribution to obstetrics was an insistence on cleanliness, upon the washing of hands. He objected that students should go directly from the autopsy room into the delivery room. He believed they somehow carried the disease from the dead woman they had been examining, or even dissecting, to the living women who next received their attention and who, in so many cases, were dead women a few days later. He ordered wash basins to be placed strategically between the two rooms, and advocated washing. The students would have nothing of his plans, or of his soap and water. Academic freedom, they said, was at stake. Who did he think they were? Midwives? Semmelweiss, never a tactful man, was content to call them murderers. On the following day he was dismissed.

All this took place, not in the Middle Ages but during the last century. Semmelweiss was born in Budapest in 1818. He qualified in 1844, and was fired from the Vienna Hospital two years later. He was dismissed despite the fact that the casualty rate for mothers attended by students was far, far higher (often with one in four dying) than those attended merely by midwives. Only the students examined the dead, and mothers implored to be left alone by them. Semmelweiss wandered sadly to Venice, but in 1847 he suddenly heard that his friend, the anatomist Jacob Kolletschka, had died, following an apparently insignificant scratch received during a dissection. At the post-mortem (and Semmelweiss luckily was in Vienna that day) all the symptoms of puerperal fever were apparent in the dead man, all the swelling and inflammation.

Suddenly Semmelweiss saw the light. 'Puerperal fever', he wrote to the Vienna Medical Society, 'is a blood poisoning caused by the poisoning that forms in the corpse. . . . It is transmitted to the pregnant women by the examining doctor.' On this occasion his ideas were listened to by at least two men who could help him. He was offered a maternity clinic to test out the theories, and therefore had to recruit some students for the testing to be exact. The midwives left and the students entered. So too did death. Mortality jumped from 9% (even the cleaner midwives had much to answer for with one mother dying out of ten) to 27%. So far, in one sense, so good. All students who had examined the dead were then forced, with Semmelweiss always on the alert, to wash their hands in a chlorinated solution. Mortality promptly dropped to 12%. This was good, but no better than the midwives. A cancerous woman provided a final clue. Semmelweiss had examined her, and then five normal women in labour. All five promptly died. The distraught doctor realised the agents of death must be present in all dead tissue. So, thenceforth, every student, whether from the mortuary or not, was forced to wash. Mortality at long last plummeted right down to less than one maternal death in a hundred.

Success? Yes, but not for Semmelweiss, and not even for his system. Obstetricians came to look, but went away amazingly unconvinced. Handwashing can do no harm, said Germany's top man, but purging and blood-letting are preferable. (Bacteriologically, these were still the Dark Ages, although their days were numbered. Louis Pasteur was then twenty-seven, and Joseph Lister twenty-two). Semmelweiss, frantic and with the symptoms of schizophrenia descending upon him, never held himself back. 'Murderer', according to one report, was the mildest expression he used. In 1849, just five years after qualifying, Semmelweiss was dismissed again, and this time for good. Broken, dazed, apathetic, indifferent and poor, he returned to Budapest.

Hungary welcomed him, and gave him a job. He married an eighteen year old girl, who was to produce five children for him, two of whom died in infancy. He was permitted to establish his washing routine, and he brought maternal mortality down to less than 1%, but his general well-being was increasingly at the mercy of his mental health. His writings, when he started them again in 1857, were generally rantings. (His most famous work, published in 1861, was poorly received at the time; but this treatise fetched £4,400 in November 1965 at a Sotheby's sale.) He also took to posting placards on walls: 'Fathers, when you summon a doctor or midwife, you are summoning death. ...' At the great Congress of Gynaecology held in Paris in 1858 its President, M. Dubois, mentioned Semmelweiss only to pooh-pooh his theories: 'It is possible these were based upon some useful principles, but the correct execution of them entailed such difficulties that the highly problematical results did not warrant their exploitation.'

Shortly afterwards, while still in his forties, Semmelweiss's schizophrenia got the better of him, and he was sent back again to a Viennese clinic, this time as patient not doctor. Just two weeks after arriving he suddenly succumbed. An infected wound, apparently the result of his last post-mortem examination, caused septicaemia. He died on August 13, 1865, from the same bacteria he had sought to conquer all his life.

The pathos is that he died young (he was only forty-seven) at a turning point in the war against bacteria. On the very day before his death, Lister began, in Glasgow, to disinfect wounds experimentally, and in the same year a Parisian doctor, Villemin, reported that TB was emphatically a contagious disease; he had succeeded in proving the tubercle's guilt. Bacteriology had begun. As Semmelweiss had known only too well, the reactionary groups would not give up easily. Fourteen years later, having isolated the streptococcus of puerperal fever, Louis Pasteur was still having a rough time in the Paris Academy. 'It is not miasma,' he shouted, 'nor anything but the nursing and medical staffs who carry this microbe from an infected woman to a healthy one.'

The tide, despite those who would not acknowledge this, had

already turned. Semmelweiss could so easily have lived to see the day. He would have been sixty-seven when Lister wrote: 'Without Semmelweiss my achievements would be nothing. To this great son of Hungary surgery owes most.'

Semmelweiss died, it should be reiterated, not in the mists of antiquity, but a dozen decades ago. And his death was by no means the end of puerperal fever. It was only the prelude to the beginning of the end. The fever as a major horror had many more years to run. For instance, in *The Lancet*'s obituary of Dr Miles Phillips, who died in 1965 and who had started to specialise in midwifery after graduation in this current century, it was written: 'Puerperal sepsis, in his time, was the scourge of obstetrics.' The scourge was therefore still rampant very recently indeed. Nowadays it is laid low. Should it occur today, despite precautions, the sovereign remedies are penicillin and the sulphonamides. Semmelweiss's battle has finally been won.

About the most shaming aspect of the whole history of midwifery is that, on occasion, more mothers died than babies. Adult women ought to be able, one might suppose, to survive the trauma of childbirth better than their brand-new and diminutive offspring squeezed so abruptly into a separate existence. Yet puerperal fever, marching triumphantly up and down the wards, found the raw wounds of the mothers far more susceptible to attack than the mewling objects which had caused them. A report from the Hotel Dieu in 1664 (the words hospital, hotel, and hostel are all etymologically the same, and took time to settle into their present meanings) stated during one epidemic that 33% of their mothers had died either during or shortly after childbirth. Semmelweiss's Viennese hospital caused maternal mortality to be almost as high. The national figures, not that such facts were collected, must have been lower as home confinements were never so dangerous as the hospitals could be. St George's Hospital, London, commented sadly in 1856, when its own maternal mortality rate was none too good, that 'women delivered in their own habitations were often living in the greatest filth and poverty and very limited accommodation, and yet these women seemed to do infinitely better than those who are removed to spacious buildings where every attention can be devised'. The same report states that there were only ten deaths out of 2,800 home deliveries (or 0·35%). The hospitals could scarcely compete. In the same year, Sir James Simpson, in his *Obstetric Memories*, wrote that among maternal deaths 'only a comparatively small proportion are the direct result of convulsions, haemorrhage, rupture of the uterus or other more immediate or primary complications and accidents connected to parturition: the great majority of these maternal deaths is produced by puerperal fever'.

'Until about thirty years ago,' Sir Dugald Baird wrote in the 1950's, 'the most urgent problem facing the obstetrician was maternal death.' Only after the 1920's did the baby start receiving something like the correct amount of attention as well. Whereas the maternal mortality rate (including abortion) has dropped to less than 0·02% (or one mother in over 5,000), the infant mortality rate (stillbirth and neonatal) is still about 110 times higher. The mother is doing fairly well these days; the child is less likely to do so.

Before leaving the mother, that percentage maternal mortality rate should be translated into less impersonal terms. It means, with 670,000 births a year in England and Wales, the deaths of 125 mothers. In New York in 1962 it was reckoned that 43% of all maternal deaths were from – mostly illegal – abortions. Yet, abortions apart, that figure of 125 maternal deaths a year in England and Wales will surely come down. A report from the Northern Ireland Ministry of Health considered that only 13% of the mothers who died between 1960 and 1963 had deaths that were 'definitely unavoidable'. All sorts of happenings were to blame: home confinements were permitted which should not have been; some women never told anyone they were pregnant; correct diagnosis of impending trouble was not made until too late; the emergency flying squad did not arrive – and over the years the errors pushed up the total. Many of the maternal deaths that currently occur are still considered to have been avoidable.

The future will inevitably reduce the death rate still further, but it cannot effect a reduction similar to that of recent years. The maternal mortality graphs indicate dramatically that many of us alive today were apparently born in a kind of mediaeval period, so far as obstetrics were concerned. Take the figure for the United States as a whole. In 1915, which was not so long ago, the number of white mothers dying was 6 per 1,000 births. (The number of black mothers dying was, and still is, at least double the white figure.) By 1930 the figures were still the same. By 1940 the white figure was down to 3 per 1,000, the black 8 per 1,000. By 1950 – a big leap – it was 0·6 per thousand for whites, 2 per 1,000 for blacks. By 1960 0·25 for whites, one for blacks. In other words the proportion of women dying in childbirth had been divided by twenty-five for whites in thirty years, and by ten for blacks. As Sir Dugald Baird put it, thirty years ago the problems were infection, haemorrhage and obstructed labour. Now antibiotics are used against the infections, transfusions against the haemorrhage, and Caesarians against obstructions. So the graph has come plummeting, and is still plummeting.

Not everywhere is like the United States or Europe. In 1930, when the white US figure was 6 deaths per 1,000, the figure in India was 23 per 1,000. Sir John Megaw, then Director-General of India's Medical Services, considered that 200,000 Indian mothers died every year in childbirth. He also reckoned this meant that one-tenth of all Indian wives would die sometime as a result of maternal labour. The modern blessings of antibiotics, blood and quick surgery are still to come in India for most of her labouring women for it is a country where most people are born, live and die without a qualified doctor's single attendance.

Infant mortality   Now to the offspring. Once again the dark ages have only just passed. At the time of the Boer War the present-day discrepancy between, say, India and Europe scarcely existed. In 1900 the Indian infant mortality rate was calculated at 232 per 1,000. In England and Wales it was then 154 per 1,000 and in the United States 122 per 1,000. By 1951 India was still in similar straits (116 per 1,000) but the English figure was 30 per 1,000, and the American 29 per 1,000.

South Africa manages to maintain such a discrepancy even within a single country and at the same time. Infant mortality figures of the 1960's were 28 per 1,000 for whites, 65 per 1,000 for Asians, 127 per 1,000 for coloureds (mixed), and 116, 250 and 397 per 1,000 for Pretoria, Cape Town and Port Elizabeth Africans. By the 1980's these figures had changed (in Cape Town) to 9·4 for whites, 20·4 for Asians, 18·8 for coloureds, and 34·6 for Africans. In general Africans do rather less well away from the big cities. Would the South Africans agree with Sir George Newman who wrote, in 1916, that 'the death-rate of infants is the most sensitive index we possess of physical welfare and the effect of sanitary government'?

Back in the genuine Dark Ages the mortality figures for infants do not exist. They only begin to arise nationally after the registration of births, deaths and marriages had become compulsory in 1837. By the middle of the nineteenth century infant mortality in towns was 200 per 1,000. This means that at least one-fifth of all babies born alive never reached their first birthday.

In those days it was emphatically best to be breast-fed. In Salford in 1904, according to A. V. Neale, the death-rate was 128 per 1,000 for breast-fed babies, 263 per 1,000 for cows' milk babies, and 439 per 1,000 (or nearly one out of two) for condensed and sugared milk babies. By the end of Edward VII's days, the national rate was 116 babies dead after one year for every 1,000 born alive. By the Great War it was 108 per 1,000. By the time peace was declared it was near 90 per 1,000. By 1923 it was 72 per 1,000. By 1928 it was 60 per 1,000. By 1935 – and the slump could not have helped – it was 62 per 1,000. By 1940 it was 57 per 1,000. By 1945, despite of, or because of, food rationing for all plus full employment, it was 45 per 1,000. By 1950, the time of Korea, it was 38 per 1,000. By 1955 it was 27 per 1,000. By 1968 it was 18 per 1,000 and by 1980 it was 12·5 per 1,000, or over 8,000 babies dying in the year before reaching their first birthday (which means almost 7,000 fewer such deaths since the late 1960's, what with a slightly reduced birth-rate and a much improved mortality ratio). Some other European countries do better, notably Holland with 8·5 per 1,000. Some do worse, such as Yugoslavia with 32 per 1,000 (but it was 80 per 1,000 in the late 1960's). The United States, which ranked thirteenth in infant mortality during the 1960's, has now improved its situation, with some 13 deaths for 1,000 born. It is therefore still behind, for example, Japan, Denmark, Holland, Sweden and Australia. Figures for the Soviet Union and China are not available.

However much the infant mortality figures may be successfully trimmed in the future there is always likely to remain a hard core. Some of the congenital malformations will see to that. The determination of a hard core to withstand medical advance is already manifest in the mortality tables, and it shows up in the first seven days of life. In 1910 there were seventy-six infant deaths (per 1,000) between four weeks and fifty-two weeks of age, a number slashed to six in this half a century. In 1910 there were fifteen deaths between one week and four weeks, a figure cut to two after fifty years. But also, in that same Edwardian year, there were twenty-four deaths in the first week, a figure modestly pruned to thirteen by the 1960's. To sum up: more children die in Great Britain of natural causes during the first seven days from birth than in the next thirty to forty years.

Hand in hand with the declining death-rate for infants has gone a reduction in the stillbirth-rate, the number of babies born dead. In 1928 (and once again a recent decade appears deeply embedded in the remote past) the proportion was 40 per 1,000 births. By 1939 it had fallen to 38 per 1,000. Then came the war, and the most dramatic fall on record of stillbirths: by 1945 it was 28 per 1,000. By 1955 it was 23 per 1,000, by 1968 only 14·3 per

1,000 and by 1980 just 7·2 per 1,000, another great leap forward. Even so this means 4,700 stillbirths in England and Wales in a year.

All this improvement, one feels, should have narrowed the gap of class inequality. It has done nothing of the kind. The social distinction remains as open as ever. In 1943 Professor Richard Titmuss pointed out that despite improvements, despite a reduction in infant mortality and the stillborn dead, the range of social class inequality in 1931 was either as great, or greater than it had been in 1911. More recently James Kincaid has told the same story. Even though the National Health Service was initiated in 1948, on the basis of equal medicine for all, the class discrepancies have remained just as constant. From Class I (professionals) right through to Class V (unskilled labourers) the death rates increase in the same sort of ratio as they have always increased. Stillbirths for Class I in Britain were 16 per 1,000 in 1949, and 13 per 1,000 in 1958: for Class V they dropped from 26 to 24 per 1,000. Death-rates for infants up to one month old fell from 16 per 1,000 for Class I in the first fourteen years of the National Health Service down to 12 per 1,000: with Class V they fell from 30 to 24 per 1,000. France reports a similar constancy between the classes. Hungary, on the other hand, does not. There, according to E. Szabady, the manual and non-manual differentiation has narrowed.

Countless factors are to blame. One is that tall women have fewer stillbirths than medium women, and medium women fewer than small women, whatever the class. The higher classes (or socio-economic groups) have more tall women. Another is that girls who grow up with many brothers and sisters are more likely on average to produce stillbirths – and the poorer classes have larger families. A third is that education helps to cut down infant mortality and Class V is less well educated. And many more. Will it be one generation or two before the Welfare State can knock down a class structure which, without the black and white clarity of the United States or South Africa, shows up inequality just as emphatically?

What of the future and that hard core of deaths? The World Health Organisation has conducted its own mortality survey. Its conclusion was that in every country 20% of the mothers constituted 90% of the risks. Each country therefore should make doubly certain that its 20% receives the necessary amount of extra care but that, as with so many remarks of what should be done, is far easier written than achieved in most of the nations in the world.

Causing death　Another hard core, nothing to do with natural causes (at least not in the strict sense), is infanticide, the killing of the baby. Only the mother can be charged with infanticide; anyone else involved may be charged with murder. The Infanticide Act of 1938 provides that, 'Where a woman by any wilful act or omission causes the death of her child being a child under the age of twelve months, but at the time of the act or omission the balance of her mind was disturbed by reason of her not having fully recovered from the effect of giving birth to the child, or by reason of the effect of lactation consequent upon the birth of the child, then, notwithstanding that the circumstances were such that but for this Act the offence would have amounted to murder . . .' she is guilty of the felony of infanticide. She may then be dealt with as if she had committed manslaughter. Some feel the law is too lax, permitting a mother to get rid of a child with minimum consequences. Certainly the felony of infanticide, according to C. J. Polson, is censured with great sympathy and understanding, for the child is assumed to have been born dead until the prosecution proves otherwise.

Certainly no judge has imprisoned a woman for such an offence recently, even though a life sentence is theoretically possible. Others feel the law incredibly harsh, for it involves the wretched mother in perhaps two visits to the magistrates' court and then, perhaps, a wait of months before a visit to the assize courts. Mr Leo Abse, the Labour M.P., feels that the woman's case should be speedily dealt with in a magistrates' court only, that it should be spared publicity, and generally treated in a less mediaeval manner. More women are charged with the offence than ever brought to trial but some are still confined to a prison hospital while awaiting trial.

At least the law is better than it used to be, and before the first Infanticide Act of 1922 which introduced the offence of infanticide. Until that time the killing of infants was always treated as murder. Between 1906 and 1921, sixty women were sentenced to be hanged for causing the death of newly-born children, and with fifty-nine of them the sentences were commuted. The exception was Mrs Leslie James, a housekeeper of thirty-nine, who was convicted on July 24, 1907, for suffocating a newborn child which she had agreed to adopt for £6. She had adopted various other children for money, and the jury in this case made no recommendation for mercy.

The basic reason for the existing leniency, and the desire for greater relaxation, is that post-natal depression is a very common result of birth. In some women it may reach uncontrollable heights. '. . . she is restless and depressed, shortly she becomes intensely emotional and impulsive, she may be suicidal or infanticidal, or both', according to an obstetrical textbook although such acute depression is less common nowadays.

105 boys: 100 girls   Back now to more natural happenings and the incomprehensible matter of sex ratio. Why should more boys be born than girls? Why are more boys born during and immediately after wartime? Why is there such a wastage of males before birth? Why is the ratio different in different areas?

In Britain 105 males are born alive for every 100 females (but the gap had been widening before the 1960's and has since been narrowing, the figure for 1968 being 105·6). The proportion varies according to the time of year, being – in 1980 – 104/100 in November, when there were fewest births, and 106·2/100 in May, the month with the highest birth-rate save for July. In the United States the ratio is about the same for whites, but 102·6 males to 100 females for black births. Greece and Korea both have 113 males per 100 females, but in Cuba it is 101 males per 100 females. After birth the ratios change because males die faster than females. Parity in Britain, or equality in numbers, is reached by the age of thirty. The United States puts the age of parity higher, in some areas as high as the age of fifty. In both cases, as males continue to die faster, the ratio is eventually reversed with more women than men being alive. By the age of fifty-five men are dying, quite suddenly, very much faster than women. By the age of ninety-five every man is outnumbered by four women. Women have started dying at a greater rate than men by the age of seventy-five merely for the reason that, by then, the population is mainly women. Men cannot then die in comparable numbers as so many men have already died.

The story is the same in the uterus. There are more males to start with, and more of them die during gestation until the ratio at birth happens to be, in Britain, 105 males per 100 females. At conception, 266 days beforehand, the ratio may well have been 150 per 100. Some scientists put it as high as 170 per 100 (or getting on for 2 to 1), while others put it as low as 120 per 100. No one knows. Certainly many more stillbirths are male than female, whether they are born at the correct time or miscarried many months earlier. A huge United States survey put the ratio at 134 males per 100 females for those stillbirths produced at term, 201 per 100 for those miscarried at the fourth month of pregnancy, and 431 per 100 for those produced two months earlier.

Unfortunately, even though these figures (of Ciocco) are

172

constantly quoted, the situation is now less clear-cut than it used to be. Perhaps the midwives and doctors, it is now suggested, confused the sex. After all the clitoris and penis, being the same lump of embryological tissue to start with, do look remarkably similar in the early stages of development. But, despite differences of opinion, no one has yet claimed that more females are conceived. It is still undeniably true that more males start life, and die off earlier all along the line.

No one knows why the males are the weaker and less viable sex from start to finish. They are bigger, and therefore more of a problem at birth, but what has birth got to do with miscarriages? They carry a Y chromosome, and not two X's like the female, but what precise benefit – if any – do those X's give over the Y? No one knows.

Nor is it known what pushes the sex ratio at birth up and down. The Boer War to a small extent, and the next two wars to a large extent, were all partnered by an increase in the number of male births. Nature, it can be said, was merely repairing the ravages of war. (She also steps up male production in neutral countries at the same time, although not so much.) If one puts beneficent Nature kindly but firmly to one side, what else can cause the wartime changes?

Again no one knows for sure. Theories which hold some water, in that they are backed up in part by some evidence in their favour, include: an increase in young marriages, which can increase the male ratio; full employment, which means greater prosperity and fewer stillbirths – hence more males; an absence of husbands except for brief periods, and therefore more conceptions by the highly fertile who might produce more males as another of their attributes; and a relative lack of intercourse and therefore pregnancy enabling the uterus to rest and be a better place for the weaker male foetus when it does come along. Different authors choose the theory they like best, and no doubt there are others.

Apart from the war story, and taking many other surveys into account, it has been found that the sex ratio is high (i.e. there are more males) amongst the offspring of mixed race marriages (notably black/white), of AB blood group mothers, of radiated fathers (as shown up by the two bombs dropped on Japan in 1945), of very old mothers of fifty or more, of United States Air Force fathers who either fly transport aircraft or who do not fly, and amongst offspring that are the first-born or who are blood group O. (None of this means that, e.g. transport pilots will produce only boys; merely that they deviate slightly from the average proportion.) On the other side girls are slightly more frequent than normal amongst the offspring of high-performance-aircraft fathers, of mothers who smoke or were radiated, of older fathers, of unskilled and manual-working fathers, and who are born either illegitimate or with older brothers and sisters. Professor Sir Alan Parkes of Cambridge summed up the heterogeneous collection of differences by stating that wherever conditions are rough, as with illegitimates and unskilled fathers, more boys will die, and therefore more girls will be in evidence. As a theory it fits – almost.

Fifty years ago everything was nicely balanced so that, with more boys dying and more being born, the sexes were exactly equal in number within the fifteen to nineteen age-group. Obviously, said everyone, it is arranged that every Jack should find a Jill when both are at the right reproducing age. A country like India gives a different picture. Throughout the whole of that sub-continent's census history, Indian Jacks have not had enough Jills. In 1921 there were 940 women for every 1,000 men. In 1951 there were 947. More males are born there (as well as elsewhere), but the infant girls are treated with less enthusiasm and are then rapidly subjected to years and years of remorseless childbearing.

One suspects that Indian conditions represent a truer historical picture than twentieth-century Britain. Therefore the sex ratio should, by this comparison, be reversed with more females being born. Also, with the human method of reproduction, Nature is being wasteful by giving one wife (or nearly one wife in India) to every man. One man could (and kings like Rameses II, Louis XIV or the musical one of Siam have done so) sire many children while his true wife is merely gestating one. The animal kingdom does not help with an explanation, but merely indicates how deep-rooted is the relative weakness of the male sex and how basic the greater number of male conceptions. Cows produce 134 males per 100 females in stillbirths, and 105 per 100 for live births. Pigs are similar, and so too are rats, mice and birds. Horses produce more male stillbirths, but more female live births. The general picture, therefore, is the same, with males dying faster, with some likelihood of parity at the reproductive age, and yet with little likelihood of any female remaining unmated whether in the mousehole, the bovine herd, the Indian village or twentieth-century Britain. The human female is even less likely to be unmated in the future. We are heading, says Professor Parkes, 'for a world shortage of marriageable females'. The cause is the greater survival of males.

We are also heading, willy nilly, for a time when the sex ratio will be adjustable. Already claims are being made for an ability either to distinguish male from female sperms in bulk (it is the male who ordains the sex) or to separate them. Some day both will be possible. Will there be a rush for male children? Certain countries today treat female offspring with something of the reverence they accord to yet another litter from the cat. Will they maintain that attitude when women are in short supply? One longs to know, but one's ignorance will do nothing to halt the arrival of an artificial sex-determining mechanism. The contraceptive pill will then be either pink or blue, or both, with either boys or girls or neither being temporarily desired.

Humanity is most undecided about the laws of chance, particularly when a baby's sex is involved. On the one hand great financial stakes will be placed at a race-track on most slender odds; on the other most families will consider themselves somehow odd if all the siblings are of one sex. An outsider's chances in a race, listed at 63 to 1, will cause many to stake hard cash. The thought of having either six boys or no boys at all from six pregnancies, each of which has a likelihood of 63 to 1 against, will strike most mothers as totally improbable.

In fact, again assuming six successful pregnancies, and disregarding the slight inequality in the sex ratio, the chances for and against particular boy/girl combinations can readily be assessed. The chances are 1 in 64 (or 63 to 1 against) that all will be boys, 6 in 64 (or 29 to 3 against) that there will be 5 boys and 1 girl, 15 in 64 (or 49 to 15 against) that there will be 4 boys and 2 girls, 20 in 64 (or 11 to 5 against – almost 2 to 1) that there will be 3 of each, 15 in 64 (or 49 to 15 against) that there will be 2 boys and 4 girls, 6 in 64 (or 29 to 3 against) that there will be 1 boy and 5 girls, and 1 chance in 64 (or 63 to 1 against) that there will be no boys.

Putting this another way round, the chances of all 6 being of the same sex are 1 in 32 (or 31 to 1 against), that 5 will be of one sex are 6 in 32 (or 13 to 3 against), that 4 will be of one sex are 15 in 32 (or 17 to 15 – almost evens), and that both sexes will be equal are 10 in 32 (or 11 to 5 – almost 2 to 1). Having 3 boys and 3 girls is therefore much less likely than having 4 of one sort and 2 of the other. Having 5 of one sex and only 1 of the other will cause many mothers to worry about themselves (or their mates). What, they ask, is causing such a one-sexed emphasis? The answer is, probably, nothing; or at least no more than the chance (6 in 32, or 13 to 3 against) of getting either 5 heads or 5 tails should a coin be tossed 6 times.

The weight of a baby, and the fact that males are heavier, has often been blamed for influencing the sex ratio. Bigger babies, assuming equal mothers, can plainly be more of a problem at birth, but the size difference can hardly be a problem back in the early days of pregnancy when males are also less viable. At the foetal age of eight weeks both male and female young weigh 5 grams – a sixth of an ounce – and sex weight differences are scarcely detectable until a month or two before birth. At birth the average male is heavier by three to five ounces, and he has spent, on average, one day less in the uterus than a girl. The weight at birth is important, partly because it can be correlated, so far as the population is concerned, with survival chances.

Present-day enthusiasm for recording birth weights is intense. Fathers announcing the news consider the offspring's poundage as exciting as its sex or general well-being. Such intensity did not exist in the past. There appears to be no mention of birth weight in any ancient Greek, Roman, Arabic, or Hebrew writings. The first mention, according to one surveyor of this subject (Cone) was in a book by Mauriceau. (This French obstetrician of the seventeenth century was the one who cleverly produced a labouring thirty-eight year old dwarf to confound Hugh Chamberlen: see page 155.) Strangely, the Frenchman put average weight at '13 livres', or almost fifteen British pounds. As if the subject were fishing, even this tall story was topped in 1747 by the English surgeon, Theophilus Lobb, who wrote that 16½ lbs. (7·5 kgs.) was 'a representative baby'. Quite suddenly reason, and better estimates, prevailed. The German Röderer of Göttingen wrote, in 1753, that past writers were 'hallucinating' as he found weights to be 6 lbs 12 ozs. (3·1 kgs.) for boys, and 6 lbs 5 ozs. (2·9 kgs.) for girls. The first proper English study (in 1785) put both weights some 7 ozs. (198 gms.) greater. And there, on average, give or take an ounce or two, they have remained.

As a temporary digression into the animal world for comparison, the human birth weight is neither particularly big nor small. Bats produce infants of a few grams. Marsupials, like the kangaroo, give birth to minute embryos half an inch or so in length. The largest marsupial, standing 6 ft. (1·8 m.) tall, is the red kangaroo; its young are born when ¾ in. (2 cms.) long and weighing less than a gram. Even some large placental mammals can give birth to very small offspring, such as the giant panda whose much sought-after cubs weigh only a few grams at the start of their external life. At the other extreme, still within the range of mammals, is the blue whale, the largest animal that has ever existed. Its calf is 22 ft. (6·7 m.) long, and weighs almost 10 tons. This huge thing is milk-fed for half a year (and is sexually mature two and a half years later when it has grown to 75 ft. (23 m.). A baby African elephant is the weight of an overweight man. A baby blue whale is the weight of a full-grown African elephant. A human mother, producing some ten young of 7½ lbs.(3·4 kgs.) each in her lifetime, is giving birth therefore to about three-fifths of her own weight during her life. A mouse-mother gives birth to a quarter of her own body-weight in a single pregnancy, and five times her body-weight in her mouse lifetime.

The mean birth weight of boys and girls together in European communities is about 7 lbs 4 ozs. (3·3 kgs.). Pygmies, the smallest human beings, produce babies of, on average, 5 lbs 11 ozs. (2·5 kgs.) according to one Zairean survey. Birth weight is almost entirely under the mother's genetical influence, not the father's. Weights range from virtually nothing at all, in extreme prematurity when the chances of survival are nil, up to 11 lbs (5 kgs.) or so when chances are again less satisfactory. Weights of 5 lbs (2·3 kgs.) to 10 lbs (4·5 kgs.) are often regarded as normal, despite the disparity of one being double the other. It is strange that this 'normal' divergence is preserved in adult life. From women of 100 lbs (45 kgs.) to men of 200 lbs (90 kgs.) is the normal range that embraces most of mankind. Although countless babies survive that are either extremely underweight or equally overweight, the optimum weight accompanied by the greatest chance of survival is slightly higher than the mean weight. Mortality is then, according to one large survey (Karn and Penrose), less than 2%. Mortality increases as birth weight diverges from normal. Babies either of 6 lbs (2·7 kgs.) or 9 lbs (4 kgs.) have a mortality of 3%. If they are 5 lbs (2·3 kgs.), or 9¾ lbs (4·4 kgs.) the mortality has gone up to 5%. Mortality then leaps up as divergence increases. It is 10% for babies of 4½ lbs (2 kg.) or 10½ lbs (4·8 kg.) and far more severe for babies either lighter or heavier still.

The European groups produce heavy babies relative to other racial groups. A world average is nearer 6½ lbs (3 kg.) than the European 7 lbs 4 ozs. (3·3 kg.). Particularly small are the Southern Indians and the Sri Lalese, which are nearer 5¾ lbs (2·7 kgs.), and those African Pygmies, which are even smaller. Unfortunately, no one knows whether these differences can be attributed to ethnic origin (as with black skin), or poor maternal diet and physique, or a wretched environment, or a mixture of the lot.

Studies on Europeans show up a few of the complexities. Birth weight rises with birth order. Second babies are heavier than the first-born by about 3½ ozs. (100 gms.). Third babies are heavier still. Thereafter the trend is less definite. Possibly the increasing age of the mother is having a counter-effect on weight. Other counter-effects are smoking, maternal height, and a small pregnancy weight gain. Cigarette-smoking mothers produce babies, on average, 5–7 ozs. (141–198 gms.) lighter. Short women also produce lighter babies, and so do light women. Finally, mothers who put on little weight during pregnancy produce slightly smaller babies than average. Much of this work was done at Aberdeen, where the Medical Research Council had for many years an excellent obstetric unit.

Prematurity makes nonsense out of traditional birth weights. Also modern methods of treating these smallest of all independent humans is making nonsense of earlier beliefs about their survival possibilities. Every year the survival percentages for each weight range improve. Miniature babies less than 2 lbs (1 kg.), in weight are occasionally being nursed into a normal existence. The current limit for survival seems to be about just over 1 lb (500 gms.), but survival is then unlikely. Such minute forms have only spent less than two-thirds of the traditional time in the uterus. They are more than half the length of a normal baby, but only a quarter of the weight.

Internationally (and even premature babies have to be defined) any birth weighing less than 5½ lbs (2·5 kgs.) is called premature. Other definitions are a length (crown to heel) of less than 18½ ins. (47 cms.), or a head circumference of less than 13 ins. (33 cms.), or any delivery up to the thirty-fifth week after fertilisation. Apart from being small, light, and badly proportioned (in relation to an ordinary baby), a premature's wide-set eyes are usually shut, its cry is weak, its temperature regulation is bad, and its nails are short. Its skin is without fat, being wrinkled like an old man's, and dull red and transparent. Its body and upper face may have some hair which falls out, and all its movements are without strength, causing a frequent difficulty in establishing breathing. About 130 premature babies are born in Britain every day, or one in fifteen of all births. (That definition of 5½ lbs (2·5 kgs.) or less confuses the picture. Some 5½ lb babies are merely light in weight for their age and maturity. Others of the same weight, the true premature infants, are nothing like so advanced.) On the basis of weight alone, 6·7% of all English and 7·6% of all American babies are premature.

The youngest, smallest and most delicate group of premature babies with any chances of survival were probably conceived some twenty-four to twenty-eight weeks beforehand. At the time of their birth, which may frequently be a very abrupt business compared with normal labour, they weigh less than a third of normal birth weight, their length is slightly over half normal, and both their head and chest circumferences are again only slightly over half normal. The difference in proportion between these and a normal baby is roughly equivalent to the difference between a normal baby and a two-year-old. They require much nursing, particularly to detect early signs of trouble. One surgeon/author recommends four nurses in each premature unit for every nine babies. The incubator has to be kept at 26·6°C (80°F) to 32°C (90°F) (the sort of temperature Britain experiences twice a year) and with a humidity of 60% to 70%. Food is usually given – by the gram – after forty-eight hours, and then through an instrument much resembling a fountain-pen filler. (Earlier feeding is recommended by some, but vomiting can be a problem; therefore intravenous feeding is sometimes given.) Weight falls initially, as with normal babies, but the pick-up time is slower. A 2–3 lbs (1–1·4 kgs.) baby will take six weeks to put on another 2 lbs to 3 lbs. Food is given every three hours, night and day. When the baby is feeding properly, and can regulate its temperature, and is no longer vomiting, then it can leave the incubator for the cot, but it must be a cot kept in hot (26·6°C or 80°F) moist (50% humidity) air.

Gradually things improve, and the premature infants attempt to catch up, both physically and mentally with the other babies, born weeks or months later, who are of the same conceptual age. It takes years for the smallest of them to reach the normal weight for age. It used to be thought that prematures had the dice loaded against them mentally partly because the abnormal number of severe defectives among the prematures (10% to 15% of them) lowered the average intelligence for the group as a whole. When these are discounted, the premature appear remarkably normal. D. J. Baird concluded a survey by saying there is 'no clear indication that . . . premature expulsion from the uterus does the foetus any serious harm'. (Some notable and historical prematures have been Napoleon Bonaparte, Isaac Newton, Charles Darwin, Victor Hugo, and François Voltaire.)

A recent statement, and from the Queen Victoria Medical Centre in Melbourne (which is famous in caring for the premature), reported a 50–60% survival rate for babies weighing less than 2·2 lbs (1 kg.) with 11% of the survivors having a major handicap (discovered within two years). For babies born weighing 2·2–3·3 lbs (1–1·5 kgs.) the survival rate was 80–90%, with 6%

suffering a major handicap. The smallest survivor at the unit weighed 1·04 lbs (470 gms.) and the most premature was born sixteen weeks early. Oddly the smallest survivor – or rather the smallest to have been well documented – was born long before modern premature-ward techniques were available. Marion Chapman was born at South Shields, England, in 1938 with a weight of 10 ozs. (283 gms.). She was fed, hourly, with a fountain-pen filler, and did so well that she had a weight of almost 14 lbs. (6·35 kgs.) by her first birthday, a more than thirteen-fold gain in those first fifty-two weeks. She died in 1983 shortly before her forty-fifth birthday.

What can do harm is to confuse a full-term but diminutive baby with one genuinely premature. It has been asserted that perhaps 30% of all alleged prematures are, in fact, dangerously under-developed full-term babies. It has long been customary to disregard maternal statements about the date of their last menstruation (and hence the onset of pregnancy) whenever foetal weight appears to belie those statements; after all some women can be gloriously vague on the subject. It is now thought – by some – that the mothers may often be more right than the doctors. There are genuine prematures, and there are also poorly developed nine-month babies. Of the two the prematures are to be preferred. A baby weighing 4 lbs (1·8 kgs.) when born after thirty-four weeks has a better chance of being normal than a baby weighing 4 lbs (1·8 kgs.) when born after forty weeks. The premature is to be preferred to the undergrown.

Some causes of prematurity are known. They may already afflict the mother (such as high blood pressure or diabetes), or they may be linked to the pregnancy itself (such as an abnormal uterus), or they may be associated with the embryo (such as malformations). By no means does each cause always have the effect of producing prematurity. Also, by no means is every cause known. About 50% of prematures happen without any cause being detected, and some women just do regularly produce premature babies. Whatever the cause, the lot of any baby cast out before its time from the snug, aquatic cocoon of the womb is being improved, year by year. Babies of 16 ozs. (0·5 kg.) or even less on occasion, have already survived. Will even such exceptional successes become the rule in the future? Will there even be a case for inducing them from women who resent the load of pregnancy, and prefer the substitution of an artificial incubator for their own bulging bellies?

The puerperium   As soon as the placenta has been delivered, the puerperium (or puerperal period) begins. It is the time of reversion while everything more or less returns to normal, and it takes roughly six weeks.

The changes in the uterus are a dominant feature of this time. Before delivery it is 12 ins. (30·5 cms.) long. Immediately after delivery (as can easily be seen during a Caesarian) it shrinks to become a pallid and wrinkled thing 7 ins. × 5 ins. × 3½ ins. (17·8 cms. × 12·7 cms. × 8·9 cms.). This weighs 2 lbs (1 kg.), or so. A week later it weighs half as much. After six weeks it weighs a mere 2 ozs. (56·7 gms.). In size it is then about 3 ins. × 2 ins. × 1 in. (7·6 cms. × 5 cms. × 2·5 cms.). No one has yet adequately explained how so much uterine tissue both shrinks and disappears, but despite this impressive shrinkage, a uterus which has produced a baby is detectably different from one which has not. There may be after-pains after birth, perhaps for two to three days, occurring mainly in mothers not having their first child. The uterus, by continuing to contract and relax, causes the pains, often when the baby is put to the breast, thereby indicating a connection between the uterus and the mammary gland.

The vagina, hugely stretched in birth, takes a similar time to revert to normal, although the first birth always leaves it slightly wider. The hymen, never completely destroyed even in child-birth, remains as persistent fragments of tissue. The abdominal wall should return virtually to normal, although flabby people can become flabbier. Striae, or small strips of skin, can remain as permanent reminders of the enormous stretching of the abdominal wall for all those months of pregnancy. The breasts change, but more about them under lactation.

Menstruation comes back slowly, and the speed of its return differs depending upon whether or not the mother is producing milk, and if it is her first child or not. On average, according to American sources, the first menstruation returns within six weeks for 25% of mothers, within twelve weeks for 61% and within twenty-four weeks for 77%. In other words, over three-quarters of American women are back to menstruation, these days, in less than half a year. The proportion would be smaller if more women were breast-feeding their babies. It is not true, despite widespread confidence, that the nursing mother cannot become pregnant. Instead it is true that, at least for the first few months, a lactating mother will not menstruate, and a pregnancy is unlikely in the absence of menstruation; but the golden rule of 'milk therefore no new baby' is by no means gilt-edged. When a new baby is conceived, milk for the old one often dries up fairly rapidly.

Other changes are that blood quantity (30% to 40% up on

normal at the time of birth, i.e. 3 to 4 pints, 1·7 to 2·3 litres up) reverts to normal within a week. The bladder is unusually active after birth, with perhaps 4 or 5 pints (2·3 or 2·8 litres) being emitted within twenty-four hours. Unfortunately it sometimes fails, and retains rather than loses liquid, at this time. This demands the use of a catheter, and more frequently than is usual with ordinary non-emptying bladders because of the abnormal quantities of urine involved after birth. Pulse rate and temperature may change slightly in the puerperium, the temperature being higher and the pulse slower. Too high a temperature or too rapid a pulse indicates fever. (It used to be thought there was such a thing as 'milk fever' with lactation somehow causing the rise in temperature. This is now dismissed as untrue but probably, in the past, it was the first indication of bacteria-induced puerperal fever.) Bacteria always do reach the uterus within forty-eight to seventy-two hours (quicker if the labour has been long) but they tend to get washed out by the blood being discharged from the shrinking uterus.

At birth the mother's weight naturally drops as she is relieved of the baby, placenta and fluid. Average weight loss caused by these three is 13 lbs (5·9 kgs.). In the next few weeks average extra weight loss is 5 lbs (2·3 kgs.) (these averages are more meaningless than most, as women can gain anything from 0 to 60 lbs (27 kgs.) in weight during a normal pregnancy and therefore have more or less to lose). Coupled with the increased activity of the bladder is the decreased activity of the colon. Defaecation is likely to be delayed for a few days. Despite the flood from the bladder, the mother of a new baby is usually dehydrated, and so drinking is important. In general, provided everything is going satisfactorily, women feel physically well during the puerperium.

A word now about words. Keen students of awkward phrases will have been quick to observe examples hereabouts, exemplified by women 'who are having a child not their first' or 'any pregnancy that is not the first one'. Medical literature uses instead some excellent and convenient words which I have been careful to resist. They are undoubtedly handy terms but ponderous, or so I considered, for the uninitiated. All women having their first pregnancies are primigravidae. For their second and successive pregnancies they are multigravidae. Birth itself is parturition with the woman being parturient at birth. After birth she is a primipara, and her uterus, etc., can be called parous. After her second child she is a multipara, and a group of such women are multiparae. After each birth they and their organs are postpartum. The puerperium and puerperal fever have already been encountered. It was sentences like: 'The multigravid thirty-eight-year-old suffered postpartum haemorrhage though when primiparous her puerperium had been . . .' that I did not wish others to encounter, however brief, accurate and to the point.

The role of the father during labour is improving. Many hospitals now welcome his presence during birth, and he is no longer such a music-hall butt. Husbands can get in the way, become distraught, urge the wrong action, faint, scream, and yet be the one person the labouring woman – if not the obstetrical staff – wishes to have nearby. There is even argument, and evidence, that he may suffer during his wife's pregnancy from symptoms which resemble hers. An article in *Discovery* quoted work which suggested that one in nine fathers suffered from minor ailments without any good physical cause during the time of their wives' pregnancies, and more so than men whose wives were not pregnant.

Known as the Couvade syndrome, from the French verb meaning to brood or hatch, the idea of a father actually suffering has a venerable history. Elizabethans quoted it, and there have been some very improbable events throughout the centuries, even to the extent of a man's abdomen swelling hugely and inexplicably to subside when he has seen his newborn child. Male lactation is also on record, with there even being a story about one man in a lifeboat who succeeded in producing milk for a baby also on board. Most paternal symptoms, according to Conlon and Trethowan, who made an investigation, reach their peak at the third month. It used to be thought, in some places, that witches or midwives could transfer labour pains from the woman to the husband. In others it was made more certain that the husband should feel pain by hanging him by his heels in the next room. Without doubt the present trend of the husband being in the same room, and helping more positively, is on the increase.

Ergot Ergometrine, often a useful drug today for the delivered uterus, deserves amplification. The history of this drug goes back into the past, and to a time when, somehow or other, its properties were accurately divined. (Many of the old remedies were undoubtedly nonsensical, but embedded among traditional folk-lore are such gems of knowledge that they make one wonder about the alleged nonsensicality of the rest.) No one knows how or when, but it was learned that ergot could be useful during the delivery of a child.

In 1808 a New York doctor, John Stearns, wrote that he had

been repeatedly importuned by a local midwife to find diseased heads of rye, and administer them to labouring women. He did so, and described the amazing results. Long before this old midwives' tale was thus reported, and subsequently made use of by the medical profession, diseased rye and ergot had had a powerful history of their own. Ergot is a fungus which preys on grasses, particularly rye. If the small black spurs it causes are mixed with the flour and made into bread there can be, and have been, great epidemics. The disease of ergot is a painful gangrene of the extremities. Fingers, toes and tips of ears rot away as the ergot disrupts the blood flow to those parts. Known as St Anthony's Fire (after the third-century Egyptian saint) there were major outbreaks in the Middle Ages, notably in France and Germany. A pilgrimage to St Anthony's shrine did have the secondary advantage of moving the patient, probably, into an area of unaffected rye and St Anthony would probably be granted credit for any cure. Ergotism reappeared in Russia in the last century, and it was even reported in this century in Manchester (1928) and in France (1951), although there is argument concerning the true nature of these two outbreaks.

Anyway back to 1808 and Dr Stearns. The magic stuff was used both to hasten labour and prevent undue bleeding afterwards, but it was abused as well, and frequently brought death rather than birth. The dosage was often grossly wrong. As soon as chemical technology was sufficiently developed, ergot was subjected to intense scrutiny. One by one its powerful alkaloid constituents were extracted – ergotoxine (1906), ergotamine (1918), ergometrine (1935). The first was later found to be a mixture, the second proved to be too slow in its effects, but the third, ergometrine – called ergonovine in the United States – is a powerful tool of the obstetrician. According to J. Chassar Moir, Professor of Obstetrics and Gynaecology at Oxford, who did the first crucial experiments with ergometrine, the drug probably provides the best single method of checking bleeding after birth, and of curtailing the third stage of labour. The ancient black heads of rye have finally found their proper place.

Who has babies?   In Britain some girls start before they have reached their teens, and some finish when in their fifties. At the lower end the age of twelve is usually the minimum. In England and Wales, for example, in 1980 nine babies were born to eleven-year-old mothers. Girls of twelve had 8 babies in the same year, girls of thirteen had 18, girls of fourteen had 191, and girls of fifteen had 1,056. Marriage for girls in Britain is permitted after their sixteenth birthday, but only 9% of British babies are born to girls who are teenagers (or less). In the ten years between reaching twenty and leaving twenty-nine girls produce 64% of their babies. In the next ten years, aged thirty to thirty-nine, they produce 25%. In the years after forty only 1% are born. The span is a short one. At fifteen it is considered too early. At forty-five it is, almost always, too late. With women living, on average, over seventy years the fertile span is therefore less than half their life time. With custom, tradition and fertility difficulties all having their say, the actual span is even shorter. In the fifteen years between the ages of twenty and thirty-five, women produce well over four-fifths (83·8%) of the babies.

Although only 1% of British babies have mothers over forty (the proportion was 3% in the 1960's), and although women over forty-five consider rightly that their fertile days are probably over, there have been some remarkable cases. Phyllis Keeble of Glamorgan, seven times a grandmother, and with her eldest child aged thirty-one, was a mother again for the eleventh time in October 1965. She was then fifty-three. Beating her into the position of oldest British mother is Winifred Wilson, of Cheshire. She was fifty-five when, in 1937, she had her tenth child. (It is easier to have later children if pregnancies have been frequent, although Mrs Keeble had not had a baby for sixteen years when she almost broke the record.) The world record seems to be held by Ruth Kistler, of Portland, Oregon. She was fifty-seven when she produced a baby in 1956. Men, on the other hand, can be, and have been fertile well into their eighties, or to the end of their days. No man seems, so far as I could discover, to hold the position of world's oldest father as a record-breaking consort for Mrs Kistler of Oregon. The ancient inhabitants of Georgia in the Soviet Union make the boldest claims although the Bible is equally bold on occasion. Sarah was ninety when she conceived Isaac, Abraham being a hundred when the child was born. Sarah died when she reached a hundred and twenty-seven, but Abraham carried on until he was a hundred and seventy-five, making his story more than a match even for most Georgians.

More than half of British babies are born to marriages less than five years old. And four-fifths are born in the first ten years of marriage. The rest are either born later or – as 12% are born to no marriage at all – earlier. Despite the numbers of pregnant brides ('it was nice seeing the three of you coming down the aisle' is inevitably shouted by the inevitable drunk), some 80,000 illegitimate babies are born in England and Wales every year. London has twice the national average. The biggest change in recent years has been in the number of babies born to teenage mothers. Back in 1971 the birth rate (per 1,000 girls/women aged sixteen to nineteen) was 63·1. By 1975 this had dropped to 45·1 and by 1980 to 38·1, and is now at its lowest level since 1935. This considerable fall can only partly be explained by an increase in the abortion rate for the same age group which, from 1971 to 1980, rose from 19·2 (per 1,000) to 24·7. The peak birth period of the year for all babies is summertime, from April to September, with the conceptions having taken place from July to December. The last quarter of the year sees fewest births, with November having fewest of all. The annual birth graph, starting from the November minimum, rises steadily until March. It stays fairly constant until October and then declines again.

British people conceive, therefore, somewhat in tune with the seasons, with more conceptions occurring during the second half of the year. That second half is when most people take their annual holidays, it is warmer, more money is circulating, the days are wetter, and anybody is entitled to choose his reason why British people conceive more from July 1, through the holiday period, through the start of the academic year, past the beginning of the shooting season, and on to the Christmas period, before the start of the New Year brings this enthusiasm for conception to a slightly lower level.

14. Lactation

Human milk/Breast-feeding/The price of milk/Polythelia/
Keeping abreast/Weaning

Human milk is usually bluish-white, tastes sweet, and weighs slightly more than water. Essentially, like all milks, it is an emulsion of fat globules (which make it white) in a fluid. The protein adds the bluish colouration. Milk production is usually greatest from those breasts which have increased most in size during pregnancy (although all rules are likely to be broken here as elsewhere) and large breasts which do not grow are generally not good milk producers. Breasts can either fail to grow at all during pregnancy, or each increase in size by 800 millilitres (a total volume increase of almost 3 pints). Average increase for each breast is 200 millilitres. For a given breast enlargement the girl having her firstborn will be able to produce more milk. Although there is usually rapid enlargement of breasts during the first two months of pregnancy, this is an increase in the size of blood vessels. After the third month the breast tissue itself begins to grow.

The content of human milk fluctuates from person to person, from birth to birth, and from the start to the finish of the feeding period, whether it is weeks, months or years. Average composition is 1% to 2% protein, 3% to 5% fat, 6·5% to 8% carbohydrate, and 0·2% of salts. The rest is water. Cows' milk has double the protein (goats, ewes and sows have even more), the same fat proportion (although different proportions for the different fats), and less carbohydrate (4·9%). Cows also have more thiamin, more riboflavin, four times more calcium, less iron and less vitamin C. Human milk will also contain alcohol if this is drunk sufficiently by the mother.

Breast-feeding in the world is extremely dependent upon custom and need. It varies from 1% to 2% (as in certain maternity units in the United States or Canada) to 99% or above (as in poor Bengali villages). The percentages can be reversed. In Calcutta 18% of well-to-do mothers fail to start lactation. In Minneapolis, after an intensive pro-breast-feeding campaign, 96% of mothers were doing so at the end of the second month, and 84% after six months. Maternal milk, although an admirable food initially, is

187

not so for all time, even if the supply is maintained. Beyond nine months or so the human child needs other foods to supplement it. (Many mammals are breast-fed for much longer: orang-utan, three to four years; some chimpanzees, two years; Macaque monkey, eighteen months; baboon, one year.) Malnutrition and prolonged breast-feeding in humans often go together. So does the belief, now confirmed by the reproduction biologists, that the prolongation of breast-feeding will deter another conception. 'Breast feeding still prevents more pregnancies in developing countries than any other form of contraception,' according to Roger Short. However, a nursing mother needs another 1,000 calories herself per day.

Ample milk production does not begin the moment a child is born. Initially the breasts produce only colostrum, a pale yellow liquid which can be expressed from the nipples long before birth. It has a little protein, and a little sugar but quite a lot of fat. In some species, such as the cow, horse and goat, valuable antibodies which could not get through the placenta are contained in the colostrum. They help to immunise the new born against disease. This boost to the colostrum is less important with humans as the placenta is less of a barrier.

Real milk production only begins after the hormones have had a chance to adjust to the post-natal situation, and proper flow usually starts on the third day. With women having their first baby everything tends to occur both later, and less gently. A new baby fasts after birth, thus permitting the delayed milk production to be on time. In one study of babies who were given as much as they would take, the first day's consumption was 25 ml per kilogram of birth weight, and even the second day's was only 50. When between eight and thirty days old they were taking 200 ml per kg, or eight times the first day's intake. The fasting inevitably leads to weight reduction, both by water loss and the breakdown of tissue. Weight loss for healthy babies is often one-tenth of the birth weight, or 11 ozs (312 gms.) of the original 7 lbs (3·2 kgs.). The human baby is remarkable for its tolerance, and many animals would perish if not more squarely nourished during the first vital days (piglets start suckling before the last of the litter is born).

During pregnancy the nipples enlarge and become more supple. The dusky ring around each of them, the areola, becomes both duskier (certainly during the first pregnancy) and larger, its diameter becoming half as wide again as it had been.

Breast-feeding can either fail to start, be suppressed artificially (by the injection of oestrogen) or continue for years. A. C. Haddon observed in 1908 that Eastern Islanders of the Torres Straits sometimes continued a single lactation for three years, and to a time when, presumably, an offspring was perfectly capable of intelligent speech and asking for it. Milk production itself can either be nil or overflowing in its abundance. In earlier times, when alternatives to human milk were scarce, dangerous or even non-existent, the wet-nurse was much in demand when the mother's milk failed to run. Other members of the family often supplied the need. David Livingstone, the exploring missionary, observed grandmothers frequently fulfilling this role in Africa. Elsewhere, as with Maoris and American Indians, the grandmothers help out because mothers are too busy on other tasks.

In the past, and before the arrival of suitable substitutes, wet-nurses could make a fair living from their personal milk production. Even in more recent times there can be, and has been, money in milk. Ronald Illingworth wrote in 1963 about two American ladies, one of whom sold 30,000 ozs in two lactation periods for $3,717, and another who sold her excess after only one offspring for $2,020. The value of their milk was therefore about three times the prevailing price of beer. It was reported from Stockholm in July 1965 that a young Swedish girl named Loborn was receiving ten Swedish crowns for every 2 pints (1 litre) of her excess milk. Mrs Loborn's child consumed $1\frac{1}{3}$ pints (0·75 litres) a day, but there were over 2 pints (1 litre) a day extra for the Orebeo maternity hospital, Stockholm, meaning extra money for the Loborn household. Yet another woman, who must surely hold one of the most useful of records, is said to have been able to keep seven babies happy on her supply. She could produce 10 pints (5·7 litres) a day, or one-third of that of a fair cow. Such stories seem to be less prevalent today, presumably due to the wide variety of milk substitutes currently on the market, and the wet-nurse has finally yielded to the tin.

Whether or not to breast-feed is not so much a perennial argument as the swing of local prejudices and convictions. Primitive places have no choice, and even if they had a choice a feeding bottle in the tropics without hygiene discipline is a lethal weapon. In places with choice the authorities are, in general and in principle, in favour of breast-feeding.

However, ordering a woman to breast-feed is a different matter. She has her own views. An article in the *British Medical Journal* made a formidable list of maternal reasons put forward against breast-feeding. They were an ability to know how much the baby was getting, the fact that bottle-fed babies do just as well, the pointlessness of starting if failure was to follow, the embarrassment of either disliking it or liking it too much, the loss of figure,

189

the nuisance of it particularly with dribbling nipples, and the pain of it associated with bulging breasts, with uterine contractions, or sore nipples, or even an abscess. As the article affirmed, it is a formidable array.

The main advantages and disadvantages can most suitably be listed. Generally accepted benefits are:

1. It is the right food in the right proportion.
2. It is cheap.
3. It is simple, readily available and correctly warm.
4. It assists in the healthy regression of the uterus, and recovery from childbirth.
5. It contains some antibodies (e.g. against the common cold) and helps to keep various illnesses at bay for longer.
6. It encourages (if all goes well) the bond between mother and child.
7. It is reasonably sterile and leads to far less gastroenteritis.
8. It will mean less soreness around the baby's bottom, even though faeces are looser.

Disadvantages are:

1. It is often a struggle to establish. Bottles are temptingly near at hand.
2. The baby cannot be fed without the mother.
3. Some breasts just do not produce enough at the best of times.
4. Worry, etc., can stop even a good flow.
5. Breast and nipples can be painful.

Despite the benefits, despite the simplicity and naturalness of it, despite the gentle prodding of the medical profession in favour of it, breast-feeding in Europe is still not common, but there is a current enthusiasm for it that was lacking in, say, the 1950's. However, the fall-off rate of those who start and then stop early is still considerable. In the United States two out of five women give their babies a chance to suckle. In Britain the proportion varies widely, and a recent attempt to sum up the situation said only that 'it is still quite commonly practised'. It is the minority rather than the majority who establish it. There have been huge follow-up surveys of breast and bottle-fed babies, but no lasting differences have been established. There is much talk about psychiatric benefits, both to mother and child, but they are hard to prove.

There is no comparison between the two methods in countries where hygiene is regarded casually. A milk bottle, unboiled and unwashed, can harbour pathogenic bacteria to a devastating degree. The milk itself may not be good even straight from the cow, or whatever animal is milked in that area. (Goat, ass, camel, llama, caribou, dog, horse, reindeer, sheep, water buffalo and cow all contribute their milk to humanity – somewhere.) Human milk, although sterile by comparison with anything that has been standing in a bottle, is not completely so. One survey showed that 93% of babies were drinking in staphylococci with their milk. Differences in the nature of cow and human milk mean a different population of bacteria in the baby's colon. Breast-feeding leads mainly to *Bacillus acidophilus*, but cow's milk leads to *Escherichia coli*. Without any shadow of doubt, the second type of colon is more likely to suffer from gastroenteritis. Even in Britain at the turn of the century the mortality of bottle-fed babies was far, far higher. The reliability and cleanliness of cow's milk is quite a recent advance, and it took huge strides between the last two world wars.

With maternal milk being such a vital commodity for previous generations, the purveyors of notions for stepping up its production had their field-day. Illingworth has amassed a few of the recommendations. They include the eating of powdered earthworms, dried goat udder, cuttle-fish soup, shrimps' heads, boiled sea-slugs, powdered silk worms in wine, and blow-fly larvae garnished with wine from glutinous rice. A more physical recommendation was tickling trout, forcing expressed milk into the jaws, and then setting the fish free. Equally physical, but demanding less sleight of hand, was the practice of a husband providing breast stimulation by sucking it himself. The so-called 'witch's milk' has nothing to do with any of these vintage recipes. Instead it is the secretion produced from infant breasts. Every foetus encounters many maternal hormones. Some of these stimulate breast growth, both in males and females because breasts are present in both, and at birth the small breasts may actually be secreting this witch's milk. The event is common.

Two final points about cow's milk concern allergy and ionising radiation. For some reason, or reasons, many children are intolerant of cow's milk. It has even been suggested that a few inexplicable infant deaths may have been caused by extreme sensitivity to some cow milk protein. These days the problem of allergy to milk is usually circumnavigated by finding some suitable, synthetic milk-substitute. Getting around ionising radiation is less easy. Fall-out falls everywhere, and did so particularly during the late 1950's and early 1960's when the three major nuclear powers were busy testing their new devices in the atmosphere. Cows, by skimming off the grassy top layer of any field, were also skimming off much of the strontium-90 which had descended upon it. Strontium and calcium are similar chemicals. Consequently, as milk is rich in calcium, it is also rich in strontium-90 when nuclear fission has added it to the grass diet. Cows' milk, due to their grazing technique of food gathering, is several times more radioactive than human milk. On the other hand humans are more adept at collecting, for example, insecticides, the chemicals being frequently lavished on foods destined for humanity. It has been said, notably in the United States, that human milk could violate local laws concerning contaminated substances if attempts were made to transport it across state lines.

Polythelia Quite a frequent abnormality is the possession of more than two breasts. They occur on what is known as the mammary line, and are said (by some) to affect 1% of the human population. This mammary line exists early in embryonic life. At seven weeks from conception it is clearly recognisable, but almost all of it then disappears leaving only a small spot of tissue on each side in the chest area. These spots become male and female breasts. Should other nipples develop they occur somewhere along the old line. If superimposed on an adult human, the mammary line resembles the traditional shape of a vase. On each side, the lip of the vase is at the armpits. The lines then travel down through the conventional breast positions, and nearly meet in the genital area. Extra or accessory nipples are somewhere along these lines, a condition known as polythelia. Should they develop into something more breast-like than mere rudiments the condition is called polymastia. Frequent sites are within or near the armpit. An Egyptian doctor in Cairo reported that they were 'common' there among Greek, Armenian and Lebanese inhabitants, less so among native Egyptians. He said that males had them more frequently, and the usual spot was $3\frac{1}{2}$ ins. (8·9 cms.) below the normal nipples. They were generally single, slightly hairy and only occasionally accompanied by a partner on the other side.

In a sense, bearing in mind the evolutionary complexity of human history, and the wide assortment of mammary arrangements among other mammals, it is reasonable to expect the human species to have occasional variations. Most primates (monkeys, apes) have just one pair of milk glands, but an insectivore called *Centetes* has eleven pairs. Usually the total is related to litter size, as with pigs, dogs and cats. Animals with a small number of glands may have them in the chest area (humans, most other primates, bats, sea-cows, elephants, sloths) or at the other end, the inguinal area (whales, ungulates – the huge hoofed group, which includes cows, horses, hippos, giraffes, rhinos, etc. etc.) or at each end (rodents) or in a row joining both ends (cats, dogs, pigs).

There are plenty of oddities. Cats have four on each side, but they do not all necessarily function at the same time, with perhaps the front two being active, the back two quiet. Bats, hanging upside down and with their single pair at the chest – or bottom end – have a pair of false teats at the other end (called pubic teats) from which the young can hang more conveniently. Whales have a single pair; each may be 6 ft. (1·8 m.) wide and 1 ft. (30·5 cms.) thick during lactation but, as with various rodents and the mole, the whale's nipples may be far from the glands.

All mammals have either nipples (as with humans) or teats (as with cows) except the monotremes (such as the duck-billed platypus). Their exuded milk merely flows on to the surface through scattered pores, and has to be licked off the hairy body by the young. The coypu, the South American aquatic rodent which invaded East Anglia so overwhelmingly, has a lateral system. The young travel on the mother's back as she swims, and the nipples are conveniently placed more towards each side of the normal ventral position. All in all, as mammals are one big related group, it seems not too unreasonable that the species man should every now and then produce something other than the traditional pair.

The traditional pair has been of dominant interest to countless humans, notably in this twentieth century. Somewhere along the line the importance of breasts as a source of natural food suddenly adopted second place to their importance as stimulants of sexual appetite. Babies have been denied the comfort of a breast for fear that the breast's shape may subsequently not be such a comfort to the mother. And some mammary girls have reached pinnacles of success solely on account of their own glandular gifts. Dr Erwin O. Strassman, of Houston, Texas, concluded after a survey of 717 childless women that, as a rule, the bigger the breasts the smaller the IQ.

All of this interest in one organ is not, of course, entirely arbitrary; nor was it chance which caused such a rumpus when creators of fashion thought that toplessness in places other than beaches or restaurants might catch on in 1964. And it was not guesswork when many predicted the failure of this attempt. The breasts, wrote Dr C. B. Goodhart of Caius College, Cambridge, do form an important part of a woman's biological equipment for courtship: 'It is a question not so much of morals as of tactics to

193

consider at what stage in the proceedings they are to be deployed to the best advantage.'

Innumerable societies other than our own use the woman's breasts both before and during intercourse, but few insist as we do in the Western world upon their concealment. Where clothes of any form are worn, the genitals are always hidden. Not so the breasts. As C. S. Ford and F. A. Beach put it, 'the bare bosom is usually not inconsistent with ideals of feminine modesty'. These authors also amassed different preferences for the differing breast shapes. Whereas nine of the societies in their investigation just wanted large breasts, two wanted them to be upright and hemi-spherical, and two (both in East Africa) wanted them long and pendulous.

Science has no kind of an answer as yet for the visible prominence of areolae, the dark circular backgrounds to the nipples. Admittedly they contain glands (of Montgomery, named after a nineteenth-century Irish obstetrician) which, by secreting a fatty substance, protect the area against the trauma of suckling, but their sign-post visibility in light-skinned people is without any explanation. One suggestion is that babies, fumbling their way through life and possessing only hazy sight, need all the assistance they can get to find that vital nipple. Certainly the areolae darken during pregnancy, but babies with black mothers, and also white babies in the dark, seem to find their way equally successfully.

Weaning begins after a few days, or after a year or two. Once again, the subject is full of controversy. Naturally the child should be weaned if the milk is drying up, but in any case by the time he weighs 16 lbs (7·25 kgs.) he is taking (at the normal rate of $2\frac{1}{2}$ ozs. a day per lb. some 40 ozs. (1·2 kgs.) of milk in twenty-four hours. Mixed diet is then considered satisfactory. Weaning is often a time of intense mortality. In West Africa the children suddenly removed from reasonably sterile milk and fed with quite unreasonably pathogenic adult food get gastroenteritis and die in large numbers. Weaning is often upsetting to mothers, partly as the breasts may hurt, but also because the break can hurt emotionally. Babies can find the process upsetting if the previous regimen of warm milk flowing from the nipple was entirely satisfactory, and powerful methods have in the past been used to make them think again, particularly when weaning has been long delayed. Once again Illingworth has a list. Breast deterrents include bitter sap, wrapping nipples in hair, or putting on tobacco, soot, aloes, garlic, ginger, and red pepper.

A final point is that no animal, despite the birth of singletons in so many species, has just one breast. For this the mothers of twins can be grateful. A satisfactory sight is of a well-organised arrangement when a pair of twins, both with their heads supported by their mother's hands, with their bodies gripped by her elbows, and their legs sticking out behind, are contentedly sucking, one to each, at the pair of breasts before them.

15. Twins and Abnormalities

Occurrence of twins/Their value/Multiple births/Twin diagnosis/Conjoined twins/Proof of separation/Why twins at all?/Congenital malformations/Their registration/The principal afflictions/German measles/Contergan and Distaval/The cause of malformations/'Maternal impressions'

Occurrence and characteristics Twins occur in Britain about once in every eighty-four births, taking into account both the identical and the dissimilar types. Many other countries have more, e.g. Belgium one in fifty-six, and many less, e.g. one in 125 or fewer still in some South American countries and among certain groups of Mongoloids. In the United States blacks produce twins more often than whites, and many African tribes have particularly high ratios. (The Yoruba of Western Nigeria are said to produce twins in every twenty-two pregnancies.) Together twins weigh more than a singleton, but not twice as much; the average British weight is 5½ lbs (2·4 kgs.) each. The maximum live born recorded weight for twins is 27 lbs 12 ozs. (12·6 kgs.). The maximum stillborn twin weight is 35 lbs 8 ozs. (16·1 kgs.). Dissimilar twins are nearer in weight to each other at birth than are identical twins, presumably because they have a placenta each instead of having to share one competitively. Twins in Britain are born earlier (on average 248 days after conception), triplets earlier still (233 days), and quadruplets even earlier (223 days). Britain had no successful quins until 1969. Triplets generally each weigh less than each twin, but the total may be more. The maximum total recorded is 23 lbs (10·4 kgs.), while the British maximum is 21 lbs 13ozs. (10 kgs.). A royal bounty for triplets (£3) was introduced in Britain in 1849 but was abolished in 1957.

Twins face greater difficulties from the start, as they die more frequently both as abortions and at or shortly after birth. Maternal mortality is also higher with multiple births. Twins have a higher incidence of mental subnormality and cerebral palsy. There is some evidence to suggest that the larger twin may be the more intelligent, and the second twin to be born is more likely to die at birth. With European groups roughly 63% of all twins are of the same sex, 37% are one boy, one girl. Twins marry less. Fingerprints of identical twins are not completely identical, but have extreme similarity.

Mothers are more likely to bear dissimilar twins as they get older; there are 6 per 1,000 births for women under twenty, and 16 per 1,000 for women aged thirty-five to thirty-nine. The rate then falls off rapidly. Also women are more likely to have twins as they have more children – quite a distinct phenomenon from the age increase. Contrary to dissimilar twinning the chance of identical twins hardly varies with maternal age. Identical twins have the same blood groups, and left-handedness is common among such twins. The difference in the twinning rate between different races and countries is almost entirely confined to the dissimilar twins. The identical twinning rate hardly varies throughout the world, and fluctuates between one in 260 and one in 340 births. In Britain, with a fairly high total twinning rate, the proportion of identicals to dissimilars is about one to three (25%). Conversely, in Japan with a low total rate, the proportion of identicals is far higher, being better than one to one (60%).

Twinning does run in families. Identical twinning does not, but dissimilar twinning does, although without any definite rules. There is evidence that it can come through the female line alone, or from both parents. A pair of dissimilar twins born at the same time can have separate fathers, although this is extremely rare. The rate of dissimilar twinning in Britain has been rising, from one in 117 births between the wars to one in 110 between 1951 and 1956. It is thought that better food is responsible. Certainly the dissimilar twinning rate fell during the war in occupied France, Holland and Norway, but not in Denmark, Britain or Sweden where there was no serious food restriction. Recent reports from Scandinavia suggest that illegitimate offspring are more likely to be twins than the offspring of married parents.

There is much argument that twins should be given a greater chance to go it alone in childhood, and not be identically dressed, etc. If one identical twin is homosexual his brother almost always is. If one dissimilar twin is homosexual his brother is also homosexual in about 50% of cases. Good dogs can distinguish between the scents of identical twins, although there is initial confusion. A bizarre rarity occurs when identical twins definitely from the same egg are born with a different sex. It is thought that faulty cell division has caused the male to be normal (XY) but the female to be immature (XO). The twins should both have been

male. (The chapter on inheritance explains the sexual ramifications of X and Y.)

Nomenclature The first problem with twins is what to call
them. Identical twins are also known as one-egg twins,
monozygotic twins, monovular twins, uniovular twins, similar
twins. Dissimilar twins are also known as non-identical twins,
two-egg twins, dizygotic twins, binovular twins, diovular twins,
fraternal twins. I shall stick to identical and dissimilar. As a
complication it should be remembered that identical twins are
only similar and not the same. Even Siamese twins show marked
differences between each member of the union.

Dissimilar twins are undoubtedly caused by the ovulation and
then the fertilisation of two ova. There is much less certainty
about identical twins. It used to be thought that all of them were
caused when the egg divided into two cells which then mistakenly
separated to form two independent, single-celled embryos. These
then divided, according to the old story, but their subsequent
two-cell stages remained correctly united and eventually grew
into one pair of identical twins. Work in the 1960's suggested
instead that identical twins are formed much later. Dr J. H.
Edwards, of Birmingham, reported to a Genetical Society
meeting in 1966 that only 30% of identical twins are formed
before implantation (and that happens six days after fertilisation);
the remaining 70% are created after implantation.

He made his deductions after noting whether the chorion and
amnion were in fact double or single for each pair of identical
twins. A singleton offspring has only one each of these foetal
membranes, but identical twins were by no means as uniform in
this respect. Dr Edwards' report made it abundantly clear that the
previous belief in just one type of identical twin was entirely false;
so too was the conviction that identical twins were always the
result of a divided cell that failed to remain united. He and others
therefore totally upset the subject, and made it far more
fascinating.

The value of twins Twins are a godsend and of prime
importance to the biologist. They enable him to get to grips with
the nature versus nurture arguments, and whether a characteristic is born or made. With any laboratory animal an experiment
can easily be planned to discover whether, for example, height is
inherited, or a matter of environment, or a complex product of
the two. With humanity always breeding at random and living at
random there is no such simplicity. Is a man's final height due to

inheritance or due to the way in which he was brought up? Richer people in Britain are taller. Is this because richer people give better food to their children, or because taller people are more likely to have taller children – and are more likely, once rich, to make them richer too? Twins often provide an answer to this sort of question, and in two main ways.

Sometimes – and this is the prime benefit – identical twins are not brought up in identical fashion. They are separated at birth, or very soon afterwards, and are brought up in different homes. In other words, their genetics are indisputably similar and their environments are not. A clear line has been drawn between the two. The second point, less of a godsend but also of importance, is that most dissimilar twins are brought up together. In other words their genetics are different but their environment is the same. Both these situations are good substitutes for an ability to breed humans experimentally. Countless statements about humanity would be impossible without the support of a twin study to provide the evidence. Sir Francis Galton (born 1822, died 1911, genius, polymath, cousin of Charles Darwin) was the first person to write about the possibilities of twins as a source of information, and published his first study in 1875. Literature and mythology had long been interested in twins, notably Castor and Pollux (dissimilar, although each had another partner in his egg), Jacob and Esau (dissimilar), Romulus and Remus (probably dissimilar), Viola and Sebastian (dissimilar); yet it was up to Galton to point out the advantage of having twins to study. Neither literature nor mythology is as interested in identical as in dissimilar twins.

Multiple births Once upon a time it was thought that a neat formula existed for the occurrence of multiple twinning. Hellin, a German, said it was a matter of squaring (for triplets) and of cubing (for quads) the actual incidence of twins. He said, bearing in mind that European twins appear roughly once in every eighty births (in Germany then the figure was eighty-nine), that triplets appear once in 80^2 (6,400) births and quadruplets once in 80^3 (512,000) births. There is, for Europe, some sort of validity for this 'law', but it is only a rough guide. In 1980 there were born in Britain 6,308 twins, ninety-one triplets, four sets of quadruplets, and one of quintuplets. Triplets were down, and quads were up by Hellin's law.

The United States always has the complexity, and greater interest biologically, of its two principal races. Blacks are definitely up for all forms of multiple birth. American black rates per million births (with the white rate in parentheses) are twins

13,423 (10,059), triplets 141 (85) and quads 1·8 (1·0). Once again Hellin's law does not fit; triplets are down for both races, and quads are down for whites but absolutely correct for blacks. Some say his law is 82^2 and 82^3 for triplets and quads, or even 87^2 and 87^3. These figures usually fit even less well. The best that can be said for the law, and the reason for mentioning it, is that it is a fair guide, a kind of reminder of multiple birth frequency. Anyone wishing to lay a bet on anyone's chances of producing more than a singleton baby should bear Hellin in mind.

Natural multiple births (formed without any hormonal interference to the mother) above four are very rare indeed. The arrival of four, although a staggering confrontation for the average family, is rare, but does not set the world on fire. Britain has at least one quad birth a year. The United States has one new black set and three white sets of quads every year. Other countries with similar medical abilities have similar figures. The arrival of quins is an extraordinarily rare event and only occurs once in twenty to forty million births. The world can expect to see about two lots a year, but much of the world cannot yet keep them alive. The most famous were the Canadian Dionne girls, who were all alike, and who all came from the same egg. The Argentine quins were, conversely, all different and from five eggs. The United States had a set of quins born in January 1964 in Aberdeen, South Dakota. Their arrival did not just cause an upheaval in the family. It created a complete change as jobs were given up, visitors were prepared for, money poured in, and earnest economists debated on television the overhaul this one multiple birth would inevitably mean to Aberdeen's commerce and financial state.

The stimulation of the ovaries, artificially and by F.S.H. (see fertility), meant many more multiple births and, though many of these multiples died, some have had happy outcomes and therefore form part of the record of successful multiplicity (although one can argue over their 'successfulness' as a singleton was presumably intended). Such a multiple birth was to a twenty-six-year-old New Zealander in 1965. She had been given hormone treatment, and by the twelfth week of gestation her uterus was the size normally reached after twenty weeks. At the eighteenth week an attempt was made, with an electrocardiogram, to count the foetuses. Counting two foetal heart beats, with their slightly different timing, is fairly easy. Counting more is far more difficult, but even the instrument failed with the quins. So one X-ray picture was taken, which showed four foetuses and room for another. The following week the mother was admitted to hospital. By the twenty-sixth week she had a waistline of forty-five inches, but that was a maximum. Labour began at the end of the thirty-third week. It was short, and uncomplicated. A 4 lbs. (1·8 kgs.) male was born first, and head first. After nine minutes, a 3 lbs. 3 ozs. (1·4 kgs.) female was born bottom first. Then, after six more minutes, came a 4 lbs. 3 ozs. (1·9 kgs.) female, again head first. Five minutes later was a 4 lbs. 2 ozs. (1·87 kgs.) female, bottom first. Finally, after four more minutes, was a head-first female of 3 lbs. 7½ ozs. (1·6 kgs.). In short, 19 lbs (8·6 kgs.) of humanity – all five dissimilar – had been born in less than half an hour. All were in good condition, save number three who needed artificial respiration for ten minutes. Having failed to provide her solitary daughter with a sibling for six years, the mother had made up for lost time by arriving home, fifteen days later, with five. It was then only the fourth time anyone had ever done so.

Diagnosing twins has its own problems, particularly when the twins are small. Often it is the mother who first has suspicions. She may know of twins in her family. She may feel larger than normal – and most twins are born to women who have already had the experience of one child. She may complain of discomfort. If the woman is indeed large, and found to have a girth of 40 inches (101.6 cms.) or over, suspicions can increase. They can increase even more if the developing form is felt to have more than a reasonable complement of limbs. A more positive method than this feeling (palpation) is listening to the heart beat (auscultation). Normally, and in a conventional pregnancy, there is just one distinct heart beat. With twins there are two. Either two men with two stethoscopes, or one man with one counting the heart rate in one area and then in another, should be able to state that there are two individuals within, if such is the case. An X-ray picture will be proof, but the current dislike for any extra radiation may weigh against it and the current procedure is to use the technique of ultra-sound. The pictures are generally less clear than those on an X-ray plate, but greatly preferable in minimising damage to the womb's contents. Certainly they are capable of showing up just how many foetuses are actually in residence. Many a midwife delivering twins has been surprised to receive a triplet. Even twins can confound the deliverer by being born at different times. In Texas, for example, Mrs Rita Castro gave birth to a boy on December 9, 1966 and then to a girl on January 8, 1967.

As soon as the probability of twins has been established during the mother's pregnancy she deserves greater care. She also needs more rest. The birth of twins carries, on average, a greater risk than normal; in one out of sixteen twin births at least one twin dies and in one out of seventy-five twin births both die. The statistics for triplets in a recent year show that all three survived in seventy-five out of eighty-seven births. This means that at least one baby died in one out of seven of the triplet births.

Conjoined twins 'But I am not me any more', said Santina Foglia in May 1965, on coming round from the anaesthetic. It was true. She had been separated from her Siamese-twin-sister, Giuseppina, and both were doing well. In 1959 the two Italian girls had been born joined together at the lower end of their spinal cord. For four years they had lived and played as one person. They then had their first operation. Two anuses had to be created, conjoined parts had to be separated, and the final separation came when they were six. It was successful, and they became two people. Three weeks later, even though they had previously fully mastered a cumbersome technique of getting about on their four legs, they had learned again how to stand up. Within a month each took their first independent steps. They had joined a very small band of less than four score people who had been born as less than two score, and who had been successfully separated.

The most famous Siamese twins of all, Chang and Eng, were not strictly Siamese as they had Chinese parents, but they were born in Bangkok in 1811. (The Siamese called them the Chinese twins.) With a shared liver, and joined together at the lower end of the chest bone, they were taken to the United States when eighteen years old. Eng and Chang, meaning Left and Right in Thai, were exhibited by P. T. Barnum in his circus. The world flocked to see them and the name Siamese twins has been indissolubly linked ever since with a conjoined birth (scientists refer instead to parabiotic twins).

Such conjoined twins existed long before Barnum jumped on the band-wagon. In 1100 the Biddenden maids were born in England and lived for thirty-four years joined from their shoulders to their hips. James III of Scotland kept the 'Scottish brothers' at his court. A pair of Hungarian girls, born in 1701, were joined back-to-back, sharing both anus and vagina. One pair of girls, called the Blažek twins and similarly joined, complicated matters still further when one became pregnant. Throughout the pregnancy the other girl continued to menstruate, but both of them produced milk when a healthy and normal boy was born.

Long before Chang and Eng were born attempts at separation surgery had been made, notably by a Basle doctor in 1689.

Not until this century, in Britain in 1912, was the world's first successful separation performed, and in the last twenty years there have been nine separations. With modern surgery, whatever Barnum might have said about the matter (and *Medical World News* suggested he would have sold tickets to the operation), Chang and Eng could probably have been divided successfully. As it was they lived to be farmers in North Carolina, they married two daughters of a clergyman, and they had twenty-two (or twenty-one) normal children. At sixty-two Chang died – of a clot on the brain, it is thought. Eng then died within a few hours of his brother. A post-mortem showed no reason why he should have died, and a doctor at the investigation wrote that the cause of Eng's death was probably 'fright'.

No one knows the extent of parabiotic twinning, or the cause. Some say that there are six such births recorded in any year (rarer than quads, commoner than quins) with a lot dying in neonatal days and a lot being quickly killed, as with many gross malformations. Others have assessed the incidence as high as once in 80,000 births. Some seventy pairs have been described in the medical literature, and more of them are female than male. No conjoined pair has lived so long, either before or since, as Barnum's twins, the famous pair who gave their name for ever to this rare consequence of a single fertilised ovum which divides to form twins, but never quite separates them.

Twin proof Quite apart from the surgical separation of Siamese twins, ordinary twins have been inadvertently separated by hospitals. With babies everywhere the mistake cannot be hard to make (although the modern custom of giving babies identifying wrist-bands as soon as possible must have cut down on the number wrongly assigned). When the error is discovered, perhaps after years, mothers are understandably reluctant to swop children merely to keep the books straight. A famous case of this kind occurred in Switzerland. One day, shortly after World War Two, the father of a pair of twins was told about a boy called Eric (the journals stick to pseudonyms even in the original descriptions). The man's two six-year-old sons, Victor and Pierre, were dissimilar, but Eric was said to be the double of Victor. And Eric had been born in the same hospital on the same day. A professor was informed, and he became convinced that Eric and Victor were identical twins, but Eric's mother refused to entertain any such notion, and clung to the boy. Eric's father was dead. The

203

doctors took blood samples and discovered that Pierre was definitely not the son of the mother he had been living with. Unfortunately, by blood group tests alone, it was not proved impossible for Eric to be the son of the mother clinging to him. Further tests had to be carried out to sort it out.

Sir Archibald McIndoe, well known by then for his wartime prowess with burnt pilots and others desperately in need of plastic surgery, was called in to help. If Eric and Victor were indeed identical twins, they would accept grafts of skin from each other. By the same token Pierre would reject them. Grafts between any two individuals will only take if those two people are identical twins. Sure enough the grafts between Eric and Victor did take. Those between Pierre and Victor did not. Eric and Victor were therefore identical twins. Informed of the powerful evidence operating against her own affections, the foster mother of Eric relinquished him and took her son Pierre in his stead.

Hospitals must make the error of muddling babies from time to time. Only under very rare circumstances, as with the three Swiss boys, will the mistake manifest itself. The rest is silence.

Why twins at all? They complicate pregnancies. They die more frequently than singletons. They cause their mothers to die more often than is the case with normal births. On the other hand, why are there not more twins? Most animals produce more than one offspring at a time (called polytocous, as against monotocous).

The reason, if it can be put so bluntly and teleologically, is a compromise, a result of the various forces of selection. The trouble with litters is that there tends to be intra-uterine competition amongst their number, with the fastest grower being the strongest and the most active at securing a nipple. The smallest piglet may not even find a nipple at once; therefore he may not survive. The selective pressure on a litter is for speed of development, for survival first and foremost over one's brothers and sisters. Even the second human twin is more likely to die than the first, mainly because the placenta detaches itself too early. There is no similar struggle with a singleton. He either makes it, or he does not. He never has to behave like a squealing piglet. He can relax. He can take his time over development, whether born weighing 5 lbs. (2·3 kgs.) or 10 lbs. (4·5 kgs.).

It is this relaxed and leisurely period of development which is important. Young mice are born after nineteen days, having competed as embryos within the womb, to face further competition outside it. The human baby lingers on, comfortably. It is born at about the time when any further growth in size would be unmanageable within the confines of a human mother. Unlike many other singleton deliveries, such as so many of the ungulates which have to run the day they are born, the human baby can continue its relaxed growth. Only at its first birthday does it totter to its feet. Only at two is it running and roughly half its height. Only at twelve does puberty begin. Only at twenty is it full grown. Such unhurried progress, so different from the sub-panic anxiety of imminent shortage with which a piglet sees the world, is a boon to a more advanced developmental process, involving greater learning time and less reliance upon instinct. Being a singleton is a necessary precursor to this lengthy period of maturation.

The lengthened growing period is common to the higher primates. It happens both before birth, and after birth. As Sir Peter Medawar has said, birth itself is a moveable feast. Animals can be born soon and immature (as with mice), late and mature enough to run (as with ungulates), or late and physically mature enough to do nothing but suckle (as with man). The more advanced the primate, the longer the gestation period. Also the more advanced the primate, the lengthier the post-natal growth period. It is two to three years with lemurs, up to seven years with monkeys, eleven years with the great apes, and twenty years with man. This is not because man is bigger than the monkeys. The apes, such as gorillas, are often far bigger than man.

Such elongation of time, growth, and learning could not have happened without the birth of singletons, and in practically all the Anthropoidea single birth is the rule. Twins are the exceptions to this rule. It is not known why they should happen, or, for that matter, why they should not happen more. It is known that they are selected against – to use evolutionary language; they are less likely to survive; the dice are loaded against them, whatever the cause. As it happens these dice are even more heavily biased in many primitive communities. Twins are frequently killed as soon as they show themselves. The woman, it is often said, must have taken another man if she has produced a second child. Or the mother, distressed by the paired abnormality, is shamed into killing them herself. Both the natural pressures of decreased viability, and the superstitions of human societies, have been levelled against the twin. Both must have had their influence upon this minority group. Both have seen to it that 99% of all births, a fraction less here, a fraction more elsewhere, are of the favoured and solitary form, namely singletons.

Inborn errors Nine months is plenty of time for doubts to fester in most maternal minds about the normality of the infant they are incubating. Will all the fingers be there? And toes? And nails on those toes and those fingers? The whole process of foetal development is so bewildering that any mind can be excused if it veers wildly from total confidence to an abject lack of it. Why shouldn't there be five toes on each foot? Why should there? The maternity authorities, books, people and pamphlets, generally attempt to bolster the confidence either by skating round the problem, or by making light of it. Consequently only when the child has been born, and inspected, does an average mother become totally convinced of its normality.

Registration and information   In the past congenital malformations, defined as gross structural defects present at birth, were relatively unimportant. So many babies died from infection that the few who died from such gross defects were statistically swamped. Nowadays, following the medical triumphs over infection, the numbers of babies suffering from malformations are more significant. They form a higher and higher proportion of infant deaths. Simultaneously more and more of them are surviving. At the turn of the century only about one-thirtieth of infant deaths were attributed to malformations. By 1980 the proportion was nearer a quarter. As the causes of these malformations are largely unknown, the proportion is likely to increase still further as babies are steadily prevented from dying for other reasons. There is no immediate promise of any big drop in the numbers of deformed babies being born. At present about two in every hundred are born with some blatant and severe malformation. Another two or so in every hundred are born with defects either less severe or less noticeable until some time has passed.

In all accounts detailing the extent of the malformation problem there is a vagueness about this extent. Surprisingly a major cause of this uncertainty is the gross shortage of information. Only in 1960, in London, did the world hold its first international conference on congenital malformations. Only in 1964 did Britain initiate a system of notification of congenital abnormalities. Until then almost all British figures were extrapolations of facts collected in Birmingham and Scotland. Even now the system does not include children of all ages, and many malformations do not manifest themselves for weeks, months or even years after birth. In the United States there are some areas from which not more than half the malformations are notified to the central authorities. Scandinavia, which has such a good record of low infant mortality, has only recently set up malformation registries. In Finland, for example, reporting of all malformations in the newborn has been obligatory just since the beginning of 1963, and Finland is a pioneer in this field.

Apart from a lack of general registration there are plenty of other causes for lack of facts. In the past no one has cared too much about any physical defects in spontaneous abortions. (Recent work indicates that more than a quarter of these foetuses have detectable abnormalities.) Also in the past, no one has cared too much about stillbirths and their defects. Autopsies on them are rare. Examination of their chromosomes, which could be highly relevant to the malformation problem, was hardly ever done, but is now being done at the major centres. A consistent classification of the various possible defects, plus their varying combinations, is still at a primitive level. Finally, with members of any one family having their babies up and down the country, it is exceptionally hard, if not impossible, for any researcher to discover familial links in any particular defect or set of defects. Lack of good and global information is certainly a dominant difficulty.

Nevertheless, despite the paucity of information, various general points have emerged. A higher standard of living does not reduce the number of malformed offspring. The world's races appear to be similarly afflicted, so far as total numbers are concerned, but each race has its own partiality for certain of the afflictions. Likewise the sexes are equally affected, but certain afflictions beset one sex or the other more frequently (for example, males more often have hare lips). Malformed babies have a poorer chance of surviving than ordinary babies. Normally 95% of ordinary babies alive at the twenty-eighth week of pregnancy will still be alive five years later; if malformed only 50% will be alive by then. With some malformations, such as anencephalus – or virtual absence of brain – death is generally inevitable. However hydrancephalics, whose brain volume is much reduced owing to the presence of excess cerebro-spinal fluid, can be kept alive if a shunt is introduced to prevent the fluid pressure from rising abnormally. Some of these individuals have achieved maturity and, with apparently negligible quantities of cortical tissue, have even achieved university degrees. To call them water-brained, as the Latin name suggests, is plainly not entirely true.

Only about half of malformations are detectable at birth, but signs are almost always apparent by the end of the first year. Some malformations are more likely to be detected at birth or shortly afterwards if the birth is in a hospital. Mongols, for example, are

less frequently reported in home deliveries, but some malformations do not manifest themselves to anyone – however expert –
until very late in life. A few are genetically determined, i.e.
inherited, but most congenital malformations are not. Congenital
means manifest at birth; genetical means determined at
conception. This clear-cut distinction of meaning is blurred by
man's current inability to observe many congenital defects at the
time of birth; but, nevertheless, there is a distinction between
something which has been caused genetically and something
which has not. German measles causes malformations on
occasion, and such causation is not genetical. However, to blur
the issue once again, one individual can, for genetical reasons, be
more prone to suffer from a totally non-genetical agent, such as
German measles.

A final complexity is associated with bad information; different
places and people have such different ideas about recording
defects. Is a mongol child who has died of pneumonia dead
because his mongolism made him more susceptible to the disease,
or not? Similarly what is a malformation? How gross does this
gross structural defect have to be? Plainly the lack of a brain, or of
an anus, or of a fused palate, or of a heart wall are all gross; but are
slightly webbed toes, or the possession of a small sixth digit, or just
the merest fragment of that small sixth digit? Are they gross? All
biological activity resents being neatly pigeon-holed, however
enthusiastic the efforts of the biologists.

Yet certain defects, without any shadow of any doubt, are
congenital malformations. They can be listed, together with their
distinguishing characteristics; so can some of the suspected and
infamous causes, notably German measles and thalidomide.

As a foreword to entering these lists it should be pointed out
that practically everything can and does go wrong with human
development. Take the eyes, for example. There can be no eyes.
Or just the development of a single median eye. Or varying
degrees of fusion of the two. The eyes can be much smaller than
normal. Or there can be distortions of the retina, of the iris, of the
cornea. Or the cornea may be opaque, in one eye or both. The
opacity may be a ring or total. The pupils may be slit-shaped, and
there may even be more than one pupil per eye. There may be far
too much pigment, or next to none as in albinism. The lens may
be minute and spherical. There are also plenty of kinds of
congenital cataract, and there are squints, glaucoma, a total
absence of tears, and blocking of the tear ducts. And so on. There
are countless ways in which some part of some development of
the human system can, and does, go awry. Yet there are far fewer

ways in which development frequently goes wrong. Some errors are common. Some, happily, are extremely rare, possibly recorded once or twice. The body is just more prone to the making of certain errors, or of having such errors inflicted upon it, during development. Of all these eye defects only one is common, and that is congenital cataract.

A further relevant point is the length of time during which a certain area of development is susceptible. As with the manufacture of some engineering structure, each part has a critical time. Assume every builder on a job is drunk for a week. The parts of the job most severely affected will be those they are working on at the time. The parts already completed will be relatively immune from their alcoholic actions; so too the sections still to be begun. It is their handiwork of that particular week that would be most disastrous. So too with human development. Any malevolent influence will be most likely to affect organs in their crucial and critical developmental stage. With the brain it is week two to week eleven. With the eye it is weeks three to eight, the heart weeks two to eight, the fingers and toes weeks four to nine, the teeth weeks six to eleven, the ear weeks six to twelve, the lips weeks four to six, the palate weeks ten to eleven, the abdomen weeks ten to twelve. In pregnancy as a whole it is the first twelve weeks after conception that is the crucial developmental period. The subsequent six months are mainly devoted to growth. Hence, with the few causes of malformations that have been found, such as German measles and thalidomide, their influence has to occur during the first twelve weeks rather than the remaining twenty-six for it to be most destructive.

Principal manifestations Thus a combination of organ susceptibility, of timing, of the powers of the causative agent, of the genetics involved, either directly or indirectly, of the resistance of the foetus and its mother – all such factors and many more must be influential in creating the list of congenital malformations which afflict man – or rather one in fifty or so of his children, or over 20,000 a year in the British Isles, or sixty a day. Of the 656,234 births in England and Wales in 1980 a total of 14,134 malformations 'observable at the time of birth' were notified. Therefore in that year such malformations affected 2·15% of all births, live or still (as against 1·9% in 1966).

The malformations not noticeable at the time of birth expand the total in the weeks, months and years after birth.

Central Nervous System   In Britain this is the commonest site for major malformations. The three principal deformities are anencephaly, spina bifida and hydrocephalus. One or other of them, or a combination of them, occurs once in every 150 births. First anencephaly. It mainly strikes female babies and it is always fatal, either before birth or very shortly after. It involves the major lack of brain tissue. From one in 500 to one in 2,000 babies are born with it in European communities.

The main malformation at the other end of the spinal column is spina bifida, where the skin often fails to cover the spinal cord completely. There are various forms, some invariably fatal, particularly when coupled with other malformations. In Britain its liveborn incidence, 750 babies a year, is more frequent than anencephaly. Even ten years earlier the incidence used to be about twice as high, and this considerable fall is also partnered, although less dramatically, by a fall in the stillbirth rate of those afflicted with spina bifida. In other words fewer such malformations are being formed as it is unlikely that the declining incidence can be explained by an increase in the spontaneous abortion rate of such defects. A better supply of vitamin A to the mothers is currently believed to be responsible for a general drop in the levels of CNS malformations, and during the 1980's the Medical Research Council undertook a major survey in an attempt to prove this point.

Thirdly, slightly less common but often found associated with spina bifida, is hydrocephalus. Sometimes called water on the brain, it is an excessive accumulation of liquid within the cranial cavity. An abnormally large head, either at birth or within the first few months of life, is the most obvious sign. It is more frequently fatal if in association with other nervous malformations.

All three of these most unpleasant of malformations vary in incidence according to social class, and age of mother, and geographical area. Information from Africa is slight, but good evidence from Uganda shows a total lack of anencephaly and spina bifida. North-west Europe is particularly prone to these two malformations, but North-west Europeans are less prone if they have emigrated elsewhere.

A totally different defect, but one with undeniable effects upon the central nervous system, is mongolism. In 1866 this unfortunate name was given to the one in 600 babies born who are often congenital idiots, and whose eyes bear a very superficial resemblance to Asiatic eyes. It was later discovered that a mongol's defects are linked to the possession of an extra chromosome. Asia in general has reasonably resented the name of mongolism, and continual effort is being made to replace the well-entrenched term by the more medical description of Down's syndrome (named after the nineteenth-century English physician Langdon Down who described the condition).

Such babies are usually born small. They are recognisable either at or shortly after birth. They occur in all races (and certainly among the Chinese where they are called white idiots). Their hands are stubby; their little fingers particularly so. Their palm prints have unique characteristics. Many of their organs, apart from their brains, are poorly developed, and they are much more likely to be born to older mothers. Their eyes do slant, but for quite a different physical reason from the normal Asiatic slant: the distinction is particularly important for the offspring of marriages with some Asiatic blood in them. The raising of mongol children is beset with problems, but the happy nature of many of these children can provide considerable compensation. By no means do they all have extremely low I.Q.'s: there is great variation in intelligence. Two-thirds of all mongol children are still alive five years later, but only one-quarter if they have other defects as well. The incidence of mongolism rises with maternal age; if mothers are twenty the risk is 1 in 3,000, if over forty the risk is 1 in 100, and may be even higher for mothers over forty-five. However, owing to the greater numbers of babies in general born to younger mothers, most mongols are born to mothers under thirty-five. If mongols have children there is about a 50:50 chance (although evidence is scarce) that their children will also be mongols.

Circulatory System   After the central nervous system, this is the second commonest site for major malformations in Northern Europe and White America. In Britain about 4,000 babies a year are born with heart defects. Perhaps half of them are discovered at the time of birth. The remainder are generally discovered before the first birthday, but some take years to manifest themselves.

Briefly, the defects can be holes between the heart's chambers, anomalies in the big blood vessels, or the failure of the foetal circulation of blood to change fully into the post-natal type, a change necessitated by oxygen coming from the lungs rather than the umbilical cord. Surgery is making great advances among the very young, but congenital heart defects still cause many deaths. In a fourteen year survey, 37% of babies possessing such defects were dead in a week, and 54% in a month. The causes of heart defects are about as little understood as brain and cord defects, but German measles is definitely to blame for some of them,

211

although how and why is a different matter. One-quarter of babies born with congenital heart defects will be alive five years later, but only 15% if there are other defects as well.

Cleft Palate and Lip   About two in 1,000 of all babies in Europe are born with one of the three types of this deformity. Hare lip (called Group I) accounts for 20% of them; cleft palate (Group II) for 46%; and combined cleft palate, gum and lip (Group III) for 34%. Sometimes the lip clefts are two-sided; but, if one-sided, the left side is the commoner. Fathers of all these children are older than average, and mothers are older than average for Group III. Birth weights of all groups are low, and about one-fifth of the victims have another kind of malformation as well. When a cleft lip/palate defect is in association with other malformations, then death is the rule within a few months. If there is only a cleft lip/palate defect, then death is rare, and 87% are alive five years later. (This repetition of viability up to five does not indicate that the fifth birthday is a turning point. It is just that five years is a traditional follow-up time, both for congenital malformations and for diseases like cancer.)

Cleft palates and lips cry out for surgery, and there is a great deal that surgery can do. Group III is most difficult, and repair work is generally begun six to eight weeks after birth. As Group II infants can swallow normally, although they cannot suckle, there is less urgency. The cleft palate is usually repaired between the first and second birthdays. Group I babies can both swallow and suckle, and therefore treatment need not start until they are fully weaned. Because speech is so important, and because poor speech can lead to poor reading, and because poor reading can mean poor progress at school, palate and lip repair is carried out, if possible, before speech begins. As one surgeon has put it, treatment should allow these infants to speak well, look well, and eat well, and in that order. Most children will develop normal speech if palate repair is done in time. If the operation is not done until they are, say, ten then most will not develop normal speech, however much the speech therapist may try to get the right sounds out of a mis-shapen instrument (although orthodontic appliances can help). Good speech also requires good hearing, and many cleft palate babies have poor hearing as well. The possibility of this double handicap should always be suspected.

Inheritance plays a part in cleft lips and palates. If normal parents have one child with a hare-lip the chance that the next baby will have one is 4%. If two children have them, the chance for the next baby goes up to 9%. If one of the parents has a hare lip also, and one baby has one, the chance for the next goes up to 17%. With cleft palates, which afflict females more than males, the odds are different. With both parents normal, and one cleft palate baby, the chance for the next baby also having one is only 2%. If both one parent and one child have cleft palates that chance goes up to 15%.

Talipes (Latin for ankle-foot)   In European communities about four babies in 1,000 are born with some form of club foot. Correction of the defect is begun very early, in fact almost immediately after birth. There are many methods, and the results are usually good. About a quarter of talipes babies, or one in 1,000 of all babies, have both talipes and some other malformation. There is, once again, a genetic link for this deformity. One deformed baby means a greater chance than normal of its siblings being similarly deformed, and greater still if the parent was a victim.

Pyloric Stenosis   About three babies in 1,000 in North-west Europe are either born with the stomach blocked at the exit end, or more frequently develop such a blockage in the first few weeks after birth. It is five times more common in boys than girls. It was often lethal before 1920, but nowadays surgery is nearly always effective. The alimentary canal can also, more rarely, be blocked elsewhere along its length. A fairly frequent condition is a blockage at the very end, called an imperforate anus. About 500 British babies a year are born with it, and skilful surgery is necessary.

Congenital Dislocation of the Hip   It affects one baby in 1,500, generally girls, by a ratio of six to one. It is commoner in breech births than head births (because of the lie and not of the birth) and among winter babies than summer babies. It is commonest in Northern Italy, among Lapps and Red Indians (making one think of tight swaddling). It is usually diagnosed by the so-called Click Test within the first week after birth. The test is negative where there is no deformity, and all positive cases should be treated within the first few weeks after birth.

So much for six groupings of congenital malformation. Though only six, about 90% of all malformations come under these headings. The remaining 10%, either more or less for different areas, include every other type of congenital defect, all those

213

defects of the eye already mentioned, all known gross errors of development. Very briefly a few of these others are:

Congenital deafness. Affects one in 3,000. Often caused by German measles.

Achondroplasia. Affects one in 40,000. Short, thick-set, square-headed dwarfs. Intelligence and fitness good. Ability to breed poor. Caesarians imperative when they give birth.

Polydactyly. Extra fingers. Affects one in 1,000 in Europe, more in Africa.

Accessory auricles. Additions to the external ear. Easily removed surgically. Conversely the external ear may be absent, and can be added by plastic surgery.

Icthyosis. Large group of anomalies characterised by scaly skin and dryness, with or without reduction or complete absence of hair follicles and/or sweat glands.

Various concluding generalisations are possible. Most races suffer a similar number of malformations. Most common sites for malformations are the nervous and circulatory systems. Most malformations are readily detectable at birth or shortly afterwards. Most hamper chances of life. Most causes are unknown. Inheritance is involved in many of them. Most malformations which commonly occur can be classified under a few headings. Most other types of malformation are extremely rare. Many of them appear in association with other malformations. Such associations are most likely to be fatal. There is little or nothing that most mothers can do to offset malformations. There is much that surgery can do. Most babies that die at birth or very soon afterwards are dying not because of malformations, but the proportion of malformation deaths is rising as other neonatal causes are attacked, and controlled. Many of the congenital anomalies regularly fluctuate in their incidence according to the season.

German measles   Now to the two most potent malformation stories of recent years. First rubella, alias German measles. Rubella is a sort of translation, or rendering, of the original German term rötheln, both words indicating redness. Confusingly, ordinary measles is called rubeola. It was in the middle of World War Two that a paper linking German measles with malformed babies was published in Australia by Dr N. McA. Gregg (later Sir Norman). He was an eye specialist and so saw many children with congenital cataract. His brilliance lay in also realising, after a German measles epidemic, that many of their mothers had contracted German measles during their pregnancies. The startling figures were: 78 cataract children and 68 German measles mothers. Once he had put two and two together the rest of the world rapidly confirmed his opinion.

The catching of German measles early in pregnancy somehow causes an offspring to possess malformations more frequently than normal. Apart from cataract, these are principally deafness and heart defects. The earlier the measles is caught during pregnancy the worse it is for the foetus within the womb. There have been countless surveys which emphasise this point. A recent one gave the malformation risk as 60% if rubella is contracted during the four weeks following the last menstrual period, 35% in the next four weeks, 15% in the next four, and 7% in the next four. So the first three-month period, or trimester, is the crucial one, and the very first month the most crucial of all.

Such loaded dice immediately raise ethical questions. Should abortions be encouraged, or even compelled, for mothers who have a 60% chance of producing a malformed baby? As a complication, mothers have been uncertain in the past if the sickness they had was German measles, or just a rash, or just a fever. The disease is not very severe. It often does not even create a rash, but it is highly infectious, and tends to come back in epidemics every seven to twelve years. (It was such an outbreak which brought the minor spate of children with cataract to Dr Gregg.) The severity of the epidemic, as judged by the affected mothers, does not seem to be partnered by any similar severity of effect upon the foetuses. Stranger still, one twin may be badly affected by the mother's catching of rubella, the other twin may be quite unharmed.

Fortunately, in 1962, and twenty-one years after that famous Australian report, the German measles virus was successfully grown in a laboratory. And less than three years later a practicable one-day test was devised, in the United States, for confirming and accurately diagnosing German measles. This research only clarifies the ethical issue. Should the woman who definitely has had German measles during the first four weeks of pregnancy be encouraged to abort her probably malformed infant? Should young girls be deliberately infected with rubella in a kind of antibody finishing school? The rubella vaccine was first licensed in the United States in 1969, some three years after claims for an effective vaccine had been made by scientists from the us National Institutes of Health. Ordinary measles had slumped from 385,000 us cases in 1963 to 21,000 cases in 1968, and vaccination had undoubtedly contributed to the fall. Hence enthusiasm in many lands to make certain, largely by vaccination, that all girls experience the rubella virus before they have a chance of giving birth.

As if to hammer home some of the important factors involved, the United States was suddenly hit by a vast epidemic in 1964. There were nearly 2,000,000 cases, including tens of thousands of pregnant women. Although all the affected babies from this single epidemic were born either later that year or in 1965, it was some time before its full effects could be finally listed. By 1968 it was said to have caused at least 30,000 stillbirths, miscarriages and birth defects. Total deafness is easy to detect fairly swiftly in a child; partial deafness is far harder and takes much longer, but it was estimated that half of the children whose mothers were affected are suffering from at least a partial hearing loss. It is easy to equate this natural disaster with the man-made tragedy of thalidomide. Fortunately, so far as the developed countries are concerned, such large scale rubella assaults upon the unborn are now improbable.

Contergan and Distaval The thalidomide disaster did have various fortunate aspects. It was lucky, in a sense, that it produced such a rare malformation. It was good that a German doctor saw, and so promptly stopped, the malevolent connection between drug and defect. It was fortunate that certain countries, notably the United States, never even sold the drug. There is even a hypothesis that the living victims of the drug might not even have survived to be born had their mothers not swallowed thalidomide and thereby prevented their deformed foetuses from being aborted naturally. A large proportion of natural abortions are malformed, and no one knows how thalidomide has its effects. Until this is known hypotheses will inevitably arise concerning the effects it did have, particularly in this relatively unexplored world of cause and effect in congenital malformations.

Germany synthesised the new non-barbiturate hypnotic in 1956. It was tested on animals, found to be satisfactory and then given to human beings. As a sleeping pill it was more than satisfactory. It was cheap, it gave a good sleep, and it would not kill would-be suicides. Above all, it was effective in reducing the nausea of early pregnancy. In West Germany, where it was first marketed as Contergan, the drug was very popular; it could be bought over the counter without a prescription. Such a profitable drug did not go unnoticed by other countries. Many of them made thalidomide. Britain did so, called it Distaval and Tensival and sold it. Australia sold it. Other countries had other names –

Softenon, Talimol. The United States made it but, to the eternal credit of the Food and Drug Administration, did not sell it. The land of intensive crop-spraying, of Los Angeles smog, of Nevada's atomic explosions, of the people who eat 300 tons of barbiturate a day before going to bed, who kill 50,000 a year on their roads and 20,000 every year at work, this same land considered that the new drug had failed its safety tests. A few Americans swallowed thalidomide, partly because one or two trials were conducted, and partly because it was available elsewhere, but they were very few. In that country, with nearly 4,000,000 pregnancies a year at the time and a highly active drug industry pushing a new, cheap and popular hypnotic and anti-nauseant, the effects could have been catastrophic. As it was, West Germany suffered the most from its new creation.

By 1959 cases of phocomelia were beginning to show up in the German hospitals. This condition, whereby small hands and feet are attached to the trunk by a single small bone, is usually extremely rare. It is one of a variety of limb defects, such as amelia (total absence of limbs) and micromelia (very short limbs). By 1960 such defects were appearing even more frequently, with phocomelia heading the list. By 1961 the frequency went up several times over. And the defects were appearing more often in other countries, in Britain, where Distaval had been available since 1958, in Japan, in Australia, and in the rest of Europe. During 1961, a mere five years after the drug's first synthesis, many German doctors were urgently trying to find a cause for the surge of phocomelia. Thalidomide, still unsuspected, had been put on the prescription list (meaning it could only be sold with a doctor's prescription) in April of that year for quite another reason: prolonged taking of the drug was being blamed for a form of neuritis, and a wasting away of certain tissues.

Dr Widikund Lenz, of Hamburg, was particularly active in asking young mothers what had happened to them during their pregnancy. Where had they been? What had they eaten? By November he suspected the drug Contergan. A noteworthy proportion of the women had mentioned it; so back he went to the others and found that a greater proportion had not even bothered to mention it, knowing its innocent reputation. At once he called a meeting, and notified Contergan's manufacturers. On November 20, 1961 he reported to the Düsseldorf Paediatric Society. Similar detective work was being done in Australia by Dr W. G. McBride, and he too notified the drug company. Suddenly, after five years of tranquillity, events moved fast for the new drug. Before November had ended the drug had been withdrawn from the German market. McBride's suspicions gave rise to urgent cables between Australia and London. The German Ministry of Health advertised in newspapers and television that women should not take the drug. And in Britain Distaval was quietly removed from the market on December 2 of that year. The sudden flurry then subsided into an uncanny quiet. The British people were almost totally unaware that anything untoward had been happening (and maybe some of the pills were still being taken after that date, medicines in cupboards being rarely thrown away and sometimes donated to a friend).

In the United States there was also silence. It was to be broken principally by a heart specialist, Dr Helen Taussig. In January 1962 she heard casually from one of her former German students about the rash of phocomelia victims in his country. The next day she impulsively flew to have a look. On leaving Germany for London she felt 90% certain that thalidomide was to blame. On leaving Britain, having heard about Australia too, she was 99% certain, and flew back to America determined to publicise her findings. This single personal mission was to break open the subject, once and for all. For some reason, about which the world's journalists should feel a certain disquiet, those German advertisements had never been noticed for the news story that lay so blatantly within and behind them. Dr Taussig spoke at the American College of Physicians, and the press at last realised what had been happening.

The British public first read about and understood the horror of thalidomide in their newspapers in the summer of 1962. This was six months after the drug's quiet removal from the approved list. There had been intermediate newspaper references in the depths of the parliamentary columns. For example, Baroness Summerskill had asked in March if the drug had been withdrawn, and had then made a few sweeping remarks about the ruthlessness of the pharmaceutical industry. Miss Edith Pitt, in May, had told the House of Commons that pregnant women should not worry unduly; the medical profession had been and would again be warned of the drug. Only during the subsequent months did editors finally give thalidomide the space it merited, and made certain that the difficult name of this six-year-old drug became a familiar household word. So, by the time the hue and cry began, the fox had already gone to ground. Nevertheless, pregnancy is nine months long. Mrs Sherri Finkbine, an American who had swallowed the drug, kept its name in the headlines by her efforts, which were eventually successful, to have an abortion in Sweden, and so rid herself of a possibly malformed child. At that time there were many other women carrying foetuses within them that might have been deformed by pills taken in 1961. Time would tell how many.

In 1964, when time had told, the Ministry of Health published *Deformities Caused by Thalidomide*. This survey was limited to live-born children. It showed that 349 malformed children were born in England and Wales between the beginning of 1960 and the end of August 1962 to mothers who had certainly, or probably, taken thalidomide. In 1964, 267 of these children were alive. Bearing in mind the difficulty of collecting all the facts, the concluding estimate of this report was that between 200 and 250 children were living in England and Wales with thalidomide-induced deformities. There had originally been fears that the numbers would be far worse. In Germany they were far worse. In November 1965 Dr Lenz, who had delivered his report four years earlier, said that there was a total of 7,000 thalidomide babies in the world; of these the majority was in West Germany. The original total of thalidomide children may well have been double Dr Lenz's 1965 total. Two other German doctors attempted in 1964 to re-examine all the children they had seen between 1959 and 1962 who had had defects typical of thalidomide teratogeny. They could not examine 45% of their former 400 malformed patients: those 45% had died.

Science is still trying to explain how thalidomide has its teratogenic, or malforming, effects; but effort was made on direct behalf of the affected children. They needed prosthetic limbs, they needed surgery. They needed special schools – Germany built eighty by 1969. They needed psychological care. They also needed money. In Britain the Distillers Company (Biochemicals) Ltd, which made the drug, agreed to pay substantial damages for phocomelic babies whose parents had successfully sued the company. In Germany, where 5,000 such children were born, Chemie Grünenthal, of Stolberg, which had been making one million of the pills a day, was also sued.

Britain was then killing 7,000 people a year on its roads, and in a thoroughly man-made manner, but there was something additionally tragic about the man-made malformations of the 7,000 living victims of thalidomide.

What causes malformations? Plainly German measles and thalidomide are guilty. Ionising radiation is also guilty. Hiroshima and Nagasaki made that abundantly clear when the pregnancies of August 1945 were delivered. Diabetic women, and women with pre-diabetic signs, also produce more defective children than normal (diabetes is itself a defect caused by something else). Vitamins in excess, hormones, measles, high altitudes, lack of oxygen, cosmic rays, gas poisoning, syphilis, mumps, polio, chickenpox, any and every sort of virus infection, quinine, antibodies, food deficiencies – the list of suspects is huge, but medical science has not found guilt an easy thing to prove.

However, unlike the law, the present attitude is to consider everything guilty even if the evidence is lacking. Give a pregnant woman nothing, and give her even less in the first three months, sums up much of the current attitude towards medication for mothers to be. A warning note inserted into *Clinical Pharmacology* stated that 'The medical profession clearly has a grave duty to refrain from all inessential prescribing of drugs with, say, less than ten to fifteen years widespread use behind them, for all women of childbearing age'. It used to be thought that the womb, nature's own sacred womb, was somehow exempt from disaster. Rubella in 1941 and thalidomide in 1961 shook that complacency. Like two enemy agents suddenly discovered in the stronghold, they have made everyone else suspect.

In the past there was simplicity. Malformations, it was said, were caused by 'faulty germ plasm'. Now there is no longer simplicity. There is solely a bewildering list of malformations, a huge array of suspects, and yet an exciting feeling that at last the subject has been opened up. Alas, but radiation, German measles and thalidomide should get most of the credit for this awakening.

'Maternal impressions' A postscript to this subject is 'maternal impressions'. Defined as 'effects on the foetus which result from similar impressions on the mind or body of the mother', there have been some uncanny examples – or coincidences. The pregnant mother catches her arm in a collapsing deck-chair, and lo and behold the forthcoming baby is without that very same arm. Coincidence? Or cause? A correspondent to the *Journal of the Tennessee Medical Association* recorded in 1910 that 'Mrs –, twenty-three years old and about two months pregnant, was one day very badly frightened by her son, aged two, nearly cutting off his left thumb, the member hanging by but a thread. She was without anyone to assist her and dressed the injury as best she could. Her mind constantly dwelt on the accident and in due time she gave birth to a boy, who, to my great surprise, had his left thumb hanging from his hand by a thin pedicle of flesh.' Such an incident is called a photographic impression, owing to the similarity between the trauma and the defect. Scientists working in this sphere are prepared to accept that certain 'impressions'

upon the mind or body of the mother have led to consequences for the foetus, despite the fact that most pregnancies punctuated by shocks of all kinds lead to normal healthy babies. But current thinking excludes the idea that maternal shocks cause 'similar' defects in the baby which resemble the original trauma. Coincidence is always possible, but nothing more: the photographic theory of maternal impressions cannot, these days, be upheld.

16. Inheritance

'Take care to get born well.'

GEORGE BERNARD SHAW

By way of introduction to this subject here is a potted version of
the mechanism of inheritance, more to outline the form the
chapter will take, than to make immediate sense of human
genetics.

In a sexual system all offspring acquire their inheritance from
both parents. The sperm and ovum fuse to form a single cell, with
all the inherited material being within this single cell. Therefore,
because of this fusing, both parents must each have a system for
presenting only half of a human cell's vital needs. The two halves
then unite in fertilisation to form a whole cell. That united cell
must then have a system whereby it divides to form more cells,
each similar, each carrying the same quotient of inheritance and
each dividing equally. Within each divided cell there must be an
arrangement for the storage of all this inheritance, so that its
information exists throughout the developing body. And, of
course, there must be a mechanism for converting all this store of
information into reality, for acting upon it, for transforming the
single cell into a multicellular adult of the correct and inherited
kind. He becomes a man, and not a mouse.

Now some names. The system for forming half a cell is called
meiosis. After fertilisation the system of normal and equal
division is mitosis. The storage of information is principally
within the chromosomes of the nucleus, each consisting of long
strands of deoxyribonucleic acid, hereinafter called – as is the
custom – DNA. Within the DNA, and according to the arrangement of chemical bases on its structure, is a code that ordains
which proteins are to be formed, and when, and how. With the co-
ordinated production of these proteins by the DNA, the development of an individual gradually unfolds. He becomes *Homo
sapiens*, and not *Mus musculus*.

Therefore chromosomes have to be described first, as their
manner and form are crucial to meiosis and mitosis, and to the

storage of information. Next comes coding, and the cracking of the most condensed blueprint of all time. Then, having established the mechanism, come its faults. First mutations, the irreversible changes. Second, the larger scale chromosomal errors when bits of them break off to cause intersexes, mongols, some cancers and inevitably much else besides. Finally, the end products of all this inheritance, whether correct or faulty: the inherent traits of man. Why do women pass on haemophilia, and yet not suffer from it? Why do blue-eyed parents virtually always produce blue-eyed offspring? Why albinos? Why not more of them? Why any?

Background Practically all the work on this subject is post-World War Two. The first correct count of human chromosomes was made only in 1956. DNA only rose to fame in the 1950's. Code cracking only began in the 1960's.

Nevertheless ideas on breeding are older than historical man. Some unsung hero must have been the first to take certain hopeful looking grasses, select the most promising, breed from them, and thus artificially develop cereals fit for man and his burgeoning agriculture. Much later, and long before biologists had anything to say on the theory of breeding, practical genetics was highly skilled. The monoglot but multifarious world of the dog is a testament to this ingenuity. Compare also the natural and wild ponies or the early Przewalski's horse of Europe (now seen mainly in zoos, or, occasionally, in Mongolia), with a Derby winner or a Shire stallion. The original British Soay sheep, now found only on St Kilda, produce a 1½ lb. (680 gms.) fleece; modern fleeces are 28 lbs (12·7 kgs.) or so. A mediaeval farmer got 3 cwt. (152 kgs.) from each strip-shaped acre; modern yields are 30 cwt. (1,524 kgs.). Radishes of 28 lbs. (12·7 kgs.) have been grown. Cows can produce 10 gallons (45.5 litres) of milk a day, not one gallon or less as formerly.

Part of all this improvement lies in better cultivation, better disease control, and better understanding, but better genetic methods are paramount. Robert Bakewell, a practical biologist who lived in the eighteenth century, and before even the word biology had been coined, had much influence on the increasing dimensions of British livestock. The average weight of bullocks sold at Smithfield market in 1710 was 370 lbs. (167 kgs.), it rose to 550 lbs. (249 kgs.) in 1732, and 800 lbs. (362 kgs.) in 1795 – or more than double in less than a century.

However, such good practical work, founded on controlled mating, was not scientific genetics. That began in 1865 when

Gregor Mendel, a teacher-priest of Moravia, sent an article on hybridisation in plants to be published in *The Proceedings of the Brünn Society for the Study of Natural Sciences*. It was not the obscurity of the journal that prevented this article from becoming famous, but a transient blindness among the appropriate scientists. Charles Darwin died in 1882 never having heard of Mendel, who died two years later. The article was the formal birth of genetics, but only at the turn of the century was this birth finally registered when three biologists from Germany, Austria and Holland encountered Mendel's work.

Briefly, Mendel had produced a theory which accurately predicted the nature of the hereditary mechanism. Before his work, again very briefly, inheritance was thought to be an alloy, an amalgam or random blending of parental properties; after him it became an assortment of discrete inherited units, the genes of genetics. Mendel's brilliance was accompanied by an apparently complete foreknowledge of the results of his experiments even before conducting them. He seemed to comprehend the system entirely without knowing anything of the mechanics. It has even been said that Mendel's gardener fixed the results according to Mendel's preconceptions. The actual hereditary mechanism, then still totally unknown, has unfolded with twentieth-century discoveries into a most positive vindication of the man who died, not as revered scientist, but revered Abbot of the Monastery of Brünn.

Chromosomes have provided the framework for almost all of this twentieth-century activity in inheritance. They were discovered in the last century, but meant little to anyone. Chromosome means only 'coloured body', and each one showed up colourfully within the cell's nucleus when the material was suitably stained. Then, in 1903, they were stated to be the carriers of genetic information. It was not mankind's chromosomes which were studied in those early days of genetics, but primarily those of a minute insect often to be seen flying above dust-bins and rubbish heaps. The fruit fly, Drosophila, was for many decades, and still is, a humble and willing servant of geneticists. Its eight large chromosomes are easy to see, and the members of this helpful species breed rapidly – like flies. Mankind's chromosomes were never so easy to detect in those early days, and for thirty years all books said we had 48 of them.

Suddenly in 1956 better techniques were to shake the genetical world and make it blush. J. H. Tjio and Albert Levan reported from Sweden that they could only count 46. Inevitable disbelief followed, but so did many other counts of 46, and 46 it will always be. Strangely, all the great apes have got 48, an inconsistency between man and his nearest evolutionary relatives which poses a problem. Other species, far less closely related, have widely differing chromosomal counts. Examples of species having more than man, apart from the apes, are the chicken (78), and some snails (54). Less than man, apart from Drosophila, are the hamster (22), the honey bee (16), the green fly (10), the frog (26), the cat (38) and the mouse (40). There is a temptation to say the more advanced the animal the more the chromosomes, but the rule is soon trampled upon by the exceptions.

The new counting techniques were possible because of greater microscopic clarity. Even before this was available it had been known that chromosomes closely resembled entwined strands of wool. Shortly before a cell divides the strands become very positive, and are in fact doubled. Then, as if pulled by invisible threads from each side of the cell, each double strand is separated. The invisible threads continue to pull from their opposite poles, and have soon amassed a group of chromosomes around them, with just as many in each group as there had been in the single group before it divided. Forty-six, in human cell division, have become 46 plus 46, and soon the cell itself divides to form two cells, each with the correct chromosomal complement. Such is mitosis (after the Greek for thread).

Reduction division, or meiosis (from the Greek for diminution), so necessary to give the sperm or the ova half the correct number of chromosomes, is a very similar process. The strands of wool become more positive, just as before, but are not doubled. Consequently, when the same polar attraction starts work upon those chromosomes to pull half each way, the 46 divide to become 23. Further divisions, to produce more sperm or more ova, then take place as in conventional mitosis, and 23 chromosomes regularly divide to produce fresh cells, each with 23 chromosomes in them. The number will only change again at fertilisation when 23 meet 23 to produce 46, when the ova and the sperm meet to fuse into the fertilised egg. Half of that fertilised egg's chromosomes have therefore come from each parent.

Not all the strands of chromosomes are equal, although the uninitiated eye finds them remarkably similar save in their length. Each chromosome is split down the middle, but is still attached at one point (called the centromere). The chromosomes are like lengths of two-stranded wool threatening to part, and the centromere is like a blob of glue holding them together. As the strand lengths are different, and the blob is unevenly placed, some chromosomes are long, with long arms and long legs, some have short arms and long legs, some are short with short arms and legs and some are medium-sized. There seems every variation on this theme, but each variation does in fact have a brother, a partner similar in appearance. Humans have 23 pairs of chromosomes rather than a straight 46.

The glaring exception to this regular pairing occurs in males. Females are all right, their 23 being nicely matched with another 23. In males there are two which do not match. These are the sex chromosomes. Whether someone is black, or big, or clever, or blue-eyed does not show up on his or her chromosomes, at least not with present-day techniques; but the fact of sex is immediately blatant under the microscope. Both males and females have 22 similar pairs, but females also have a similar 23rd pair, both of which have been named X chromosomes. Males have a dissimilar 23rd pair, one X and one named Y. The Y is far smaller than the X. With females the division during meiosis of 23 pairs into 23 single chromosomes is straightforward. The female eggs are then all 22 plus X, all alike. With males the division of 22 of the pairs is straightforward, but the XY 23rd pair is less so. The X goes one way, the Y the other. Hence, males produce during their meiosis two types of sperm, a 22 plus X type and a 22 plus Y.

It is for this reason that the male is the arbiter of an offspring's sex. During fertilisation the female's 23 chromosomes (X plus 22 others) will meet a sperm bearing either X plus 22 others or Y plus 22 others. If the former the sperm's X will pair with the egg's X to form XX, i.e. a female. If the latter, then Y will meet X to form XY, i.e. a male. The Y makes for maleness, the lack of it makes for femaleness; but why this happens is for the future to say. Somehow the genes on that short Y chromosome cause the glands to develop which, in their turn, cause the male organs to develop. Within that Y are coded all the necessary instructions for determining maleness. (Plainly the time will come soon – see page 175 – when parents can choose the sex of their future offspring. Dr Augustus B. Kinzel, writing in *Science* in 1967, warned that we could 'expect sex predetermination by 1980'. We have not got it yet, but the technique cannot be far distant, even though society may manage to forbid its use.)

Coding   First the size of the problem. Spies have sometimes used microdots for passing on information. Within a comma or an exclamation mark is the shrunken image of some document.

When suitably enlarged this small image will yield up its secrets. Such refined expertise is as nothing compared with the minuscule scale of the genetic code. Human cells each contain about six millionths of a millionth of a gram of DNA, and there are 28 grams in an ounce. As there are 4,000,000,000 people now living on earth, and as each individual egg cell which initiated all these people contained no more DNA than those millionths of millionths of a gram, this means that all the DNA information for all those people weighed 0·024 of a gram. (I am indebted to the geneticist Theodosius Dobzhansky for this arresting calculation.) Taking the arithmetic still further, and bearing in mind the population growth of recent centuries, it could well be that the total inherited and informative DNA of every individual born since the time of Christ has weighed no more than one gram. The world of microdots is gargantuan by comparison.

The history of DNA has been short. Only during the last war was the idea first promoted that the substance responsible for hereditary messages was DNA. The structure of this extraordinarily complex molecule was only determined in the early 1950's, and the code only started to crack in the 1960's. The rapid progress in this science of molecular biology was achieved when scientists of various disciplines worked together to solve the biological problem. Of the two Nobel prizes given in 1962 and shared by five men for work in this field only one, J. D. Watson, was trained as a biologist. Two were physicists, F. H. C. Crick and M. H. F. Wilkins, and two were chemists, M. F. Perutz and J. C. Kendrew. (Watson was also the only American in this otherwise British achievement.)

Long before the code began to be cracked there was a basic understanding of the mechanism involved in the progression from one cell to one adult, from the store of facts within DNA to the realisation of all that information in the end product of a developed human being. Basically the order of events was known. First make the proteins, and these proteins would then make the body, partly by forming the structures of it, and partly by making the enzymes which control its chemical reactions. It was also known that only twenty different substances – the amino-acids – were needed to make the proteins, and thus the enzymes, and thus the body. The amino-acids are consequently called the building blocks of development. The order in which these blocks are picked up to make each protein ordains what kind of protein it is, what kind of enzyme and structure it makes, and thus what sort of development results. It had been known for a long time that whatsoever controlled this order controlled the development, and

227

it then became known that DNA, somehow, did the controlling.

Having simplified this issue it is now necessary to complicate it to give an indication of the scale of things. There are only those twenty kinds of amino-acid, but an average protein may contain 150 different amino-acid molecules, all amassed in the right order. As amino-acids are themselves fairly complex molecules, the proteins are even more so. Each protein molecule contains many thousand of atoms, whereas some simple molecule like sugar has forty-five, penicillin forty, and sulphuric acid seven.

Within each cell there are probably a few thousand different proteins at work, producing different enzymes, causing different chemical reactions. Taking the body as a whole there must be many times that number of large protein molecules at work. And all of this is ordained by the information contained within those wispy threads, the long tenuous molecules of DNA. These DNA molecules possess somewhere between 100,000 and 10,000,000 atoms. As molecules they are big; but, even so, they are small in comparison with the chromosomes that contain them. Chromosomes are small enough, demanding good microscopes even to see them, but the thinnest of them are one hundred to two hundred millimu across (with a millimu being a millionth of a millimetre). A DNA molecule is only about two millimu across.

Now to the code. Although long and thin, and although giving rise to such multitudinous activity, the DNA molecule is made up of only four types of chemical units. Just as proteins are made from the twenty building blocks of amino-acid, so DNA is made from a repetition of just four chemicals. There are many tens of thousands of these in each molecule, but only the four types. They are called adenine, guanine, cytosine, and thymine. They are all chemical bases, and hereafter are called A, G, C and T. Somehow their ordering along the DNA chain controls the order in which the right amino-acids are picked up to make a protein. Plainly the order in which just these four kinds are arranged would be inadequate for twenty different kinds of amino-acid. If A meant one amino-acid, G another, T another and C another, there would be sixteen left over that could not be called upon. The code was obviously more complex. It was then thought that the order of each *pair* of bases might be the controller of a single amino-acid, and could ordain which amino-acid was to be taken from the pool. Unfortunately there are only sixteen possible pairings of A, G, C and T (namely AA, AT, AG, AC, TA, TT, TG, TC, GA, GT, GG, GC, CA, CT, CG and CC) and these are still less in number than the twenty amino-acids. These sixteen types of arrangement could not therefore select from twenty amino-acids. So it was inevitably wondered whether the magic number was three because two was plainly inadequate.

If A, G, C and T are thought of as threesomes there are sixty-four different ways in which they can be arranged – AAA, AAG, AGA, GAA, AAC and so on. Sixty-four is therefore ample for a selection from twenty. In the 1960's it was proved that this triplet arrangement does form the code with some amino-acids being chosen by more than one triplet, and some of the triplets acting as punctuation to mark the end of the formation of a particular protein. The first amino-acid to be correctly associated with a definite triplet was phenylalanine. This happened in 1961. As with the deciphering of any code, the first crack is the important one. Thereafter the secrecy falls apart. After 1961 it did so very rapidly, and by 1965 the code was known. The controlling sequences of bases, whether one triplet or several, had all been identified with the twenty amino-acids. Just what controls which had been unravelled.

The work went so quickly that, within the same year, the code was actually being utilised. Such work immediately posed many questions for the future, and there was considerable unhappiness that genetic engineering, as it came to be called, would lead to all manner of infringements upon liberty and life. Sir Macfarlane Burnet, the Australian virologist, confessed in 1966 that he had been fascinated by molecular biology 'since before some of the great molecular biologists of today were born. . . . But, as a man, as a doctor of medicine, it worries me intensely. At the human level, where is it leading us? What good can it do?'. The feeling in the 1980's is altogether more relaxed. Chromosomes have been manipulated, and for the good of many individuals, enabling them, for example, to create various substances they were pre-viously, and wrongly, unable to manufacture. Presumably all advances in science are capable of good or evil, and there is no built-in compunction that this particular form of tinkering should lead to evil. There will have to be effective safeguards, but there is nothing new in that requirement. Besides, genetic engineering and its capabilities will not go away. It is simply up to us to use it and them for the greatest good.

Mutations Nature has been knocking off atoms, and adding them here and there ever since life began. The essence of life is stability, but the essence of evolution is the ability to change. Mutations are the main mechanisms for permitting change, and are therefore desirable. They are also most undesirable, for they often kill and they generally harm. Their changes are irreversible, and they are very rare.

It is not known how rare. It just happens that, from time to time, cells suffer hereditary alterations. For ordinary body cells the change means, provided they are still capable of division, that the cells stemming from them will thenceforth carry the new characteristic. The change may be for the good, for the bad or inconsequential. Anyway it dies out when the individual's body dies. But this is not so if the change has been in his or her germ cells, the producers of sperm or ova. The future descendants will also carry this change. And this change will remain a permanent characteristic until it too is changed in some fashion by yet another mutation. And so on. Species will be replaced by quite different forms. And they in their turn will be replaced. Long before the science of genetics began, P. B. Shelley wrote: 'Man's yesterday may ne'er be like his morrow; naught may endure but mutability.' He was absolutely correct. (As a coincidence Shelley drowned the very year that Gregor Mendel was born.)

The delicacy implicit in the whole coding system makes it quite plain why any alterations are likely to be harmful. After all, mutations are random changes, and only the very smallest have the greatest likelihood of being acceptable. Hence, in a sense, their rarity. Were they frequent no organisms could survive such a plethora of unwelcome change. Experimental evidence suggests that one germ cell in 100,000 or possibly only one in 1,000,000 carries a recent mutation. Lots of effects can increase this rate. Radiation is one of them, a fact discovered in 1927. Many chemicals can induce mutations, but no way has yet been found of reducing them, and such a reduction system would be welcome.

Mutations get talked about familiarly as if everybody knows precisely what has happened in any one of these changes. In fact no one knows the course of a single mutation in forms of life higher than bacteria. Is the change which causes, for example, achondroplasia (the thick-set, large-headed form of dwarfism) a single change in a single part of the DNA code, the picking up of one wrong amino-acid, or many? Or is such a mutation a combination of mutations which only manifest themselves if they involve a number of changes, a quorum of alteration?

The molecular biologists unravelling DNA will inevitably sort out a few mutation answers in time, this being the same side of the coin as genetic engineering, occurring naturally rather than induced. Allowing for the traditionally rapid rate of discovery in this new branch of science, the time should not be very long.

Chromosome mistakes and intersexes   Klinefelter's syndrome is the awkward name given to an unfortunate form of sexual development. The sufferers are male, they are always infertile, their testes are exceptionally small, they scarcely ever have to shave, their breasts may grow, and they are likely to be of poor intellect. Many examples of this abnormality had been recorded in the past, but no explanations. And various other forms of physical sexual abnormality like Klinefelter's syndrome were known. Some were called pseudo-hermaphrodites, some pseudo-males, but all were unexplained. Suddenly in 1959 the whole area of intersexuality was burst wide open.

In that year Patricia Jacobs and John Strong, both working in Edinburgh, reported their examination of the cells of a Klinefelter man. It was only three years after the revolutionary Swedish work which had made chromosome study a more precise act, and had fixed the human number for all time at 46. The Klinefelter chromosomes were counted in Edinburgh, and they came to a total of 47. One extra chromosome was responsible for the developmental mix-up of this form of intersex.

Mutations, whatever they are precisely, are changes that are microscopic by comparison. They somehow alter the hereditary units, but on a minute level compared with the relatively gross aberration of the addition of a whole chromosome to the standard complement of 46. The effect of this major alteration is sufficient, with Klinefelter sufferers, to prevent any likelihood of the abormality being inherited.

Also in 1959 it was discovered that several mongols had an extra chromosome. It was a different one from Klinefelter's, and no mongol has since been found without this additional chromosomal fragment. Therefore, somehow, its presence must cause the poor intelligence, the inabilities, and the physical abnormalities of the mongol. As with Klinefelter's the extra chromosome was found to exist in every body cell examined. Therefore, the chromosomal error must have been present at conception. As the parents of both types of abnormality are usually perfectly normal, the error must have been in the chosen sperm or egg.

The mongol chromosomal addition is an extra chromosome of the pair known as pair 21. The Klinefelter extra is another X chromosome, i.e. one of the 23rd pair. Normal females have two X's, and normal males have one X to partner their Y, but Klinefelter's have two X's and one Y. It becomes immediately apparent why their masculinity is both trampled upon and accompanied by female characteristics. XXY contains both the coding for male (XY) and for female (XX); hence the muddle; hence the intersex.

The 1959 disclosures opened the gates. Turner's syndrome, long observed like Klinefelter's, but with opposite effects, was found to be represented by only 45 chromosomes. (Both Henry Herbert Turner and Harry F. Klinefelter were Americans, the first a gland specialist, the second a physician.) In this case there was no Y, and just one X. As XX is the customary female condition it is therefore thoroughly understandable that the Turner sufferers (called XO) are all immature females, each with a diminutive uterus, small genitalia, and sometimes without an ovary at all. They are also very short people. Other wrong numbers found include XXX females, XXXX females, XXXY males, XXYY males, and even XXXXY males with 49 chromosomes instead of 46. The medical journals regularly recorded details in the 1960's of this spate of new discoveries, and caused one *Lancet* reader to complain in doggerel fashion: 'Wouldn't it be nice, say, just once or twice, to open this journal and not find pictures of peculiar, hieroglyphic creatures which look like worms, but are called chromosomes?'

One of the last to be discovered had been notably rare, namely XYY. As XY is the normal male, and Y makes for the maleness, it might therefore be construed that XYYs were doubly male. An investigation was made in a Scottish hospital for difficult criminals and rapidly the rare XYY anomaly became almost common. A few XYYs had been previously described; then nine men were found in this one hospital, and most of them were taller than average. After that first examination of a state prison-hospital some ordinary prisons were examined. At Nottingham, for example, one-fifth of the men over 6 ft. (1.8 m.) were found to be XYYs. It was first thought that the extra Y would possibly cause, apart from tallness, more aggression, more sex crimes, more wild behaviour. It now seems as if the association is with modest crime, with pilfering, with stupid stealing. Society does not mind about the height, but it locks the men up for their behaviour. The extra Y, it seems, is the real villain of the piece. Therefore, should courts of law hear cytological evidence before locking a man up in the fairly idiotic hope that mere punishment will rid him of the insidious influence of one more Y chromosome?

With 46 chromosomes controlling human development it is scarcely surprising that the possession of 47 sometimes causes such an upheaval that death follows. It is more surprising that the possession of 45, 47, 48 or even 49 in other cases has not caused more of an upheaval in development and viability. The two chromosomes principally in charge of sex determination seem to be modifiable with less disruption to the body as a whole than if the modifications were applied to the other 44. The sex is disrupted because intersexes of one sort or another are formed; but the individual frequently survives.

Sometimes, as a variant to this theme, there are mosaics. Instead of every cell in the body being uniformly and distinctly deformed, some are and some are not. Or some are deformed one way, and some another. Presumably, in mosaics, the deformity arose not before conception but shortly after the multiplication of the original cell had begun.

To sum up. Mutations are inheritable changes, either for the subsequent body cells or for subsequent offspring. They can, by radiation and chemicals, be made to occur faster but not slower. They can be lethal. If not, they are probably harmful. Chromosomal changes are much cruder, and the traditional number of 46 chromosomes can range from 45 to 49 with varying results, often lethal, often damaging, and almost always associated with reduced fertility.

Attempts to summarise any aspect of human genetics inevitably lead to over-simplification. Be warned that a work exists called *Humangenetik*. Described as *Ein kurzes Handbuch in fünf Bänden* the idea of anything short in five volumes is forbidding, and a reminder of that international elephant story. The British, Americans, and French described the sport, size, and love-life, respectively, of this species. The six German volumes were entitled *Das Elefant, ein Vorwort*. Human genetics is also a huge subject, and see *The Human Pedigree* (Allen & Unwin 1975) for my personal attempt to make the subject both compact and legible. At least it is only in one volume.

Inherent traits   The DNA carries the information. The chromosomes carry the DNA. Mutations are changes in this information, and inter-sexes, mongols and so on are manifestations of changes to the chromosomes. The inherited information, whether mutilated or not by change, leads on to the development of the individual's characteristics. The information received is different for each and every one of us. People are most emphatically not born equal.

The most useful, overworked and misused word in the subject of genetics is 'gene'. There is a vagueness about it, a lack of precision which is treacherous. It is defined as a factor of inheritance, a chromosomal unit, the carrier of heredity. The

233

genes are presumed to lie along the chromosomes (and elsewhere) with each one having a finite responsibility. They are the agents of Mendel's 'particulate inheritance', the separateness which he demonstrated and which is now the basis of genetics. Yet, for all that, there is an imprecision about this concept of a gene. How many do we have, for instance? Professor Curt Stern, of the University of California, wrote that 'the true figure is probably not less than 2,000 or more than 50,000'. He hazarded 10,000.

So, bearing the roundness of this figure very much in mind, and being correctly suspicious of its validity, the number 10,000 is a convenient assessment of the human genetic inventory. At once the number should be doubled, and be considered as 10,000 pairs. Each individual, as is well known, receives a nearly equal contribution of genes from each parent. (The slight inequality is associated with those 23rd, or sex, chromosomes.) The child gets 10,000, according to this current assumption, from his father and 10,000 from his mother. It is correct to write of 10,000 pairs rather than 20,000 genes, for each of the 10,000 is coupled at fertilisation with a similar partner from the other 10,000. They are like two sides of the same coin.

The coin analogy is a helpful one. Imagine having two sacks of coins. In each sack there are 10,000 different coins, and each coin has both heads and tails to it, totalling 10,000 pairs. Each coin in each sack has its partner in the other sack. Now take one of the sacks, shake it up, and pour its contents on to a table. Each of the 10,000 coins will show either heads or tails. This assortment can be considered as the chance contribution of a single sperm. Do the same with the other sack, and it represents the equally chance contribution of one ovum. Pair up each coin with its partner, and these 10,000 pairings form the genetic material for the offspring. With each coin from each side having two possibilities the number of possible combinations is seemingly infinite. Hence -- quite apart from other factors, such as mutations – the uniqueness of the individual.

Now imagine one coin out of the 10,000 plus its partner. Each has a head and a tail, and so there are four possible combinations – head head, head tail, tail head, and tail tail. These four varieties give a clue to the mechanism of dominant and recessive genes. A character is said to be dominant if it manifests itself whenever it can, even when it forms only half of the final pair. Suppose that heads indicates dominance. Out of the four combinations three include heads, and only one does not. Therefore, if a dominant gene is involved in the situation, its effect will be three times as common as those of its recessive partner.

Brown eyes and blue eyes are an example. The brown-eye gene is dominant to the blue. Babies may have inherited from their parents either brown and brown, brown and blue, blue and brown, or blue and blue genes. In the first three cases the baby's eyes will be brown, and only in the last case will they be blue. If such a blue-eyed person mates another blue-eyed person all their children should be blue-eyed because they have inherited no brown-eyed, and dominant, genes. The same is true for albinos as albinism is similarly recessive, with pigmented skin and eyes being the dominant form. Most people inherit genes for pigment from both parents. Some of us, one out of seventy, inherit a gene for pigment from only one parent, and nothing – or albinism – from the other. However, the pigment gene is dominant, and so normality results. Only if albinism is inherited from both parents, which occurs once in every 20,000 fertilisations, does an albino result, with the characteristic white hair, pink eyes, and general lack of pigment.

With dominant characteristics the rarity of the gene is also important. Brown eyes are dominant to blue and are common. So brown-eyed people, although they may be carrying blue-eyed genes, are likely to marry brown-eyed people, and therefore extremely likely to produce brown-eyed children. Achondroplasia, the massive form of dwarfism, is a rare dominant character. Assuming such a dwarf marries a normal person the chances of an achondroplasic child are only evens. Assuming the dwarf marries a similar dwarf the chances are still only three to one for a child being achondroplasic.

Unfortunately such clear-cut precision is exceptional. A single gene for a single characteristic is rare, and not the rule. Even the blue-eye/brown-eye situation is more complex with possibly three gene pairs involved as two blue-eyed parents can (in about 2% of cases) produce brown-eyed children. Other characteristics are, like eye-colour, known to be caused by more than one gene, and some genes are associated with more than one characteristic. Qualities like intelligence (which is no one thing), height, build and colour are controlled by more than one gene. Inter-marry some short, intelligent, thick-set black with some tall, stupid, thin European and the offspring will be a bewilderment of variation.

Inheritance is a hotch-potch, as well it might be. The mutations which have permitted evolution to occur, and which have permitted man to be evolved differently from the monkeys, have happened in no set fashion. Consequently our inheritance is no orderly system but a muddled assortment of interacting genetic factors which, somehow or other and via that coding system, have led to the development of distinct human beings. Knowledge about human genes is still sparse partly because of the impossibility of breeding human beings experimentally and the general lack of information from past generations. But, however knowledgeable humanity becomes, the arrangement of genes will always be seen as a hotch-potch. Take the eye, for example. Little enough is known now, save that a single gene for the eye will never be discovered. It is too big an organ and too complicated for such simplicity. Probably hundreds of genes are involved, each controlling, each playing its part or being dominated, each the result of all those years of evolution since the first light-sensitive areas developed into the direct ancestor of the human eye.

It is humanity's fringe characteristics, such as eye colour and albinism, which are better understood genetically than its main properties. Dominant traits (with the recessive condition in parentheses) which have been well studied include:

Brown eyes (blue); pigmented skin (albinism); Roman nose (straight nose); black skin (white skin); curly hair (straight hair); dark hair (fair hair); non-red hair (red hair); short-sightedness and long-sightedness (normal vision); early baldness – males only (normal hair); free ear-lobes (attached ear-lobes); long eye lashes (short eye lashes); large eyes (small eyes); achondroplasia (normal stature); blood groups A, B and AB (Group O); more than five fingers or toes (normal); webbing between fingers or toes (normal); very short fingers (normal).

It is an odd list, and gives little idea either of humanity or of human variation. Not every human characteristic is either controlled conveniently (one pair of genes is most satisfactory) or is suitable for study. The manifestation of the characteristics must not be blurred by too many other genetic influences, and must be obvious. Something like free or attached ear-lobes is ideal; something like the shape of a forehead is not. Another difficulty is the timing of the gene's action. Most effects, such as blood groups, are present when the child is born. Some, such as hair and eye colour, are not definite until shortly after birth. Others, like early baldness, show up much later, and a few of the heritable diseases are later still. Glaucoma, which can cause blindness due to increasing pressure within the eye-ball, is an example of a late manifestation.

Nevertheless, despite the great complexity involved when two individuals fashion at random one other individual, despite the great blurring of most genetic effects, despite the unscientific manner of human breeding, despite the twenty-five years between one generation and the next, despite this and despite that, there are already some fascinating insights into human inheritance. Once again they make an odd and apparently arbitrary list. From it I have made my own choice, the first of which is the disease of bleeders, most strangely called the 'love-of-blood' disease.

Haemophilia   Wrongly considered to be caused by inbreeding, it has achieved notoriety in modern times mainly because it has plagued the royal families of Europe for three generations. Almost always the victim is a man. All men receive it from their mothers, the so-called 'carriers'. The few women who do actually suffer from the disease, instead of merely passing it on, have the ill-fortune not only to have a haemophiliac father but also a mother who is a haemophiliac 'carrier'. With haemophilia being a rare disease anyway such a double circumstance is extremely rare.

The more common and traditional pattern is for a woman who is carrying this defective gene to marry a normal man. She may not know she has this defect, but if she has children the defect will manifest itself over the years. Half of her sons will, on average, be bleeders. Half of her daughters will be similar carriers to herself, and half of their sons will be bleeders. None of the affected men can pass it on to their sons, but all of their daughters will be carriers. And, once again, in the subsequent generation there will be the same fifty/fifty chance for the boys to be bleeders and the girls to be carriers.

The picture in any family is therefore not clear for many years. With most families the clinical details of any forbears are poorly recorded and generally forgotten, but not so with royalty. It was therefore of enormous importance when, on April 7, 1853, assisted by the new gas chloroform, Queen Victoria gave birth to her eighth child. The sickly Prince Leopold proved to be a bleeder. He was the first indication that Queen Victoria, whose descendants were to percolate throughout the royal houses of Europe, was herself a carrier of classical haemophilia.

The disease had been known for centuries. The excessive bleeding of the haemophiliacs had even resulted in special rules in the Talmud. The Jews had written, in the second century AD, that boys need not be circumcised if two older brothers had already died from blood loss following the operation. More significantly, they added that the sons of their sisters were also exempted. They knew that the disease came from the mother and was passed on through her daughters. Early in the nineteenth century, several decades before Prince Leopold was born, several men, notably Professor C. F. Nasse of Bonn, described the pattern of haemophiliac inheritance. Queen Victoria's diaries do not indicate that she understood the pattern as expressed in 'Nasse's Law', but the full understanding of this pattern did not come until the discovery of sex chromosomes in the twentieth century.

The defective gene, which is responsible for the defect in the blood's clotting mechanism, is carried by the X chromosome; it is what is known as a sex-linked characteristic. The sex chromosomes cause a person's sex, but they also possess genes which cause characteristics irrelevant to sex, such as colour-blindness, webbed toes, muscular dystrophy – and haemophilia. A man receives his X chromosome from his mother, never from his father. Hence he can only get this defect of the X chromosome from his mother, never from his father. And women do not suffer from the defect because they have two X chromosomes, only one of which is defective and one of which is normal; the normal one gives her an effective clotting mechanism. To each daughter and to each son she passes on one of her X chromosomes. It may or may not be the defective one. Hence the precisely fifty-fifty chance that her sons will be haemophiliacs and her daughters will be carriers.

Queen Victoria's case is an excellent example of haemophilia in action. She had nine children, and the family gene was passed on to one son, and two or perhaps three of her daughters. Four out of nine is as near fifty-fifty as the unevenness of the number nine will permit. As Victoria's antecedents had not in any way been victimised by haemophilia, a recent mutation must have been present in one of her X chromosomes. Perhaps she had received the faulty X chromosome from her elderly and long-time bachelor father, who died the year after Victoria was born, his only legitimate child. (Various families claim that their ancestors are in fact illegitimate children of the Duke's, but none of them claims that these illegitimate ancestors were afflicted with haemophilia.) Or perhaps he was guiltless, and the mutation occurred within herself.

In any case Leopold was the first indication of things to come. He bled from trivial injuries, and his baptism was postponed for three months. Leopold's childhood was beset with illness, and even when he was an adult of twenty-six Victoria refused to let him visit Australia. Either his health would suffer, she wrote, or hers would for worrying about him. Nevertheless she let him marry three years later. He had time to be the father of a girl, and to make his wife pregnant with a boy before he died, aged thirty-one, as a result of a minor fall and a major haemorrhage.

Before that happened, and before he passed on his affliction (inevitably making the girl a carrier, although the boy was safe), his mother had had further indications that, as she wrote, 'Our poor family seems persecuted by this awful disease'. Her daughter Alice, born ten years before Leopold, married in 1862. Two of her daughters were themselves to be proven carriers, and one son was a haemophiliac who died from a haemorrhage, after falling out of a window at the age of three, eleven years before the death of Leopold. Another of Victoria's daughters, Beatrice, married in 1885. Of Beatrice's four children (the Battenberg/Mountbatten famly (one was a girl carrier and two were boy victims. Both boys died without issue – one after surgery and one in action. Victoria's daughter Victoria may or may not have been a carrier. She gave birth to the Kaiser of World War I, who certainly caused others to bleed for him, but was himself exempt. Two of his brothers died early, at two and eleven, and haemophilia just may have been present.

So much for the first generation, with one sick son, two carrier daughters, one suspect daughter, two normal sons, and three normal daughters. King Edward VII was one of the normal sons, and thus the present British Royal Family, who descended from him, has no fear of the disease.

Not so the descendants of Victoria's carrier daughters, and of Leopold's daughter. Various royal males have testified, since the Queen's death, to the persistence of this defective X chromosome. Prince Henry of Prussia, grandson of the carrier Alice, died in 1904 aged four. The Russian Alexis, another of her grandsons, was sickly and Rasputin was frequently summoned. However, the boy died of a bullet and not of his haemophilia when, together with his sisters, the family was shot following the Russian Revolution. Gonzalo, a grandson of Victoria's daughter Beatrice, died aged twenty after a modest injury suffered in a car smash. Alfonso, an elder brother of his was also a haemophiliac. He too died only four years later, after breaking his leg and suffering a few cuts in another motor accident. Both victims were without issue.

Nevertheless the mutant gene probably still exists, despite the lack of issue and despite the early deaths. According to Professor V. A. McKusick, of the Johns Hopkins University, who studied the royal tree in detail, there remain three females each of whom has a fifty-fifty chance of being a potent carrier. Two of them are the sisters of the unfortunate Alfonso and Gonzalo. Both are married, and both have had sons and daughters. The sons are apparently not afflicted, but the daughters might be quietly carrying this unwelcome feature of Victoria's inheritance. Although Leopold was born over a century ago, and died so summarily as a result of haemophilia, he did pass the gene to his daughter; therefore the disease did not necessarily die with him. Time will tell whether the unwelcome gene will crop up again, or whether, aided by early death and childless victims, the gene is no longer around to plague Queen Victoria's descendants. After all, whether Victoria's father did give it to her or not, the defect has already been reigning for a century and a half, and it has caused much misery in that time. But, concludes Professor McKusick, the prospect that it will reappear among some of Victoria's male descendants, notably among some of her great-great-great grand-children, is a 'real one'.

Queen Victoria's family, so beset with haemophilia, gives a depressingly false picture of the extent of the disease. Only about one in 20,000 of the population of Europe is afflicted by it. In Britain there are 2,000 males without the globulin factor in their plasma that is necessary for the normal coagulation of blood. There is also the Haemophilia Society which operates for their welfare from 16 Trinity Street, London S.E.1. As yet there is no sign of a cure for the haemophilic disorders, but the missing factor can be temporarily infused. The effect of such an infusion is short; half of it has gone in half a day. In the past, bleeding episodes have often led to permanent disability, adding positive injury to the insult of haemophilia. Nurseries are now designed with special care for the haemophiliac child, with padding, low furniture, non-skid floors and a preference for soft toys, but it is also argued that over-protection can be harmful. With no cure foreseeable at present, save via the manipulations of genetic engineering, it will obviously be a little time before the world can rid itself of the hereditary disease that Queen Victoria called 'the worst she knew'.

Tasters and Non-Tasters   There is a group of chemical compounds which have been of bizarre interest to geneticists ever since 1931. In that year it was discovered that some people could taste them while others could not. The compounds included substances like phenylthiocarbamide and phenylthiourea. To possess an ability to taste a remote and unimportant chemical like P.T.C., as the former is called, is a curious characteristic, but it is most emphatically an inherited trait. Where both parents are non-tasters all the children will be non-tasters. If one parent is a taster and one a non-taster the children will be either, but probably tasters. If both parents are tasters the children also will probably be tasters. The situation, which is of practical interest in paternity disputes, is similar to the blue-eyed/brown-eyed story. Blue eyes and non-tasting are both recessive characters. Both are controlled by an inherited single pair of genes, but both stories can be slightly complicated by the modifying actions of other genes.

In Europe, and hence in North America, about 70% of the Whites can taste P.T.C. Both Arabs (63%) and Aboriginal Australians (51%) are less able in general to taste the substance, while Chinese (over 90%), Negroes (95%) and American Indians (up to 98%) are more able to do so. Amongst animals, who should never be forgotten in any discussion of mankind, most primates seem to be tasters. The evidence is limited, but twenty out of twenty-eight chimpanzees in British zoos showed, in unambiguous fashion, an ability to detect the bitter unpleasantness of P.T.C. So why this ability, and why its variation between species and races? The only clue so far is a link between it and certain forms of thyroid disease, or nodular goitre; those who get this disease are more likely to be non-tasters.

A casual link between one disease and a curious ability to taste does not have the compelling power of the haemophilia story, but both are invaluable material for researchers into human genetics. Both are detectable, neither effect is too confused by too many other genes, and both are clearly inherited. These characteristics, and others, ranging from the Hapsburg lip to colour blindness, have all added their quota to the general knowledge of genetics. They are all examples of straws at which geneticists have had to clutch. The stream of inheritance which flows every time a new life is created is too great, too complex by far for detailed study. Straws are clutched whenever possible, and the following are all further examples of them.

The Hapsburg Lip   Any group of women around any pram, or push chair, or papoose bundle, will be identifying inherited characteristics. Out of the myriad nose, forehead and jaw shapes in the world, it is remarkable how similar father and son can be, or even mother and son and father and daughter. The shapes are impossible to define, but easy to observe. Sherlock Holmes, when investigating the Baskerville family, was himself astounded by the similarity between an ancestor and a modern. After mentally stripping the whiskers off a family portrait he was then quick to unmask a living descendant of the family posing in another guise. A famous example of a dominant family characteristic is also preserved in the Hapsburg line, a family rich enough to have had their portraits painted through the centuries and wily enough to have stayed in power and prosperity.

The Hapsburg lip is probably the work of a single dominant gene. The protruding and ugly lower lip is accompanied by a narrow jaw and often by a slightly open mouth. Fortunately, painters and engravers did not belittle the deformity as it is depicted fairly consistently up to modern times. Notable possessors of the lip were the Emperor Maximilian I (born in the fifteenth century), the Emperor Charles V (sixteenth century), Maria Theresa of Austria and Archduke Charles of Teschen (eighteenth century), the Archduke Albrecht and Alfonso XII of Spain (nineteenth century). Whenever such a rare variant keeps on turning up in a family, and is passed on only by the affected members, it is probably the work of just one dominant gene. Whosoever chances to inherit it – the chances are 50:50 – will manifest it and have an even chance of passing it on to each of their descendants.

Most of us have nothing equivalent to a Hapsburg lip, but there is a general and curious ability associated with the tongue. Some people can curve their tongue into a U, with both sides pointing upwards; some cannot. The distinction is most positive. The genetics of this characteristic have not been thoroughly worked out, and the advantage of being able to manoeuvre the tongue in this manner, although the spittle can probably be fired further, is even less clear.

Albinism 'She became pregnant and brought forth a child, the flesh of which was white as snow, and red as a rose; the hair of whose head was white, like wool and long; and whose eyes were beautiful.' The baby in question was Noah, later to build the Ark and survive the flood. The description of him is from the Book of Enoch, allegedly written a hundred or so years before Christ. The details are those of an albino. As Noah was subsequently to populate the planet one might expect a greater abundance of albinism amongst us. It is caused by a recessive gene and therefore, assuming no selective disadvantages, one quarter of us should be albinos. As it is the incidence is far fewer, but all races are afflicted. It can be called an affliction because the total absence of pigment results in weak and astigmatic eyes, an intolerance of the sun and, of course, a rare disfigurement. The European albinos can walk with reasonable equanimity in a region speckled with blond and pale people; not so the white blacks, Japanese, or Red Indians.

The customary ratio quoted for Europe is one in 20,000 or approximately 2,500 in Britain and 9,000 in the United States. In darker communities albinism is commoner. Nigeria has one in 3,000 and they are easy to spot in any Nigerian town. Among a group of Indians (the San Blas) in Panama the ratio is one in 132. The recessive gene of albinism is possessed by one in seventy Europeans, but only shows up when such a person (or heterozygote) marries another such heterozygote. When both parents are albino all the children will be albinos. When only one parent is albino the children are heterozygote just like one in seventy of us, and look normal. There is no way of knowing, except after the birth of an albino child, which of us is the one out of seventy. And no way of telling which possible mate is one girl in seventy. If one albino has already been produced in a family the chance of a subsequent sibling being albino is one in four.

Finally albinos can be piebald, and the lack of melanin is probably due to the lack of the enzyme tyrosinase. This catalyses the first stage of turning tyrosine to melanin. The pinkness of the eye is not a positive pigment coloration, but instead a lack of pigment permitting the blood and its red colour to be seen. In bright light an albino's eyes can become very inflamed, and dark glasses are sensible. No cure exists for those who, like the patriarch Noah, have flesh as white as snow.

Height Unlike haemophilia, Hapsburg lips, blue eyes and an albino's skin, the inheritance of height is different in two distinct ways. For one thing it is polygenic; many genes are involved. And for another the population is not either tall or short, like blue-eyed or brown, haemophilic or not; instead there is a continuous variation from the lowest to the highest. There are very few at either end, and a bulge in the middle. The distribution pattern is what is called the 'normal curve'. It is bell-shaped, with the lip of the bell flattening off to take in the very short and the very tall. The dome of the bell incorporates everyone else, the 95% or so who are neither short nor very tall but whose height is slightly more than or less than the average. As with intelligence, which follows a similar pattern, or with coloration, or weight, or virtually any physical measurement, the normal curve indicates the distribution. Most people are very near average; just a few are extremely clever or extremely stupid, very heavy or very light. There are so many genes involved that, however distinct they are individually, the resulting effect is a blur or mixture of their activities. Mendel's 'particulate inheritance' still exists, but it is far harder or impossible to detect. Too many genes are at work.

243

Nevertheless the hereditary mechanism is still most positive. Like still produces likeness, tall produces tall, and short produces short. However, assuming that both tall and short men marry average-sized women, the sons of the tall men will be shorter than their fathers, and the sons of the short will be taller than their fathers. There is, assuming such constant affection for average women, always a regression to the mean. In fact, many tall men can be embarrassed by short women, and would like to have a hearing aid lodged in their umbilicus with which to hear them better. Similarly, short men will not be readily partial to relatively gigantic women. Like do marry like as well as beget like.

The average British male height is about 5 ft. 8 in. (1·7 m.). The average female height is about 6 ins. (15·2 cms.) shorter. So an average man married to an average girl could expect average-sized children, i.e. 5 ft. 8 ins. (1·7 m.) if male and 5 ft. 2 ins. (1·5 m.) if female. If both parents are 5 ft. 8 ins. (1·7 m.) tall the girl is exceptionally tall and they could expect sons taller than the father, but girls shorter than the mother. A 5 ft. 8 ins. girl is equivalent in height to a 6 ft. 2 ins. (1·9 m.) man because of that 6 ins. (15·2 cms.) differential between the two sexes. (Some say the equivalent male height of a woman is reached by adding one inch for every foot of her height.)

In many ways height prediction for offspring is a matter of arithmetic. For boys add the father's height to the mother's height plus 6 ins., and then divide by two to get the boy's final height. Assume a 5 ft. 10 ins. or 70 ins. father and a 5 ft. or 60 ins. mother. The son is 70 + 60 + 6 all divided by 2. The answer is 68 ins. or 5 ft. 8 ins. For girls, assuming the same parents, those six inches of differential should be subtracted. Therefore the equation is 70 − 6 + 60 all divided by 2. The answer is 62 ins. or 5 ft. 2 ins. In this family the girls will (probably) be taller than their mother and the boys shorter than their father.

However, predictions are by no means the same as results. A coin has a 50:50 chance of being head or tails, but its perverse behaviour can confound all predictions. Just as it can produce ten heads in a row for no particular reason, so can a human family make nonsense of forecasts. However, taking the population as a whole, the predictions make more sense. Family sizes then behave as they should, and conform to the average pattern, the normal curve, the traditional outline of inheritance. A coin may be heads ten times in a row. It will not be heads a thousand times in a row, but will be quite near to 500 heads and 500 tails. Large population numbers are necessary for good predictions in polygenic inheritance.

Small family numbers can make nonsense out of forecasting. Further nonsense, particularly with regard to height, is associated with the environment. Poor health, bad food, or wrongful up-bringing can make the final stature less than it should be genetically. Similarly, because of environmental difficulties in their youth, parental height may be less than it should be, genetically speaking. An average-sized couple who were short of food in their growing period may produce a series of taller children. Once again the predictions will have been proved wrong.

Strangely, the final height of people these days is not markedly different from earlier times. Suits of armour in museums appear small, and cottage doorways smaller still, but skeletal remains indicate a remarkable constancy. Today's average male height in Britain is from 67 ins. (170 cms.) – 68 ins. (172 cms.), varying slightly from area to area. Old Stone Age man in Western Europe was, on average, 69 ins. (175·2 cms.). New Stone Age or Neolithic Man was 66 ins. (164·7 cms.), Bronze Age man was 68 ins. (172 cms.), Iron Age Celts 66 ins. (164·7 cms.), Anglo-Saxons 67 ins. (170 cms.), Mediaeval British 66 ins. (164·7 cms.). Perhaps those suits of armour, perched crustacean-like up on their stands, are arranged more squatly than they should be, and perhaps mediaeval man cracked his head on the lintel in his cottage just as frequently as modern man.

What has changed notably in recent years is the speed of growth. The average height of eleven year olds in London schools has gone up by 4 ins. (10·2 cms.) since the start of this century. Both puberty and full adult height are now being reached earlier. Consequently soldiers, for example, are now theoretically taller than they used to be because their heights are measured on recruitment; boys of eighteen are now much nearer their final height than they were in the past. Improved environment has speeded things up, although the genetics have remained the same.

Just because a baby is long there is no reason to believe it will be tall. Neither the height nor the weight of a baby is markedly affected by its genetic constitution. There is a belief that a child has reached half its final height by its second birthday. This is not true, but it is true that a child of exactly two is about half the height of an adult. Work carried out at Aberdeen suggested that good predictions of adult height could be made at the third birth-day. For boys exactly three years old their final stature was likely to be 1·27 × present height (in inches) + 21. For girls it was likely to be 1·29 × present height (in inches) + 16·1. Thus a boy of three measuring 3 ft. 2 ins. will be slightly over 5 ft. 9 ins., while a girl aged three of similar height will end up 5 ft. 5 ins. tall. The trouble with average formulae of this type, which are good for working out the average height of the population, is that individuals will be capable of proving them wrong and their final height may be two inches on either side of the prediction.

Colour is another example of many genes at work, and most of the work on it has been based on black/white matings. The situation is complicated by the fact that whites are not without pigmentation, and blacks are not pitch black. Moreover, the skin colour of Africans varies much more than that of whites. The actual skin colours can be assessed fairly accurately by matching them with spinning discs on which colours have been painted in certain proportions. If a disc is spun fast enough all its painted colours merge into one shade. A particular 'white' skin of a normal European was once matched by a disc painted 5% black, 34% red, 15% yellow, and 46% white. Similarly one ordinary black's skin was matched by 75% black, 13% red, 2% yellow, and 10% white. The white man is even less white than the black man is black.

When black marries white the mulatto offspring are, as is well known, a colourful mixture, being neither white nor black, but various shades in between. When these hybrids intermarry the second generation is predominantly a similar mixture, but there are then some offspring much resembling the original black and white grandparents. Also, for some reason, the first generation mulattos are, on average, slightly closer to the white than the black parents in their pigmentation. There is a popular belief that the marriage of a white to another white who, due to some earlier mixing, has a black gene or two in his or her system, may suddenly produce a very black child, or throw-back. Neither the term nor the likelihood is particularly acceptable to geneticists. In general in such a marriage no child should be darker than the off-white parent, the one with the tar-brush's touch in his or her genetical make-up. However if the couple both possess dark genes the offspring can sometimes be darker than either parent. Many investigations of alleged throw-backs uncover the wayward behaviour of a parent, and consequent illegitimacy, rather than any wayward behaviour of the genes.

Panmixis is a term for total interbreeding. Consequently it describes the situation when all racist barriers come down, and mates of all colours are chosen at random. Within the United States barriers are still up, but Professor Curt Stern has calculated what will happen to the colour of subsequent generations when the Negro genes flow without restriction throughout the fifty states of the Union. He has assumed ten gene pairs to be involved in pigmentation, with pure whites having no black genes and pure blacks having all ten black.

Assuming panmixis, and assuming blacks to be a tenth of the population, the resultant colouring will be very largely white: 48·4% will have only white genes, 36·4% will have nine white genes out of the ten, 12·3% will have eight, and 2·5% will have seven. This total of 99·4% is classifiable as 'light types'. The remaining 0·6%, the 'dark types', will have four, five, six, seven, eight, nine or ten of the black genes. In other words almost half the population would be just as white as the whites are today, over half would be slightly tanned by their one or two dark genes, and only half a per cent would actually have half a black's dark complement of genes as their inheritance. Such a panmixis would in fact lead to fewer really dark faces than exist today. Not one man in ten, but one man in 200 would then have the traditionally black face of the Negro. At long last his migration would have been absorbed into the nation of immigrants.

Baldness well exemplifies the difficulty of studying the inheritance of a particular human trait. At first sight it is easy. Do the parents show baldness? Do the offspring? At second sight it is far more complex. Baldness can start at the back, like a monk's tonsure. It can creep up the parting. It can start when a man is in his twenties. It can start much, much later. It can become total over the top. It can cease merely as an extreme thinning. It is, in short, no one thing.

Nevertheless it is probably inherited. It occurs predominantly in men, but is not a sex-linked gene like haemophilia or colour blindness. It merely affects men more frequently than women – which is not the same as sex-linkage. A bald man will often transmit the characteristic to about half his sons, but exceptions abound. It appears that there are two main types of baldness, both genetically different. In one the thinning starts before thirty, and is extensive before forty; in the other it starts later.

'No boy ever gets bald, no woman and no castrated man', noted Aristotle. The link between the three is the male hormone. The genes for baldness do not act if the level of androgens is low. The alleged link between baldness and virility, sometimes claimed hopefully by the bald, is only true in so far as bald men are more virile than boys, women and eunuchs. A final confusion, irrespective of gene inheritance, irrespective of male hormones, is that hair thins with age in both women and men. This thinning is distinct from baldness, although women can go bald, and so can eunuchs.

The initial simplicity of bald/non-bald is therefore confused by its age of onset, its type of recession, its association with male hormones, its link with old age, and with more than one pair of genes involved in its inheritance. Simplicity yields to confusion, and predictions about the hairiness of future offspring are rarely hazarded. However, early and emphatic baldness does act as if a dominant gene is involved. It is only dominant in the male, and so a victim of such early fall-out will, on average, cause half his sons to be similarly affected and none of his daughters. In the next generation half of those sons and half of those daughters will cause half of their sons to go bald with the same rapidity.

Colour-blindness is another anomaly that is generally inherited. Total colour-blindness is rare, it is from a recessive gene and it affects both sexes equally. Partial colour-blindness affects about one person in thirty, is from sex-linked genes, and is much commoner in men than women. Brushing aside all exceptions there are certain basic rules. When a normal woman marries a colour-blind man their children will probably be normal. When a normal man marries a colour-blind woman the sons will be colour-blind and the daughters will be normal. Taking the first case a generation further, the normal daughters of that marriage are carriers of their father's colour-blindness, although unaffected by it. Hence half their sons will be colour-blind, half will be normal, and half of the daughters will be carriers like themselves. If carrier daughters marry colour-blind men the situation will be aggravated; half of their sons will be colour-blind (just as if the father were normal) and half the daughters will be carriers, but the remaining 50% of daughters will be colour-blind. Finally, if a colour-blind woman marries a colour-blind man all their off-spring will be colour-blind.

As with haemophilia the explanation of these ramifications is the X chromosome. A boy can only get his single X chromosome from his mother. A girl has two X chromosomes, and gets one from each parent. If a girl has a colour-blind father she will have inherited his X chromosome and hence his defect, but she will have inherited a normal X chromosome from her mother. One defective X and one good X result in good vision; but she is a carrier. Half of her sons will receive the good X and half the defective X; thus half will be colour-blind and half normal. Give or take a number of exceptions this is the mechanism of colour-blindness inheritance.

Colour-blindness varies in its incidence between the races. Europeans have a lot of it, but Aboriginal Australians and Eskimos, for example, have much less. It is assumed that natural selection is still discriminating against the colour-blind, and making it harder for them to survive in the more natural and primitive communities. In Europe about 7% of males, and 0·5% of females, are either colour blind or colour-weak. In communities with less colour-blindness as a whole, the proportion of affected women is an even smaller fraction of the male number.

As an addendum to this tale it is not coincidental that both haemophilia and colour-blindness are linked with the X chromosome rather than the Y. Almost all of the sex-linked characteristics, such as these two, are associated with the X chromosome. The reason is straightforward. The X chromosome is quite large, with plenty of room on it for genes. The Y chromosome, the chromosome peculiar to males, is a relatively stunted thing. Only a few characteristics are known to be linked with the Y chromosome apart from maleness, and one of them is 'hairy ear rims'. It is pleasing in a sense that the complex world of genes should be associated with such a matter-of-fact item as space and accommodation upon a chromosome.

As a subject, the inheritance of human traits has only achieved a scratched surface. Just a few characteristics have been studied because only a few have been amenable to study. The majority of our inheritance is poorly documented, to say the least, and most unamenable. What facts are there about the inheritance of liver size, of memory, of virility? Also, as Kenneth Mather put it in his book *Human Diversity*, the inborn and the early inculcated are often hard to disentangle. As parents know to their cost, children can blame both inheritance and upbringing upon the same people. Are the children stupid because their parents led stupid lives or, like their children, had stupidity thrust upon them? How much influence did the genes have, and how much did the surroundings?

Once upon a time no one cared too much about the environment. Peasants were considered ignorant clods and incapable of improvement because they were sons of peasants and not because they were brought up in ignorant homes. R. L. Dugdale in 1875 studied a particularly ill-fated pedigree of a family called Jukes. Old Max Jukes, long dead, had given rise by then to 709 descendants. These included 76 convicts, 128 prostitutes, 18 brothel-keepers and 200 paupers. The theory was that a bad seed, rather than a slum existence, was at fault. Similarly, and at the same period, Oliver Twist's essentially inbred goodness shone brightly through the wickedness of his foster-world. Literature is packed with examples both of 'good breeding' and of inborn villainy. The racists have followed suit, forgetting the circumstances, and assuming instead an inherited lack of ambition, laziness, love of corruption, and poverty of mind. Nowadays circumstances are achieving a more realistic consideration in the mechanism of inheritance. The genes exist, but they are as nothing without an environment. Put at its most basic, an individual needs food for development, however packed he may be with the best of genes.

I shall give the geneticist Theodosius Dobzhansky the last word: 'It has been said before, but it will bear repetition, that the genes do not determine "characters", such as proneness to criminality or smoking habits; the genes determine the reactions of the organism to its environment.'

17. Inbreeding

The churches' attitudes/Cousin marriages/Forbidden matings/
Effects of inbreeding/Incest/Genetic recessives

It seems that society in every age has had taboos which have discouraged inbreeding. Probably these laws, which prohibited incest ('sexual commerce of near kindred', *Oxford English Dictionary*), were primarily to promote favourable relations within the camp. Father/daughter and uncle/niece matings could cause great disunity. It is unlikely, or so most anthropologists think, that the harmful effects of consanguinity would have been such a spur for the taboos, had they been noticed, as the immediate disharmony caused by close matings. The religions emphasise this point. Most of the forbidden marriages are not based upon sound genetics. Before the recent revisions the Church of England's *Book of Common Prayer* used to list thirty types of relations forbidden as partners for each sex in its Table of Kindred and Affinity. Of this number only ten made genetic sense, in that some inbreeding would then be taking place. The *Koran*, as will be seen, is about as restricting and only partly concerned with the harmfulness of consanguinity.

In fact there is current argument in many places that the old laws should be brought more into line with modern knowledge. Ought outbreeding, and more distant mating, now so much easier, be actively encouraged? Should first cousins still be permitted to marry? Should modern science be allowed a say in the matter to try and reduce the toll of, for example, those congenital malformations more likely to be observed in the offspring of near relatives?

To set the scene it should be remembered that every group is inbred to some degree. Just as there has been antagonism to incest there has also been antagonism to outbreeding. The girl next door has been the ideal. It is not just parents in the Southern States, or Hillbrow, Johannesburg, or Jewish mommas whose children are pressed to marry 'one of us'. Mating has rarely been far-ranging. Geography, social barriers, and custom have seen to that, but even if it had been more venturesome, with two parents for everyone and four grandparents, and two parents for each of them, the time soon comes when there must be overlap.

251

Six generations ago, about the time Queen Victoria started on her reign, we each had sixty-four direct parents alive. Sixteen generations ago, at the time of the Armada, there were, in theory, 64,000 direct parents (or ascendants) for each one of us today. Obviously, there must have been considerable overlap, with the actual number being much smaller. (Even the offspring of first cousin marriages have already 'lost' some ancestors, as they have six, not eight, great grandparents. The loss is caused by having some common ancestors.) By the time of Agincourt the theoretical situation is yet more ridiculous, with 4,000,000 direct ascendants for each one of us today. As the population in Britain was then probably less than 4,000,000, it is not a question of: 'What did your father do in the Great War?', but 'What did your entire army of fathers, every manjack of them, do at Agincourt?' Barring recent immigrants to Britain, barring those of Henry V's men who died without issue, and barring lineages which stopped, every single man at Agincourt was a great-great-and-so-on grandparent many times over of every single one of us in Britain today. As 1415 was only some twenty-two generations ago we are, it can be argued, an inbred lot.

Cousin marriages   In fact the situation is much more severe. Marriages are local affairs. A study in Northern Italy showed that a man used to find his mate at an average distance of 600 yards (540 m.). Then, after the invention of the bicycle, the distance leapt up – to 1,600 yards (1,460 m.). In Britain, despite inter mingling and the invention of the motor car, six out of a thousand marriages were still between first cousins during the 1960's. In the United States it was six out of ten thousand. Urban Austria, urban Brazil, and urban Spain were all similar to Britain, but studies elsewhere have shown up far more inbreeding. The number of first cousin marriages per hundred were: Spain 4·6, Nagasaki 5, rural Japan 7, Swiss Alp village 11, Parsees in Bombay 12·9, Brazilian village 19·5 and Fiji Islands 29·7. Hence in Fiji the occurrence was fifty times more frequent than in Britain and five hundred times more so than in the United States. Occasionally high prevalence is influenced by choice and social preferences (as with Indian castes and Japanese families) but generally such inbreeding is force of circumstances, such as isolation, and many of those countries have become less isolated in recent years. Hence the numbers of cousin-cousin marriages must have dropped. They still occur, but in fewer numbers probably everywhere.

Inbreeding has sometimes been actively promoted. Royalty in

Ancient Egypt and the Incas of Peru were prominent examples. Rameses II had at least fifty daughters and he married quite a few of them. The Azande of Africa encouraged their chiefs to mate with their own daughters. Generally, first cousin marriages are as near inbreeding as society allows. The Egyptians at their most permissive (among the Ptolemies) and the Incas allowed brother/ sister unions; but nowhere, according to C. S. Ford and F. A. Beach in their excellent *Patterns of Sexual Behaviour*, are such matings permitted to the general population.

Marrying one's relations sets up genetic trouble because it increases the chances of harmful recessives manifesting them- selves. (There is more about dominant and recessive genes in the previous chapter.) Every human being carries potentially harmful recessive genes, but they can only manifest themselves when both parents not only possess the same harmful genes, but pass them on, and the harmful manifestation then occurs in their offspring. Just as a child can only be blue-eyed when both parents possess at least one blue-eyed gene (although this recessive character does the child no harm) so can a child only suffer from phenylketonuria, amaurotic idiocy, or alkaptonuria, for example, if both parents happen to possess the recessive genes for these complaints. Only recessive PLUS a recessive can show up. Recessive plus anything else does not. Each of us inherits an assortment of harmful recessives, but each of us is likely to inherit a different lot. Hence the chances are that one person's assort- ment does not match another's; when they marry there is then no pairing up of harmful recessives, no phenylketonuria and the like.

The apple cart is upset when like mates with like. Brothers and sisters have, on average, half their genes in common. Fathers and daughters (also mothers and sons) *must* have half their genes in common. (Whether inheritance 'must be' or is so 'on average' is important, but the difference can be forgotten in discussing populations.) The fractions become smaller as the relationships grow more distant. Granddaughters must have one-quarter of the genes of each of their grandfathers, and a further quarter from each grandmother. Nieces have, on average, one-quarter of their uncle's genes; and nephews have one-quarter of their aunt's genes. For still more distant relationships, such as a cousin, the proportion is one-eighth. A cousin is, say, a mother's brother's daughter, hence the proportion is $\frac{1}{2} \times \frac{1}{2} \times \frac{1}{2}$, or one-eighth. A second cousin is, for example, a mother's mother's brother's son's son and the compound proportion here is $\frac{1}{2} \times \frac{1}{2} \times \frac{1}{2} \times \frac{1}{2} \times \frac{1}{2}$, or 1/32. Considering the embargo that society of every kind places on father/daughter matings (where half the genes are shared) and usually places on uncle/niece matings (where one-quarter are shared) it is remarkably casual about cousin matings (where one-eighth are shared).

Forbidden matings Until the Marriage Act of 1949 the State and the Church of England forbade marriage with thirty different kinds of relation. Of this number ten made genetic sense in that there existed, either necessarily or on average, shared genes. The new Act struck off ten from the original list, none of which had shared genes. In 1949 these ten were called 'statutory exceptions', and they permitted a man's marriage with the sister, aunt or niece of his former dead wife, or with the widow of his brother, uncle or nephew. In 1960, as a further change, came the Marriage (Enabling) Act. This was much the same as the 1949 Act, but a man's wife need no longer be dead, only divorced, for him to be free to marry her sister, aunt or niece. Similarly he can now marry the former (and divorced) wife of his brother, uncle or nephew; there is no need for her to be a widow. Taking these new Acts into consideration, as well as the genetic relationships, an amended Table of Kindred and Affinity is on page 256. (The text of the *Book of Common Prayer* of 1662 is Crown copyright and extracts used herein are reproduced with permission.)

Other countries and other religions have different rulings. In the United States, for example, a minority of States permit uncle/niece, nephew/aunt marriages, and yet a third of the States prohibit first cousin marriages. The Roman Catholic Church forbids both second and first cousin marriages without special dispensation. (The unfortunate Anne Mowbray, who died aged nine and whose coffin was unearthed accidentally in London in the spring of 1965, had been married in 1478, when five years of age, to Richard, Duke of York, then four. This had necessitated a Papal dispensation, but not because of their ages: Anne's great grandmother and Richard's grandmother were sisters.) Jewish law permits uncle/niece marriages, but not aunt/nephew unions. However Jewish law bows down on such matters to the local law of the State. The only Jewish uncle/niece partners in Britain must have been married elsewhere. Mohammedan law is surprisingly similar to that of the Church of England, considering that a Muslim is permitted to be polygamous. 'Forbidden to you are your mothers and your aunts both on the father's side, and your nieces on the brothers' and sisters' side, and your foster mothers' and your foster sisters' and your mothers of your wives and your stepdaughters who are your wards, born of your wives to whom ye have gone in: (but if ye have not gone into them, it shall be no sin to you to marry them); and the wives of your sons who proceed out of your loins; and ye may not have two sisters; except where it is already done. Verily, God is Indulgent, Merciful!' (From J. M. Rodwell's translation of the *Koran*.) The Muslim can therefore, like the Jew, marry his first cousin.

Before moving on to the harm done by marrying near relations, the convenient word 'cousin' should be defined. Basically, of course, it means the son or daughter of someone's aunt or uncle. There are in fact ten different types of cousin – or cousinships – from this basic formula, depending on whether the cousins are boys or girls, or one of each, and their parents are sisters or brothers, or one of each. In some other countries this relative straightforwardness is thrown overboard. Perhaps a man's father's brother may marry the man's mother's mother, and therefore present him with young cousins who are also his half-aunts and half-uncles. Professor Curt Stern quotes a Navaho Indian of America whose parents were simultaneously first cousins, third cousins, and first cousins once removed. Working out the likelihood of genetic similarity among their offspring requires great devotion. Professor J. B. S. Haldane did it after emigrating from England to the familial bewilderment of India. He listed, named and calculated the degree of genetic relationship for hundreds of possible cousinships.

Effects of inbreeding Without doubt the offspring of near relatives who marry have a rougher time in life than those from unrelated couples. The nearer the relationship the worse the time, the greater the chance of premature death and the more likely the presence of severe abnormalities. Some reports have illustrated the situation. With first cousin matings in the United States the percentage mortality among their offspring for the first ten years was 8·1% as against 2·4% for unrelated marriages. In France the figures for neonates (less than one month) were 9·3% as against 3·9%. In Japan, for children aged one to eight, the figures were 4·6% against 1·5%. Similar work has been done on disease (TB) and defects (malformations) in both groups of children. The percentages of those affected, according to a Swedish report, were 16% as against 4% for the offspring of unrelated parents. In the USA comparable figures were 16·15% as against 9·82%, in France 12·8% as against 3·5%. (These reports cannot be compared precisely with each other as differences exist in the compilation of the facts, but they do all indicate the same unfortunate trend.)

Certain congenital malformations show up particularly badly

KINDRED AND AFFINITY

A man could not marry his	Common name	Permitted after 1949	Common genes? (whether 'must be' or 'on average')	A woman could not marry her
1. Grandmother		No	$\frac{1}{4}$ (must be)	Grandfather
2. Grandfather's wife		No	—	Grandmother's husband
3. Wife's grandmother		No	—	Husband's grandfather
4. Father's sister	aunt	No	$\frac{1}{4}$ (average)	Father's brother
5. Mother's sister	aunt	No	$\frac{1}{4}$ (average)	Mother's brother
6. Father's brother's wife	aunt	Yes	—	Father's sister's husband
7. Mother's brother's wife	aunt	Yes	—	Mother's sister's husband
8. Wife's father's sister		Yes	—	Husband's father's brother
9. Wife's mother's sister		Yes	—	Husband's mother's brother
10. Mother		No	$\frac{1}{2}$ (must be)	Father
11. Step-mother		No	—	Step-father
12. Wife's mother	mother-in-law	No	—	Husband's father
13. Daughter		No	$\frac{1}{2}$ (must be)	Son
14. Wife's daughter	step-daughter	No	—	Husband's son
15. Son's wife	daughter-in-law	No	—	Daughter's husband
16. Sister		No	$\frac{1}{2}$ (on average)	Brother
17. Wife's sister	sister-in-law	Yes	—	Husband's brother
18. Brother's wife	sister-in-law	Yes	—	Sister's husband
19. Son's daughter	grand-daughter	No	$\frac{1}{4}$ (must be)	Son's son
20. Daughter's daughter	grand-daughter	No	$\frac{1}{4}$ (must be)	Daughter's son
21. Son's son's wife		No	—	Son's daughter's husband
22. Daughter's son's wife		No	—	Daughter's daughter's husband
23. Wife's son's daughter		No	—	Husband's son's son
24. Wife's daughter's daughter		No	—	Husband's daughter's son
25. Brother's daughter	niece	No	$\frac{1}{4}$ (average)	Brother's son
26. Sister's daughter	niece	No	$\frac{1}{4}$ (average)	Sister's son
27. Brother's son's wife		Yes	—	Brother's daughter's husband
28. Sister's son's wife		Yes	—	Sister's daughter's husband
29. Wife's brother's daughter	niece	Yes	—	Husband's brother's son
30. Wife's sister's daughter	niece	Yes	—	Husband's sister's son

in first cousin offspring. They are phenylketonuria (causing a rare form of mental deficiency if not detected soon after birth),[1] alkaptonuria (a metabolic disorder), amaurotic idiocy (which starts with blindness), and various others, including possibly albinism. Definitely not on this list, despite widespread belief, is haemophilia. Phenylketonuria is most certainly on it and between 5% and 15% of affected individuals have related parents. With alkaptonuria the proportion is 30% to 42% (according to different countries). The figures appear alarming but are considerably less so when balanced against the actual incidence of the two diseases. The first disease affects one child in 40,000, the second one in a million. In fact, the rarer a disease, the more alarming the incidence appears to a first cousin offspring. With alkaptonuria it so happens that one in 500 of the general population carries its recessive gene. If two cousins marry, one of whom has the gene, there is a one in eight chance that his or her mate will also have the gene. If the affected cousin marries anyone else there is one chance in 500 of the mate possessing the gene. Odds can also be calculated for the offspring of these marriages. In the cousin–cousin marriage the probability of a child being affected by the disease is one in thirty-two. In the cousin/anyone else marriage the odds lengthen to one in 2,000.

Playing this kind of genetic roulette with one's future offspring is obviously unsatisfactory. Unfortunately, no one knows all the genes he or she carries, and therefore no one has any idea whether the marriage is genetically sound. Both he and she, if cousins, can only know that a greater risk exists for their children. The very rare recessive diseases will be slightly less rare for them; the common ones will be slightly more common. In the *British Medical Journal* the anonymous expert once wrote in his weekly column: 'My own practice with first cousin couples who plan to marry is to explain the additional risk and to tell them that, if they really want to marry, it is a very reasonable risk to take.' If one of their children is then afflicted with an autosomal recessive disorder (meaning that both cousins had the bad luck to possess and pass on the same harmful and recessive gene) there is then a one in four chance (or three to one against) that their next child will be similarly affected. The odds do not vary however many children they have.

Some notable cousin–cousin offspring of the past have been

1 It is phenylketonuria which is primarily responsible for the routine testing of new-born babies' nappies/diapers which is carried out in many areas. Testing the urine will determine whether the child is a victim, and therefore whether it must have a diet very low in phenylalanine during the critical early months of life.

Charles Darwin, Edward Fitzgerald, Toulouse-Lautrec, and John Ruskin. Although a brilliant collection, Darwin was also, according to most, a hypochondriac, Fitzgerald a homosexual, Toulouse-Lautrec a cripple, and Ruskin impotent. However, Darwin (after lengthy correspondence in the medical journals of today) just may have been suffering from Chagas' disease, picked up when in South America (although he lived on until aged 73); Toulouse-Lautrec's legs were probably the result of a dominant gene for osteogenesis imperfecta (though several falls in his childhood did not help, his father had similar trouble). And no one knows why or if Ruskin was impotent or Fitzgerald homosexual. Certainly none of their complaints can be laid at the door of the cousin marriage which created them. And neither, for that matter, can their brilliance.

Undoubtedly there will be fewer cousin–cousin marriages in the future. America's mixed and mobile community (with only six cousin marriages per 10,000) is an indication of the modern trend. In Utah in 1870 such marriages used to be 1% of the total; by 1890 they were 0·25%, by 1910 0·1%, and now they are next to nothing. In France (Loire-et-Cher) there were 6% in 1918, 3% in 1932 and 1% in 1952. Nowadays not only are people getting out of their valleys, but families are smaller. Fewer brothers and sisters in one generation mean fewer cousins in the next. (My own family is a good example of this point. My father was one of nine siblings – he had seven sisters and one brother – and hence there were twenty-three first cousins for me and my brothers. My mother was an only child: hence there was not a cousin in sight on her side.)

Everything said about cousin matings is multiplied for closer relationships. It has already been pointed out that cousins have, on average, one-eighth of their genes in common. Because of this fact cousin matings are more likely to result in faulty offspring. Uncles and nieces have one-quarter of their genes in common; therefore the percentage of malformed children is likely to be greater still. Fathers and daughters, or mothers and sons, or brothers and sisters have half their genes in common (the parental relationships are certain to be 50%; the siblings have 50% on average); consequently the mortality risk for children from these closest of all incestuous relationships is four times greater than for the children whose parents are first cousins. Detailed studies of the offspring of close incestuous unions are rare, but the Institute of Child Health in London has been examining the case histories of thirteen such children. Of this number only five are both physically and mentally normal.

Emile Zola said incest was 'so stupendously vile that I cannot decently contemplate it'. The eighteenth chapter of Leviticus is also most adamant. Incest is still customarily vilified by most of the population, although not so categorically and with a twinge of curiosity about the relationship's causes. It is presumably less common than before, and is presumably being influenced by the same factors that are reducing cousin–cousin matings, but the number of cases known to the police is actually growing. Whether for reasons of increased police diligence or not, the pre-war average was less than a hundred cases a year. The law in England states that males over fourteen and females over sixteen can be charged with incest, and the maximum penalty is seven years' imprisonment. If the man's partner in the incestuous relationship is less than thirteen he can be imprisoned for life. Both in England during Oliver Cromwell's time and in Scotland until 1887 the maximum punishment for incest was death. Public opinion on punishment has changed much since those days. A Liberal member of the Swedish Parliament even announced in January 1967 that he would try to change the existing law so that brothers and sisters could marry, but they still cannot do so. Previously a Swedish court had ruled that a man aged thirty-four and his younger half-sister (both had the same father) could continue to live as a married couple; they have a normal child of eight. Public opinion may be changing, but geneticists are likely to be inflexible about the hazards of excessive consanguinity.

However, as long as cousins are permitted to marry (with their common genes), there will be pressure of a kind that all the non-genetic matings should be permitted, namely ten out of the twenty relationships currently forbidden from marriage. In 1984 a British lorry-driver and his former mother-in-law made history when the British parliament agreed to consider special legislation to permit them to marry. Without doubt the 1949 Act, amended slightly in 1960, will be further amended soon to permit former mothers-in law and others to be future wives, provided they have no more genes in common with their prospective spouses than the normal gamut of available mates. Even a Church of England group, appointed by the Archbishop of Canterbury, recommended (in May 1984) that marriages between stepparents and stepchildren over twenty-one should be permitted. The majority of the group urged that legal restrictions on the freedom to marry of those related only by affinity should now be as few as possible. In which case the pre-1949 ban imposed on thirty degrees of relationship, and which now covers twenty such degrees, may be whittled down to ten degrees of relationship before so very long.

Recessives and the future   There is a final sting in the tail of this inbreeding story. The fact that such marriages are on the way out does not magically remove the problem of all those harmful and recessive genes lurking within every one of us. It merely postpones it. When two cousins do have the misfortune to beget a malformed child, and that child dies, the two recessive genes it carried die with it. Nothing is passed on to the next generation. If cousin marriages stop, such swift elimination also stops – for the time being. As the years, and the generations go by, the lack of elimination means the accumulation of a larger and larger number of recessive genes among us. Hence the greater chance of unfortunate pairings from unrelated marriages. Exclusion of consanguinity in one generation, as Curt Stern puts it, merely transfers the load of affected individuals to later generations. Cousins who do decide to marry now may take a little cold comfort from the fact that, by accepting the risks today, they are doing something for all our descendants, for posterity. The fact becomes truer, and the comfort colder still, should a recessive defect chance to kill one of their offspring.

18. Blood Groups

All human beings belong to one of four blood groups. These groups are distinguished by the reaction between the red blood cells of one individual and the serum (the blood's liquid) of another. Either the blood cells are distributed evenly in the serum, or they club together (known as non-agglutination and agglutination). Red blood corpuscles are said to possess antigens called agglutinogens; there are two agglutinogens (A and B) and two agglutinins (anti-A and anti-B). Blood corpuscles may contain either A or B or both or neither. These form the four blood groups. Blood serum may contain either anti-A, or anti-B, or both or neither. Corpuscles containing A are agglutinated by serum containing anti-A. B corpuscles are similarly treated by anti-B. People possessing, for example, A corpuscles do not possess anti-A serum, but do possess anti-B. O people have serum containing both anti-A and anti-B. AB people have neither. Trouble following the wrong blood transfusions (A blood for a B person or vice versa) is almost entirely caused by the agglutinins in the recipient reacting against the introduced corpuscles. The agglutinins of the donor's blood generally do much less harm.

Karl Landsteiner   Just as Sir Winston Churchill's actions seem to cover more than the life span of a single individual, so does the name of Karl Landsteiner occur relentlessly in the history of the development of blood group knowledge. Born in 1868, the son of an Austrian journalist, he became a doctor in 1891. At the end of the nineteenth century he was working in Vienna on the curious clumping together of blood from different individuals. In 1900 he described three blood types. In 1902 a fourth was added, thereby completing the A, B, O and then the AB blood classification of mankind. Even at that time Landsteiner foresaw the importance of the different blood groups, but it was not until the Great War's spilling of blood – and the subsequent urgent need for transfusion – that the importance of his work received general recognition. In 1922 Landsteiner emigrated to the United States, and became an American citizen as soon as he could.

Five years later, together with Dr Philip Levine, he discovered yet another blood grouping; they proclaimed all human beings to have either M, N or MN blood. This secondary classification is irrelevant to blood transfusion, as it does not matter what kind of M or N blood each person receives, but it is scientifically exciting and of extreme genetical relevance in cases of doubtful paternity. Accused and unhappy males can be found not guilty of fathering some child if blood is examined and compared; but more about that later.

In 1930 Karl Landsteiner, then sixty-two, received a Nobel Prize. Seven years later he started work which led in 1940 to yet another discovery. Working this time with Dr Alexander Weiner, he hit upon the famous rhesus factor. It may seem valueless to care, let alone announce, that if a rabbit is injected with some blood from the monkey, *Macacus rhesus*, the rabbit then makes an antibody which clumps together the red blood cells of 84% of New York's white people. Yet the work was crucial. Without this breakthrough the 16% of women who have so-called Rh-negative blood, and who, until then, frequently produced jaundiced and dying babies, would not find their condition treated with such routine care in maternity units today; but also more about this later. Landsteiner finally died in 1943 when still in harness (he had a heart attack while at work in his laboratory). More than any other man he had made blood transfusion possible.

Transfusions   These began, as can be imagined, erratically and, curiously, with the assistance of Sir Christopher Wren. In 1665, Great Plague year, he suggested to the Cornishman, Dr Richard Lower that blood might be passed from one animal to another. Apparently the transfusion worked. Two years later Mr Arthur Cogan, a Londoner, capped this tale by passing into himself blood from a pig, and legend claims success. It seems that Lower, who became Court Physician to Charles II, let things be, but one of Louis XIV's doctors, Jean Denys, gave a patient a transfusion of lamb's blood. There was, reasonably, a violent reaction; but, less reasonably, the patient lived. In 1668 another patient died after similar treatment. (The purpose of the transfusion – with calf's blood – had been to transfer gentle and bovine characteristics to a philandering husband.)

Thenceforth blood transfusions in France were forbidden. Not until 1818 did anyone attempt a transfusion of human blood, when Dr James Blundell, of London, tried it. The patient died. One trouble, which had nothing to do with blood group clumping, was the clotting of the blood. Many different techniques were tried to keep it fluid, and were soon successful, but then came the brick wall of blood groups. Some transfusions were entirely successful. Some were equally disastrous, and the whole of the nineteenth century had to pass before Landsteiner solved the riddle.

Now transfusions are a commonplace. Over 2,000,000 Americans receive them in a year. Britain uses over a million bottles of blood and plasma annually. Blood banks are every-where. There are even banks of extremely rare blood, as at the Chelsea Naval Station, Massachusetts (where, for example, Rh-null is kept, reported so far in only a handful of people). Swimming against this tide are, amongst others, Jehovah's Witnesses. Unlike Christian Scientists, who ignore medical healing but are not definitely forbidden to take blood, the Witnesses accept hospital treatment but not blood. They quote the Acts of the Apostles Chapter XV ('. . . that ye abstain from meats offered to idols, and from blood') and Genesis ix, 4 ('But flesh with the life thereof, which is the blood thereof, shall ye not eat') as authority for their principle, and since 1945 have ruled out transfusions. There are over 50,000 active Witnesses in Britain, and they achieve world-wide publicity when the enforced lack of transfusion precedes a death. Mr Walter Stevens, for example, of Adelaide, refused permission in June 1965 for his wife to receive blood during a difficult delivery of twins. She died shortly afterwards, and the newspapers were quick to pounce.

Distribution of the groups All mankind, whether New Yorker, Eskimo or Aborigine, is either O, A, B or AB. The letters seem unnecessarily complex, and I, II, III, and IV would seem a much better and simpler method of classification. In fact Roman numerals were used originally, but these gave way to letters because the letters are more meaningful. A and B refer to the two clumping factors (technically these are two lipopolysaccharides, with both being known as agglutinogens) which human blood contains. It is these factors which cause all the trouble. Without them blood transfusions would be relatively plain sailing. Blood type A (42% of Western Europe) has one of them, blood type B (9%) has the other, type AB (3%) has both and blood type O (46%) has neither. Group O are often called universal donors, and in World War Two only group O blood used to be sent to the front in many areas; but it is a misnomer. For various reasons not all group O blood can be pumped into A, B or AB people. Some group O donors are more universal than others. (It all depends upon the amount of anti-A or anti-B agglutinins in their serum. Some O's have a lot. Most do not.) Similarly the AB people, the rare 3% of us, are not the 'universal recipients' they are frequently said to be.

To offset this nuisance value in transfusion the ABO groups are intriguing geographically. All races of man have them, but differently. During that war, for example, and during blood collection in Northern Wales, someone noticed that Welsh names (Jones, Evans, Williams) produced more O and B blood than English names. In fact, all Celts are now known to produce less A blood than the descendants of those who caused them to retreat. South-east England produces much more A blood even from its inter-mixed inhabitants today.

In fact, a history of Europe, albeit a short one, could be written from blood groups alone. The A people, whoever they were – but with a high A group percentage, seem to have occupied Europe with particular entrenchments in Scandinavia, Spain and Turkey. The O people fared less well and either preferred or were forced into the corners such as Ireland, Scotland, the Pyrenees, Iceland, Sardinia, Corsica and the Eastern Black Sea. Meanwhile the B's, never so plentiful, just get more so the further they are from Europe, whether up into Russia or down into the Middle East. The B's look like a second invasion force (the Mongols?) who were unable to oust the A's.

Globally the picture is more complex. South American Indians are 100% O. North American Indians are also rich in O but have more and more A towards the north. They have virtually no B. Maoris and Australian Aborigines are, like the Eskimos, half A and half O. West African blacks are mainly O (52%) but equally A and B (21% and 23%). Top B people are the Ainu of Japan (38%), the Asiatic Indians (37%) and Tartars (33%). Top AB's are the Congo Pygmies (10%), the Japanese (11%) and the Egyptians (10%). The O group is the most persistent. Some people have no A, like those South American Indians; some have no B, like many of the Indian groups further north; and some have no AB, like almost all the New World Indians; but no group anywhere has no O. Thirty per cent is about the lowest figure for O's, possessed by some Pygmy, Chinese, Russian and Tartar groups.

The blood group distribution therefore seems an awful mix-up. Why do the Basques have ABO proportions like the Australian Aborigines? Or why do the Greenland Eskimos and the French? Why do the Mongols, who are supposed to have supplied the ancestors of the Red Indian, have blood rich in B, while the Indians have next to none of it? The short answer is that advantages must exist for certain blood groups under certain conditions.

Group advantages In recent years some hints of differing benefits have begun to show. Professor Ian Aird, the surgeon, found O English people were more likely to get duodenal ulcers. More A people get stomach cancer (about 20%), pernicious anaemia (25%) and even diabetes mellitus than O's and B's. Such slender information, which by no means implies that the blood groups actually cause the diseases, is a long way from explaining, for example, why all the B's died out as the Asiatic groups, so rich in B blood, entered the New World via the Bering Straits – which they almost certainly did.

Extra facts about the ABO system are: the blood group of an individual is established irrevocably probably by the end of the second month of pregnancy; which group it is (useful in paternity trouble) can be found out by the twelfth week of pregnancy if the shed cells in the mother's amniotic fluid are examined; the muscle tissue of 4,000-year-old mummies has been blood grouped; bones long buried soon will be; dissimilar twins are more often of the same group than mere chance would indicate; and O mothers married to A fathers produce more miscarriages (certainly in Japan, and probably elsewhere) than A mothers married to O fathers.

All other groups While the ABO groups are vital in transfusion they are by no means the only groups. Second to be discovered was the M, N, MN system. Two genes control this. English people are 32% M, 19% N and 48% MN. To be N both parents cannot be M. To be M both parents cannot be N, and to be MN means that both parents cannot be exclusively M or N, all of which is crucial evidence for the paternity courts. Once again the geographical distribution is odd. Blacks, Pacific Islanders and Aboriginal Australians are rich in N and weak in M. Eskimos, Red Indians, Welsh, Chinese and Japanese are rich in M and weak in N (the Welsh are never good at conforming). There is slightly more sense in the MN distribution than with the ABO's, but that Bering Straits migration is still awry. Asiatic Mongoloids have many N's among them; American Mongoloids hardly any.

Since Landsteiner started the blood group avalanche, more and more have been found, such as the MNSs, the Rh-HR (the famous Rhesus factor), the P groups, the Kell, the Lutheran the Duffy, the Kidd, the Lewis, the Diego, the Hunter and Henshaw, the

Sutter – most of which are named after the patients in whose blood they were first found. Therefore a man is not just O or O and M, but – to quote the commonest type of Englishman – O, MSNS, P_1, CDe/cde, Lu^bLu^b, kk, Le^bLe^b, Fy^aFy^b, Jk^aJk^b. London's Blood Group Research Unit estimates this lot occurs once in every 270 people, and there are 303,264 possible combinations of blood groups in Englishmen. Consequently do not leave fingerprints or blood at the scene of the crime, particularly at some foreign scene. Any European's blood can be distinguished from a West African's ninety-five times out of a hundred, thus narrowing the field of suspicion abruptly.

It is reasonable to wonder where all these blood groups came from. Consequently tests have been done on the higher animal species. Chemically there are differences but many more similarities. Most monkeys are A or B, rarely both. The gorillas seem to be either A or B, the chimpanzees are mainly A and sometimes O, the orang-utans and gibbons A, B or AB. It is strange that O, the most frequent group in man, is the rarest (bar 10% of the chimpanzees) in the apes and monkeys. The evidence is insufficient to provide any powerful pointers towards the origins of man. Perhaps the less well-known groups, with perhaps less evolutionary influence, will provide better clues when comparative work has been done.

Rhesus If you are a Rhesus-negative girl (one-sixth of British girls are) marry a Basque, or, better still, a Walser of Switzerland. But think twice about Pacific Islanders, Australian Aborigines and American Indians. You may be making life marginally easier for yourself and your obstetrician, and considerably easier for your babies, if you choose a Rhesus-negative mate. Thirty-six per cent of the Basques are eligible, only 16% of Englishmen, but no Red Indians, Papuans or Aboriginal Australians. Also hardly any Chinese or Japanese are Rhesus-negative. If you do choose a Papuan, or even a Rhesus-positive Englishman, there is a chance (about 72%) that your baby will be Rhesus-positive, like his father, unlike you. This could mean trouble, but such a marriage with such a baby is only troublesome in 2% to 5% of cases.

Reasons were at first poorly understood, but whether or not there was foetal bleeding at birth, causing the mother's blood to meet the foetus, is now known to be highly relevant to the fate of the next child. So are the ABO blood groups because, if both parents have different ABO groups, the new babies are less likely to suffer from Rhesus trouble. This stroke of good fortune is linked with the massive resentment between one ABO blood group and another blurring the minor resentment between the two Rhesus groups. Anyway if you, a Rhesus-negative female, do mate with a Rhesus-positive male, there will be fewer Rh-incompatible foetuses if you are O and he either A, or B, or AB. Things are less well from the Rh point of view if you are A and he is either O or A.

This trouble, this Rh-incompatibility of the foetuses, has many names such as Haemolytic disease of the Newborn, Erythro-blastosis foetalis, Kernicterus, and Icterus gravis neonatorum. Whatever the name it is basically a destruction of the baby's red blood cells. The mother (having produced antibodies against her Rh-positive foetus) is doing the destroying. So the baby becomes anaemic and jaundiced and, were it not for transfusions, would die seventy-five times out of a hundred, sometimes before birth. Recent treatment was fairly straightforward. The sooner the baby was born the better (within reason), for the sooner it had left its mother's destructive (haemolytic) influence the better. Then its blood was exchanged to get rid of all those maternal antibodies and in 1963 a foetus's blood was even changed *before* its birth.

Although such ante-natal transfusion is an extremely modern technique, it was outmoded almost at once by an even better system for protecting the victims of Rhesus incompatibility. Professors C. A. Clarke, P. M. Sheppard and others at Liverpool University, after working with Rh-negative male volunteers, devised a way of preventing Rh-negative mothers from producing antibodies against their Rh-positive offspring by using gamma globulin with what is known as a high titre of Anti-D. This new system, being one of prevention rather than cure, has enabled one in 200 of all British babies (roughly 3,000 a year) both to survive more frequently and have an easier time of it than formerly. The period, therefore, between the first comprehension of Rhesus incompatibility and the discovery of an effective counter-measure was less than three decades.

The clue which led to the unravelling of this story at the start of World War Two was that virtually no first babies were affected, with one vital exception. If the Rh-negative mother had ever received a transfusion the chances were (with 84% of us being Rh-positive) she had received Rh-positive blood; hence she could have become sensitised, hence the preparation of Rh-antibodies, hence her firstborn's blood being cruelly attacked. Nowadays no Rh-negative woman before her menopause is given Rh-positive blood. The situation is reasonably in hand.

A fascinating upsetting of the situation came from Sweden in

October 1965 and was published in *The Lancet*. A girl, Rh-negative, aged twenty-four, had been a virgin when married in 1961. Consequently she had had no earlier children or abortions and no chance to get sensitised from Rh-positive blood. Also she had never had a transfusion. But her first baby was born jaundiced, and she was full of Rh-antibodies. Her second baby was similar. It was all a mystery until she admitted receiving blood, when aged nine, during a girls' 'brotherhood' ceremony. The amount transfused was presumably small, but Lund University doctors think it could have done the trick and sensitised her against her first child born a dozen years later. Anyway, they sought out and found her blood brother. Sure enough she was Rh-positive.

Although the ABO groups have been linked with disease in a few cases, the advantages or otherwise of being Rh-positive or negative are still totally obscure. As Sir Peter Medawar puts it (and he rarely misses the opportunity of a happy phrase) the Rh blood types are associated with nothing 'except the unqualified incubus of transfusion accidents'. It would seem better for the human race if all of us had similar Rh blood (and similar Kell blood too, for the same kind of story exists with that grouping). Why, therefore, the dissimilarity? There is still no explanation nearly half a century after the Rh discovery. 'It is not known,' says Medawar; 'it is merely being groped after.'

Paternity and the courts Blood groups may be irksome in transfusions, but they are the breath of life in paternity cases. Many a male, harassed by accusations, has had cause to be grateful to his grouping, to hers, and to that of the child.

Can a man prove he is the father of a child? Can he prove that he is not? Briefly, the answers are 'No' and 'Probably', respectively. The key is blood. There are three ABO blood group genes, and four ABO blood groups. No child can acquire a gene, and consequently a blood grouping, if it is not possessed by either parent. If the wife is group O, the husband group O and the baby group A, that husband has no reason whatsover to believe the baby is his. He can regard quizzically those of their friends who are A or AB. (The British courts accept blood group evidence and do now have the power to order tests in paternity cases; bringing Britain into line with, say, Scandinavia and some United States states, such as New York. The procedure starts becoming really crucial when illegitimate children are to be given as much of the father's wealth as his legitimate offspring.)

If the mother is	and the child is	the father can be	but not
O	O	O, A or B	AB (3% of European population)
O	A	A or AB	O or B (55%)
O	B	B or AB	O or A (88%)
A	O	O, A or B	AB (3%)
A	A	any group	
A	B	B or AB	O or A (88%)
A	AB	B or AB	O or A (88%)
B	O	O, A or B	AB (3%)
B	B	any group	
B	A	A or AB	O or B (55%)
B	AB	A or AB	O or B (55%)
AB	AB	A, B or AB	O (46%)

The percentages are important. It is obviously easier to settle the paternity question if 88% of the population can be excluded rather than 3% or even 0%. (It's a wise mother who knows her lover's ABO blood group before she begins complicating the issue.) Taking the population as a whole, the probability of freeing a wrongfully accused man, using just the ABO groups, is only about 20%. So other groups are used, such as the MN and Rh groups.

If the baby is MN and the mother is M		the father cannot be M
MN	N	N
N	any group	M
M	any group	N

The MN test frees a man from responsibility in about 18% of cases. For the Rh test (if both parents are Rh-positive, the child cannot be Rh-negative; if both parents are negative, the child cannot be positive), the maximum chance of exclusion in Western Europe is about 25%. With all three blood tests being used any man has at least a 50% chance of disproving his paternity, and it could be more (75%) if further blood groups are tested. At present most men wrongfully accused of fathering children can prove their innocence by blood group tests. In the world of fact – and the courts – fewer get off. Many women must, therefore, be telling the truth when they point the finger.

In Germany one hundred children plus their correct fathers were once examined by an expert independent panel. (Such paternity experts have a high status in Germany and Scandinavia, not so in Britain.) Using customary methods (and merely looking at parents plus offspring without any recourse to blood testing) the examiners had to decide whether it was probable, improbable or not determinable that the children were correctly assigned. The results were emphatic. Ninety-three of the children were 'probably' the sons of their fathers (varying from 'more probable than not' right up to 'probable to a degree bordering on certainty'); seven of them were not determinable; but none were said to be definitely not the sons of their fathers. In other words, like produces likeness – an adage more correct, biologically, than like producing like. Like producing like implies duplication, much like a printing machine. In biology, certainly above the most primitive level, the offspring is only similar to its parents, not the same.

Apart from blood and resemblance there are genetic extras like eye and hair colour. When both parents are blue-eyed all their children ought to be blue-eyed, but occasionally an incomprehensible brown eye creeps – legitimately – in. Two red-haired parents ought to have all red-haired children. And the same sort of thing goes for parents both tall, or both with attached ear lobes, or both with straight hair, or both blond. They are more likely to have tall, attached ear lobe, straight haired and blond children. Tall parents may produce short children, but on average they do so less frequently than short parents. Some characteristics obey definite rules, notably when only one pair of genes is involved (as with blue eyes). Some obey or follow probability, as when many genes are involved (like height and hair). Both types have a part to play in linking children with the parents who produced them.

In a blond, blue-eyed, tall, fair-haired community these particular characteristics are valueless in assessing paternity, and the infrequent ones are more important. In Norway a normal mother produced a brachyphalangic (short-fingered) child, a rare condition. The court asked the accused to hold up his hand. His fingers were short, and he therefore had to pay. Other men have been caught by the presence of hair on the middle digit of one of their fingers. This characteristic also cannot be passed on if it is not possessed. The courts can be wrong in coming to a yes or no answer, but it is possible to calculate – given all the facts about relative frequencies of certain characteristics – how probable or improbable a wrong decision may be (the postman of that Norwegian couple may also have been brachyphalangic, but this is unlikely). Curt Stern has called it 'an exact numerical evaluation of the probability of correctness of a paternity judgement'. The law already possesses a better phrase – beyond all reasonable doubt.

As a tailpiece to paternity there are occasional cases of double-sired twins. In the first to be officially recognised the mother (blood group O and M) produced dissimilar twins, a boy (B, M)

270

and a girl (A, MN). One man could have produced both, by being AB and MN; but no such man was in that mother's life. She had known, as the Bible uses the word, only two men, one being A and MN, the other B and M. The first could not have produced the boy, for where did the lad get his B from? The second could not have produced the girl, for where did she get her A or N from? Both men must have played their part.

19. The Growth of Babies

The rate of growth/Survival/Cot-deaths/Battered babies/
Developmental progress/First reflexes/Changes in lungs, brain,
sleep, skeleton, muscle, fat, heart, kidneys/Spock/Swaddling/
Child law

*From the child of five to myself is but a step, but from the new-born
baby to the child of five is an appalling distance.'*

LEO TOLSTOY

The human baby, already 266 days old at birth, is incapable of
survival without help. Countless mammals who have spent less
time in the uterus are able to stand and run within minutes of
being born, whereas the human baby will stagger for the first time
into a walk some twelve months after birth. Most mammal
offspring find and feed from the life-supporting nipples with great
competence; the human baby needs all the assistance it can get.
By the time many other large mammals have grown up to rear
their own families, the human infant is still slobbering around,
falling down stairs, eating anything that comes its way, crying,
getting lost, defaecating at random, and generally behaving in a
witless fashion. At what age, one wonders, would human infants
be able to survive if removed from parental care and placed
instead on some sumptuous island, rich in foods of every kind? At
birth they would die virtually at once. At the age of one they
would fare little better. At two they might get by, but apathy and
precipices would ensnare them in the end. At three or four the
mortality rate would still be tremendous, even assuming a
disease-free island. *Homo sapiens*, the wise one, the pinnacle of
evolution, certainly takes his time to achieve his wisdom.

Nevertheless, that is his strength. His growth is slow. The whole
pace is slow. The time before breeding is long. The days for
acquiring and testing individual skills are seemingly endless.
During the first eighteen years of life a baby's stature, from top to
toe, is multiplied less than four times, and although mothers may
think otherwise, as nipples yield to bottles, and bottles yield to
plate after plate of food, a child's gain in weight is pitifully slow by

272

comparison with most other animals. A mere 5 lbs. (2·3 kgs.) in the first three months, or 14 lbs. (6·3 kgs.) in the first year, or 38 lbs. (17·2 kgs.) in the first six years of life – such increments are most modest. (For comparison, a bacon pig weighs 3 lbs. (1·4 kgs.) at birth, and 200 lbs. (90 kgs.) at 6 months; a Friesian calf weighs 100 lbs. (45 kgs.) at birth, and can weigh 900 lbs. (400 kgs.) ten months later.)

Even during the adolescent growth spurt when boys (and to a lesser extent girls) eat hugely both at meals and in between meals, when they gather food towards them much as magnets gather iron, the rate at which they put on flesh is small. A gain of 12 lbs. (5·4 kgs.) in one year is a fair pubertal increment. The average height increase for boys at this time of most rapid growth is about $3\frac{1}{2}$ ins. (9 cms.) in a year, and $2\frac{1}{2}$ ins. (6·3 cms.) for girls. Aunts and uncles express amazement at the weed-like change in stature, and rows of outgrown clothes give a more objective assessment; but in fact there is nothing biologically remarkable in $3\frac{1}{2}$ ins. (9 cms.) a year, or less than a third of an inch (8·5 mm.) a month. Mankind does take its time. A boy born 20 ins. (50 cms.) long and reaching 6 ft. (1·8 m.) after eighteen years is growing less, on average, than 3 ins. (7·6 cms.) a year. From his first to his eighteenth birthday he is only growing about $2\frac{1}{2}$ ins. (6·3 cms.) a year, and from the second to the eighteenth birthday only $2\frac{1}{4}$ ins. (5·7 cms.) a year. No other creature of similar bulk is quite so leisurely.

In a sense growth has practically stopped by the time of birth. At the twelfth week of pregnancy the baby's increment per month is 600%. Four weeks later it is 220%, four weeks later still 120%, then 90%, then 50%, then 25%, then – in the last four weeks – 20%. Most of the original growth power has gone by the time of birth; a small percentage remains to finish, in eighteen years, the job that began so fervently nine months before. Growing up is – for humans – a slow business, particularly after birth, and is also a dangerous business.

The child is more prone than the adult to infection, and any country which achieves an improvement in its child mortality figures is generally doing it by better control and treatment of infection. However, even in modern countries, many children still die. In Great Britain about 25,000 children (0–14 years) die every year. Of this not negligible number about 16,000 have failed to reach their first birthday. And of these short-lived infants about 13,000 have not even survived for a month. That first month is made hazardous primarily by birth itself, by the trauma of the event, by actual birth injury, by the possibility of oxygen lack, by the problems of prematurity and by unfitness for the sudden change. It is also the time when a large number of congenital malformations have a telling effect upon the statistics; severe deformities frequently lead to death in the first few hours of life.

In many poorer communities, even today, it is the exception rather than the rule for babies to survive into adulthood. Dysentery and the host of allied enteric disorders have probably been the greatest killers of all time for mankind; they are still formidable barriers to survival, notably at weaning time. Not so long ago in England more babies died than lived. In London the births of 315,456 babies were registered between 1730 and 1749. Of this number 74·5% died before the age of five. By the end of that century, and in a similar nineteen-year period between 1790–1809, the births had risen to 386,393 while the deaths up to age five had fallen to 41·3% of the total. To a very large extent this solitary pair of facts explains the sudden surge in Britain's population. This percentage was even better than seventeenth-century mortality figures for the upper classes. Of the 32 royal children born between the reigns of James I and Anne only 10 lived until they were twenty-one.

Medicine in the eighteenth century was still archaic, towns were still foul, hygiene was still poorly attended to, poverty was extreme – and yet about twice as many babies struggled success-fully through their first five years at the end of the century as at the beginning. By the time of World War One 87% of babies were surviving until five, and today about 97% are surviving until their fourteenth birthday.

Control of infection among the community at large was primarily responsible for the improvement. Even since 1925 there has been huge change. In that year there were 2,774 deaths in England and Wales from diphtheria, 988 from scarlet fever, 5,337 from measles and 6,058 from whooping cough. By 1960 the figures for these four diseases were 8, 3, 27 and 152 respectively, or 1·25% of the 1925 total. Some diseases and childhood ailments have proved more stubborn. Pneumonia and bronchitis dropped dramatically in the 1930's and 1940's, and have since remained steady, but congenital malformation deaths have stayed steady since the 1930's, thus forming a higher and higher proportion of child deaths. Deaths from birth injury, asphyxia and prematurity dropped in the 1940's, but have since stayed fairly constant. Accident deaths have fallen slightly since the 1930's, with the rising number of actual accidents being just offset by the gradual improvements in surgery and nursing. Malignant disease in children has been slowly rising since the 1930's (although one immediately suspects that more accurate diagnosis has had much to do with the rise). The current order of severity for the causes of child death is: birth injury (including asphyxia and prematurity), infection, congenital malformations, accident, malignancy, and then the various other relatively minor hazards. All in all the first part of life is still the most perilous until the individual reaches advanced middle age.

Which part of that early life is the most serious? In advanced countries the first month is definitely the most severe hurdle. In some primitive countries the second year can be the worst. Their first year is relatively care-free, with a lot of breast milk (therefore reasonable nourishment, little contact with bacteria, a close relationship with the mother's own immunity system), and not much opportunity for grubbing around and picking up infection. The second year child is no longer carried so readily, nor fed so well, nor protected so happily from pathogenic bacteria. Infection and malnutrition then form a deadly alliance. The death rate in primitive countries for that lethal second year can be fifty times higher than in advanced countries.

It is the combination of poor food plus infection which is so disastrous. The same measles virus exists in the United States, Chile, Guatemala and Ecuador. Its strength, or virulence, is roughly the same, but its killing power is totally different between the one advanced country and the other three. In the United States children do not, in general, die from an attack of measles. In Chile similar attacks kill 138 times more frequently, in Guatemala 189 times and in Ecuador 418 times more frequently than in the United States (figures from the Pan American Sanitary Bureau).

Diarrhoea can be equally devastating, often killing – in Guatemala – within 24–48 hours after it has begun. Well-fed children do not, in general, die with such promptitude from similar infections. Attempts to discover which particular bacteria are killing Guatemalan children afflicted with post-weaning diarrhoea have often proved unsuccessful; no known bacterial pathogen can be isolated. The conclusion is that many such lethal diarrhoeas represent a reaction to bacteria which would not ordinarily be pathogenic to well fed children. It is not just the bacteria, it is not the lack of good food; it is the crippling effect of the two together that still prevents so many millions of weaned children from reaching their second birthday.

Cot-deaths Abrupt child death is not only a feature of poor countries. The so-called cot-death of the advanced countries (crib-death in America) can be equally sudden. Its principal characteristics are that it is totally unexpected and generally inexplicable. United States' estimates vary from 10,000 to 20,000 cot-deaths a year. Some coroners prefer to blame a virus rather than unknown causes for these deaths; consequently, according to various pathologists, the US figure may be as high as 30,000. British estimates are somewhat lower, but still large. Some say such deaths occur 14 times in every 10,000 babies; one town (Hartlepool) gave a rate of 35 per 10,000 babies. The 1982 figures for England and Wales were 2,139 boys and 1,503 girls dying during their first year from SIDS, the sudden infant death syndrome. A further 14 boys and 8 girls died from this cause, or multiplicity of causes, when over one year old. About half of these die at home, indicating a certain suddenness, and about 80% of the sudden home deaths come into the cot-death or SIDS category. The remaining 20% are caused by entirely explicable accidents or disease. Of the 80% which are classed as cot-deaths a definite cause of death is not found for most of them even after extensive post-mortem. The highest death rate is within the second to fourth months of life, and most such deaths occur in winter at night. The cot-death syndrome, by whatever name, is not to be found in the scientific literature until after World War Two. It is now the biggest killer of babies in most industrialised societies.

Of course there are plenty of theories about these sudden deaths. In the past bedclothes and pillows were often blamed, or the breathing in of vomit. More recently, particularly as child deaths from other causes are decreasing and raising the proportion of cot-deaths, greater attention has been paid to this problem of the suddenly dead child, and the possible causes are listed as: allergy to some of the proteins in cow's milk; virus infection; the effect of abnormal parathyroid glands; the result of a maternal illness during pregnancy; the lethal presence of ammonia arising from wet nappies/diapers; a failure of the carotid body within the brain that monitors and regulates breathing; and even cold. The wintry increase in the cot-death rate made people think of cold, although there seems to be no correlation between cold nights and cot-death incidence. An engineer recently took the trouble to write to the *British Medical Journal* with the idea that the babies might even be suffocating themselves with their own expired air. With a deep cot, and no side ventilation, he thought the resulting accumulation of carbon dioxide could be silently fatal. Whatever the cause, or causes, the thousands of abrupt cot-deaths in the night are a reminder that a baby's existence is fraught with danger even in the well nurtured, well doctored western world.

Battered babies   Quite apart from the modernity of the 'cot-death', or at least of its term, there is another title of similar age: the battered baby syndrome. Some babies are literally battered to death by one or other of their parents, and many more suffer violent physical pummelling and mutilation. The British are kind to dogs but cruel to children, says the old maxim, and there is still truth to it. The National Society for the Prevention of Cruelty to Children investigates cases involving tens of thousands of children annually. Roughly half the cases are of child neglect, and roughly 10% of actual ill-treatment. In a recent year, according to the Home Office, 275 men and 265 women were found guilty in magistrates' courts, and seven men and three women in higher courts, of wilfully assaulting, ill-treating or neglecting children in a manner likely to cause them unnecessary suffering or injury to health. And, for one reason or another, due to various forms of family breakdown, 11,213 children in England and Wales were received into care by local authorities during that year. Although cruelty is old, and the NSPCC itself is a Victorian foundation, the battered baby name was only coined three decades ago.

An American radiologist, Dr John Caffey, gave the classical description of the new syndrome in the *British Journal of Radiology* in 1957. The article, plus the subsequent publicity and discussion about it, brought many more cases to light: a one-year survey of all United States hospitals revealed 749 cases. This is a frequent development following new descriptions, but the battered baby accounts are complicated by the fact that all such pummelling is caused by other human beings. Therefore both ethics and the law are involved. It is not a matter of describing, say, a case of measles caused by a virus but one of positive ill-treatment caused by one particular person against a small child.

Eight years after Caffey's article describing the battered baby syndrome had been published, the first conviction in England was brought against a man for murdering his children in this fashion. The jury could not accept the man's statements that the two children had received their injuries accidentally. This case's details well exemplify both the problem *and* the difficulties of detection.

In December 1963 a four-month-old girl, Susan, was found dead in her cot, having been 'off-colour' for a few days. No doctor had attended her during this final illness, and her death was reported to the coroner. A pathologist discovered many bruises, a

4½ in. (11·4 cms.) skull fracture, and a ruptured liver. The father mentioned that Susan had struck her head on the cot three days before her death, and he said the other injuries were possibly due to his attempts at artificial respiration. An open verdict was returned. Ten months later there was another death in the same family. Michael, who would have been Susan's younger brother had he and she survived, died at five weeks. (Michael must have been conceived within a very few days of Susan's death.) Upon examination the infant Michael was found to have nineteen bruises and a ruptured liver. This time the father said his knee had accidentally caught the baby in the stomach. After this death there was a trial. On January 19, 1965, Laurence Michael Dean, aged nineteen, was convicted at the Old Bailey of the murders of both Susan and Michael.

Both before and after this case it had become clear that other similar deaths have obviously not been recognised in their true colours. With many accidental deaths happening naturally, with victims too young or injured to speak for themselves, and with parents only too happy to distort the truth, Laurence Dean must have had many other quite undetected predecessors. Professor Keith Simpson, the forensic scientist, wrote after the Dean case: it is hoped that doctors will 'sharpen their perception of the battered baby syndrome' for the crime is probably widespread and can only too easily escape detection.

In the United States, where the problem is also serious, there is much current controversy over the reporting of such incidents. Should the doctor, or a neighbour, or any suspicious passer-by tell the authorities? Some states, such as California, insist upon the reporting of all injuries inflicted by violence. Others do not, and keep such mandatory laws for violence from, for example, guns. During the 1960's those states without some form of child abuse law, demanding that such cruelty be reported, brought themselves up-to-date with this unwelcome necessity. It is not presumed that more children are being battered these days. Instead society is more determined that such misconduct should be brought to book.

One final complexity is that some children can look for all the world like battered babies, in that they are covered with scars and bruises, but they can be victims of inherent weakness in their connective tissue. The result is a skin that is easily cut and bruised, as well as the possibility of a totally false diagnosis.

Developmental progress   Assuming survival, a baby's physical development is a bewildering assortment of achievements.

Parents may consider each stage to be an eternity in itself, an endless drooling, a perpetual repetition of ma-ma, a ceaseless and apparently hopeless determination to crawl, to stand, to use a spoon without a total spilling at the very moment of intake, to achieve the impossible, to grow up. Childhood is interminable – and yet each stage is over in a flash. It is a time of continual change – and yet lasts for eighteen years. It is the development of a mature human being, with sexual maturity suddenly and strangely arriving some two-thirds of the way along the long route to adulthood.

At birth a baby is 7½ lbs. (3·4 kgs.) of potentiality, but little ability. He (or she) can suck, he can swallow, he can salivate, he can cry, he can smell, taste and hear. He can yawn, hiccough, sneeze, cough and stretch. If on his face he brings his knees under his stomach. If held by his feet his head hangs down. He has a very long way to go, despite being over a quarter of his adult height. He is also a bundle of reflexes, some of which are extremely short-lived and pay temporary lip-service to his evolutionary past:

The Moro reflex A rapid change in position can initiate it. The arms suddenly fling outwards, with the hands open, and then come together as in an embrace. The reflex is harder to elicit after two months.

Grasp reflex Stimulation of the palm leads to a closing and gripping of the hand. For two months this grip can be strong enough for the whole weight of the baby to be lifted in this fashion. The same reflex exists for the foot, but the foot cannot grip. After about two months the foot grasp (always hard to elicit) disappears, and the toes then tend to turn upwards and spread apart when the sole is stroked. This so-called Babinski reflex then disappears in normal children at about two years, and is replaced by a turning downwards of the toes. (Should the Babinski reflex persist, it is an indication of some defect in the nervous system.)

Rooting reflex Touching of the cheek or corner of the mouth elicits head turning in that direction. Later on the lips will also protrude when this is done. Initially the lower lip is lowered, and the tongue moves to the touched side.

Blink reflex Stimulation of the eyelashes causes blinking, whether awake or asleep.

Sneezing Irritation up the nose or even bright light can cause sneezing.

Walking reflex   A baby can be made to walk by holding him vertically with his feet on the ground, and then moving him forwards. The co-ordination and rhythm are good, but this reflex has gone by six weeks.

Tonic neck reflex   When lying down the face will be turned sideways. If the head is then turned to look the other way, either voluntarily or if forced, the legs and arms will also change position. Those flexed will straighten, and those straight will flex. The reflex may not be present at birth, but goes after three to six months.

Dolls eye reflex   Eye movement lags behind when the head is rapidly rotated. This is a very short-lived reflex lasting only a few days.

Crossed extension reflex   If one leg is stretched and the sole of the foot stroked, the other leg bends and then stretches. Usually this reflex has gone after a month.

Galant's reflex   Stimulation of the trunk between ribs and hip on one side causes a curving in of the trunk on that side. This reflex disappears after the second month, but reappears much later.

The development of a child follows a well-worn path, but every child takes a unique route of its own along that path. Read any book about child growth, with its generalisations of progress by each birthday, and they only fit a particular child in a very general way. Such attributes as watching a dangling toy, as squealing with pleasure, blowing bubbles, disliking certain foods, seeing a mirror image, placing one brick upon another, jettisoning the pram's contents repeatedly, saying 'Why?' no less infrequently; all these and countless others are manifestations of progress, but their timing is highly variable. They are milestones along the way from being a pathetic, toothless, hand-flailing collection of reflexes, crying without tears and excreting without restraint, into the adult human form, a creature more of reason and sense than of reflex and instinct.

Such progress is by no means even. Thomas Carlyle is alleged to have said, as his very first words, 'What ails wee Jock?' to a fellow one-year-old suffering vehemently in the next cot. He then spoke only in sentences. Other children, who grow into adults of no lesser stature, are content to drool and speak meaningless absurdities until deep into their childhood. The milestones are indeed passed, but without too much concern for the calendar.

280

The importance of human development lies in what takes place, rather than when it takes place.

Nevertheless a punctual and triumphal procession through all the myriad stages of childhood does indicate normality, even though all normal children are relatively forward in some respects, relatively backward in others. If a baby lying on its stomach can lift its head momentarily at four weeks, readily at six weeks, recurrently at eight weeks, if it can bear weight on its forearms at twelve weeks, and raise itself at sixteen weeks so that it can look directly forward, then average development is occurring. At sixteen weeks rattle-shaking begins, at twenty weeks the deliberate grasping of objects, at twenty-four weeks the bottle is grasped, at twenty-eight weeks biscuits – and much else – are soddenly sucked, at thirty-two weeks unsupported sitting is possible and supported standing, at thirty-six weeks nearby furniture is suitable support, at forty weeks there is strength enough for the child to pull itself into a standing position, at forty-four weeks one leg can be lifted, at forty-eight weeks a crab-like shuffling is effected, and at one year a walk is possible, either supported with one hand or even, however short-lived, with no exterior aid whatsoever.

All such activity and aptitude indicates a correct development of various organs, for balance, muscle, bone, brain and vision are each involved. The outward and visible signs of progress are merely manifestations of all the organic changes taking place within. Nothing is possible without suitable development of the appropriate organ, or organic systems. They come first.

Lungs   Before birth a baby has no air in its lungs. The alveoli – the countless bulb-like endings of the lungs – are mostly collapsed. The major airways of the bronchial tree plus the nose and mouth are full of liquid, the amniotic fluid of the uterus. At the first vital breath the alveoli open up, and air reaches the lungs. Amniotic fluid is then absorbed through the alveoli, and by the third day all the alveoli have expanded. The first breath can make use of about half the potential lung space. By the third day, breathing at 30–80 times a minute and taking in about 20 cc. of air per breath, all available lung space is being used. Premature babies breathe faster, and take in less air per breath.

Within a year the respiratory rate has gone down to 20–40 per minute and each intake up to 48 cc. By two years the intake is 90 cc., by three years 125 cc., by ten years 320 cc., and by twenty years 500 cc. Breathing rate is then 15–20 times a minute.

Brain　This more than doubles its weight in the first year after birth – 12 ounces (350 gms.) to 32 ounces (910 gms.). Growth continues to be rapid, and by three a child has already accomplished three-quarters of its post-natal brain growth. By the age of seven the brain is almost completed (2 lbs. 10 ozs./1·2 kgs.) and by ten it is virtually adult-sized (2 lbs. 14 ozs./1·3 kgs.). Less is known about the actual development of the brain with regard to nervous pathways, etc.; but certainly such pathways do develop, at least until the age of four. Physical changes to account for intellectual development after this time have not yet been discovered. (There is much more about the brain's development in this book's companion volume, *The Mind.*)

Brain plus nervous system form one-tenth of the total body weight at birth. By five they are one-twentieth, and by adulthood one-fiftieth of the total. Naturally no amount of teaching, coaxing or practice will cause a baby or child to perform some skilled act until the necessary mechanisms in the nervous system are mature.

Sleep, more closely associated with the brain than any other organ, is extremely variable from child to child. Dr Arnold Gesell, the expert in pre-school development, has said sleep can be one of the easiest aspects of child-life to overcome or one of the most difficult. A difficulty lies in the amount considered – by the child – to be sufficient. A study of 783 Minnesota children aged two and a half showed their average sleep to be 12·9 hours, but they ranged from eight hours to seventeen. Many a child needs no more sleep than the average adult, and many children can bounce back into wakefulness in a totally non-adult fashion. From a dormant lump of flesh wracked with weariness they can leap up again thirty minutes later ready for hours more play – if their former sparring partners are not incapable.

The newborn baby cannot help going to sleep. Within a year the process can be inhibited by the child. Between nine and twenty-one months going to sleep is often partnered by some fearful head-banging and bed-shaking, or gentle rocking and finger sucking. Waking up often sets the two- and three-year-olds crying. Boys are more restless in sleep than girls.

Skeleton　At first this is connective tissue, like the soft fontanels of a baby's skull; then it is cartilage, and finally it is bone. The process is not complete until the age of twenty-five.

The skull bones are soft at birth. During birth the head is often moulded or distorted, but such effects usually disappear in a few days. However the skull is still soft, and moulding can continue, caused just by the weight of the head as it rests on the bed-clothes. Remarkably, the flattened skulls of three to four months sort themselves out, and become symmetrical by the first birthday (except for those for whom genetics and family characteristics have ordained a flat-sided head).

There are in fact six fontanels (or fontanelles) at birth, with the anterior fontanel being the most noticeable and longlasting. The posterior fontanel is also in the midline at the top of the skull, but two inches further back. Two others are on either side, roughly where the temples are, and the final two are even lower down the sides of the skull and further back. The anterior fontanel, which pulses visibly and makes any baby seem yet more fragile and vulnerable, actually enlarges for the first two months after birth. It then shrinks, and finally closes. The actual time of its closing is notably variable; anywhere between four and twenty-six months (in one survey of 1,677 babies).

The proportions of the skeleton itself change markedly. At birth the neck is short, the shoulders are high, and the chest is round. Later on, particularly between the ages of three and ten, the neck lengthens as the shoulders are lowered, and the chest broadens and flattens as the ribs slope increasingly downwards. Children's arms look normal long before the legs do. E. E. Price has written that 'a very casual observation of the posture of normal children will refute the suggestion that the legs should be straight'. Bow-legs and knock-knees are common among the three-year-olds. About a fifth of all normal children at the age of three are knock-kneed to the extent of having two inches or more between their ankles when they are standing upright with knees together, and many a bow-leg does not straighten until the child is four or five. Flatfootedness is also a normal feature of infancy, and many children toe-in their feet to an astonishing degree. As with flattened skulls it is quite remarkable how seemingly deformed a normal child can be.

It should never be forgotten that the upright human is a recent occurrence. It is particularly relevant to remember this point in connection with skeletal development. For geological epochs man's direct ancestors were four-footed creatures with a horizontal spine. The change to bipedalism had, and still has, far-reaching mechanical consequences. 'The erect posture that helped man master the animal world also makes him vulnerable to low back pain' states an advertisement which draws attention to this point. Just because the body is muscle, bone and tissue does not nullify the engineering consequences of up-ending the structure.

The Body

Admittedly the human body has changed many of its characteristics and proportions since that 90° switch to two-leggedness, but its entire ancestry lies in a four-legged past. The changes since then have been amendments, not creations; they have been modifications to an existing form. There has never been anything equivalent to a scrapping of the old design plus the sudden initiation of a brand new one. The skeleton has had to continue its supporting role without a break, both during and after the up-ending process. From being horizontal to vertical, from being four-legged to two-legged and two-armed, it has had to adapt itself and make use of a structure primarily suited for another purpose in another age.

Muscle At six months the volume occupied by muscle is very small, and by bone is very large, relative to an adult. In the first three years of life, muscle grows faster than bone. X-ray pictures show muscle widths to be increasing at twice the speed of bone width. By the age of three or four the two rates of growth are similar, and bone and muscle continue to grow equally. Then, at adolescence, there is a sudden leap forward in both bone and muscle growth.

Generally speaking, the increase in a certain skill involving a muscle means that the part of the system in nervous control of that muscle has developed sufficiently for the skill to be achieved.

Fat 'Nothing is known', Professor J. M. Tanner has written, 'of the physiology of the wave of fat which threatens to engulf the one-year-old.' This wave starts at about the thirty-fourth week of pregnancy. Subcutaneous fat is then laid down for the first time, and the process continues to produce the bicycle-tyre one-year-old. (Of course some babies never swell in this fashion, and are presentably slim throughout their lives.) The fat inflation customarily recedes after that first birthday, and continues to decrease until the age of seven. It then increases again, although slowly. At adolescence the increase may stop for boys, and the total fat may even diminish. It will not be replaced until adolescence is over, and the boys are out of their teens. The traditional picture for girls is that there may be a very slight halting in the deposit of fat, but it is nothing like the masculine decrease. By the time their adolescence is over the girls have the correct feminine curves, which are formed and softened by the fat lying beneath their skin. If not beauty, then shape is more than skin deep.

It is not known whether the balloon-like one-year-olds are

more likely to grow into fat people. If they are still plumper than average at the age of three they will probably be consistently fatter than most throughout their lives. A child's abdomen can be, and often is, quite naturally, vast. Apparently distended beyond all hope of retreat, mainly by a disproportionately large liver within, the bulbous shape does in fact recede eventually. By his fourth year the child's outline is no longer a caricature.

Heart   The foetal heart beat is 130–160 a minute. The pulse rate of a newborn baby is, like its breathing, very fast. An average is 140 heart beats per minute, falling to 130 at one month (although it may actually rise in the first few weeks), to 115 at one year, to 110 at two years, to 95 at eight years, to 85 at twelve years, and to 82 at eighteen years. The heart beat of girls is consistently higher, by about five beats a minute, throughout this slowing down. Conversely, blood pressure steadily rises. The systolic pressure may only be 70 (mms. of mercury) at birth, and less in prematures. By six months it is about 93 – although blood pressures are notoriously unwilling to be associated with averages. By six years the pressure is about 100, by ten years 110, and by sixteen years 120. In other words blood pressure is doubled from infancy to maturity while the heart rate is almost halved. In the meantime the heart itself has grown twelve-fold.

The newborn heart, despite its 140 beats a minute and its pressure of 70 mms./Hg., is very small. It weighs less than an ounce (24 gms.). By one year it is 1·6 ozs. (45 gms.). By six years it weighs $3\frac{1}{2}$ ozs. (99 gms.), by fourteen years $5\frac{1}{2}$ ozs. (156 gms.), and by maturity between 10 ozs. (283 gms.) and 11 ozs. (311 gms.).

Kidneys and bladder   At birth the kidneys are of course small (the pair of them weigh less than an ounce) but in fact they are large in relation to adult proportions. They grow rapidly after birth, doubling their weight in six months and trebling it after a year. By five years old the two kidneys weigh over 4 ozs. (113 gms.), and by adulthood 11 ozs. (311 gms.) (i.e. the same as the heart).

The bladder may even contain urine when a baby is born, but the production of urine is initially modest. Some babies do not produce any for at least twenty-four hours. Such urine as is produced helps to create the post-natal weight loss. By the end of the first week urine secretion is anywhere between 50 ccs. and 300 ccs. every twenty-four hours, but for the bulk of the first year the daily output is about 450 ccs. (or four-fifths of a pint). This is remarkably high, bearing in mind the total daily water loss of an adult which, via lungs, evaporation and excretion, is only about four times as much. By the age of five the daily urine output is about 750 ccs., or 1·3 pints.

Bladder control, a subject that has generated fervour, remedies and advice for centuries, does not become voluntary until, on average, fifteen to eighteen months. (Once again an average is a guide, not a rule. Voluntary control can be much earlier, and very much later.) Control can be simulated beforehand by skilful maternal anticipation, but the first true stage of voluntary control is a child's awareness of passing urine at the moment of its passing. Later the alarm is given seconds before the actuality, still giving no time for action by others. Finally, by the second birthday – or later – or earlier, mothers are given time, and then the child itself will gradually take over.

The agony of enuresis (or lack of control) can be prolonged if the ability to sleep is stronger than the ability to control. The child-care books offer help, but the problem is immense. A leader in *The Lancet* on the treatment of enuresis estimated that 15% of boys and 10% of girls aged five were afflicted by it, and still 5% of children by the age of nine. Buzzers, which operate as soon as urination starts, are the currently favoured devices for treatment because they alert the child into a contraction of his sphincter; but they are not always successful. Some alarms wake everyone else except the soundly-sleeping bed-wetting one who is failing to make the necessary excursions. Happily, for the child's and the problem's sake, the age-old remedies for enuresis, like applying stinging nettles to the penis, or hedgehog flesh, or goat's claws, are no longer considered valid. (Old wives who dreamed up tales never seemed short of blood-curdling curative recipes.)

Dr Spock If human growth were solely a matter of increasing body size, of improving kidney performance, of increased lung volume, and so on, it would be a matter primarily of measurements; but it is not. It is the development of a human being, a unique individual brought up in a unique fashion by young and amateur parents who have a mixture of love, sense, impulse and tradition to guide them. Consequently, extra guidance is regularly sought after, particularly in societies where tradition is less powerful or not universally known and accepted. Why does this one cry so much? Why does this one not cry enough? Children never conform. This one loves its play-pen; this one finds it hell. He loves company; he hates it. He loves toys, cots; he hates toys, cots. He likes the new baby, new people; he considers them anathema. He eats well; he will never eat. He sleeps. He wakes. He vomits. He defaecates. He is loving. He is fat. He is spoilt. He sucks his thumb. Is your child a problem child? ask the advertisers – confidently. Is he sensitive? Are you doing the best for him?

The triumph of Dr Benjamin McLane Spock occurred because his guidance proved more satisfactory than anyone else's had ever done, and countless other authors have tried and are trying to be equally helpful. By the mid-1960's his *Baby and Child Care* had sold more than 17,000,000 copies in 145 printings of one edition. The current figure is nearer 30 million. No other book first published in the United States has done so well. It is still doing well but less phenomenally, mainly because so many other books are covering similar ground. (There is nothing like a best-seller for causing other publishers to think along similar lines.) The sociologists now talk of a Spock-reared generation and, as the book came out in 1946, Spock babies are already having Spock babies of their own. Some mothers have a Spock in every room 'for instant readiness', a practice he has deplored.

In an interview the author explained that 'the book set out very deliberately to counteract some of the rigidities of paediatric tradition, particularly in infant feeding'. 'It emphasised the importance of the great differences between individual babies, of the need for flexibility, and of the lack of necessity to worry constantly about spoiling. ... I didn't want to encourage permissiveness. I wanted to relax rigidity.' Many have charged that in Spock they will find whatsoever guidance they are looking for, as he talks both of parental rights and of the baby's rights. Undoubtedly, more than any other author, he has relaxed the previously powerful rigidity, the sort which could not accept that some $2\frac{1}{2}$-year-olds need seventeen hours sleep, some only eight. His huge sales emphasised a general acceptance of the facts that babies are indeed different, that parents are different and that nothing is simple about the complex development of each new human being.

Despite today's talk of freedom for the child, and a general discarding of old traditions, babies are still cared for traditionally. It can also be argued that we constantly acquire new traditions. Take swaddling. Today's western mother would throw up her hands at the idea of the baby not being able to throw up his. Yet swaddling, used in the Bethlehem stable and by the Jews, Greeks and Romans long beforehand, was employed in Britain until the end of the eighteenth century, and there is a lot to be said for it. Many Russians, Yugoslavs, Mexicans, Lapps, Japanese and North American Indians still do it. The promoters of tight and restrictive envelopment for the child have various theories about its values, but there is general agreement that the child is quieter, and not just because of *force majeure* with all the blanketing. Swaddling does create less fretful infants. E. L. Lipton and his colleagues put this ancient practice to scientific test. They examined total swaddling, partial swaddling (with arms free) and total freedom. With cardiotachometers, strain gauges and thermistors to measure the infants' general activity, they finally concluded that swaddling does indeed reduce both heart-beat, breathing rate and crying, and that it increases sleep. One can imagine many Russians, Yugoslavs, Mexicans, Lapps and all the rest, plus our pre-eighteenth century ancestors, being aghast as today's infants thrash themselves so regularly into a flailing, unfettered fury.

Child law Putting today's attitude to the rights of a child to one side, there are also legal rights for children. Primarily, no child in England less than eight can be guilty of any offence. Keith Simpson in *Forensic Medicine* refers to a two-year-old who deliberately suffocated a four-month-old brother with a cushion, but the law considers all under eight are incapable of forming sufficient malice aforethought to be guilty of a criminal act. Between eight and fourteen, to quote again from the same source, 'there is the presumption of innocence which is rebuttable except in respect of certain sexual crimes'. All people over the age of fourteen are subject to the criminal law. (Hanging, when it used to be carried out in Britain, was only for those over eighteen.)

The problem of overlaying used to be much more serious in the days of worse overcrowding, and there are laws to protect the infant. It is a criminal offence if anyone over sixteen, who is drunk and who shares a bed with a child under three, causes that child to die of suffocation. In 1911 there were 1,157 cases of death caused by overlaying. There were 255 in 1937, but 517 after the war in 1947, a time of great housing shortage. The numbers since then have decreased, with total deaths (in England and Wales) from 'accidental mechanical suffocation' being about 170 a year.

Finally, anthroposophists and others believe that each human spirit has some kind of choice in deciding which parents will be blessed with its birth. The spirits who choose poor parents are certainly the most masochistic because poverty will weigh the balance against them in almost every society; in the poorest it can mean the difference between food and no food. In the rich countries poverty is still chronic, and even in Welfare Britain the poor have a greater relationship with the hazards of life. They have more abortions, more still-births, more premature babies, a higher perinatal mortality and infant mortality, more congenital malformations of the brain and nervous system, more deaths from respiratory and many other diseases, more accidents, more bed-wetting, more speech disorders, more delinquency and, as a final testament to their poverty, the children become smaller adults.

'Why should the wives of unskilled labourers give birth to children who are ten times more likely to die of malformations of the central nervous system than are the children of professional men?' asked Dr John Apley in a lecture to the Institute of Child Health. Is there one reason, or many? Is it to do with the father's job, or the house, or the air at the poor end of town, or what? How does poverty strike? 'Man is a multiple amphibian', wrote Aldous Huxley, 'living in many worlds at once. There is his heredity, and his environment; his family, his community and his society, and there are multiple causes for the multiplicity of effects in the development of a human being.' 'There is one condition', wrote Cicely Williams, 'that affects or has affected every single member of the population, and which in some regions is associated with a mortality rate of over 50%. This condition is childhood.'

For those who survive all the vicissitudes of the early years, who acquire scars and learning and phobias and skill, there is yet another and more specific condition to which all are subject. Coming long before adulthood the phenomenon of puberty makes its premature addition to the complexities of growing up.

20. Puberty

Puberty is not adolescence. It occurs as part of adolescence, and marks the changes to sexual competence. Adolescence, the time of growing up, lasts much longer and embraces the whole decade from childhood to adulthood. It is important to the human species that sexual maturity is achieved long before intellectual maturity. It is also achieved before physical adulthood. Growth continues, particularly with the male, long after the reproduction system has proved itself capable. Moreover the age of puberty is gradually becoming younger, both for girls and boys, and head teachers of primary schools state that, since World War Two, they have had to provide facilities for their menstruating pupils.

Because girls experience the dramatic incident of the first menstrual loss, much more is known about female puberty dates than male. The average age of girls on this all-important day now varies between thirteen years two months and thirteen years four months in Britain, according to the region. In the United States the average is just below thirteen years. In hotter countries, despite a widespread belief possibly engendered by the frequency of early marriages, the age of puberty is no younger, and there is some evidence that it occurs later. The actual range from which the British averages are taken extends from nine to seventeen years, with just as many girls experiencing their first menstruation on either side of the average age. This average age is certainly lessening. Back in 1890 it was about eighteen months greater, with fifteen being the age of menarche, of that first menstruation. The minimum school leaving age then was twelve. In 1973 it was raised from fifteen to sixteen. In other words ninety years ago puberty used to occur, on average, three years after most girls had left school. It is now occurring, on average, three years before the minimum school leaving age. Therefore what used to be a minority problem is no longer so.

There is an argument that the reduction in pubertal age may only be an apparent reduction. The removal, for example, of extremely delayed pubertal dates, caused perhaps by disease or extreme malnutrition, can lower the average date; hence an apparent shrinking of the average age at which girls experience

290

menarche as soon as those reasons for delay have been eliminated by improved living conditions and medicine. Despite this argument, which is equally applicable in many other average findings, it is generally realised that the pubertal age is going down, and is still going down at just the same rate. An enthusiast of extrapolation once calculated, bearing in mind today's rate of change, that Juliet, for example, would not have experienced menarche until well into her twenties and a decade or so after her well-chronicled affair with Romeo. Obviously the current rate of change is exceptional. It is even suggested that the present age recession is merely a restoration of the situation which existed before the appalling living conditions brought about by an industrial revolution and squalid urbanisation. At present there is no evidence that the trend towards earlier development has stopped. Nevertheless, as Professor J. M. Tanner put it, 'commonsense dictates a stop in the foreseeable future'.

Although menarche arrives on a definite date there is no such clear-cut delineation with other changes of female puberty. Moreover that first menstruation does not indicate a sudden fertility. It usually precedes the ability to rear offspring by a year or more. The earliest menstrual cycles occur without an egg being shed from the ovary.

The first sign of impending female puberty is customarily the appearance of breast-buds between the ages of nine and eleven years. Yet the breasts will not reach maturity for another eight years or so. An average age for nipple pigmentation is twelve to thirteen years. Pubic hair starts growing at about eleven years and is well developed by fourteen years, but it may start earlier and end later. Axillary or armpit hair comes a year or two later: it probably starts between twelve and fourteen. The date of menarche, important as it is, is bounded on both sides by all the other manifestations of sexual maturity. Nevertheless it almost always comes after the peak of the height spurt, the rapid growth in stature that is another feature of the age of puberty. Both puberty and the growth spurt occur earlier in girls than boys. Consequently girls of twelve to fifteen are often both taller and heavier than boys of equal years. Boys not only have their growth spurt later, but carry on growing for a longer period. It is all over with girls by the time they are twenty or twenty-one, but boys frequently go on getting taller until twenty-three to twenty-five. (This point is often neglected by official organisations like the Army. A recruit's height at seventeen or eighteen is assumed to be his height for life.)

Boys are just as variable in their puberty dates as girls, if not more so. In any normal group of thirteen- to fourteen-year-olds there are boys who are sexually mature and those who are virtually immature. Such disparity provides a fertile field for feelings of inadequacy or abnormality. Penis development, for example, starts on average at thirteen, and lasts for about two years; but its development can start at eleven or fourteen and a half, and it can end at thirteen and a half or seventeen. So some are completed and mature before others begin. Usually the first sign of male puberty is an acceleration in the growth rate of the testes. They start their period of rapid development at about twelve and end it at about sixteen, but some boys start it at ten while others end it at eighteen. Pubic hair begins to grow between ten and fifteen, and will have completed the bulk of its growth anywhere between fourteen and eighteen. Generally, axillary hair begins a couple of years after pubic hair has started, but it can even reverse the situation by starting its growth first.

A girl of eighteen has lost the childish voice she had before puberty, but the pubertal voice change is more emphatic with boys. The male larynx starts its growth spurt at about the time the penis is completing its period of accelerated development. Customarily voice change is gradual, and may extend over several years. Sometimes, during the time of laryngeal growth, boys may croak haphazardly in speech as their voices break. However ungainly the noise, nothing is in fact disintegrating; it is uncertainty in muscular control over the growing larynx which causes the hoarse warbling. Male breasts are not totally unaffected by puberty. The areola, that button-like background to the actual nipple, may enlarge and darken. In a third or so of boys a lump of tissue may actually form beneath the areola, only to recede at some later time in adolescence.

Finally, much to the disgust of almost every adolescent, the period is one of extreme susceptibility to skin trouble. Acne is a form of skin eruption that is definitely linked with the sexual changes of the adolescent. Eunuchs do not suffer, but will do so if given sex hormones and they will lose their acquired acne when the hormone treatment is stopped. For some reason the sebaceous glands do frequently become plugged during adolescence, and are then subject to secondary infection – hence the acne. The complaint is so common that it is said to be physiologic rather than pathologic, normality rather than disease.

Whereas male puberty demonstrably starts with an enlargement of the testes, and the age of this acceleration is about twelve but may be ten, mature sperm are not produced until the age of fourteen to sixteen. Even so, as with the female, sexual maturity is reached long before physical and physiological maturity of the body as a whole. This single fact is of enormous importance and influence in human society.

21. Size

'How long should a man's legs be?
They ought to be long enough to reach the ground.'

ABRAHAM LINCOLN

Having started off as a single united cell each human being grows until he or she consists of some 500,000,000,000 cells. That first cell is larger than any of its subsequent offspring – until the manufacture of further ova. Ordinary cell size varies from 200 to 1,750 cubic micra (a micron being one thousandth part of a millimetre) but each human egg is about 1,400,000 cubic micra. Most mammal eggs are about the same size but there is a wide range in adult mammal size. The smallest is the shrew and the largest is the blue whale. Isaac Asimov, who calculates such things, has estimated that a man is 45,000 times as massive as a shrew, while a blue whale is 1,300 times as massive as a man. Whereas the largest whale weighs 130 tons the largest land animal alive today is the seven-ton African elephant.

It is highly relevant to the size of any land creature that weight increases as the cube of the length, breadth and height. If one land animal has twice the dimensions of another it does not have twice the weight, or even four times the weight, but eight times the weight. Human beings exemplify this law of volume. A child of two is approximately half the length of an adult, and has approximately half his chest and hip girth, and yet he weighs about one-eighth as much. (On his second birthday I measured my son. His height, chest, waist, neck, foot, hand, and arm girth measurements were all almost exactly half mine, but his weight was less than 2 stone (12·7 kgs.). At the time, with a chest of 42 ins. (106·7 cms.) against his 20 ins. (50·8 cms.), with a foot 11½ ins. (29·2 cms.) long against his 5¾-inch (14·6 cms.) foot, I weighed 15 stone or 95·2 kgs.) Martin Wells once took this problem of weight, gravity and mass a little further by comparing possible planetary life. The bigger the planet the smaller the animal. Earth visitors to Mercury should carry a grenade or two, he pointed out, and to Jupiter a fly swat. In any case the science fictional giant spiders on spindly legs just could not exist.

The law of volume to length is also influential in the normal adult population. Discounting dwarfs and giants, almost all the human population can be included within a 2-foot (61 cm.) span.

Size

Virtually everyone has a stature between 4 ft. 7 ins. (1·4 m.) and 6 ft. 7 ins. (2·0 m.). Therefore the tallest are less than half as tall again as the shortest. No such similarity exists for weight. It frequently occurs that a normal man is considerably more than twice the weight of a normal girl. Men are, of course, customarily heavier than women, but the difference is small between men and women of equal height. American women of 5 ft. (1·5 m.) are, on average, only 5 lbs (2·3 kgs.) lighter than 5 ft. (1·5 m.) men. The average difference is only 3 lbs. (1·4 kgs.) if both are 5 ft. 4 ins. (1·6 m.), only 4 lbs. (1·8 kgs.) if both are 5 ft. 8 ins. (1·7 m.), and only 6 lbs. (2·7 kgs.) if both are 6 ft. (1·8 m.). Consequently most of the traditional weight disparity between men and women is due not to broad shoulders and muscles, but to the height differential.

The average European and American male is about 5ft. 8 ins. (1·7 m.) in height and weighs 168 lbs. (76·2 kgs.). The average such female is 5 ft. 3 ins (1·6 m.) and weighs 142 lbs. (64·4 kgs.). But in one American survey (of 6,672 people) it was found that, although the averages held good, the weight of 90% of the males ranged between 126 lbs (57·1 kgs.) and 217 lbs. (98·4 kgs.), and 90% of the women ranged between 104 lbs. (47·2 kgs.) and 199 lbs. (90·3 kgs.). Therefore, although the word 'normal' customarily applies to the characteristics of 95% of a population, the normal weights – of 90% – of these people varied between 104 and 217 lbs. (57·1 and 98·4 kgs.).

Men reach a maximum average weight of 172 lbs. (78 kgs.) when middle-aged (between thirty-five and fifty-four). By the age of seventy-five the average weight is 150 lbs. (68 kgs.), presumably aided by the fact that many fat ones have already died. Women reach a maximum average of 152 lbs. (69 kgs.) when even deeper into middle age (between fifty-five and sixty-four). By seventy-five this average has dropped to 138 lbs. (62·6 kgs.) – by precisely 1 stone.

Dwarfs and giants are almost always the result of some malfunction of the pituitary gland. This diminutive egg-shaped organ, a mere half inch long and attached by a stalk to the base of the brain, is primarily responsible for the abnormal lengths of both the very short and the very tall. Over-secretion or under-secretion of this gland's growth hormone both lead to spectacular results. In fact these hormones are probably the most spectacular of all human hormones in their effects when produced either excessively or inadequately.

There are two main types of pituitary dwarf. In one of them, the Lorain type, the hormone deficiency results in just a very small human being, well proportioned – although with the measurements more of a child than an adult. He may or may not be sexually competent, he is probably intelligent and certainly not ugly. In the second type, called Fröhlich's, the results of pituitary abnormality are very different. This fully grown dwarf is fat, sexually underdeveloped, often stupid, lethargic and sleepy, and less physically attractive than the Lorain type. Other forms of dwarfism are caused by thyroid inadequacy (which also leads to cretinism), certain forms of diabetes, and various other abnormalities, including of course the lack of food of the right quantity or quality. Early rickets, causing a high forehead and great warping of the leg bones, can cause adult dwarfs. The United States has an estimated 7,000 midgets. The Little People of America is a society whose members are all under 4 ft. 10 ins. (1·5 m.), and every year there is a midgets' annual meeting to further their interests. The most hopeful line of research is the prevention of pituitary dwarfism by the administering of growth hormones as soon as their lack has been detected. Until the 1980's the only significant source of growth hormone was the human pituitary gland, and these glands together with their crucial growth product could only be obtained from the dead. Fortunately biochemistry brought this practice to an end when it learned how to synthesise the hormone, and this started to become available at the start of this current decade.

Conversely there are all the forms of giantism (or gigantism). Principally these indicate an over-production by the pituitary gland, but there is a great difference whether this hypersecretion occurs before or after the normal growth time has ceased. If produced before the end of adolescence the extra growth hormones will cause extra growth, and the resulting giant may be up to 2 ft. (0·6 m.) taller than a normally tall man. Their proportions are normal, and they hardly ever seem to get taller than 8 ft 6 ins. (2·6 m.). If the pituitary over-production occurs after adolescence the hormone does not succeed in increasing stature but causes acromegaly (Greek for large extremities). The ungainly appearance of this disease is the result of the extra growth hormone acting upon those parts of the skeleton still insufficiently mineralised, and still capable of further growth. The main growth areas are the hands, feet and jaw.

Both these forms of giantism draw attention to the big unknown associated with growth. Why is it that a human being stops growing? He or she is still continuing to produce the growth hormone, and yet something happens to inhibit its effects at a certain time. Sex hormones, arising in abundance a few years before the end of growth, do have an inhibiting effect, but castrated animals with a grossly depleted supply of sex hormones stop growing just as emphatically, although the stoppage may occur slightly later. Even though animals vary in size from shrews and smaller to the giant blue whale, and even though their proportions vary throughout this size range, with some having huge feet and small tails or vice versa, there is a steady consistency throughout the animal groups: the size of the pituitary gland almost always bears a direct arithmetical relationship to the size of the animal. The bigger the gland the bigger the animal. This is interesting, but it still does not explain why a certain quantity of pituitary gland leads to a certain size of animal, particularly when added to the extra dilemma that growth stops even though growth hormone is still being produced.

22. Physical Ability

Fitness/Athletics/Somatotyping/Man versus woman/The future

'The preservation of health is a duty.'

HERBERT SPENCER

'Attention to health is the greatest hindrance to life.'

PLATO

Fitness 'Early to rise and early to bed makes a man healthy, and wealthy, and dead', said James Thurber. 'Keep fit', shouts the sergeant. 'Yes, fit to drop', mutter his men. 'Fitness', according to Sir Adolphe Abrahams, Olympic medallist, then Olympic medical adviser, 'means a satisfactory adjustment to one's environment.' It therefore means being able to run away from predators. It also means being able to sit at an office desk for eight hours without running anywhere, or wanting to. And it means getting wet through, or thoroughly exhausted, without succumbing to the first infection that hits at a weakened body. There is evidence that poliomyelitis invasions leading to paralysis often follow periods of strenuous activity. The magnificent physiques, with muscles arching and flowing over a broad frame, do not win the longest races. Marathon men are short and thin. Weight-lifters have short legs and short arms. World record holders of track events often have awkward, gangling and even mis-shapen bodies. Michelangelo's broad-shouldered, long-legged and well-muscled David would not have had a hope in any race longer than a medium sprint. The quarter-mile would probably have been his running event.

Mankind's athletic abilities are a great complex of abilities. In each of them he is completely outclassed by at least some members of the animal kingdom. Man can run, briefly, at 27 m.p.h. (43·5 k.p.h.) on a smooth surface. Practically every mammal as large or larger, including heavyweights like the rhino and hippo, can catch him up, rough surface or smooth. Man can run a mile at 15 m.p.h. (24·2 k.p.h.). Most mammals of similar size or larger could do far better. Good high jumpers can get over

298

a 7 ft. (2·1 m.) bar. A bushbaby the size of their fists can beat them. And so can countless ungulates, and kangaroos. Man has jumped 29 ft. (8·8 m.) into a sandpit. Animals do not jump into sandpits, or jump once and then fall over; but, if somehow persuaded, jumpers like the impala and the springbok could plainly do better. Man's great weakness is the possession of only two legs. For most of the time neither leg is exerting any thrust. A four-legged system is almost always faster. Man has sacrificed two legs to have arms instead. Parallel-bar gymnasts and trapezists can do wonders with them, but think of a brachiating gibbon hurtling through the tree-tops, or watch a monkey do wonders with just one arm.

Man instead is the great all-rounder. He can swim a bit, climb a bit, run a bit, and jump a bit. He can also exert himself for reasons of his own, without either chasing or being chased as a necessary stimulus. It has been claimed that a man can outwalk a horse. George Littlewood, for example, walked 531 miles (854 km.) in six days in 1882. Man is not much of a swimmer, in that sharks, whales, seals and penguins must consider his top speed of 3·7 knots virtually stationary. Nevertheless he has plodded on to swim the English Channel both ways. He cannot fly, but he is trying to; the enthusiasts are pedalling their man-powered-flight-machines ever faster and higher.

Athletic competitions do not in general test mankind's all-roundedness; they test individual skills. There are pentathlons, and even decathlons, but most events do not set out to discover 'Athletic Man'. Such a victor ludorum would make more sense, were he to compete in every event, as a physical test for mankind. Today's specialisation demands a special type. The shortest 400m. runner in the Olympics is taller, by and large, than the tallest marathon runner. The heaviest marathon runner is 10 lbs. (4·5 kgs.) lighter than the lightest 400 m. runner. Men who jump high, or put the shot, or throw the discus almost have to be taller than 6 ft. (1·8 m.) to win. The longer the race over 400 m., the shorter the man who, on average, wins it.

All of these human abilities are being improved all the while. The shot-putt record has gone from 51 ft. (15·5 m.) to nearly 73 ft. (22·2 m.) since 1909, a 42% increase in slightly over half a century. The 17ft (5 m.) pole-vault height, the 7ft (2 m.) high jump, and the 4-minute mile, all previously considered to be obstacles near the human limit, have all been triumphed – substantially. Ten years after Roger Bannister had run (in 1954) his world-famous mile in six-tenths of a second less than 4 minutes, forty-four other athletes had run it in less than those 240 seconds,

299

and the current barrier to be broken is 3 mins. 45 secs. Without doubt the mile record will be changed again and again in the future. Long-range forecasters are predicting that 3 mins. 30 secs. is the ultimate time for mankind to run a mile, and 3 mins. 41 secs. will be achieved by the year 2000.

Why the steady and unflinching improvement? First, the athletic catchment area is far greater. The 285 participants in the 1896 Olympics were drawn from 50,000 athletes in training. The 10,000 participants at Tokyo (in 1964) were representing about 100,000,000 possible competitors. And the games of 1988, so the forecasters say, will have the representatives of at least double that number. Much of this increased representation is an invasion of talent from the world's less well developed continents. Almost all the athletes sent to the games now from Australasia, the New World, the Soviet Union or Western Europe are of European stock. Africans and Asians have been poorly represented but are gaining all the time.

In athletics the races are not born equal; neither are they brought up in equal fashion, nor in equal climates or altitudes. The differing potentialities of black and brown will have much influence upon future records. Even now the American blacks do three times better in collecting Olympic medals than their proportion of the United States' population would indicate. One wonders avidly what might happen should the Watusi of Africa enter the arena. With an average male height of 6ft. 5 ins. (1·9 m.), some of them close on 7 ft. (2·1 m.), and a national fondness for jumping, they should sweep the field. Jumps of 7 ft. 4 ins. (2·2 m.) have been recorded under circumstances far from ideal. They do not use a carefully balanced bar, but a firm rope between two trees. The Masai, also of Africa, are phenomenal walkers. An American team visited Masailand with a fiendish 'treadmill' whose moving platform increased its slope by one degree every minute. Two of the Africans immediately beat the record of an American champion who had trained on the device for six months. The Masai pulse rates had gone up, but not their respiratory rates. Professor Ernst Jokl, who ran for Germany and then emigrated to America, has written of the superior athletic abilities of the black, and quotes work assessing the order of 'athletic efficiency' as first Negro, second Asian, and third European. Only now are the black and brown countries fully entering the competitive territory formerly reigned over so casually by the whites.

Better training methods, and longer periods of training, must also have improved abilities. Nearly all the top Australian and

American swimmers have started their hard training by the age of twelve or so. Although instructors have plenty of theories concerning the production of an athlete's top performance on a certain day, there is a scarcity of fact about the physiological and anatomical changes which take place during the training. When muscles get bigger are the muscle fibres becoming larger? Do the blood capillaries multiply in the training of a muscle? And do changes happen in the nervous control of muscular contraction? It is thought that the capillaries increase and the muscle fibres just get larger; but most training methods are empirical, based on experience and not on physiology.

At present record-breaking occurs relentlessly, but the sprint records are already harder to beat than those of the longer races, partly because training cannot do much to improve the 'oxygen debt'. All the oxygen used up during a 100 yd. sprint is replaced after the race is over. Some sprinters do not even bother to breathe during their 10-second dash; some only gasp once. A mile runner has to breathe in half his oxygen needs for the race during the (usually less than) 4 minutes of the race. Consequently training can do more to improve the replacement. This partly accounts for the improvement of over a quarter of a minute on the mile record since 1921. In that year Charlie Paddock ran the 100 m. in 10·2 seconds: since then only 0·3 of a second has been knocked off his time. The limit of mankind's sprinting ability is obviously very near. (The former custom of measuring performance in tenths of a second became insufficiently discriminating, and during the early 1970's it became customary – and technically possible – to record all times in hundredths of a second.)

The first scientific attempt on a reasonable scale to find out about Athletic Man was carried out during the Rome Olympics of 1960. Some measurements of stature had been made at earlier meetings, but only sketchily. Dr J. M. Tanner's survey at Rome was the first comprehensive study of the physique of the Olympic athletes. He and his team measured, X-rayed, photographed and classified all the male athletes they could get hold of, and then drew their conclusions. They also somatotyped them.

Somatotyping is one system of identifying physique, and it considers shape alone, not size. It is based upon the work of William Sheldon in 1940. He tried sorting out the photographs of 4,000 American college students, and finally surfaced with the opinion that there were three extremes of body shape. The three were then awkwardly called endomorphy, mesomorphy and ectomorphy. The endomorph is essentially rounded, with a

301

round head, a bulbous stomach, a heavy build and a lot of fat; but he is not necessarily a fat man. When short of food he does not shrink to become a mesomorph, let alone an ectomorph: he just becomes a starved endomorph. The mesomorph is the sculptor's model, with a large head, broad shoulders, a lot of muscle and bone, not much fat, relatively narrow hips. When fattened up he does not become an endomorph; he is then a fat mesomorph, for these three characteristic shapes are basically quite distinct. The ectomorph is the thin one, all sharpness and angle, with spindly legs and spindly arms, with narrow shoulders and still narrower hips, without much muscle or fat, but with a large skin area relative to his diminutive bulk. Even when fattened up he is still the ectomorph.

Each one of us, according to Sheldon, has a bit of all three in his frame. Even the three extremes, the spherical man, the Hercules, and the thin one, are all assumed to have a small fraction of each other's characteristics. Sheldon judged the amount of each characteristic possessed by each of his 4,000 pictures, and he rated these amounts from one to seven. The three extremes, the round one, the Hercules, and the thin one, were called 7–1–1, 1–7–1, and 1–1–7. The 7 measured the predominant quality. The 1's measured a token possession of the other two qualities. A medium man, with an equal amount of all three, was labelled 4–4–4. A man with some shoulders (of the mesomorph) and some roundness (of the endomorph), but with much more of the narrow angularity of the ectomorph was pronounced 3–3–5 or perhaps 2–2–6. Despite the apparent difficulty of this type of subjective assessment, different assessors do in fact produce nearly identical classifications of a group of men. Judgement is much assisted by the use of standard photographs of all the different somatotype ratings.

If these assessments are then plotted upon a triangular graph, with the three extremes at the three corners, the picture is soon dotted all over like a target of an erratic rifleman. College students, who are chosen for their brains and not for their shapes, produce just such a random picture. However, students at a military academy, who are chosen mainly for brains and partly for shape, produce a modified arrangement of dots, like a bad rifleman with a consistent fault. An academy does not accept either the extreme endomorphs or those with too much endomorphy about them. The Olympic athletes produce an even more compact picture. Of the 137 men measured by Dr Tanner all of them were in one half of the graph. No one was at Rome who scored more than four out of seven for endomorphy. The men were either ectomorphs or mesomorphs, or combinations predominantly of these shapes.

In other words, however determined, the 5–4–1's, the 4–3–3's, the 3–2–6's of this world, and many more still plumper and rounder, are not to be encountered at an athletic meeting. They should forget the whole idea of being an Olympic star. School boys and soldiers, told hoarsely that it is just 'guts' or 'drive' which wins races, should examine their shapes seriously, and then yield if blatantly on the wrong side of the somatotype graph for that particular event. They just were not born for it. All the shot-putt men, for example, are found to have their ratings clustered together in a bunch near the 4–6–2 area. The 50 km. walkers are similarly grouped, but in the 2–4–4 area. These are thinner men, less powerful, and even less endomorphic. The 2–4–4's do not weight-lift, or shot putt, or throw discus, javelin, or hammer. They run races, and would run circles around any 4–6–2 who dared to compete with them.

Moving from shape to stature it is equally wrong to be ambitious for the wrong event. No shot-putter measured at Rome was under 6 ft. 1 in. (1·85 m.). Their arms were long, while those of the discus men were longer still. The high jumpers were all over 6 ft. (1·8 m.). Part of the answer is straightforward mechanics; a shot will travel further if put from a greater height, and a discus will be given more momentum if whirled round a wider circle. Another point is that runners shorten as the race lengthens. The average heights (to the nearest inch) for the 400 m., the 800/1,500 m., the 5,000/10,000 m. and the marathon were 6 ft. 1 in. (1·85 m.), 5 ft. 11 ins. (1·8 m.), 5 ft. 8 ins. (1·72 m.), and 5 ft. 7 ins. (1·7 m.). The tallest marathon runner and the shortest 400 m. runner measured at Rome were both 5 ft. 10 ins. (1·78 m.). The 100 m. race does not fit into this gradation. Its runners were, on average 5 ft. 10 ins. (1·78 m.).

Weight mimics height precisely. The longer the race the lighter the runner, with the 100 m. race being the exception. The 400 m. average weight was 12 st. 1 lb. (76·7 kgs.), and the marathon average was 9 st. 8 lbs. (60·9 kgs.) with the heaviest marathon man 11 lbs. (5 kgs.) lighter than the lightest 400 m. man. The 100 m. average was 11 st. 7 lbs. (73 kgs.). Weight was also an important distinguishing feature between those competing in the running and jumping events and the throwing events. Every runner, hurdler, jumper and walker measured at Rome was lighter than the lightest discus, javelin, shot or hammer man. Discounting one particularly light-weight javelin thrower, the lightest throwers were 29 lbs. (13·1 kg.) heavier than the heaviest of the running, jumping, and walking men. Men using their arms to win are thus in a quite different weight class from those who use their legs to win, who have to carry their bodies with them. The heaviest man of all was a shot-putter of 18 st. 4 lbs. (116 kgs.) who was 6 ft. 4 ins. (1·9 m.) tall. The lightest man was a marathon runner of 8 st. 4 lbs. (52·6 kgs.) who was 5 ft. 5 ins. (1·6 m.) tall. With one being 2·2 times the weight of the other, and nearly a foot taller, they might almost be considered different species. As it is they help to exemplify the physical specialisation of the Olympic games, and to illustrate the impossibility of determining an optimum physique or 'Athletic Man'.

Age is extremely relevant. A generality is the longer the race the older the men running it. The average age of the 100 m. runner at Rome was twenty-three. By 400 m. it was twenty-four. At 1,500 m. it was twenty-five, at 5,000 m. twenty-six, at 10,000 m. twenty-seven, 20 km. walk twenty-eight, marathon thirty, and 50 km. walk nearly thirty-one. The youngest marathon competitor ($25\frac{1}{2}$) was less than one year younger than the oldest 100 m. competitor. All track and field athletes are, reasonably enough, young men: the oldest was thirty-six (a 50 km. walker) and the youngest was nineteen (a high jumper). In the games as a whole, taking into account all the events including those the ancient Greeks never thought of, the age distribution spans many decades. At Helsinki in 1952 the youngest competitor was thirteen (a swimmer) and the oldest (a clay-pigeon shooter) was sixty-six. The youngest winners are invariably amongst the swimmers.

Normally altitude is considered of little consequence. Most large towns, and therefore the sites of most Olympic games, are only marginally above sea-level. The International Olympic Committee decided to hold the 1968 games 7,400 ft. (2,255 m.) above sea-level at Mexico City. 'Foolish, astonishing, absurd', wrote Roger Bannister. Many current athletes also protested, mainly because lengthy acclimatisation was considered helpful. At such a height the sprinters had a good chance of beating existing world records – and did. The long distance runners had little hope of doing so – and did not. The sprinters have all their oxygen within their systems before the race, and could therefore benefit in speed from the 23% difference in the air density. It is easier running through thinner air. For long races this benefit is outweighed by the fact that there is 8% less oxygen in Mexico City's atmosphere than at sea level.

Dr L. G. C. E. Pugh, Himalayan mountaineer and medical physiologist, made an exploratory foray to Mexico's high altitude stadium. On his return he stated flatly that athletes would not be permanently injured by the experience, but performance in all endurance events would inevitably be reduced. Others calculated from his figures that times would deteriorate roughly in accordance with the logarithm of the distance. For the 800 m. this suggested 2·6% longer, for the 10,000 m. 15% longer. The British athletes who went with Dr Pugh were about 6% slower over three miles, even after four weeks of acclimatisation, but they had been 8% slower on arrival. When compared with sea-level races such running would put them way behind the rest of the field. Any runner with a superior technique for acclimatisation, and with a good ability to mimic sea-level performance in a higher altitude, would clearly romp home at Mexico City. Conversely, any future sprinter trying to win records at sea-level will have difficulty emulating those achieved in Mexico's upper air.

In the event, and as predicted, many of the quick burst events at Mexico did create new Olympic records, such as the 100 m., 200 m., 400 m., 400 m. hurdles, 4 × 100 m. relay, 4 × 400 m. relay and long jump (all for men), and the 100 m., 200 m., 80 m. hurdles and 400 m. relay (for women). Certainly not predicted was the smashing of the Olympic 1,500 m. record by Kipchoge Keino, of Kenya. He both lives and trains at altitude, and behind him came Jim Ryun of the US, who was 2·2% slower than his fastest time. The forecasters had predicted a 3% drop in speed for this event, but few had said that the major problem for the athletes would be gasterointestinal rather than respiratory. Mexico's traditional welcome, known as Montezuma's revenge or just as turista, felled many competitors far more effectively than oxygen lack could do.

Women will never equal men in athletics – so say some of the experts. A sport like swimming makes for good comparisons. In those Mexico games the men's times at 7,400 ft. (2,255 m.) were still better than the women's times at sea level, but women are improving their times at a faster rate than men, and some of today's records for women are better than men's records for the year 1900. Although men are physically more proficient generally, women athletes are certainly more proficient than most men. Mary Rand long-jumped over 22 ft. (6·7 m.) at Tokyo. The male record is over 29 ft. (8·8 m.), but how many normal males could hope to beat Mary Rand's leap, which was the length of four average-sized men lying head to toe?

Women have a lighter and weaker build than men. On average, their bodies have a smaller percentage weight of muscle, and more fat. Their legs are shorter and less muscular. Their bones are lighter and smaller. Their total strength and their individual muscle strength is less. Their shoulders are narrower and they have a lower centre of gravity, good for standing firm in the shot-putt, bad for jumping. They have a lower general metabolism, with the male/female ratio being 141/100. They have a smaller heart in proportion to body size, and smaller lungs.

Many societies do not enter female competitors in the Olympic games, presumably going along with Pierre de Coubertin, founder of the modern Olympics. He said that 'the primary role of women should be like in the ancient tournaments – the crowning of the male victors with laurels'. Every possible argument seems to have been lodged at some time against women athletes – they will look like men, bulge with muscle, fail to marry, fail to have children, overstrain themselves, die young.

The Venus de Milo may have been the Greek ideal of woman-hood, but undeniably, with or without arms, she would have fared badly on the track. Athletic women do not look like her. A French delegate to a recent sports medicine congress said that 'although the predominance of male characteristics in female athletes is not rare, sports are not the contributing cause. It is rather these inherent anatomical attributes of the female which enable her to excel in sports'. She too, like the male, has to be the right shape and size to win.

She certainly has been winning recently in a manner unthought of in the first half of this century. Only in 1928 were women allowed to compete against each other over 800 m., but as several of the runners collapsed after the race it was decided that they should never run further than 200 m. again. However, the 800 m. was reinstituted during the 1960's, and the 1928 time (slower than the men's by 25 seconds) became – by 1980 – only 8 seconds slower than the male record. Without doubt the women have been improving their times far faster than the men have been able to do. In 1976 the marathon record for women was 28 minutes behind that of the men. By 1984 the men had cut their time by 1 min. 42 secs., but the women had slashed theirs by 14 minutes. The winner of the women's marathon at Los Angeles in 1984 achieved a time one minute slower than was achieved by Emil Zatopek in 1952 when he won the Helsinki marathon. Some argue that the gender gap for speed events will level out at 10% or so, but the margin for endurance events may be less or even zero. Others contend that women will equal the men even in the speed events, suggesting equality in the 400 m. by the year 2020 and in the 3,000 m. by 2003. However, women cannot – yet – catch up in eight of the twenty-seven Olympic sports for the basic reason that they are barred from taking part in weight-lifting, wrestling, judo, soccer, bobsleigh, boxing, ice hockey and the biathlon.

The inherent attributes of sporting women have been questioned. At the European Athletics Championships held at Budapest in 1966 certain female competitors were asked for the first time to submit to examination to prove their femininity. Although no one seems to have said so outright they were being asked to prove that they were neither a male nor some form of hermaphrodite, an intersex combining femininity with masculinity (see the chapter on inheritance). Not everyone is emphatically either man or woman, and certain forms of intersex combine a theoretically feminine sex with many virile characteristics. If these characteristics help them to beat less virile females they have an unnatural advantage, and the Budapest examinations were conducted to rule out such intersexual unfairness. Significantly, some competitors withdrew from the Budapest games rather than suffer the questions and tests.

Later the 100 m. Polish runner Ewa Klobukowska was banned from international races by six East European doctors on the grounds that she had 'one chromosome too many'. Her totally feminine reaction was to burst into tears, and to provoke others to ask whether there should be the chromosome Olympics, exclusively for XY, XX, XO, XXY or whatever. With certain males being an XYY (see p. 232) should all those tall male throwers and jumpers be carefully examined from now on? Although some women have had to drop out owing to their uncertain femininity, no males have yet had to withdraw owing to their double masculinity.

What of the future? Presumably the athletic catchment area will grow until all of humanity has a chance of competing. By then the actual numbers of humanity will also have grown. Drugs may or may not be able to step up performance. Jokl says bluntly that 'medical science does not know of any substance whereby a well trained athlete's performance can be improved pharmacologically'. The British Association of Sport and Medicine, in issuing a ten clause policy statement denouncing the doping of athletes, also mentioned that, in its view, there is no known chemical agent which will both safely and effectively improve performance in healthy subjects. Over the years quite a few athletes have most positively shown their disagreement with such an opinion, not only in taking certain drugs but in providing positive results after the obligatory tests.

Perhaps the future will witness some positive eugenics in the

307

world of sport. At present mankind breeds locally, at random, and without much genetic regard for the next generation. No expert in animal husbandry would think of employing such a casual system. Maybe the future will be less casual. Indeed the American netball team were said to have been chosen partly by computer for the 1984 Olympics. Some extremely tall and suitable girls were selected, and then asked if they would like to take up the sport. The American team eventually lost to the Chinese who may also have been using a computer or the more ordinary fact that China possesses four times as many women from whom to choose.

The difference between a Derby winner and a basic run-of-the-mill, randomly-bred horse is considerable. The casual horse is completely out-classed. So, perhaps, the athletes of today will be outclassed by the thoroughbreds of tomorrow. Height, for example, is good for high jumping. Systematic breeding of tall men and tall women would guarantee exceptionally tall offspring in a generation or two. After all, the bullocks sold at Smithfield market doubled their weight in the eighteenth century alone, when fenced enclosures permitted controlled mating. What size could man become? What weight and height? And what then would be the distance that he could put the shot, or throw a discus?

Today he is like a mediaeval cow, with no selective breeding except the natural sort. Tomorrow the methods of the breeder, developed by man for meatier cattle and faster horses, may even be applied by man to make himself more meaty (for weight-lifting) or faster (for races) or more agile than today. The Olympics are already over-shadowed by undesirable national-ism. This unwelcome attribute would itself pale if set beside the enormity of breeding people specifically for the games.

23. Old Age

What is senescence?/How long is life?/Record old age/Animal comparisons/Changes in the old/Their causes

---

'Others merely live; I vegetate.'

The Unquiet Grave by PALINURUS

'The old . . . are positive about nothing; in all things they err by an extreme moderation.'

ARISTOTLE

'Men of age object too much, consult too long, adventure too little, repent too soon and seldom drive business home to the full period, but content themselves with a mediocrity of success.'

FRANCIS BACON

'It is a man's own fault, it is from want of use, if his mind grows torpid in old age.'

SAMUEL JOHNSON

'Everybody wishes to live a long time. Nobody wishes to be old.'

SIR ADOLPHE ABRAHAMS

'There's a certain moment in life when you realise you're born with a deadly disease which is life.'

JEANNE MOREAU

Every child is suddenly confronted one day with the disturbing fact that he or she will eventually die. It is so worrying a realisation, and so distasteful to accept before being even full grown, that the confrontation is often linked with a determination to be the first exception. Later on, having progressed through childhood, having left adolescence for adulthood, and having met the first positive deteriorations associated with senescence, the determination not to die becomes more confused and confusing. At what age should one not die? At Peter Pan's? As Chekhov's perpetual student? Or with the worries and wisdom of a John of Gaunt? ('Old Gaunt indeed, and gaunt in being old' died at fifty-nine.) Plainly, being too old is too late, for King Lear hated him 'that would upon the rack of this tough world stretch him out longer'. Like Tithonus, whose wife acquired immortality for him from Zeus but forgot to ask for eternal youth as well, the actual prospect of a perpetual existence is as forbidding as death, unless its gift were partnered by innumerable other assets.

In fact the prospect is as remote as ever. Despite our morbid interest in the subject, next to nothing is known about the mechanism of the ageing process, and medical science has done next to nothing to achieve any prolongation of the traditional life span. (It has merely enabled more humans to experience more of this life span.) Certainly no one knows whether or how we may be able to control our rate of ageing. Put at its crudest, the advances of medicine are enabling more and more of us to achieve senility. The man who dies naturally today at a grand old age is experiencing a most unnatural form of death. Such an end out in the wild is highly improbable, because predators plus the other hazards of existence tend to kill off creatures long before they have lived sufficiently to attain even a moderate degree of senility. Small birds can live for twenty to thirty years; but, with at least half of each small bird population being killed off every year, the avian individuals who achieve anything approaching longevity must be very rare indeed.

Senescence has various paradoxical definitions. Firstly, it begins at birth. (Think of wound-healing. This procedure is most efficient at birth and declines thereafter.) Secondly, as Charles Minot was the first to point out, ageing is faster in younger than in older animals. Thirdly, as Sir Peter Medawar defined it, senescence merely 'renders the individual progressively more likely to die from accidental causes of random incidence'. Finally even that word 'accidental' can be disputed because all deaths are really accidental. No one dies solely from the burden of his years; no one dies from old age.

On this last point a clinical pathologist (in *The Lancet* in 1938) once tried to recall from his experience any patient for whom the burden of years had become excessively burdensome, and who had died solely for this reason. The nearest person to suit these requirements was a man of ninety-four whose life had seemed just to fade away. Unfortunately it had done nothing of the kind. At autopsy a lobar pneumonia of four days standing was discovered. The pneumonia, and not just those ninety-four years, had killed the old man.

More commonly the cause of death in the very old is an assortment of causes. One actual cause may have been the coup de grâce, but there were several others waiting in the wings with their own rapiers had the first coup failed to kill. One case quoted is of a lady of eighty-three who died of haemorrhage due to a duodenal ulcer. She also had trouble with her oesophagus, a hiatus hernia, gall stones, a faulty colon and duodenum, diseased lungs, a fatty heart, calcified valves, an atheromatous aorta, an ovarian cyst, a tumour of the uterus, a soft spleen, arthritis of the knees and two clots on the brain. Her doctor knew of many of these deficiencies, but had been presented with no evidence of the others. In any case he would have been hard put to it to predict the one lethal ailment of her unfortunate assortment; but it was certainly not her four score years and three that killed her by themselves.

However, there is one quality of mere years that does great damage entirely on its own. Modern society, although permitting its politicians and its judges to care for our well-being until they are steeped in old age, likes to evict most of its citizens from their customary employment at sixty or sixty-five. Quite suddenly men and women are discarded. They may well have lost initiative, and be finding their work rather more arduous than formerly; but on one day, a birthday, they suddenly lose high office, or at least medium office, and find themselves with no office at all. The daily rate of retirement in Britain is 1,000 persons. (Of course a minority have had to stop work before retirement age. Of those who are in the Civil Service at the age of twenty-five one-quarter will have retired or died before reaching sixty.) Among the mass of people, wrote V. S. Pritchett, 'retirement is often a fatal assault on the ego, and calls for an unusual gift of courageous adjustment in the person concerned'. To be cast out in this fashion just because of a birthday, without reference to actual deterioration, can be both a physical as well as a mental assault. Removed from routine the change can be physically disastrous – although, once again, there is always a specific cause of death. No death certificate can give 'He had to retire' as a fatal cause any more than it can give 'Anno domini'; but a sudden banishment from useful activity can accelerate a person's decline with outstanding haste.

How long is life?   One of the most remarkable biological statements in the Bible is that the days of our years shall be three score years and ten. In those disease-ridden days, with so much biblical talk of plagues and afflictions, of leprosy and sores, of pestilence and hunger, the life span of man was reckoned to be seventy years. Now, in these days of welfare and medicare, with talk of antibiotics and injections, of surgery and immunisation, the average life span is still seventy years. In the United States it was 70·2 years in 1961, and 69·9 years in 1964. There have been minor fluctuations since then but the male survival rate is as near the classical three score and ten as makes no difference.

Whereas the new-born Hebrew delivered in Judaea had a slim chance of attaining his rightful old age, the new-born American has a chance better than 50:50 of doing so. The biblical span indicates man's potentiality; today's span indicates man's actual and average life-time. Modern chances are better than 50:50 because so many infants die; one death aged one brings the average span rocketing down.

Recent figures for Britain show the life expectation for a boy at birth to be 69·2 years, and for a girl to be 75 years. United States figures are made more complicated (and more revealing) by splitting the races as well as the sexes. It is particularly interesting that the advantage of being white is, from the longevity point of view, about the same as the advantage of being female. White males have a similar life-span to non-white females, but the difference in expected span at birth between white females and non-white males is thirteen years.

In one sense it might appear from the 'expectation of life' charts that mankind is immortal. Each new age reached stretches forward the likely date of death. For example, in Scotland in 1982, newborn babies could expect to live for 69·2 years (if male) and 75·2 years (if female). On their first birthdays those infants could expect to live for a further 69·1 and 75·0 years, a gain of almost one year as reward for having survived the perils of the initial year. By the age of fifteen the children could expect to live for an additional 55·4 and 61·2 years, totalling 70·4 and 76·2 years, a further slight gain for having survived childhood. By the age of 45 their expectation is 27·1 and 32·2 years, totalling 72·1 and 77·2 years, another gain of a year. By age 65 expectation is 12·2 years for men and 15·8 for women, totalling 77·2 for men and 80·8 for women. The difference between the two sexes has narrowed owing to the fact that many more men than women have already died. It is like the argument that the hare never catches the tortoise, because the tortoise has always advanced slightly during the time taken by the hare to reach the tortoise's last position. However, despite the average prospect of new life at each new birthday, the process is not infinite. All will die in time. The hare will always reach the tortoise. The system has never been known to fail.

Although the Bible may speak of those seventy rightful years, and the modern statistician now speaks of them as the seventy average years, it took a long time for the average to creep up to the biblical right. It has been reckoned (although with slender evidence) that life expectancy in Britain in the Iron Age was eighteen. Later estimates, of increasing validity, indicate expectancies of 22 (2,000 years ago), 33 (the Middle Ages), 33·5 (1687–91), 35·5 (1789), 40·9 (1836–54), 49·2 (1900–02), and 66·8 (1947). Nowadays people in large areas of countries like India have a better life expectancy than people in Britain had a hundred years ago.

The poorer countries are still pushing up their average age of death, but the richer countries have already met a kind of limit. According to the Metropolitan Life Insurance Company an American male of sixty-five in 1950 could have hoped for another 12·8 years. By 1962 a man of sixty-five could hope for 12·9 years, marking a mere five week improvement in longevity in a dozen years. American women, on the other hand, went marching on. In 1950 the US female of sixty-five could have hoped for fifteen more years; in 1962 she could expect sixteen years. In those dozen years she had gained a dozen months of life.

The steady pushing back of the time for dying means a steady rise in the numbers of old people. In 1961 there were 5,750,000 people over the age of sixty-five in England and Wales, a sizeable number roughly equal to the population of Scotland. By 1981 there were 7,500,000 old people in England and Wales, or 13% of the population, or 87,000 more each year. (As women acquire their pensions at age sixty rather than the male age of sixty-five this means that 17% of the British population were then on a pension, a formidable price for the state to bear. As a further 6% were unemployed, and some 25% were either children or students of some sort, the total load of dependants was therefore about half the population. Third world countries are similarly burdened, but almost entirely by the young as a 3% growth rate in population means 50% under the age of sixteen.)

Should the medical profession reach its goal, said Sir George Pickering, 'those with senile brains and senile behaviour will form an ever-increasing fraction of the inhabitants of the Earth'. This Regius Professor of Medicine at Oxford thought it a 'terrifying prospect'. Dr Alex Comfort, the gerontologist, admits an increasing preponderance of old people, and does not expect any radical change in the immediate future years. 'The best we can hope for from medicine and hygiene *alone* is that the average life-span will increasingly become 75–80 years'; the future will of course bring 'palliative possibilities ... but there is no graft, hormone or other preparation known at present which is capable of producing more than a limited reversal of a very few senile changes in human beings'.

Today, even though there are relatively fewer old people than there will be, the problem of the aged is enormous. Of the

7,500,000 British people over sixty-five at least half of them depend wholly or primarily on cash and other benefits from the State. Four per cent, or 300,000, live in institutions; and severe incapacities, such as bad feet, the need for delivered meals, bad eye sight, bad living conditions or even a totally bed-ridden condition, affect hundreds of thousands of others. The old are not only retiring earlier and living longer, but they have had fewer children and therefore have a smaller chance of unmarried daughters to care for them.

Bernard Cohen is an American very much in favour of a quantitive approach to risk. He considers that facts should be set against each other if, for example, the risks inherent in nuclear power are to be properly appreciated. On 'Loss of Life Expectancy' he has compiled a list, which includes: Being male rather than female – 2,800 days lost; being unmarried – 2,000 days; working as a coal miner – 1,100 days; being poor – 700 days; driving small cars rather than standard – 50 days; fire, burns – 27 days; diet drinks (one per day) – 2 days; if all electric power in the US were nuclear (as calculated by Union of Concerned Scientists) – 1·5 days; if all electric power were nuclear (as calculated by Nuclear Regulatory Committee) – 0·03 days. It should also never be forgotten that the step-ladder is one of the most lethal pieces of equipment around the standard home.

Record old age is a subject for extreme fantasy. Stories come out of Russia, for example, with tall tales to tell about longevity. Georgia, and its minute population, has been alleged to include 2,000 centenarians now living. More than one grain of salt is necessary in assessing extravagant claims, particularly as most long lives have their infantile origins back in an illiterate and unrecorded past. In 1966 the Russians reported the death of Shirali Mislimov, allegedly aged 160, said to have been the oldest resident of the USSR and to have fathered children until 1936; but one wonders about the recording of his peasant birth back in Tsarist 1806. (They also reported his death again in 1973 at the age of 168, and added the fact that there were then in the Caucasus eighty-four people aged over 100 for every 100,000 inhabitants. According to the *Guinness Book of Records*, which takes a particular interest in such things, 'no single subject is more obscured by vanity, deceit, falsehood and deliberate fraud than the extremes of human longevity' and scientific research has revealed that 'the correlation between the claimed density of centenarians in a country and its regional illiteracy is 0·83 ± 0·03'.)

It is more likely that 115 is the authentic limit for mankind, and at least two people are believed to have reached these five score years and fifteen (although the *Guinness Book* considers that the greatest 'authenticated' age is the 113 years 124 days lived by Pierre Joubert, a French-Canadian bootmaker who died in 1814). Ages beyond 110 are likewise extremely rare, but centenarians are almost common. Buckingham Palace sends telegrams on significantly ancient birthdays, notably the 100th, if reliably informed of the event. Some 300 such telegrams are despatched annually. However, the Palace's telegrams do not just record the 100th birthday, and they have to rely upon individual informants; therefore the numbers are none too reliable. There are currently about 1,300 people in Britain aged over 100, more than ever before. The most famous of them is Lord Shinwell, born in 1884. He became a member of parliament in 1922.

The Registrar-General's figures show that about twenty men die every year who are over 100, and about one hundred women. Women traditionally survive longer than men, but the ancient stories from Georgia in the USSR reverse the situation; it is the Georgian men who thrive deep into a second century while their women scarcely reach their first. The better survival of men used to be the feature also of primitive communities. Presumably the hazards of obstetrics were greater than the hazards of war, despite the additional and over-all male weakness. In England and Wales roughly a quarter of a million of each sex die annually, and the chances of anyone putting one hundred candles on a cake are about 12,500:1 against for men and 2,500:1 against for women.

It was during World War Two that the claims of oldest inhabitants quietly became verifiable. British birth certification had started officially in 1837. There had been parish registers for centuries beforehand (back to 1538) but in 1943 the first woman died who was not only a centenarian (she was 106), but the possessor of a birth certificate. In 1945 the first such man died – at 105. Apart from a lapse in 1948 when a woman died – at 115, having allegedly been born four years before certificates were introduced – current British claims for longevity are now made in association with birth certificates.

Mankind may not think so, as old age and death hurtle along with comet-like rapidity, but the human species is nearly the oldest-living species in the entire animal kingdom. Man certainly lives longer than any other mammal. Alex Comfort, and S. S. Flower before him, have sifted through zoological records, scientific literature and countless claims to select the most ancient faunal inhabitants. With mammals the Indian elephant

315

may reach the 60's or 70's, one such elephant having died in California in 1975 with a known age of seventy-eight, but otherwise only an occasional horse, hippopotamus, rhinoceros and ass either pass or nudge up towards a fiftieth birthday. The authentic bird record is held by an eagle owl of sixty-eight, but some extravagant claims of greater antiquity have been made for cockatoos, vultures, geese and parrots.

Certain reptiles undoubtedly do beat mankind, notably many tortoises. Some are known to have reached 100 and to be still going strong. One or two have reached 150. Crocodiles do less well, with fifty being very old for them. Snakes may also live long, but they do less well in zoos and barely reach thirty years. The amphibia have a reputation for longevity, with some toads living to thirty-six and even frogs well into their teens; but no amphibian reaches man's age. Fish have an even greater reputation, and long legends are told about carp. Without doubt some sturgeons have reached their 70's and 80's, and a 10-foot halibut was once landed at Grimsby whose age – as shown by scale examination – was over sixty. The carp stories may well be true, but no fish has yet been convincing, following scientific scrutiny, that it has lived longer than a long-lived human. No invertebrates, despite some languid molluscs, have yet proved a life of more than two or three decades; although, once again, age assessment is difficult.

The conclusion, whatever humans may think, is that man is a very long-lived animal, beaten into first place emphatically by the tortoises and possibly by a few other contenders. Womankind does even better than mankind, although even she cannot compete with the tortoises.

Changes in the old 'Fear old age for it does not come alone', said Plato. It certainly does not. Old people are a plethora of change. They have lost an inch of stature. Their hair has thinned from their heads, yet it may sprout thickly from nose and ears. Exposed skin, dry and wrinkled, may be blotchily pigmented. There are often dark flat warts on the body, and more frequently there are small, red spots on the abdomen. Limbs may shake. The jaw or head may also quiver, and walking may change from a stride into a shuffle. Hands and feet, in particular, may get cold.

Internally the brain becomes lighter as it atrophies, while the grooves of the sulci become deeper and wider. Less oxygen is used by the brain. The ability to feel both heat, cold and pain, may all lessen. Body temperature cools, and a thermometer in the mouth of very old men may read 2°F (1°C) less than normal. (Therefore

316

normal temperatures may indicate fever.) Certainly the ability to smell decreases as the olfactory fibres diminish; even the added pungency of domestic gas is undetectable by many of the old. The lungs shrink, the joints stiffen, the reflexes decrease, the pupils become lazy, the liver lessens its bulk, and the basal metabolic rate gradually falls – to about 12% of normal. Teeth are lost, and both jaws shrink, although the chin tends to jut out. Hearing suffers, characteristically for the higher tones, and so does vision. For females the uterus and ovary may atrophy into minute fragments of tissue, while for males the prostate may enlarge from chestnut-size to apple-size. Bodily composition varies between the young and the old. Between the twenty-fifth and the seventieth birthdays water content goes down from 61% to 53%, cell solids down from 19% to 12%, bone mineral from 6% to 5%, but fat goes up from 14% to 30%. Consequently an old man, priding himself upon no weight increase during his middle-aged years, may merely have exchanged his youthful muscle for elderly fat.

Incontinence of the bladder is frequent, but incontinence of the faeces is rare although constipation is a common complaint, possibly because the intake of food and drink by the old is much reduced. Cramp is common. The pulse rate normally decreases with age, but can rise again with the extremely old; the range of some Chelsea pensioners was found to vary from 44 to 108 times a minute. Hearts can shrink or enlarge with age and some hearts of the over-eighties can be almost three times the weight of others. Blood pressure rises with age. The speech of an old person becomes not only tremulous but metallic. And of course there are all the chronic mental disorders, like senile dementia. Trevor Howell (in *A Student's Guide to Geriatrics*, 1963) gives a disturbing list of ten questions to check a patient's degree of orientation and intellectual impairment. The disturbing point is their simplicity, for they include: Where are you now?, What year is it?, What is your birthday?, and How old are you? Finally, to quote Marjory Warren, 'Existence to the old often becomes a pathetic attempt to kill time, before time at last kills them.'

What causes ageing? What indeed! Whatever does happen is happening right from the start of birth, and it never fails to happen. As a system, mechanism or occurrence it is perfection: it always succeeds. No one has ever failed to die. Progeria, the condition of premature ageing, may first show its severe symptoms when children are starting school, and the victims die on average at sixteen, often from coronary thrombosis. The opposite condition, that of Dorian Gray or Rip van Winkle, does not arise. Those who die very old look very old and are physiologically very old. The process of ageing is unfailingly effective. It is not difficult assessing a person's age possibly to within a year or two, certainly to within a decade. The passing of time produces too many clues for us to work on.

There are three main theories about the controlling factors of our rate of ageing. The first concerns loss, the loss of too many cells or the loss of irreplaceable parts. Brain cells, for example, undoubtedly die off in their hundreds of thousands and they cannot be manufactured again after a very infantile stage in each human life. Some get killed, as by natural radiation, and some just die, but none is ever replaced. Unfortunately for this theory, although large parts of the brain or other organs can either be heavily damaged or partially lost, there is no effect on the rate of ageing. It neither speeds up nor slows down in consequence. Moreover, animals have totally different ageing rates but suffer cell destruction at similar speeds. Mere cell destruction cannot be the sole determinant.

Theory number two concerns mutations. A dividing cell does not always divide correctly and provide two cells quite as satisfactory in every way as their single predecessor. All sorts of errors can creep in, aided and abetted by natural radiation. The list of agents is long which have been blamed for causing cancer, for causing the berserk behaviour of excessive proliferation; and the list is just as long, or longer, of agents which could cause faulty replication of cells in cell division. Alex Comfort has pointed out that any theory which is linked to cancer research is a sure grant-winner and therefore has something to be said for it. This is a reminder of two other faintly sick scientific aphorisms: 'Cancer is God's gift to biologists' and 'There are more people living off cancer than dying from it'.

This theory does demand a fearful number of mutations for them to be so powerfully effective throughout a person's life. Perhaps the changes do not just put cells out of commission, but make the mutated cells harmful. Or perhaps the small loss of mutated cells has a more powerful effect with certain types of cell, such as the endocrine glands or some of the constituents of blood. It does not seem – at present – as if mere changes of this kind are sufficient to be the sole determinant of the ageing process. Nevertheless there was the interesting announcement that 10% of the cells of very old women had lost an X chromosome. Theory number two may well prove to be the correct theory, but only when the world's increasing number of gerontologists come up with some equally enlightening information as that 10% loss of a whole chromosome.

A third theory, now falling swiftly out of favour, is concerned with the accumulation of unwanted chemicals. It could perhaps be that some vital substances can only be replaced at cell division, and a general decline in the rate of cell division could lead either to a lack of needed substances or an excess of unwanted ones. The theory is gaining less support these days partly because it assumes too static a state of affairs in the living cell, whereas cells are dynamic systems. Another point in this theory's disfavour is that ageing occurs also in animals which continue growth and a generalised cell division throughout life. Mankind stops growing, but fish, for instance, grow and grow – until death. In the meantime they too have aged.

Spare parts and the storage of human tissue do not really influence our ageing. Some of us are uneven, in that one part needs replacement prematurely; but ageing is an increase both in the number and variety of faults. Unlike Stanley Holloway's weapon in the Bloody Tower, which had had a new handle and perhaps a new head but was a real old original axe, the human body is more complex. More like an old car than an old axe there is always another part waiting to fail.

I shall give the conclusion to Alex Comfort. 'To the question "Can the effective human life-span be prolonged artificially?" the most probable answer ... would appear to be "Yes". To the further question "By what factor?" no meaningful answer can be given until we know more of the nature of the predominant processes which determine human senescence.'

24. Death

When is death?/The signs of death/Its causes/Its secrecy

'Qu'est ce que la vie?' 'La vie, c'est la mort.'
CLAUDE BERNARD

'Who dies and, dying, does not protest his death, he has known a true old age.'
LAO-TZU

'The death which a man suffers by one sudden means or another is usually infinitely more comfortable for him than his birth was for his mother.'
WILLIAM JOYCE (Lord Haw-Haw, who was hanged in 1946 at Wandsworth)

'Die, my dear doctor; that's the last thing I shall do.'
LORD PALMERSTON, on his death-bed

'Lazy beggar would never work when he was alive. He can do summat now he's dead.'
A Lancashire widow, quoted by J. B. PRIESTLEY, who had put her husband's ashes in an egg-timer

Once again there is a paradox. Everyone knows he or she will die; but the medical profession does not believe that most people are aware of the fact when they are actually dying. Similarly the more proficient mankind becomes at controlling death and prolonging life the harder it is for anyone to state when death had definitely occurred.

For centuries there was no problem. Death occurred when movement ceased, when breathing stopped, and when no heart-beat could be detected for two to three minutes. The balancing of a glass of water on the chest, or the holding of a mirror or a feather over the lips, were immemorial aids. Nowadays there is no such simplicity. For one thing it is now known there are various forms of suspended animation, caused perhaps by electrocution, by drowning or drugs. The signs of life can be absent, and the signs of life can then reappear. It is not without precedent for a mortuary attendant to observe that one of his charges has plainly arrived prematurely.

However, the old-fashioned simplicity is rendered problematical by today's inability to define death. Both respiration and circulation can be maintained mechanically, and they frequently are during operations; but this is different from the mechanical maintenance of lifeless lives, those for whom no hope exists, those which are merely technical prolongations of the biological process. As one American neurosurgeon put it, 'the look of life is maintained in the face of death'. But when, asked another neurologist, 'do you pull the plug out, and make the expensive equipment available to someone who might live?'

Although it was the departure of the human spirit which doctors have been habitually checking by listening for signs of heart-beat, it is only recently that attention has been paid to the brain at death. With the brain more the spirit's home than the heart, the absence of life within the brain has recently been regarded as a surer sign of death. An electroencephalogram, or EEG, can record the constant electrical flickering of a living brain; it can also record the absence of any such activity. When the EEG has produced a flat tracing for quite a time, and when the patient has neither reflexes, nor a heart beat nor respiration of his own – only the pumping of the machine – then is the time, by today's rulings, for the machine to be stopped, for the plug to be pulled out, for death.

In May 1966 the French took a positive lead in establishing this point. The National Academy of Medicine, in Paris, decided that a man may be ruled dead if it can be proved, despite the artificial beating of his heart, that his brain will never again be able to resume control of his vital functions. The principal objects of the French recommendation were partly to clarify the situation and partly to enable organ transplantation to be carried out more effectively. Such organs can save other men's lives, but they cannot do so if they have been permitted to deteriorate grossly during the prolonged death of their original possessor. A failing circulation will cause a failing oxygen supply to those organs; hence their deterioration. This need not be the case if the body is pronounced dead, due to the lack of EEG readings, and then maintained under conditions of artificial survival as a store-house of living organs; but for how long should the brain and the EEG be quiet? For forty-eight hours or less, said Professor de Gaudart d'Allaines, head of the Commission which presented the report that was later adopted by the French Academy.

France was therefore the first country to have faced death in this fashion, and to have regarded the situation squarely by taking action at a national level. However, since it took that step, other countries have taken significant strides, and Finland was another pioneer. As EEG's are not possessed by every hospital, and as the need is often great for someone to be pronounced dead, many other criteria have been established. There is, as yet, no world concensus over the definition of death, but opinions are more uniform than they were a decade ago and legislation is catching up. (Once again there is a fuller account of this story in *The Mind* as it is more appropriate to that volume.)

Signs of death   The skin becomes pale, and loses its elasticity. There are changes in the eye, such as the lack of circulation in the blood vessels of the retina. (Some ophthalmic experts say they can see the red cells roll to a stop.) Muscles lose their tone, and the body permanently flattens in those areas which are taking the weight. Breathing stops, although it can confusingly start again after a lengthy pause. The pulse is undetectable and must be so for at least five minutes, although survival from longer pulseless periods is not uncommon. (Film characters, who rush up with their own pulses beating wildly and, after feeling a second or two beneath the coat of some alleged victim, emphatically pronounce him dead, are being outstandingly hasty and brilliant in their assessment.)

Later changes are yet more definite. The body cools, and at a reasonably constant rate, i.e. about 2·5°F (1·5°C) per hour for the first six hours, if clothed and under temperate conditions. Later cooling is slower, perhaps 1·5°F (1°C) per hour after six hours. After twelve hours the body will feel cold, and after another twelve it will – probably – have reached ambient temperature. A naked body will cool about half as fast again as a clothed one, and a body immersed in water will cool about twice as fast. Due to lack of blood circulation there will be pale areas where the body has been resting, and livid areas all around them into which blood has drained.

Rigor mortis is caused by a stiffening of muscle fibres. Its time of arrival is highly variable, but it can be seen first in the face some four to seven hours after death. A few hours later it has affected all parts of the body. Then, having lasted for perhaps half a day, it retreats very gradually, leaving the body flaccid once again. Within thirty-six hours this cycle of stiffness is probably completed, although rigor mortis is casual about its timing. For instance, what is called a cadaveric spasm, certainly a kind of stiffness, can occur at the very moment of death.

Decomposition, the final and most positive sign of death, causes a green tinge on the abdominal skin after about forty-eight hours in a mild climate. The bacteria responsible for putrefaction come mostly from the intestines. As new-born infants are relatively sterile, having no large bacterial population within their digestive system, they frequently mummify rather than putrefy. Dry hot air can do the same both for children and adults. Putrefaction is also prevented to a large extent in those individuals whose bodies have been immersed in peat bogs after death. Corpses, many centuries old, and discovered in Denmark, Germany and Holland, have been preserved in a remarkably life-like and telling fashion. Apart from such exceptions as embalming and incineration, the putrefying bacteria do their work with a distasteful and effective greed, though their effectiveness is hindered by burial or immersion in water. They achieve their biological entropy most surely between 70°F (21°C) and 100°F (38°C). Below 50°F (10°C) putrefaction is markedly slow or non-existent, and above 100°F (38°C) some form of drying or mummification is likely.

Causes of death   In Britain, by law, a doctor does not have to see a body after death, and only about two-thirds of all the dying are seen after their deaths by medical men. Nevertheless, a death certificate has to be completed promptly by the doctor who had been in attendance upon the dead man, stating the cause of death.

The Registrar-General has certain rulings about the completion of these certificates, and about likely causes. No doctor can just write: 'He got sick and died.' (The Nebraska State Medical Journal reprinted some causes of death from earlier days: 'Went to bed feeling well, but woke up dead.' 'Don't know. Died without the aid of a physician.' 'Had never been fatally ill before.') By no means are death certificates always accurate even today as to the cause of death. Recent surveys of presumed causes before and after necropsy have indicated great fallibility. Death certificates 'are not worth the paper they are written on' according to a 'professor well known to *The Lancet*', and yet great faith is pinned on them whenever they are consulted for mortality statistics. Perhaps necropsies should be commoner, as in Scandinavia. Whenever deaths are violent, suspicious or unexplained the problem of investigation is handed over to the coroner (whose other remaining tasks concern deaths in prison, treasure trove, and the removal of corpses from Britain).

The British are publicly ambivalent about causes. Everyone asks, but the facts are rarely printed. Newspapers occasionally mention the cause in their obituaries, but neither the *British Medical Journal* nor *The Lancet* does so. American newspapers are more forthcoming. *Time* magazine makes a point of stating the cause of death. However the American ballyhoo of what to do about the corpse is certainly not mirrored in the British Isles where the subject is more hurriedly and less expensively hidden. (*The American Way of Death* by Jessica Mitford, which drew attention to the costs of the funeral industry, made £40,000 for its author within two years. She also had a cheap kind of coffin named after her.) Our remote ancestors and even our Victorian forebears would be astounded at the revelation that many in Britain today achieve their own ends without having seen the dead body of a single other human being. The British are also making certain that corpses are more speedily disposed of than by traditional burial. By 1966 the cremation figure was 47% of the total, and increasing steadily, being 64% by 1979 and approaching 70% by 1983. Crowded churchyards, or even full ones, are part of the reason for favouring a form of disposal that occupies no space, save for the busy crematoria. Orthodox Jews do not cremate, but the Roman Catholic church has recently been relaxing its objections, and soon some four-fifths of the total British dead will be burned at death.

For every thousand people alive in Britain at the start of any current year, eleven or twelve of them will die during the year. The figures indicate some 640,000 deaths a year, or 1,750 deaths a day. Despite infant mortality and accidents and all human ailments, over half these deaths are caused by heart disease, cancer or stroke.

Figures for the extreme past are notoriously unreliable; everyone seemed to die from the ague, flux, fever, or the vapours. However it is only necessary to go back to 1900, and well within the history of reliable records, to find a totally different pattern in the causes of our deaths. There is a distressing simplicity about the change. Practically every form of dying has become less common than the three modern and triumphant horsemen – cancer, heart disease and stroke. In 1900 cancer killed 4%, and eighty years later 26%; heart disease killed 12·8% and this rose to 31%; vascular lesions affecting the central nervous system – strokes – rose from 7% to 14%. As compensation there have been equally dramatic slumps. Tuberculosis fell from 10·4% in 1900 to 0·2% in 1980. Accidents, suicides, ulcers and many others have remained reasonably constant. Both stillbirths and the deaths of infants under one year have been falling, but their two percentages of the total deaths may actually have risen on occasion due to the rising birth-rate.

Lung cancer also has its own emphatic tale to tell, as practically all the over-all cancer increase for males in recent years can be accounted for by the increase in deaths from lung cancer. Of males who are killed by cancer, 37% are killed (in 1982) by cancer of the lung; of females killed by cancer the percentage from lung cancer is 14%. (In 1966, for instance, the female proportion was only 9%, but women are determinedly attempting to catch up men in this regard with their considerable enthusiasm for smoking cigarettes. In the younger age groups, and in Britain, they are already smoking as much, with 31% of the males smoking aged 16–19 as against 30% of females, and with 41% of males aged 20–24 as against 40% of females in the same age groups. A decade earlier the respective figures had been 43% as against 39%, and 55% as against 48%. The young girls are therefore catching up the young men by being relatively more loyal to the habit.) The ban on television advertising of cigarettes became effective in Britain during 1965, and the United States ordered all cigarette cartons to be labelled with a warning shortly afterwards. It appears that cigarette smoking is too deeply engrained a habit to be unduly influenced by such measures, and the authorities proclaim but do not prevent – despite harrowing announcements. 'It is possible to prevent about nine-tenths of the cancer of the lung which will otherwise occur among our younger generations if we can persuade them not to smoke cigarettes', according to the Chief Medical Officer of the Ministry of Health. 'We could save more lives by persuading people to stop smoking than by any other step I can think of,' said the Minister of State for Health, '. . . Over 50,000 deaths a year in England and Wales are attributed to cigarette-smoking.'

A remarkable feature of our changing mode of dying, or rather of succumbing to various diseases, is the near-totality of defeat for the ancient diseases. Dreaded names, still with great power to them, no longer wield their fearful authority. In 1964, for example, there were no deaths from diphtheria in England and Wales, only seventy-three deaths from measles, none from anthrax, seventeen from dysentery, four from poliomyelitis, none from smallpox, forty-four from whooping cough, 2,484 from tuberculosis (but only forty-one of them were of people under twenty-five), none from cholera, none from plague. All the deaths from this frightening collection of names did not add up to the number dying that year from traffic accidents, the 7,000 deaths – or 1·3% of the total – that succumbed to the internal combustion engine. Neither do they add up to the current, and smaller, number of traffic deaths, now running at just over 5,000 a year. Nevertheless, as we have to die from something, quelling the ancient diseases merely opens up possibilities for the modern ones. And in the future, when medical science has triumphed over today's fatal assortment, there will always be – to use Dr Comfort's expression – the next layer of the onion.

Yet there is one sure sign of progress. Although we will die just as emphatically, the chances are that future doctors will be able to identify accurately more and more of the ailments assaulting us. In those days of agues and humours the medical men were frequently at a loss for a description, let alone a cure. Even today in Britain about 1·3% of the population die from 'unknown causes'. This percentage will surely drop. It is substantially higher in many other parts of the world, such as Greece (18%), Yugoslavia (25·1%), Sri Lanka (22·4%), and El Salvador (28·3%). The South African situation reflects something of the schism of that country. Whereas the deaths of 4·8% of the whites have an unknown cause, the deaths of 21% of the urban blacks are similarly unknown. The figures are higher still for the Bantu living in the country.

The secrecy of death　Should anyone, such as doctor or relative, inform a dying person of his condition? Should he or she be told that a 'terminal stage' in the disease has been reached? The most famous and frequently quoted pair of facts relating to the question serve merely to underscore the problem. Apparently 69% to 90% of doctors questioned in various studies favoured keeping the patient in ignorance, but 77% to 89% of the patients wished to know. The doctors hold the whip hand, in that they are the experts, they have watched the disease's course, and already over 50% of American and British deaths occur in hospitals; while the patient – inexpert and sick – is obviously less aware of the actual situation, unless clearly informed. One patient reported to the enquirers: 'I don't want to be denied the experience of realising that I am dying.' And at a symposium on the subject in Russia an American psychiatrist said, 'We do not even permit the dying person to say goodbye to us.'

The most famous personal case of this kind in recent years was that of J. B. S. Haldane. Here was a true scientist, alert, and constantly fascinated by new information. He was no desultory dowager, ignorant of medicine and unwilling to understand. He was undeniably capable of absorbing facts, had he been told them. At the age of seventy-one he was operated on for cancer of the rectum. This he knew, and he even had a poem published by the *New Statesman* called 'Cancer can be fun'. However he was not told the whole story, neither was his wife, nor other relatives. Eventually after he had initiated several new investigations involving colleagues, he did learn his prognosis, and that he had secondaries. 'Now I shall have no time to rest before I die,' he wrote to a friend. As soon as he had died (in December 1964) Naomi Mitchison, his sister, wrote to the *British Medical Journal* deploring the secrecy and asking why the surgeon had withheld the vital information. The ensuing correspondence was extremely forthright, notably from John Maynard Smith who attacked the custom of keeping quiet: he added that he personally felt he could accept 'bad news with fortitude' but not 'a continuing fear that the truth was being withheld'.

The Lancet had a leading article, entitled '*Timor mortis conturbat me*', on the duty of doctors to their dying patients. 'A positive, courageous, and realistic approach to death and dying has been common, at least as an ideal, through much of history, but we seem to have lost it. . . . Could we not now usefully persuade ourselves to take a new look at death and dying?' Lyndon B. Johnson, when the United States's president, highlighted mankind's current inability to regard death as death when he cabled to the very old, very sick, dying and unconscious Sir Winston Churchill: 'Whole world prays for your speedy recovery.'

Finally, at least in England, you do not legally own your body after you are dead. You may give instructions for its disposal during your life, but it is up to your next-of-kin to decide whether your wishes should be carried out. Should you wish to bequeath your body to medical research the man to write to is H.M. Inspector of Anatomy, Department of Health and Social Security, Alexander Fleming House, London S.E.1. The Government will respond, as on so many earlier instances in life, by sending the appropriate form – for death.

25. Suicide

Society's attitudes/Suicide history/Who kill themselves?/Rates elsewhere/Students/Children/Methods/Suicide and murder

'The thought of suicide is a great consolation: with the help of it one has got through many a bad night.'

NIETZSCHE

'Sufferings and hardships do not, as a rule, abate the love of life, they seem, on the contrary, usually to give it a keener zest.'

WILLIAM JAMES

'Each advancing age period of life shows a steady and consistent rise in suicide frequency.'

LOUIS I. DUBLIN

'To my friends: My work is done. Why wait?'

GEORGE EASTMAN, shortly before he shot himself in 1932, aged seventy-seven

Before mentioning facts about the where, when, who and how of suicides and attempted suicides it should be pointed out that such facts are notoriously unreliable. Different countries, religions and people have disparate attitudes towards self-inflicted injury, and their differing codes colour the statistics. Is it likely, one wonders, that figures from Manhattan, Dublin, Rome, Lagos and Tokyo are comparable? Furthermore, a society's attitude and its actual laws have frequently reached a high level of disunity. Throughout this century the conviction has been growing that suicidal acts are a form of sickness; yet in England and Wales until 1961 both suicide and attempted suicide were criminal acts, although rarely treated even as a misdemeanour, and until 1962 churchyard burials of suicides were carried out without any form of burial service.

At least, that was the theory. In practice both burial services and consecrated ground were often used, depending upon the coroner's verdict and upon the bishop's feelings. Prayers were summarily altered if they were inappropriate. Coroners, pulled both ways by the ludicrous situation, constantly recorded verdicts of 'suicide while of unsound mind', thus absolving the victim. The law did not like the custom because actual evidence of

328

unsoundness of mind – apart from the wish for self-destruction – was frequently totally lacking. It was tantamount, wrote John Roy, 'to making a psychiatric diagnosis without having met the patient'. Before 1961, therefore, those who succeeded in killing themselves were judged insane, while those who failed in the act were charged as criminals. Before 1962 the Church of England's dictates were equally confusing.

The history of society's attitude towards suicide has of course helped to cause the muddle in which today's statistics are hidden and from which today's society is wishing to extricate itself. Suicide is definitely not an entirely modern disease; some primitive peoples knew of it, others have laughed at the idea when questioned about it. Shame has always been a common cause of self-destruction, and another has been the need to accompany someone already dead, such as the practice of suttee in the East – mainly for widows. The Romans and Greeks did not, in general, regard suicide as a sin, except that the loss of a soldier was always to be deplored. Even the early Christians accepted the traditional attitude, frequently killing themselves. The Apostles did not condemn suicide, and the Old Testament makes no direct case against the act, although it is possible to construe various indirect indictments of it. There were four important Old Testament suicides: Samson, more of a kamikaze than a hara-kiri; Saul, whose armour-bearer then followed suit; Abimelech, for shame that he had been wounded by a woman; and Ahithophel, who made more of the classical suicide's preparations before hanging himself. The most famous historical case of Jewish suicide occurred when some 960 people killed themselves in the besieged fortress of Masada in AD 73 rather than suffer under the Romans.

St Augustine was the first Christian to denounce suicide, even though several earlier victims of their own hands had already been canonised. Within a few centuries suicide was considered a crime as well as a sin. Jewish law had started to denounce suicide much earlier and, traditionally, despite high rates at times of extreme persecution, Jewish suicide numbers have been low. The Koran emphatically forbids self-destruction, saying in one place that suicide is a graver crime than homicide. Thomas Aquinas said much the same thing for the Christians; whereas murder kills a body, self-murder kills both a body and a soul. The Middle Ages, not famous for its tolerance of deviants, dragged suicides through the streets, removed hearts, drove stakes through them, and either failed to bury the corpse or interred it in some unholy place. If buried near a church, a later practice, suicides were

329

grouped with stillbirths, criminals, the excommunicated, and all other outcasts. Eventually, in England in 1823, a month after a London suicide named Griffiths had been buried at the junction of Eaton Street, Grosvenor Place and Kings Road, the custom of burying such victims beneath the highway was banned by Parliament. (It was presumably the inconvenience of such road-works, as well as the distastefulness of the practice, that led to the prohibition.) Not until the Convocation of the Church of England in October 1962 was it agreed that a modified form of burial service could be used for suicides.

Legally, therefore, England had statutes against the suicide for ten centuries for it was first called an infamous crime, a species of felony, in the tenth century. In 1961 the Suicide Act was passed, and no longer is suicide or attempted suicide a criminal act, although complicity in the suicide of another is still a felony. In the United States, despite its inheritance of so much English law, suicide has never been a crime, but attempted suicide is a crime in six states and Arkansas holds that a person aiding and abetting a suicide is guilty of murder.

Who kill themselves? In Britain about 4,700 a year do so, or thirteen a day. The figure is slightly less than the road toll and slightly more than deaths attributed to accidental falls. (There has often been assertion that mechanical failures causing car accidents are murder attempts. One wonders how many other failures of machines are really suicide attempts. The Americans have even coined a word for them – autocides.) In the United States the straight suicide figure is about 22,000 a year, or less than half the road toll, or one every twenty minutes. At once it should be reiterated that the official figures are almost certainly under-statements, some say by 30%. A self-killing is still frequently hushed up. W.H.O. world figures indicate 365,000 suicides a year, and 3,000,000 attempts. No suicide facts are published by either China or the Soviet Union.

Despite imperfections in the statistics many generalisations are possible. Males kill themselves more frequently than do females. Attempted suicide is probably ten times as common as the successful act, therefore indicating some 40,000 a year in Britain and 200,000 a year in the United States. Suicide attempts vary from harmless (the pills are taken only when footsteps reach the house) to near-lethal (only something unforeseeable prevents death). Most attempts involve some danger to life (although psychiatrists often say that many self-poisoners are trying 'to alter their life situation, not to die'). Among American whites the suicide rate is five times the homicide rate; among the blacks it is the other way around. More women than men attempt suicide in most countries. Successful suicide figures increase with age; unsuccessful attempts reach a peak for both sexes at about thirty. During the first sixty years of this century the number of male suicides increased slightly in England and Wales – by 31%, but the number of female suicides rose sharply – by 171%. The male–female ratio was then 1.47/1. Both rises may be influenced by the decreasing pressures not to disclose suicide as a cause of death, but this does not explain the changing male:female ratio unless it had previously been considered even more shameful for a woman, the bearer of children, to kill herself.

The suicide numbers in Britain have changed dramatically in recent years, dropping from a total of about 5,500 in the mid-sixties to about 3,500 in the mid-seventies. The fall was all the more remarkable because it was mirrored nowhere else in the rest of Europe. Indeed the continental rates were largely rising, and people were quick to attribute the British improvement to the activity of the Samaritans, a charitable organisation that, like similar bodies elsewhere, is permanently ready to talk and offer friendly comfort to those in need. The presumed correlation between suicide prevention centres and the fall in suicide figures did not show up when examined scientifically and, although the Samaritans may well have helped, the British reduction has yet to be understood. Unfortunately, since those mid-seventies, the numbers have started to rise again, and the annual total is now nearer 4,500, a mid-way point between the high of the sixties and the low of the seventies.

Peak periods for suicides are when unemployment figures are high, when it is peacetime, when it is spring or summer, and when it is afternoon rather than morning. Both world wars lowered the suicide rates a year or two after the fighting began. The Australian figures do not mimic too well the peak spring-time levels of the northern hemisphere, for they are more erratic. Urban areas have more suicides than rural areas. London's three worst zones are Chelsea, Holborn and Hampstead – in that order. America's worst is San Francisco. Professional and managerial people kill themselves more frequently than the semi-skilled and unskilled.

Suicide is rare among children, a possible hazard of adolescence, a disproportionate problem at universities, and an increasing factor with increase in age. For Americans it is the tenth leading cause of death, and for teenagers it ranks fourth. The black American is much less likely to kill himself than the white American. The American figure of ten suicides per year per

100,000 people has remained remarkably constant since 1900. The current annual British figure is about nine per 100,000. Marital status is important: from married to single to widowed to divorced the proportion of suicide victims goes upward. Roughly one male doctor in fifty kills himself, and each British family doctor has on his list between ten and twenty patients who, sometime in their lives, will attempt to do so. According to Dr Norman L. Farberow, of the Suicide Prevention Center in Los Angeles, the doctor is 'in a unique position to be the most effective agent in suicide prevention . . . as 50% of all suicide victims visit a doctor during the last month of their lives'.

Michael Frayn once wrote in the *Observer* that Swedes are believed to have 'the highest rates in the world for divorce, alcoholism, delinquency, illegitimacy, abortion, pacifism, road deaths, suicide . . . and sin. Only a tiny minority of cranks, such as sociologists and Swedes, would care to dispute these deeply held convictions'. The proportion of Swedes who kill themselves does not exceed that of all other countries. According to World Health Organisation figures the Swedish annual rate is 16·9 suicide deaths per 100,000 people. More suicidal countries are Switzerland (18·2), West Germany (18·7) with West Berlin (37·0), Japan (19·6), Finland (20·6) and Austria (21·9). The high figure for West Berlin probably reflects the age of its inhabitants because young people tend to leave the city; all towns with an elderly population tend to have high suicide rates in consequence. Countries with consistently low rates are Holland (6·6), Spain (5·5), Eire (3·2) and Egypt (0·2). Practically every underdeveloped/developing country with a modest gross national product has a modest suicide rate. Nevertheless, to repeat the point a third time, all suicide statistics should be viewed with suitable cynicism, such as that Moslem Egyptian figure. Whereas the figures for England and Wales have been 10·2, 11·3, 11·5, and 11·3 in four post war years, the equivalent figures for Scotland have been 5·4, 7·7, 8·5 and 7·9, and for Northern Ireland have been 4·1, 3·3, 4·1 and 5·0. No one really believes that the Scotsman and the Ulsterman are so dramatically different from the Englishman and the Welshman; it is much easier to accept that the registration procedures for suicides were different in these different sections of the United Kingdom during those four years.

Student suicides are unhappily common. A recent survey of undergraduate deaths at Cambridge University listed 103 fatalities; of which forty-one were due to accident, twenty-seven to disease, and thirty-five to suicide. It has been estimated for Oxford University that the suicide rate among its students is eleven times higher than that of a similar age group among the general population. Oxford and Cambridge are particularly prone to suicide. In an investigation by Sir Alan Rook he disclosed that eleven other universities all had rates considerably lower than Oxford and Cambridge; but, even so, these other universities had rates twice as high as the average figures for young men. The figures for University College, London, fall between the rates for the two old and all the younger universities in England and Wales. Female undergraduates only very rarely kill themselves.

Apart from students, the incidence of suicide among adolescents and very young adults is low. In the United States it is 3·9 per 100,000 for the age-group 15–19. The rate is then doubled for the age group 20–24, and it continues to rise with age until it reaches a maximum of 27·9 per 100,000 for the age group 75–84. Although women do not kill themselves in such numbers, their suicide rate rises with age too, but reaches a maximum in the 55–64 age group.

The fact that there should be any suicides at all below the age of fourteen is so disturbing that their relatively few numbers provide little comfort. Only one in 400 of all American suicides is by a child under fourteen, but this means fifty a year. Most of these are at the older end of this age group, but about two a year are in the five to nine group. Once again the male:female ratio is still dominant; three-quarters of the children who kill themselves are boys. In the United States the most frequently used method of infantile self-destruction is the firearm, with hanging or strangulation coming second, and the swallowing of a corrosive substance third. The ratio of unsuccessful to successful attempts at suicide is always very high in the very young. Comparable British figures for suicides among small children are not available. In Britain there were four deaths of boys and three of girls under fifteen that were coded as suicide during 1982. Presumably children are less prone to leave suicide notes and more prone to perform acts (switching on the car in the garage, falling off a cliff) that are considered in their case to be accidental rather than suicidal. The figures for young suicides are therefore likely to be greater in any year than the seven for 1982.

Methods The American affection for firearms is reflected in the suicide statistics. During one five-year period, for example, the methods used were: firearms and explosives 47·1%, poisons and gases 20·8%, hanging and strangulation 20·5%, drowning 3·7%, jumping from high places 3·5%, cutting or piercing instruments

2·6%, and all other means 1·9%. There is not too much difference in proportion in the methods used by the American child and the American adult.

The British, with firearms less readily to hand and when a large proportion of their houses were piped with poisonous gas, chose this gas for self-immolation more frequently than any other method. In a survey carried out over thirty-three years at Leeds (and published in 1962) the methods used were: coal gas poisoning 49·5%, drowning 14%, poisoning other than carbon monoxide 12%, hanging 11%, cut-throats 6·3%, and all other methods – including firearms 7·1%. The methods used altered as the highly toxic coal gas was replaced by the far less lethal natural gas, as ovens were adjusted so that they could not deliver their major flow of gas – of whatever kind – unless that gas was being burned, and as barbiturates were freely prescribed by doctors. These sleeping pills quickly occupied the number one position formerly held by gas. Fortunately new pills then came on the market that made death from an overdose much less feasible. By 1980 the largest category for males was 'hanging, strangulation and suffocation', accounting for nearly one-third of suicides. For females 'poisoning by solid and liquid substances' is the most favoured, accounting for over half the female suicides.

It is disturbing that the British suicide rate is currently ascending, despite the gas improvements, the drug alterations, the work of the welfare agencies, and the better hospital treatment for attempted suicides. The increase has been almost entirely among males of working age, and the current recession is a natural suspect for causing this rise, what with job difficulties and even unemployment. Unfortunately there is no foreseeable improvement in the unemployment rate, and the last major suicide peak occurred in 1933, a time when 2,800,000 were out of work, a number slightly lower than today's all-time high.

There are some classical features of self-inflicted injuries. Fatal cuts, for example, are frequently preceded by very many tentative incisions before the final fatal wound. The drowned man will often have placed his accessory clothes, such as hat and gloves, in a neat pile beforehand. The firearm victim has almost always chosen the right temple (or left, if he is left-handed), the centre of his brow, the roof of his mouth, or the area over his heart for the fatal shot, having characteristically bared his chest beforehand. The gas victim will often have sealed the room, and will have made himself or herself comfortable with cushions. The hanged man scarcely ever breaks his neck. A judicial hanging used to require a drop of 6½ ft. (2 m.) to 7½ ft. (2·3 m.), according to the subject's weight, and the neck was broken in the middle of the cervical vertebrae. Most hanged victims have caused their deaths by asphyxiation, but some have killed themselves merely by blocking the neck's blood vessels. According to C. J. Polson the jugular veins are closed by a rope tension of only 4·4 lbs. (2 kgs.), the carotid arteries by a tension of 11 lbs. (5 kg.) and the actual windpipe, or trachea, by a tension of 33 lbs. (15 kg.). Thus, as the majority of hanging victims choose sites which enable them to die with both feet off the ground, they succeed in closing their tracheas because their weight always exceeds 33 lbs. (15 kg.); but successfully fatal hangings have been carried out from door-knobs, and even by people lying in bed. The human body is, to say the least, vulnerable to its own attempts upon its life.

Traditionally suicide is thought of as the single act which causes death. Today, bearing in mind the maltreatment meted out by some to their bodies, by heavy drinkers who drink heavily until their premature death, by coronary victims who do not mend their ways, there is talk of 'sub-meditated suicide'. There is no reason why suicide should be instantaneous, or nearly so.

Suicide and murder are common partners. One survey a few years ago showed that a third of the murders in England and Wales were followed by suicide. The partnership is less conspicuous in the United States which has a similar suicide rate per unit of population but a homicide rate 15 times greater than in England and Wales. Suicide and the highest buildings are also partners, although not so emphatically as the inevitable publicity would have us believe. The Eiffel Tower in Paris was the means to the end of 341 men and women in its first three-quarters of a century of existence (two survived the fall); but this average of five a year is eclipsed by the 500 or so suicides committed annually in the Paris region. The Golden Gate Bridge of San Francisco was the jumping-off place for 278 people in its first twenty-eight years, but ten deaths a year are soon swamped by the figures for California as a whole. London used to have many deliberate falls from the Monument in the City until wire-mesh made this impossible, but there was a suicide from the new Post Office Tower even before it was completed and made suicide-resistant.

26. The Brain

'In creating the human brain, evolution has wildly overshot the mark.'

ARTHUR KOESTLER

The human brain is neither the largest brain in the animal kingdom nor is it the largest by comparison with body size. An elephant's brain is about four times heavier than a man's, and many monkeys have a body–brain ratio of 20:1 whereas the human ratio is nearer 50:1. Nevertheless, even though mankind has been judge and jury in dubbing itself sapiens, in referring to itself as the pinnacle of evolution, in distinguishing between man and the animals, there is nothing quite like the human brain. (It receives but one chapter in this book, somewhat equivalent to its bulk representing 2% of the human body, but it receives much fuller treatment in *The Mind*, companion volume to *The Body*.)

As ever there are innumerable conundrums. Brain bulk is undoubtedly related to brain ability. The animals and mankind's evolutionary ancestors bear out this generalisation; but, within the species man, brain bulk seems strangely unimportant. One brain may be twice the size of another without showing any apparent difference in ability. The largest human brains, twice average size, are those of idiots. Machines can be made to operate far faster than any brain and mathematical problems of mind-boggling proportions are mechanically calculated before a brain can absorb them, let alone provide an answer; but no machine has a memory with a capacity in the same class as the human brain. Many humans say that they never forget a face, one more configuration of the same old theme of two eyes, a nose and a mouth, and yet the memory of one new telephone number spoken by that face is instantly elusive. Surely a face is more complex than a small finite series of figures? Louis Pasteur suffered a cerebral haemorrhage which caused moderate one-sided paralysis but did not prevent him from doing some of his best work. At post-mortem, years later, the injury was found to be so

336

extensive that he was said to have been living on half a brain. The brain is undoubtedly sensitive to wrong cuts by the surgeon's knife, and gloved fists cause at least a dozen deaths a year in the boxing ring, but Phineas Gage achieved fame when he had a $13\frac{1}{4}$ lbs. (6 kgs.) tamping iron blasted into his skull and his brain. He lost consciousness briefly, walked to the surgeon's office and recovered physically. Admittedly he became a drunkard and lost jobs thereafter, but he still had the wit to sell his skeleton, cash in advance, to several different medical schools, and the pushing of an iron bar with the diameter of an old English penny through his brain, the delicate and sensitive brain, demonstrated remarkable cerebral resilience. (In fact he was to die only twelve years after the accident when aged thirty-seven.)

Wars and surgery have, since Mr Gage's accident a hundred years ago, amply demonstrated man's ability to lose substantial portions of the brain without undue suffering. In 1935 a London conference was informed of the stabilising effect upon a chimpanzee caused by the surgical severing of much of its frontal lobes. In 1936 a Paris conference was yet more fascinated to hear of a similar operation upon a human being. Since then thousands of the mentally unbalanced have suffered the deliberate severance of the foremost part of each cerebral hemisphere, and many have benefited psychologically from this deliberate destruction.

Finally, although the brain is anatomically symmetrical, and although many of its functions are entirely bilateral, some others are quite one-sided. One half is called dominant and its loss is the greater loss. Had Pasteur suffered his haemorrhage on the other side there would have been no question of good work; either death or a pathetic existence would have followed. For those parts of the brain which are bilateral and unaffected by dominance an extra complexity is caused by the fact that each half tends to organise the opposite half of the body. Neither dominance nor this crossing over simplifies comprehension of man's fantastic neurological anatomy.

Anatomy   The human brain is a soft lump of some 14 thousand million cells, it weighs slightly more than 3 lbs. (1·4 kgs.), and is so full of water that it tends to slump like a blancmange if placed without support on a firm surface. Predominantly its structure is the huge pair of cerebral hemispheres which sit on top of the 10–15% of remaining tissue. This smaller fraction is much more varied, being composed of several entirely different portions of brain tissue, and fits into such space as exists in the cranium between the top end of the spinal cord and the huge convoluted growth of the two cerebral hemispheres. The complexity has been accentuated by man's upright stance. From being a tetrapod creature whose spine and head were both horizontal, to becoming a bipedal creature whose spine was vertical and whose head continued to look horizontally, the central nervous system of spinal cord plus brain has had to suffer this 90° flexure. Small wonder there is confusion at the point where the spinal column joins the brain.

In the fishes everything is simpler. In even less advanced creatures the central nervous system is merely a tube running the length of the body which swells modestly at the front end where sense organs are predominantly situated, and where sensory impressions are predominantly received. By the time of the fishes the brain is in three parts, three separate swellings each linked to a sense organ. The front swelling receives impulses from the olfactory organs, the middle one from the eyes, and the back one from the organ of balance, later to be the ear as well. These three, the fore-brain, mid-brain and hind-brain, persist with their three roles in more advanced forms, but the expanding fore-brain is to take over all co-ordination from the other two.

It may be an over-simplification to say so, but if one part of the brain is to be expanded it had better be the front part. To enlarge elsewhere is plainly a greater problem. Simplification or not, this is what happened. The paired lump of fore-brain became increasingly important as the centre of sensory correlation. To react suitably is not to react according to stimuli from one sense organ, but to sift through all available information from the exterior in one central control. By the time of the mammals the fore-brain was dominant in this regard, and the importance of the mammalian sense of smell was intimately bound up with the swelling olfactory lobes. When the primates were evolved smell was less important, but the fore-brain was by then firmly established as the centre for interpretation and correlation. Therefore, again to simplify, but with good reason, if greater correlation and association are both required and advantageous, the fore-brain needs to be expanded still further. Besides, unlike the mid-brain and hind-brain, its expansion continues as before to present less of a space problem anatomically.

During evolution the brain tissue doing all this correlating moved increasingly towards the outer surface of the fore-brain. Called the pallium, or cloak, or cortex, this correlating tissue made up the most advanced portion of the brain. Its area expanded when fissures and grooves were formed, as in a walnut, and a brain should be judged for its level of advancement more on

338

the area of its cortex than on the body of brain tissue. In human beings the cloak of cortex is only about one-eighth of an inch (3 mm.) thick or less, but its area, enlarged by all the folds, convolutions, gyri and sulci, is 400 sq. ins. (2,580 sq. cms.). Beneath that cortex lies the maze of fibres and pathways which lead to and from the thin crucial skin of cortical tissue.

The human brain is the most exaggerated form of this development. The original fore-brain, now so heavily grooved and with four paired and major lobes of its own, is over five-sixths of the human brain. Known as the cerebrum, with its two halves being the cerebral hemispheres, it performs all the higher and highest roles of human activity, the correlation of sensory impulses, of memory, of thought. The remainder of the human brain is tucked underneath the huge mushroom of cerebrum, and it also consists of four parts. These are the:

Midbrain. Short and narrow. Still linked with vision in that it receives impulses from the retina, and still acts as a centre for visual reflexes; but it mainly consists of fibres going from elsewhere to elsewhere.

Pons – Just below the midbrain. Dominantly tracts of fibres.

Medulla oblongata – The continuation of the spinal cord where it first begins to broaden. Mainly tracts. Contains much of the crossing-over of fibres causing the left cerebral hemisphere's association with the body's right hand side, and vice versa.

Cerebellum – Lies behind midbrain, pons and medulla, and below rear regions of cerebral hemispheres. It is also in the form of two hemispheres and, like the archaic hind-brain, is associated with the inner ear. Main cerebellar activity is to assist muscular co-ordination. Victims of a poor cerebellum move jerkily and without accuracy.

The cerebrum Forgetting these four other parts, and returning again to the cerebral hemispheres, a further division into four is necessary because the four lobes of each hemisphere have distinct roles, far more distinct than the lobes themselves.

Frontal lobe The 'motor lobe'. Situated at the front of the head. The precise motor area, which controls muscular movement, is only a narrow band of cortex at the back of this lobe. The band moves from the side of each lobe up to the top and down into the central fissure between each hemisphere. Each region of the body is linked with a part of this band. Stimulation of any portion of the band will always cause a muscle, even a particular muscle, to twitch in response in the appropriate bodily region. Progressing along the band from each side of the brain to the centre, the bodily areas involved are the mouth, including speech (but more about speech later under left-sided dominance), the face, eye, neck, thumb, fingers, hand, wrist, elbow, shoulder, trunk, hip, knee, ankle and toes – in that order. The amount of band involved is certainly not proportional to the size of the body controlled by it: the hands and fingers use as much of the band as the remainder of the arms plus the legs and trunk. Digital skill is correspondingly more advanced.

In front of this motor band lies the pre-motor area. Its stimulation does not ever lead to the twitching of a single muscle, but to a more co-ordinated movement of a whole area. Presumably this area affords general control whereas the motor band affords greater precision. The remainder of the frontal cortex is not only far larger but far more mysterious. Called the silent zone, great quantities of it have sometimes been cut away or severed, as in tumour operations, to leave the patient remarkably unimpaired. A small nick in the motor cortex and there is immediate partial paralysis: a huge carving up of the silent cortex and, although personality is involved, notably mood, criticism, drive and concentration, functional ability is scarcely modified.

Parietal lobe   Separated by a deep fissure from the frontal lobe, but possessing a band at its front border adjacent to the frontal motor band, this second band is sensory. It receives sensations of warmth, cold, touch and general movement from the body. The order along the band is virtually the same as the order for the motor band: the mouth, including taste, is low down on each side and the toes are in the central fissure.

Temporal lobe   The lobe nearest to the ears. Reasonably it is linked with hearing, and the lobe on each side receives impulses from both ears. The area actually linked with hearing, as with most lobe areas positively linked with any function, is quite small. There is believed to be an association between memory and this lobe.

Occipital lobe   The lobe nearest to the back of the head. Linked with vision. A small wound at the back of the head has often caused no harm other than total blindness. If only the visual cortex of the left lobe is destroyed, vision will still remain good for the right half of each eye's retina, and vice versa.

Whereas so much of the brain's volume is functionally vague, being allied to intelligence but with no precision, there are portions of the cerebrum apart from those already mentioned

340

which are quite the reverse. Below the cortex, or grey matter, the bulk of the tissue is white matter – the pathways of fibres. Nevertheless, at the base of it all, completely overshadowed in bulk by the great mass of the cerebral hemispheres, there exists some more grey matter.

Part of this tissue is the hypothalamus, a minute portion of the total brain – about one three-hundredth of it – which seems to have a disproportionate share of duties. 'Is there any pie in the vertebrate organisation into which the hypothalamus does not dip its finger?' asked a *Lancet* leader once. This fraction of brain tissue is involved in the body's water metabolism, in temperature regulation (and therefore in perspiration and shivering), in appetite, in thirst, in the levels of sugar in the blood, in growth, in sleeping and waking, in emotions such as anger and pleasure, in the cycles of the reproductive system, and it is also the main centre of the autonomic nervous system. Probably it will be found to be invested with yet more bodily authority when research-workers have probed further into its vital lump of grey matter but, considering that it is smaller than a finger joint, the list is already impressive. Naturally, with such a small but all-important organ, minor damage – even the overcrowding presence of a tumour nearby – causes drastic alteration to the body's self-regulatory abilities. Similarly, with such a complex list of duties, there is no need to presuppose a tumour to consider the occasional hypothalamus at fault in itself. Compulsive eating, the tendency to eat too often or too long, may merely be the fault of a hypothalamus whose appetite control centre is either inadequately strict or faulty in its assessment of need.

Left-sided dominance   The fact that the cerebrum is completely divided by the superior longitudinal fissure to form the two separate cerebral hemispheres is highly relevant to bodily control. In effect this division causes two regions, each capable of administration. Having two such regions of cerebral cortex incurs the possibility of a disunited command. Messages from one half of the brain can reach the other half, but the human brain obviates the need for such constant cross-reference by having one cerebral hemisphere dominant to the other. The damage to Pasteur's brain, already mentioned, was to his non-dominant half. He could not have suffered such damage to the left-hand and dominant side of his brain. Should a child receive damage to its left hemisphere before it is too late, some say before six months, it is possible for the right hemisphere to take its place. Later on the chain of command is too firmly embedded, and the dominant hemisphere cannot have its higher and more authoritative role taken over by the other inferior half.

Almost always the left hemisphere is the dominant one. It possesses the speech centre as well as wielding unspoken power over the right hemisphere. Why the left should be so dictatorial in this fashion is quite unknown. Allied to this ignorance, often unhelpfully so, is the problem of right- and left-handedness. With the left hemisphere controlling the right half of the body, and with the left hemisphere nearly always dominant, these facts might seem to explain the overwhelming majority of right-handers; but, alas for the presumption, left-handed people frequently have a dominant left hemisphere. Roughly 5% of the world's population (all areas, rich or poor, black or white, are equal) are left-handed to a greater or lesser degree, and yet the proportion of people whose brains have been discovered after disease or surgery to be right-sided is much smaller. Similarly it is a great rarity for speech to be controlled from the right-hand side, but again this is information which is only encountered by chance, and there is plenty of disagreement in the literature over the occurrence of left-handers with right speech, right-handers with right speech, and so forth.

The *Economist* once had a buoyant article on the Declaration of Human Lefts. Why, it asked, are those of the left camp gauche and sinister, whereas the right-handers are not only right but dextrous and righteous? Whence this prejudice, which also stamps radical politics as leftward of conservatism? The sinistrals, as left-handers call themselves, have reason also to resent the designs of things which favour right-handers, as well as the traditional association of left with wrong. One ardent sinistral even totted up biblical references to rightness: he discovered 1,600, most of which were hostile to the left. Although animals, such as the higher apes, tend to favour either one hand or the other, and are not markedly ambidextrous (another loaded word, meaning right on both sides), they do not show a particular preference for their right side.

Man has been prejudiced ever since he has been able to record the fact. Most hand imprints on cave walls are of the left hand, indicating an Aurignacian preference for working with the right. Military burials after the early battles consist of bones more grievously injured on their left side, indicating right-handed attackers. Castle bastions and keeps possess staircases which spiral upwards in a clockwise fashion; right-handed swordsmen find them easier to defend. Scripts used to make use of the ox-turn system, ranging back and forth alternately, but most now favour right-handers as they run from left to right. Hebrew and Arabic are exceptions, and for thousands of years China was impartial by being vertical. Chinese custom has now been overturned, and the new cultural rule is to go from left to right. Most countries drive on the right, and most people walk on the right of the pavement irrespective of the driving laws in their country. Presumably both are linked with the preference for keeping the right arm in the clear.

The pressure to conform, a strong and insidious force on its own, used to be abetted in the classroom by a reluctance to entertain left-handed writing. There is some evidence that such compulsion often went hand in hand with stammering, but there is also more recent evidence that it does not. Perhaps there are changes in the definition of left-handedness which have influenced the evidence and made it contradictory. Left-handers make use of their right hands more frequently than right-handers make use of their left, and plainly some people are only a little left-handed. Jack the Ripper, whoever he was, worked with his left hand, so did George VI, who also stammered: so did Horatio Nelson, although unwillingly for his lost right arm had been his major arm, and so do an estimated 200 million in the world today, a huge minority group that suffers from the effects, as Thomas Carlyle put it after he lost the use of his right hand, of 'probably the very oldest institution that exists'.

Clockwise, the way the sun goes in the northern hemisphere, may have had its influence in the preference for the right. Widdershins, anti-clockwise, and sitting on the left have always been inferior, often devilish, and generally sinister.

Intelligence   Man's cunning is lodged in his cerebrum, but there is no centre for intelligence. This depends upon the cortex as a whole. Previous theories that the frontal cortex, being the most forward bulge of the central nervous system, was the foremost spot for intelligence have not been supported by the frequent surgical excision of large portions of its substance. Besides, even though the measurement of intelligence quotients, and whether someone is above or below the average of 100, gives an aura of unity to intelligence, it is by no means one thing. Therefore, it is unfair to expect it to have one location on the cortex, as if it were control of the little toe.

Dr George Stoddard, in *The Meaning of Intelligence*, listed its multiplicity. He said it is 'the ability to undertake activities that are characterised by 1. difficulty, 2. complexity, 3. abstractness, 4. economy, 5. adaptiveness to a goal, 6. social value and 7. the emergency of originals, and to maintain such activities under conditions that demand a concentration of energy and a resistance to emotional forces'. Similarly Theodosius Dobzhansky said, 'What is measured by the I.Q. is not necessarily the same as what is referred to in everyday language as intelligence, cleverness, aptitude or wit. Still less does the I.Q. give an estimate of the value or worth of the person.' It has also been said that a chimpanzee will only start scoring well on intelligence tests when another chimpanzee has set them.

On learning, Alfred North Whitehead wrote: 'It is a profoundly erroneous truism, repeated by all copy books, and by eminent people when they are making speeches, that we should cultivate the habit of thinking what we are doing. The precise opposite is the case. Civilisation advances by extending the number of important operations which we can perform without thinking about them.' Sir Peter Medawar, in *The Uniqueness of the Individual*, took Whitehead's statement a stage further to produce its converse truth. 'Learning is two-fold: we learn to make the processes of deliberate thought instinctive and automatic, and we learn to make automatic and instinctive processes the subject of discriminating thought.'

The intellectual abilities of others are equally hard to comprehend at both ends of the scale. In one issue *Penguin Science Survey* had articles exemplifying both these two ends, the abnormally incapable and the abnormally skilful. One was dedicated to the several thousand spastic children born each year whose brain damage makes it difficult for them to draw a diamond. Normally a child can copy a circle by three, a square by five, a diamond by seven. Two years of development therefore separate each achievement, all three of them seemingly equally elementary to the adult and yet major milestones to the struggling child, with one of them being an unattainable milestone to the average spastic.

At the other end of the scale, when normal problems shrink to nothing, was Professor A. C. Aitken of Edinburgh University. He was asked to turn $\frac{4}{47}$ into a decimal. After four seconds he answered, giving one digit every three-quarters of a second: 'point 0851063829787234042553191 4'. He stopped there – after twenty-four seconds, discussed the matter for one minute, and then started up again. 'Yes, 191489, I can get that'. Five second pause. '361702127659574468, now that's the repeating point. It starts again at 085. So, if that's forty-six places, I'm right.'

EEG Appreciation of the mechanics of intelligence might take a step forward when current obscurity is diminished about the origin of electro-encephalogram waves. It is already sixty years since these waves were first discovered. Towards the end of the nineteenth century various men had shown that the brains, notably of dogs, possessed changing electric potentials, but it was up to Dr Hans Berger of Jena University to open up the subject.

For five years he applied electrodes to the human skull and he studied the fluctuating patterns of the potentials which his delicate galvanometers succeeded in recording. This psychiatrist then recorded his results: *Über das Elektrenkephalogram des Menschen*. For the next five years, while the world disregarded his claims, he continued to work, to record his results, and to use the same title in writing them up in the German journals. He was particularly suspect for his repeated contention that he could distinguish between different wave patterns. Recording the waves at all was improbable; differentiating the wave patterns yet more so.

Suddenly he was vindicated. Professor Adrian (later Lord Adrian) also recorded waves; the Physiological Society saw them; the 'Berger rhythm' was applauded, and electroencephalography – electric brain writing – has not looked back.

Sometime after a child's eighth birthday the dominant rhythm, called by Berger the alpha rhythm, changes from its childish speed of 4–7 cycles a second, and starts increasing to its adult frequency of ten cycles a second. Basically it is an idling rhythm, for it is most manifest when eyes are closed and minds are empty. Its casual tremor is abolished the moment anything happens, whether the eyes actually open to see something interesting or visual images are merely imagined so that the brain records greater excitement.

With the arrival of sleep different wave patterns appear, and an EEG – the advanced form of Berger's equipment – can inform when someone is asleep, dreaming, unconscious, anaesthetised or waking up. This synchronisation of waves is of course a problem. Brain cells might be expected to discharge at random, but instead there is unison. Are there pace-makers in the brain regimenting the discharges? Of what use is the rhythm? And why are different wave patterns associated with an epileptic fit? The complexity of the rhythms involved, so cleverly detected by the assiduous Dr Berger, are still just as complex, and the job of deciphering patterns in terms of varying stimuli is being handed over to computers. Therefore, although the human brain is far more cunning and compact than any computer, and although computers were devised by human brains, it may be computers which can inform the brain what makes it tick so cyclically, and why. It was a

United States Air Force scientist who was first to teach himself to alter his alpha rhythms, switching them on and off at will. He could speak his mind, as it were, without moving a muscle.

Concussion   Picture the scene. (Practically every film director has already done so.) The victim walks into the room, and glances about him. The villain, whose earlier glancing had supplied him with a vase, poker or paper-weight, then knocks out the victim with one cruel blow. The victim slumps, stays unconscious a while, then comes round, shakes his head, takes in the situation and rushes out of the room. It is all easy. It is bogus.

The causing of unconsciousness is genuine enough; it is the instant recovery that is improbable and misleading. Concussion is most likely to be caused if the head is rapidly accelerated: a threshold speed of twenty-eight feet per second is sometimes quoted. The speed of the blow is therefore important, whatever the size of the blunt instrument. It so happens that a rabbit punch on the neck and a straight left to the jaw are both good at producing acceleration; but a short neck with thick muscles is more able to resist efforts to accelerate its head. Any subsequent unconsciousness may be so fleeting that the victim is scarcely aware of it; it may last less than thirty seconds, as with most boxing knock-outs; or it may last for days and weeks, as in many car accidents.

Generally speaking, the unpleasantness of the return to consciousness and the extent of amnesia surrounding the event are proportional to the degree of unconsciousness. Nausea, headache, unsteadiness and amnesia are probable if the concussion has been at all prolonged. The instant recovery and awareness of the screen victim are both highly improbable, and give the impression that such a mishap is annoying but trivial. Dr Macdonald Critchley, specialist in the effects of brain mistreatment, has said no unconsciousness can ever be regarded as trivial, that blood clots could form and enlarge as a result of the injury, that evidence exists of association between tumours and severe concussion, and that blows not even causing knock-outs can cause damage to brain cells.

Boxers still die regularly. So do participants in virtually all other sports; but, unlike most other sports, the aim of boxing is calculated violence. In 1965 when 'Sonny' Banks died of a blood clot on the brain, aged twenty-four, three days after he had been knocked out by Leotis Martin, he was the sixty-fourth American boxer to die in five years of ring injuries. Dr Milton Helpern, then New York City's chief medical examiner, performed autopsies on dead boxers. *Medical World News* quoted him after he had conducted studies on four knocked-out victims: 'When burr holes were made in the heads of the unconscious boxers, brain tissue oozed out like toothpaste from a tube.' The four men had survived unconscious from fifty-five hours to nine days after their knock-outs. Two of them had hit their heads on the mat when they fell: two had just slumped to the ground.

A report from Finland published in the *Lancet* during November 1982 gave results of a survey concerning fourteen boxers who had been Finnish, Scandinavian or European champions. It concluded that 'modern medical control of boxing cannot prevent chronic brain injuries but may create a dangerous illusion of safety. The only way to prevent brain injuries is to disqualify blows to the head'. In the same year the American Medical Association recommended that doctors should have the right to interrupt a bout in order to examine a boxer and, if need be, stop the fight. Currently doctors have this power only in Michigan.

Memory 'If we remembered everything we should be as ill as if we remembered nothing', said William James, and for a long time that just about summed up mankind's ability at comprehending mankind's memory, or indeed the memory of any creature. R. W. Gerard lamented in 1949 that current understanding of the mind 'would remain as valid and useful if, for all we knew, the cranium was stuffed with cotton wadding'. B. D. Burns, reviewing the whole subject of memory theory eight years later, concluded that no hypothesis had proved 'a wild success'.

'It may well be that we are at last on the way to solving . . . how memories are stored in the brain', said Lord Adrian in 1965. Perhaps the future did look brighter then, for no one is so optimistic now. The chemistry of memory has advanced, in that much more is known than previously, but the progress has been in the nature of hillside walks; the accomplishment of each knoll merely gives a better, and more realistic, view of the miles and miles that lie ahead. Finding the chemical basis is only half (or less) of the story. Memory has three ingredients: there are the three R's of registration, retention and recall. Everyone registers more than he can retain, and retains more than he can recall. If RNA is the chemical which, by having its molecular pattern altered during registration, is the card index basis of memory this fact does not explain how the card index is either maintained (retention) or used (recall). What is the system that does either or both?

Similarly there is short-term memory – looking up a number, dialling it, and forgetting it – and there is long-term memory. Are these different or merely two aspects of the same phenomenon? It is relevant that memory difficulties of the aged have been demonstrated to be dominantly difficulties of recall: short-term retention is scarcely influenced by advancing years.

Nerves 'On old Olympus' topmost top a fat-eared German viewed a hop.' So runs one of the non-obscene jingles for remembering the brain's cranial nerves. Obviously the central nervous system does not exist in isolation: it must be connected to every part of the body. In all there are forty-three pairs of nerves joining the C.N.S. with everywhere else. Of this number twelve pairs go to and from the brain itself, and thirty-one pairs go to and from the spinal cord. The nerves may be either entirely sensory – passing information only inwards and towards the brain; or, more frequently, they may be a mixture of both incoming and outgoing fibres. Each of the forty-three has specific functions; hence the necessity for knowing which does what; hence the jingle which stands for the twelve names: olfactory, optic, oculomotor, trochlear, trigeminal, abducens, facial, auditory (ear), glosso-pharyngeal, vagus, accessory and hypoglossal.

Once again everything makes much more immediate sense in the more primitive forms of animal life. The fishes, for example, have both cranial and spinal nerves (although only ten cranial nerves, and a varying number of spinal nerves) but everything is simpler. The brain and spinal cord lie in a straight line, the nerves branch off along the length of this central nervous system much like rungs along a ladder, and they tend to be associated with an organ lying nearby. The first one receives impulses from the nose, the second from the eye, and so forth down to the tail. The human being acquired this same system but only after it had been evolved for the differing shapes of the amphibia, the reptiles, the mammals and the primates. The result is that No. 1 cranial nerve still receives impulses from the nose, No. 2 from the eye, and so on, but neither nose nor eye nor brain nor any other organ is sited in the same manner as in the earliest fishes. Consequently the regular rung-like simplicity has gone. Instead the cranial nerves run much more awkward courses in their atavistic efforts to join the same organ to the same bit of brain even though both brain and organ have radically altered proportions, sizes, shapes and locations.

The spinal nerves are also much more complex in the human being than in the fishes. The ladder-like regularity exists to a certain extent, but the human spinal cord does not even run the entire length of the human backbone. It starts where it joins the brain and then only continues downwards for 18 ins. (46 cms.). After the small of the back and for the last 10 ins. (25 cms.) of the spinal column there is no spinal cord, only a horse's tail of fibres, tracts and occasional nervous complexes. Fibres leading from this spinal cord are always both sensory and motor. Even though the cord is quite long, and is connected to thirty-one pairs of spinal nerves as against twelve pairs of cranial nerves for the brain, and even though it is such an important trunk route for impulses, the spinal cord is far, far less bulky than the brain. It is about ½ in. (13 mm.) wide and only weighs an ounce (28 gms.), a fiftieth of the brain's weight.

When a neuron dies it cannot be replaced. The nerve cells are so specialised that they have lost the ability to make new cells, and neurons are being lost all the time. Should an axon be cut sufficiently far away from its mother cell the peripheral part of the axon will slowly wither and die. The neuron itself will not necessarily die, although part of the axon on its side of the cut may perish. Sometimes the remaining axon will grow afresh from the region of the cut towards the peripheral area that it used to serve. Growth may be slow, but some higher animals have demonstrated an ability to grow new axon at about one-eighth of an inch (3–4 mm.) a day. This rate of about an inch (25 mm.) a week means that quite a time may elapse after the axon's severing before it arrives again at its destination and establishes a functional connection once more. Should the nerve be severed within the central nervous system of brain and spinal cord the connection is broken for all time; but this does not always create such a severance as this seems to imply. Often the body can find a way round such a problem. Other pathways are made use of, and developed with use.

The all-or-none transmission of a stimulus along a nerve fibre means a system even simpler than the Morse code. There are not even dots and dashes, just dots. The dots are all equal dots, none louder than the rest. Therefore the only way to vary the intensity of such a stimulus is to alter the frequency, the number of dots per second. Even though a nerve amplifies a stimulus as it travels along the fibre the whole system is back to normal very rapidly, so much so that several hundred impulses can be impelled along a given fibre every second. Frequencies higher than 1,000 impulses per second have been experimentally measured in some mammal fibres, but human fibres usually conduct at frequencies lower than 100 per second.

The many thousands of millions of neurons within the nervous system of each human being give some idea of his or her potential and excellence when compared with the modest neurological equipment of an insect. The bee and the ant, apart from being objects of veneration for their wisdom and diligence, lead complex social lives, and construct cunning homes for themselves. The bee has about 7,500 nerve cells, and the ant even less. Mankind, not always wise or diligent, has two million times more neurons at his command than the bee, whether he uses them more skilfully or not. In the main they are not used, as so much of the brain can be removed without apparent effect, and as individuals (such as hydrancephalics) with but a minute portion of the normal brain bulk have even achieved university degrees. It would seem, as Alfred Russel Wallace phrased it, that 'an instrument has been developed in advance of the needs of its possessor'.

27. Sleep

'Where unbruised youth with unstuff'd brain
Doth couch his limbs, there golden sleep doth reign.'
FRIAR LAURENCE in *Romeo and Juliet*

It appears to be a rule that the more important the subject the deeper is human ignorance upon that subject. What initiates birth? What makes the brain more capable as the childish years pass by? What causes old age? And what happens in sleep? Why is it necessary, bearing in mind the hazard of closed eyes in a hostile environment, and why can the brain not work without partially shutting down some of itself for a third of life?

Another generalisation about big, important subjects is that the few crucial facts which are known have only been discovered recently. The first detection of the brain's electrical rhythms, the minute flickerings which change so radically when a person goes to sleep, was only made in the 1920's. (Dr Hans Berger published his findings in 1929.) It was only in the 1950s that the two principal kinds of sleep, orthodox and paradoxical, or non-dreaming and dreaming, were first worked out: Dr Nathaniel Kleitman published his paper in 1953. Only since then has it been discovered that both kinds of sleep are crucial to anyone's well being; to be deprived of dreaming is as bad, or possibly worse, than deprivation of orthodox sleep: Dr W. C. Dement did his experiments on dream curtailment in 1960. Finally, even though the interpretation of dreams is as old as the hills, modern thinking is largely based upon a book which appeared together with this century: Sigmund Freud published *The Interpretation of Dreams* in November 1899 to earn immortality, ridicule, fame and £41 in royalties.

Like all big subjects sleep is equally afflicted with misconceptions. Sleep research suggests that no one, however tired, sleeps like a log; that everyone dreams, whether their powers of recall are good or not; that dreamless sleep cannot be called the best kind, because times of dreaming should always punctuate sleep; that dreams are lengthy affairs, possibly totalling two hours a night – they are not over in some sort of cerebral flash; that old people do not lose any desire for sleep as they grow older; and that conventional talk about deep sleep, or one hour before midnight being better than two hours afterwards, does not fit in with modern notions concerning the rhythms of a normal night's sleep.

Some elemental facts about sleep are: that its average length (according to a large French survey) is 7 hours 20 minutes; that this duration is remarkably constant from 30–80 years; that it is not strongly influenced by sex, intellectual ability (except for very low I.Q.s who usually sleep longer) or cultural background; that exercise taken during the day does not cause longer or shorter sleep; that sedentary workers are ten times more likely to take drugs than highly active people; that women take sleep-drugs more than men; that about ten days is the record for staying awake by a normal person; and that the body is relatively unmoved for at least the first few days of such a procedure – weight, heart rate, blood pressure, reflexes, stay normal, and the bodily temperature has its ups and downs equivalent to the traditional rise and fall of the daily and nightly routine. It is the brain that is so drastically affected by sleeplessness.

There have been 'wakeathons' (long bouts without sleep), notably in the United States. The traditional picture is of slightly deranged behaviour by the third day, such as laughing unreasonably or taking offence irrationally. Normal cyclical rhythm is not forgotten, despite its interruption, because the desire to sleep during the second night is greater than during the following day. After the third day mental derangement becomes more exaggerated. There are wilder imaginings, a belief that others are destroying the experiment, that the floor is heaving, that food is poisoned, and all such delusions and hallucinations suddenly come to an end after 200 hours or more with the permitted arrival of sleep. In a famous wakeathon Peter Tripp kept awake off Times Square for 201 hours 13 minutes, and then fell into a normal sleep lasting 13 hours and 13 minutes.

Quite the most significant discovery so far is the realisation that sleep has a cycle to it. A person going to sleep first falls into orthodox sleep. An electroencephalogram, measuring the minute voltage differentials playing about his brain, detects the changes inevitably occurring as a person slowly sinks from wakefulness into sleep. Some ninety minutes later the brain's electrical rhythms revert to the sort of pattern indicated by the EEG when the person had been in the act of falling asleep. At the same time the eyes start moving about behind their closed eyelids. It is tempting to call this time a period of light sleep, but the description is unworthy because, despite the activity recorded both on the EEG and by the eyes, there is also great depth to this kind of sleep: muscles, for example, are then even more relaxed than during the initial and orthodox sleep. Both types of sleep are vital, both occur alternately, and neither depth nor lightness is a valid term in this regard. During a typical night's sleep a person will oscillate from orthodox sleep to the rapid eye movement kind some five times. Orthodox sleep is always longer, and may consume 80% of the night. The remaining 20%, indicated by the quite different pattern on the EEG and by eye movement, was first known as paradoxical sleep. It arises after gaps of ninety minutes. Each burst only lasts for ten to thirty minutes at the beginning of the night, but may last for an hour by the end of it.

Once again the remarkableness of a discovery lies in its simplicity. In the late 1930's E. Jacobson drew attention to the association between dreams and eye movement. Later Nathaniel Kleitman, Bill Dement, Eugene Aserinsky and others working in Chicago experimented with this idea. Sure enough, those woken at the time of eye-flickering could in general speak of dreams; those woken during orthodox sleep could scarcely ever recall them. If anything is recalled from this time it tends to be simple cogitation of the day's events, not the spicy, glamorous, fraught, inconsequential, hostile and abject sequences typical of the dream world. Considering the number of hours mankind has spent watching either womankind or any sleeping person it is remarkable that no earlier scientific notice was taken of these jerky eye movements, still less of their association with dreams.

An experiment then took place in 1960 to discover the importance of dreaming. As soon as the EEG moved away from orthodox sleep, and as soon as the eyeballs started their blind activity, the volunteers were woken. Soon they were allowed to sleep again, but they then slept in orthodox and not dream sleep. They were woken whenever dream sleep began to manifest itself, and for several nights the regimen of no sleep with dreams was rigidly applied. Then, for five blissful nights, the men were allowed unprovoked sleep. There were three findings. First, there had to be more awakenings as time went on: on the first night the men only had to be woken from a dream state five times, but on the fifth night the numbers of necessary wakenings were 20–30. Secondly, during the five uninterrupted nights, the EEG and eyeballs recorded double the normal dream-time, suggesting that a back-log was being made good. Thirdly, despite the wakenings, each man got seven hours sleep a night; but during daytime, for all that, they behaved irritably, tensely and with many of the characteristics of men deprived of the greater part of their sleep requirements.

In other words mankind has to dream, or at least to have the

353

kind of sleep generally accompanied by dreams. When Peter Tripp finally slept for thirteen hours, having waited over eight days to do so, he dreamed for 3 hours 46 minutes (or 28%) of the total sleep time. When still awake, and when he had been suffering his hallucinations, their bouts had come roughly every ninety minutes as if they were the attempts of a befuddled brain to follow its traditionally cyclical pattern of sleep. Customarily those deprived of sleep take more orthodox sleep than normal at their first sleep; but, if pep-pills or similar stimulants have been the cause of their prolonged wakefulness, they will take more dream-sleep than normal when sleep is eventually permitted to catch up with them.

Babies and infants sleep, as is well known, for a large part of their neo-natal day; but babies and infants, as is less commonly appreciated, have quite different ideas over the amount of sleep they require. An average requirement is sixteen hours, not twenty-two as in some of the literature distributed at maternity centres. One study, which involved close observation of the babies to determine whether they were really asleep or merely quiet, discovered a remarkable sleep range among infants in the first three days of life: they were awake from one out of the twenty-four hours up to $13\frac{1}{2}$ hours. Some of them were therefore sleeping more than twice as long as others. Similarly, although the average length of the maximum sleep period throughout the twenty-four hours was 4·4 hours, some of the infants could sleep for ten hours without waking while others never achieved a time longer than 2 hours 6 minutes. Mothers and maternity homes may have invented neat rules for the offspring; but the offspring have had nine months to themselves in which to fashion quite distinctive personal attributes and behaviour patterns of their own.

A new-born baby, sleeping its sixteen hours a day, or a lot more, or a lot less, spends about half of it in the dream kind of sleep and half in the orthodox kind. As a young child the dream-time drops to 30%–40% of the total, and it further falls to 20% by the time of adulthood. As premature infants are thought to spend an even greater proportion of their sleep in the dream state, it is presumed that the foetus both sleeps longer than the new-born child and also spends a still greater proportion of that sleep in the dream state. As it is hard to visualise, and impossible to prove, that a child *in utero* is dreaming, it is perhaps better to say it is experiencing a type of sleep which will later be accompanied by dreams.

Most certainly a foetus has established a rhythm of sleeping and waking before it is born. As an adult sleeps once a day for a third of that day, and as a child sleeps twice a day for a total of half that day, and as a new-born infant sleeps half a dozen times a day for a total of two-thirds of each day, it is assumed that the elderly foetus both sleeps for a longer proportion of each twenty-four hours and has bouts of wakefulness even more intermittently than a new-born baby.

Mankind tends to regard sleep as the abnormal and wakefulness as the normal state. (There is not even an easy English noun for the state of wakefulness.) It is now generally considered, having taken note of the premature infant's addiction to sleep and its occasional lapses into a more wakeful state, that sleep is the prime condition invaded later on by increasingly lengthy periods of wakefulness. The distinction also becomes important in theoretical evaluations of the subject: can a sleep centre be postulated in the brain which sends a man to sleep, or can an awakening centre be proposed which irritates him out of his normal, passive and sleeping condition?

Animals do not, at least not yet, help to unravel the sleep problem. Undoubtedly many go to sleep, notably the mammals and birds, and there is also some justification for saying that fish which lie on the tank's bottom at night are asleep. Only mammals seem to exhibit the two kinds of sleep, dream and orthodox. Animals do to some extent explain the apparent vulnerability of sleep by proving themselves remarkably adept at waking instantly. Dr H. Hediger, of the Zürich and Basle zoos, used to glide on tip-toe towards his elephants, but they always heard him coming. Only in the disturbance and din of a circus was he able to catch some elephants sleeping and to assess their sleep periods as $2\frac{1}{4}$ hours a night. Mankind, living in his own man-made circus of noise and battered sense organs, and having to claw his way back from sleep into something approaching consciousness as the alarm pulverises him into activity, has plainly lost 99% of his jungle sensitivity. It is even harder for him to imagine a central nervous system so finely balanced, as in the animals, between sleep and instant readiness.

This knife-edge between the two forms of consciousness does not simplify one's comprehension of sleep. Just what are the animals gaining from the short period when they appear to relax, but with a hair-trigger attached to every sense organ? And why does a sheep only sleep in the dream state for 2% of the total time whereas a cat is in that state for 20% to 60% of the sleep time? Why does an infant need more sleep than an adult: are both brains equally taxed and equally in need of whatever sleep provides? And why does an adult in his thinking prime need no more and no less sleep than some hapless dotard shuffling wearily through yet another forgotten day?

Some random sleep facts are: that a person loses from 1–1·5 ozs. (28–42 gms.) in weight per hour of sleep (or about 11 ozs./312 gms. per night); that body temperature falls during normal sleep time, whether the person sleeps or not; that less urine is produced per hour of sleep; that people move some forty times in a night (certainly no logs) with roughly thirty seconds of movement per hour; that babies are only six times more active when awake than when asleep; that girls sleep more soundly than boys; that almost everyone can remember dreams when woken up at times of rapid eye movement; that almost no one can remember them if woken up ten minutes after the eye movements have ceased; that eye movements do not happen with those blind from birth, and most of their dream images are auditory; that times of rapid eye movement are regularly associated with penile erections; that some highly abnormal people have not slept for years; that learning during sleep is impractical and probably impossible; that those who say they feel worse after sleep are right, because tasks involving dexterity are better done towards the evening; that the architect R. Buckminster Fuller once trained himself to have a half-hour's sleep every three hours totalling four hours' sleep in every twenty-four, for a year; that computers may be made to 'dream' in order to review and reject redundant material; that narcolepsy is a rare condition which gives rise to irresistible and unwelcome attacks of falling asleep; that one of a pair of conjoined twins can fall asleep without the other doing so; and that men have been acquitted of somnambulistic homicide, or killing when they were asleep.

Yawning and snoring Half way between a reflex and an expressive movement, the yawn has defied attempts to explain it. It lasts a few seconds, it is sometimes partnered by stretching, it happens before sleep and after sleep, it accompanies weariness and boredom; it occurs in many mammals, notably the apes and monkeys, and it is necessary to be conscious in order to yawn. It is said to be contagious, but probably conditions which promote yawning in one person are also promoting yawning in those nearby. The act of yawning is often accompanied by increased tears, either because of direct pressure on the tear gland due to the yawn's facial contortions or because of nervous impulses partnering those which trigger this bizarre, open-mouthed and vaguely vocal form of behaviour. The act, which is linked with a slight and fleeting increase in heart-beat plus some constriction of capillaries, may be a physiological attempt to get more blood to lungs and brain; but blood flow is frequently boosted on many other occasions without the bewildering grimace of a jaw-cracking, ear-popping yawn.

The business of going to sleep is accompanied by other quaint traits. The eyes glaze over, for instance. This is thought merely to be the slowing down of the customarily swift jerks of the eye, as it leaps from subject to subject – or just leaps. Similarly there may be a wavering of the eyes when they eventually do focus on their object. When the eyes are finally covered over with their own eyelids the floating feeling of going to sleep may be followed by the horrific, heart-stopping drama of apparently total spasm. Known in the books as a start (and feeling more like a convulsive stop to everything) the muscular contractions may occur in an arm or a leg or the whole body. They are more common in adults than children and, despite their brutal interruption of sleep, do not usually interfere with it during the rest of the night.

Snoring does not interfere with the snorer's sleep, unless acts of retribution are taken by others. It is described as 'sounds made by vibrations in the soft palate and posterior faucal pillars during sleep'. When children snore, which is not common, it is commonly due to swollen tonsils or adenoids. When adults snore, which they tend to do increasingly with age, it is frequently caused by a loss of tone in the palate muscles. This loss of tone has been blamed upon obesity, fatigue, smoking, over-work, ill-health or just old age. Both sexes snore, and they tend to do so lying on their backs with their mouths open. There is evidence that snoring decreases during the dream periods of sleep, which can therefore give temporary respite to others. Snoring is harmless to the sleeper, despite the thunderous treatment of the tissues nearby, but surgery has often been attempted in an effort to reduce the noise. Other *ad hoc* remedies include hairbrushes or stones to keep a person off his back, pillows to keep his head more vertical, splints and pegs for keeping his mouth closed, and a wealth of inventions filed with the Patents Office.

In summary, therefore, sleep is a vital need but its reasons are quite unknown. Dream sleep is equally vital, and equally incomprehensible. Without sleep a normal man will die quicker than without food. Knowing that such ignorance over the main issue of sleep exists, it seems only fair that ancillary factors like yawning keep us equally in the dark about their reasons.

28. Speech

'Language most showeth a man; speak that I may see thee.'

BEN JONSON

'Truth is never where men shout, and scarcely ever where they speak.'

LEONARDO DA VINCI

At birth a child knows nothing of speech, and until that moment has not made a sound. At the age of one it will say a few words. At two its words will be sentences – of a sort. At three it will talk incessantly. At four, still assuming good hearing, a sound intelligence, no speech impediments and a normal willingness to learn and communicate, the child will have very nearly mastered the entire complex and abstract structure of the English language. As Professor David McNeill of Michigan put it: 'In slightly more than two years children acquire full knowledge of the grammatical system of their native tongue.' It is a matter of appearing, suddenly, abruptly, into a world of sounds, of imitating these sounds, making sense of them and then making use of them. At birth a child not only knows no speech but, unlike an adult arriving in Peking or Ulan-Bator without a phrase book, has no concept of speech, no preconceived notions whatsoever, let alone refinements of grammar such as 'bite dog' being so different from 'dog bite'. It is a stunning intellectual achievement, unthinkable by most adults, and yet carried out by every child even before it reaches school.

The ages for the various aspects of this achievement are extremely variable. Single words generally come at the first birthday, when children are also tottering into a walk, but the first word range is at least from eight months to $2\frac{1}{2}$ years. Phrases begin at eighteen months, but the normal range is certainly from ten months to $3\frac{1}{2}$ years. At two a child may have few words, or over 2,000. Some children complicate the issue further by speaking intelligibly, or intelligibly enough for friends and relations, and then relapsing into unintelligibility for a year or two and for no apparent reason. An experiment by L. C. Strayer seems to indicate that speech training is effective but pointless. Twins $19\frac{1}{2}$ months old were just about to talk. One was given training, and his word power was boosted from one to thirty-five words in five weeks. His twin was then still wordless, but was immediately

358

given training and he learnt faster than the first. The experiment was stopped when he had reached thirty words. Soon the children were both equally adept once again, and both took their own time to acquire a vocabulary. Speech comes, in other words, when the child is ready to speak.

Some generalisations are that speech comes later for twins than singletons, that girls speak before boys, that dada is usually said before mama owing to a preference for d and not necessarily for the father, that babies use far more vowels than consonants (adults use one vowel to 1·4 consonants), that a baby's first vowels are those which adults form in the front of the mouth, that a baby's first consonants are formed at the back of its mouth, and that children learn to speak at different speeds according to their environment. G. A. Miller said the most precocious child orator is a blind girl, the only daughter of wealthy parents. The slowest speaker is a hard-of-hearing twin born into a large family with poor parents who speak two or three languages. Whereas 100% of a child's jargon is meaningless even to a parent (and to the child too?), although a general sense may be conveyed, quite a high proportion remains unintelligible even when speech has begun. D. McCarthy reckoned that only 26% of an eighteen-month child's talk was comprehensible, 67% at two, 89% at $2\frac{1}{2}$, 93% at three and 99% at $3\frac{1}{2}$. An infant deaf in both ears does not learn to speak without special training. An infant partially deaf in both ears may learn the sounds he can see being made – b, f, w – but not g, l, r.

No one knows how man's speech began. Of course other animals convey a lot of information in sounds, but no other form of communication is comparable with mankind's talk. (The dolphin's chatter has still to be understood.) There is a theory which holds that a better valve was developed in the human windpipe to lock air more completely within the chest cavity, thereby providing a firmer base for the arms. There is another theory which holds that mankind's first words were based upon the grunts, puffs and groans that are the inevitable accompaniments to physical labour of particular kinds, each type of grunt being descriptive of a type of action. Be that as it may, man now has an effective larynx at the top of his windpipe. All air to and from the lungs goes through the larynx, but normally the laryngeal muscles are relaxed to leave a wedge-shaped opening, 0·7 ins. (17 mms.) long in men and 0·5 ins. (12 mms.) long or less in women. Making two sides of this wedge are the vocal cords.

The vocal cords are not cords. They are membranes, more like reeds than any possible kind of cord, and a sound is made after the two of them have been brought together to close that wedge-shaped opening. If air is then blown through the larynx, preferably from the lungs, although noises are possible – particularly to the ventriloquist – when air is drawn towards the lungs, a sound will be made because the valve of the vocal membranes is constantly forced apart by the pressure behind it. Each forcing apart causes the escape of a puff of air, and then the coming together again of the two folds. The vibration, which happens many times a second, occurs on more than one frequency at once. The basic tone of the human voice, despite there being other frequencies, is about 125 cycles a second.

Making a sound is not making speech: no one can talk through an oboe. It is necessary to articulate, to join together, before the basic note or notes are transformed into talk. The so-called mobile articulators are the lower lip plus its jaw, the various parts of the tongue, and the vocal cords. They have to move towards something, and the fixed articulators are the upper lip, upper teeth, the ridge of gum supporting the teeth, the hard palate at the front, the soft palate at the back, and the sides of the larynx. Say any word, and feel the inevitability of the articulation involved. Say 'pop' without the lips touching, say 'that' without pressing the tongue's tip to the upper teeth, say 'she' with the tongue wrongly in either cheek.

There are five types of articulation:

1. Plosives, stops. Breath has to be stopped to make them. As in pie.
2. Fricatives, spirants. Breath is forced through some narrow groove. As in fine, thin, yes, she.
3. Laterals. Breath has to avoid the mouth's centre, and come round the sides. As in lip, low.
4. Trills. Vibration of one of the articulators, such as the tongue. No trill in English, but Scotsmen roll 'r's'.
5. Vowels. Whole air passage is more open than in first four, which are all consonants.

These different types of articulation, plus the different positions of each type (fine, thin, she, yes, for example, all use different positions), enable a number of different sound elements to be made. These verbal elements are the basis of any language. English can be written down phonetically, not as the twenty-one consonants and five vowels of the alphabet, but as forty elements or phonemes. Some fourteen or fifteen are vowels, and the remainder are consonants. Of the forty some are rarely used, some frequently, and only nine of them are used in half of what we say, namely the consonants n, t, r, s, d, l, and th (as in then), the vowels i as in it and e as in river. However, the phonemes are insufficient to make a language of their own. It is the pattern of their use rather than the sounds themselves which enable us to communicate.

The 4,000,000,000 people in the world speak roughly 2,000 languages (this assumed total fluctuates wildly for there is eternal disagreement over the extent of barriers caused by mere dialects) or 2,000,000 people per language. There are about 160 languages with more than a million people making use of them as their prime means of speaking. English is a strong contender for the role of world language, and only about 2,000 words are necessary to speak it quite well, with some 400 million people using it as their prime language. (Mandarin Chinese is spoken by more people, and Russian by 250 million, but neither of these languages is much of a *lingua franca* outside its native area.) The importance of English is that a German uses it to check into his Peking hotel. When the Dalai Lama escaped from the roof of the world he was greeted by India's prime minister. 'How are you?' said Nehru, in English. 'Very nice,' replied the visitor from Tibet.

G. A. Miller has pointed out that, although the fifty commonest words make up 60% of those we say and 45% of those we write, the limited vocabulary of Basic English does lead to inevitable losses: 'blood, sweat and tears' becomes 'blood, body-water and eye-water'. Normal English uses several times the number for Basic English, and the extra words are important. In 1930 some telephone conversations were recorded and analysed. The result was a total usage of 2,240 different lexical units (give, gave and given are one lexical unit) and possibly 5,000 different words. A good fiction vocabulary is said to be 10,000 words, and a medical student reputedly adds 10,000 words to his vocabulary at medical school. James Joyce used 29,899 words in *Ulysses,* and the *Oxford English Dictionary* records 500,000 lexical units.

There are three properties of any sound and human speech is also associated with these three, with loudness, pitch and quality. In the English language changes in loudness and quality are each more important than changes in pitch.

Loudness depends upon the energy with which the cords vibrate, and this depends upon the pressure with which the air is impelled past them. Fairly loud singing means the passage of as much as 12 cu. ins. (200 cc.), of air per second over the cords. Trained singers can be more economical with their reserves of air, getting the same loudness from a fifth as much flow past their cords. The power of speech can be measured in watts just as any other power source. A whisper is 0·001 microwatts, faint talk is 0·1 microwatts, and very loud talk is 100 microwatts. All vowels radiate more power than any consonant; consequently loud shouts for help linger on the vowels and not on the consonants. (The English 'help' is not as good as the Italian 'aiuto' or the French 'm' aidez'.)

Pitch is determined mainly by the length and tightness of the cords. To some extent pitch can be altered, and trained singers can move happily from one octave to another, but human growth causes conspicuous changes in pitch. Both boys and girls are born with cords about a quarter of an inch (6 mms.) long, and these grow in length much as the body itself grows. By two they are a third of an inch (8 mms.) long, by six they are 0·4 ins. (10 mms.), by ten they are half an inch (12 mms.) and the boy's cords are then showing signs of growing longer even though the height of girls at this age may actually be taller than the height of boys. At fourteen, on average, a girl's cords are just under half an inch (12 mms.) long, a boy's just over half an inch (13 mms.). At twenty a boy's are nearly an inch (24 mms.), and a girl's are only just under two-thirds of an inch (16 mms.). At thirty a man's cords are $1\frac{1}{4}$ ins. (30 mms.) (a fivefold increase since birth) and a woman's are just over $\frac{3}{4}$ ins. (20 mms.) (slightly over a threefold increase). The change in pitch is not something which happens once in adolescence: it is changing throughout the first three decades, and it changes again at the end of life when cords shorten and voices become higher. Sometimes, as with eunuchoids, the voice of an adolescent remains strangely unbroken; but this can usually be remedied rapidly by suitable treatment, and a normal male voice soon results. A true eunuch's voice does not break provided castration is performed before puberty.

Quality, or timbre, depends upon the form and size of the various resonating chambers, the throat, the chest and the mouth with its cavities. The difference between the vowels is made by alterations to these chambers, without much recourse to the articulation devices. A whisper is possible solely by adjusting articulation and cavities; the vocal cords are not made to vibrate. Consequently there is difficulty in sexing a whisper because the major clue of cord length and fundamental frequency is lacking.

Nevil Shute, Somerset Maugham, Aneurin Bevan, King George VI, Arnold Bennett, and roughly one in a hundred of the general population are in the same speech category; their voices have all been subject to a stammer. To stammer or to stutter is the same, with the British being prone to stammer and the United

States more frequently stuttering. There have been efforts to distinguish the two – stammer being the hesitation, stutter the repetition – but the words are generally considered synonymous.

Two types of stammer are customarily realised. The first affects 2–4 year olds during normal learning of speech, as the child stumbles along in his efforts to be coherent and when the breathless urgency of the message is infrequently matched by the skill to impart its information. The second, the persistent stammer, first affects the 4–7 year olds, and just occasionally the first type merges into the second. Persistent stammer hits boys far more than girls, it frequently disappears in the young on its own accord, it occurs more often with late speakers and in children of low intelligence, it sometimes runs in families even when mimicry can be ruled out, and the traditional belief in an association between stammering and left-handedness, notably when right-handedness is made compulsory, is not borne out by recent studies. The several theories that stammering is purely psychological in origin are not assisted by the overwhelming frequency with which boys are afflicted, but a review on the subject by the *British Medical Journal* suggested that stammering is probably best treated by a combination both of speech therapy and psychotherapy. There are few cases of completely successful cures, although stammerers can be made to lose their stammers in the presence of suitable noise, a noise which tends to make normal people stammer. Some stammerers can be helped by having miniature metronomes beating out the time for them.

Whereas earlier generations were entirely familiar with stammering, and with the deaf-mute, there is a form of muteness which is entirely modern. Total laryngectomy, caused by the discovery of a malignant larynx, is an operation which removes voice along with the cancer. There are about 30,000 people in the United States without a larynx, and suddenly not being able to make a sound has been a bewildering shock to many of them. However the majority of laryngectomised people can regain voice of a sort even though they have to breathe through a hole in their necks and have no vocal cords.

Finally, when did speech arise? It is not just the major attribute to distinguish mankind from the rest of the animal kingdom but, arguably, the only such attribute. Perhaps it arose during the lengthy evolution when pebble tools were used or, which seems much more likely, at the start of the past 30,000 years. Artistic, technical and sociological progress has been so swift since then that speech *must* have been involved. Progress beforehand was so slow that the gift of speech could not, or so it is easy to assume, have been partner to such indolence. The certainty, as judged from now, is that no one will ever know when this unique form of communication arose. Or where. Or why.

29. Temperature

Normal/Why warm bodies?/Chilling/Heat/Hibernation/
Hypothermia

The heat of a body is the balance between heat gained and heat lost. It is gained primarily by metabolism, secondarily from the environment (normally cooler than the body), thirdly by shivering (involuntary activity of skeletal muscles) and finally by the consumption of hot food. The basic metabolism of a person at rest produces about 1,700 calories of heat per day for men, 1,500 for women. This is sufficient heat (assuming there is no loss) to raise the temperature by nearly 2°F (1°C) per hour. Moderate activity may produce 3,000 calories, extreme activity 6,000 calories, or enough to raise the temperature by over 6°F (3°C) per hour. Without heat loss a fatal heat stroke would therefore occur within a few hours.

Heat is lost primarily by radiation, convection and conduction, these three totalling perhaps 72%, but ambient temperature and the amount of clothing worn will affect the percentages; secondly by evaporation from the skin, which is 15% under normal conditions; and lastly by evaporation from the lungs (7%), by warming the air taken into the lungs (3%), and by the expulsion of urine and faeces (3%). The normal daily temperate zone sweat loss varies between 0·9 and 1·6 pints (500 ml. and 900 ml.) indicating 300 to 530 calories of heat lost (the latent heat of the evaporation of sweat is 0·59 calories). The maximum sweat rate may reach 3 pints (1·7 litres) in each hour, but with a maximum about 21 pints (12 litres) in 24 hours, indicating a loss of over 7,000 calories – four times the basal metabolic rate. Metabolism produces heat because chemical changes frequently liberate much energy as heat. During muscular activity only 25% of the energy produced is converted into work; the remaining 75% is heat. Temperature regulation, the achievement of the balance between heat lost and heat gained, is under the control of the hypothalamus in the brain. It is less efficient in the very young, and in the old.

Normal temperature The word normal, as applied to human body temperature, is a splendidly misleading description. The little arrow on the clinical thermometer, doggedly directed at

365

98·4°F (36·9°C), gives extra support to it and the exactness of that point four of a decimal emphasises it even further. The awkward figure is the fault of Gabriel Daniel Fahrenheit, the German physicist. Isaac Newton, a duodecimal enthusiast, had wanted 12 units between water's freezing and boiling. Fahrenheit's new thermometer could do better than that, permitting more units. He liked the idea of the lowest temperature (made with salt and ice) being 0, of water's freezing being 32, of the body being 96, and of water's boiling being 212, all of them nice mathematical numbers. Greater accuracy subsequently, plus an equal determination to keep freezing and boiling 180 degrees apart (another nice number) meant that something had to give – and body temperature did so.

Normally the majority of us do not have a 'normal' temperature. Infants have a higher temperature; older people have a lower one. Ordinary adults wake up with their temperature below normal and go to sleep with it above normal. Women are about 1°F (0·5°C) hotter after ovulation. The armpit (or axilla) is 1°F (0·5°C) cooler than the mouth and the groin, and cooler still in thin people. The rectum is almost a degree higher. A British publisher once said to scoffing visitors: 'I have a fever. I am 98·4°F. You see normally I am below normal.' While 'normal' in Britain is 98·4°F, it is 98·6°F in the United States. In Europe it is 37°C, which is also slightly higher than the British normal. The British Standards Institution recently suggested it was high time that 98·4°F was forgotten because the warmer 37°C is around the metric corner.

The story of temperature-taking really began in the middle of the nineteenth century with Carl Wunderlich (1815–77), a German doctor at Leipzig. Despite clinical thermometers being a foot long, and taking twenty minutes to register a man's temperature, he – more than anyone else – initiated the regular practice of recording a patient's thermal fluctuations. There is current argument that he did his work too well and that nurses now spend far too much time routinely recording temperatures; but it is also said of him that he found fever a disease and left it a symptom.

As warm-blooded (homoeothermic) creatures go, man is cool. Among domestic animals (according to H. Dukes) temperatures are: stallion 99·7°F (37·6°C), mare 100°F (37·8°C), beef cow 101°F (38·3°C), dairy cow 101·5°F (38·6°C), cat 101·5°F (38·6°C), dog 102°F (38·9°C), sheep 102·3°F (39°C), pig 102·5°F (39·2°C), rabbit 103·1°F (39·5°C), goat 103·8°F (39·9°C), and chicken 107·1°F (41·7°C). Human beings would be near to death or dead if kept at chicken heat. Mammals and birds are all warm blooded, but some are better at regulation than others. Monotremes (like the duck-billed platypus), marsupials (like the kangaroo and koala bear) and bats are relatively inexpert. The lower vertebrates (fishes, amphibians and reptiles) are frequently considered to be cold blooded, or poikilothermic. In fact, as Professor A. S. Romer of Harvard University says, there is some sort of regulation even among these forms, notably the reptiles. Not only do they seek out hot or cold places, light or shade, but there are also 'internal mechanisms of which we know little'. The old fishermen's story that tuna often feel warm has recently been proved by marine biologists; the fish can maintain a temperature 25°F (14°C) warmer than the water they swim in.

Why do we keep warm? Why at 98·4°F (36·9°C) or thereabouts? It is a compromise. On the one hand is the fact that chemical reactions are speeded up (roughly two or three times every 20°F or 11°C) with heat. On the other is the fact that enzymes, the catalysts vital to the body's chemical reactions, can be destroyed by too much heat. It is no good living like a torpid lizard on a cold day. Similarly, it is unsatisfactory to cook one's enzymes (like the white of an egg being irreversibly cooked by heat). It so happens that most enzymes are inactive at freezing point, and they have been inactivated – quite a different thing – by boiling point. Animal enzymes work best at 100°F (37·8°C), or when slightly warmer than the human 'normal'. It has even been argued that humans do better when marginally feverish. Certainly athletes perform better in the latter half of the day when they are warmer. However, there is a limit, and an emphatic one, to the desirability of increased heat. By 111°F (43·9°C) or so the destructive effects of heat (on the enzymes and other proteins) are equal to or greater than the ability of the body to repair them. The rate of destruction is then increasing perhaps forty times for every degree rise in temperature, and death becomes inevitable.

Human beings start life much hotter than they end it. Babies are poor at regulation at first. Cast out from their cosy uterine incubator, and rarely able to shiver, they drop 3°F (1·5°C) in the first hour (according to one survey) despite a room temperature of 80°F (27°C) to 85°F (29°C). A maximum drop of 4·6°F (2·6°C) was recorded at the third hour. The return to 'normal' is always a slower business, but a spate of crying (thereby increasing metabolism 180%) can speed things up. With premature babies heat regulation is even less perfect. Environmental temperature of 84°F–90°F (29°C–32°C) with a humidity of 65% (somewhat like midday at a hot Mediterranean resort) is best, according to E. H. Watson and G. H. Lowry. As the baby grows into an infant, and then into a child, the average temperature first rises, then drops. Rectal readings under basal conditions have been recorded as:

	°F	°C
3 months	99·4	37·4
6 ,,	99·5	37·5
1 year	99·7	37·6
3 years	99·0	37·2
5 ,,	98·6	37·0
7 ,,	98·3	36·8
9 ,,	98·1	36·7
11 ,,	98·0	36·6
13 ,,	97·8	36·5

Girls tend to stabilise at fourteen, and boys at eighteen.

To confound the picture of 'normality' still further there is a daily cycle. Babies acquire it quickly, and then follow the traditional pattern of a peak in the evening and a trough at dawn. An average between maximum and minimum is 2°F (1°C), but individuals can go up and down by 3°F (1·5°C) or more. The menstruation of women is a further complexity; during it the temperature is low. This increases slightly with a small but pronounced rise (perhaps as much as a degree) during ovulation when an ovary releases an egg. A hot bath, reasonably enough, sends the bather's temperature up to 100°F (38°C) or so. Exercise does even better. After three miles a runner can read 105°F (40·5°C) rectally, but less than normal on his skin due to all the sweating there. A long cool drink reduces mouth temperature by 3°F (1·5°C) and so may a blocked-up nose, as cooling air keeps oral temperature down. Infants can induce fever temperatures by quite mild bouts of rage and can drop 7°F (4°C) in a cold bath. A hot climate pushes the entire range of temperatures up by about 1°F (0·5°C), but locals (certainly in India) are neither hotter nor cooler than acclimatised visitors.

Chilling Only quite recently was the medical profession made forcibly aware of the perils of hypothermia in the elderly – or the fact that people in many 'temperate' countries quietly freeze to death each winter. A special report published during November 1964 in the *British Medical Journal* was among the first to stress that the condition, although unspectacular, had a 'very high mortality rate, very much higher than is commonly supposed'. One doctor even estimated 20,000 to 100,000 deaths each winter.

Doctors who carry only the conventional clinical thermometers, which record temperatures above 95°F (35°C) (up to 110°F/43°C), can be unaware of the actual and far lower temperature of their old patient. (Hospital temperature charts usually have no space to record sub-normal temperatures.) Dr Geoffrey Taylor, of Somerset, reported that the temperature of all old people falls below 95°F (35°C), often below 90°F (32°C), if their room temperature is below freezing. In the annual mortality statistics over 300 people are generally listed as dying from hypothermia, with about two-thirds of the total being women, presumably because more women than men are wrestling with the problems of considerable old age. The annual total is about the same as deaths from 'homicide and injury purposely inflicted by other persons'. Although the two forms of death are so very different, it is arguable that letting someone die from cold is no less callous but just more drawn out than hitting him on the head.

One trouble is the insidious nature of the chilling process. H. Duguid recognised three phases. From 98°F–90°F (37°C–32°C) is a feeling of cold and shivering occurs. From 90°F–75°F (32°C–24°C) the subject is depressed, and usually does not complain of feeling cold. Pulse slows down, blood pressure falls, and all shivering has stopped by 85°F (29°C). Below 75°F (24°C) is the downward path, with survival being rare. The temperature regulating centre ceases to function, and the body progressively cools until it reaches the level of the atmosphere. Circulation fails at about 70°C (21°C). To make matters worse, consciousness becomes clouded at about 86°F (30°C) and clothes may be taken off rather than put on. Another feature is torpidity and a desire to sleep. Those who can still walk, despite increasingly stiff muscles, may be giddy, and suffer from impaired vision and look for all the world like a wandering drunk, but anyone below 90°F (33°C) requires immediate treatment. (Considering the fact that many old people do die of cold, it is worrying to read frequently in books that 'most deaths by freezing occur in abandoned infants and alcoholics'. One wonders how many of those 'alcoholics' are really victims from cold suffering from impaired vision and balance.)

The *British Medical Journal*'s anonymous expert has recommended slow warming. One degree (Fahrenheit) rise an hour is sufficient and can be achieved by keeping the patient well covered in a room kept at 70°F–80°F (21°C–27°C), and by monitoring the return to a warmer-blooded and healthier state. For younger people, like sailors or mountaineers who are rescued in a cold state, the warming-up process should be as rapid as possible.

They have tougher constitutions than the elderly, and therefore a bath at 111°F (44°C) will heat them up with a minimum loss of time. Naturally anyone falling overboard should be retrieved with even greater alacrity. According to the Institute of Aviation Medicine five minutes immersion in water at 40°F (4·4°C) causes a body to become rigid, and incapable of making the effort to keep the head above water. Death comes perhaps after two hours, perhaps after thirty minutes. And the water around Britain is at 40°F for half the year.

Obviously a thermometer recording the conventional range of 95°F–110°F (35°C–43°C) is unsatisfactory. The *British Medical Journal* report recommended that 75°F–105°F (24°C–41°C) was necessary for general use. A range of 75°F–110°F (24°C–43°C) has also been suggested, in order to cope with the heat stroke side of things as well. One English doctor in the winter of 1964–5 (not a very cold one), who was equipped with a low-range thermometer, took mouth temperatures of old people who were not ill. He found the average was 94°F (34°C), and the lowest 89°F (32°C). One particularly severe case of accidental hypothermia was that of a young black girl. Her temperature was 64·4°F (18°C) when she was pulled out of an American snowdrift. She lost some of her extremities from frost-bite but did live. (Russians frequently report such incidents, but omit details.) That girl had gone half way to freezing point from 'normal' and then back again. The *Guinness Book of Records* gives the names of two females (is this sex more able to cope with refrigeration?) whose temperatures dropped to 60·8°F (16°C) when they were aged twenty-two and two years respectively, one being found in an alleyway and the other in an unheated room. Particularly chilly comfort is the fact that no one immersed in water can actually freeze. Sea water freezes at 28·5°F (–2°C), a temperature less than fresh water's freezing, but body tissues freeze at about 27·5°F (–3°C).

Frostbite is basically a blockage of the blood vessels; they constrict so much that blood cells cannot pass through them. At the same time the vessels suffer injury, and consequently, when the limb is thawed out, the blood vessels leak and there may be clotting. Even at temperatures above freezing the same kind of damage can be done, although far less speedily. It is then given such names as immersion foot, immersion hand or trench foot.

Heat Trouble from too much heat is more complex, partly because the body is more sensitive to increases than decreases in temperature, and partly because there are so many types of heat

370

disorder. The book *Exploration Medicine* (in a chapter by C. S. Leithead) lists them. Briefly they are:

1. Heat syncope. Generally giddiness and fatigue. Sometimes fainting. Rest (not alcohol) will cure it.
2. Heat Oedema. Swelling of the extremities. 'Deck ankles'. Unimportant.
3. Water-depletion heat-exhaustion. Sweating output not balanced by water intake; hence thirst, fatigue, giddiness, fever, then delirium. Death likely when water debt reaches 20% of total body water (20% of average total equals 2·75 gallons/22 pints/12 litres).
4. Salt-depletion heat-exhaustion. Salt loss in sweating not balanced by intake. Hence fatigue, giddiness, nausea, vomiting, cramp. Generally occurs only in unacclimatised. Seldom fatal. Normal daily diet contains ⅓ oz. (10 gms.) of salt. Sweat may initially contain 4 grams per litre. As sweating may rise to 1 litre per hour, obviously depletion can occur. But body acclimatises itself in a few days. Less salt appears in urine and sweat, and normal daily salt diet can be as before.
5. Heat cramp. Not necessary to visit tropics to get heat cramp, as stokers, miners, firemen and athletes know. Disorder extremely painful (with knotted muscles) but never fatal. It is a very rapid form of salt depletion.
6. Prickly heat (Miliaria rubra). An irritating, inconvenient and common rash occurring on skin wet with sweat. Not a hazard, but rashes can become infected.
7. Anhidrotic heat exhaustion. No sweating in affected areas, but blisters instead. Little understood and rare.
8. Heat stroke. Heat regulation may suddenly go haywire, sweating stops, temperature goes up (above 105°F/41°C), and delirium, convulsions and coma occur. Rare but very dangerous. (*Exploration Medicine*'s recommended treatment makes startling reading: 'Ideally the patient is stripped, placed in a good stream of cold, dry air, and sprayed with water at 50°F . . . or in a bath filled with ice chips and water. . . . Convulsions or shivering should not be allowed to delay the treatment. . . . Cooling should be stopped when temperature has fallen to 102°F . . . chances of survival seldom better than 65%.')

As might be expected the old and the very young are most severely affected by heat. Weak hearts in the old and vulnerability to dehydration in the very young are mainly to blame. Taking both heat and cold together the human body has a comfortable temperature range under normal conditions of some 3°F (1·5°C).

This can be stepped up to 15°F without considerable alarm. Below 90°F (32°C) (if not before) there is urgent need for treatment. Above 105°F (41°C) (if not before) there is an even more urgent need. The range is small. Yet it has been extended from 60°F (16°C) with those frozen girls, to a maximum of 111°F (44°C) or so under extreme conditions of heat stroke, a range of 51°F/28°C (or the difference between the day for skating on a lake and basking in it).

The girls, cold as they were, still had the temperature of a warmish room. They were hot by comparison with a hibernating animal. Humans cannot hibernate, but great efforts are being made to induce an artificially cool state. Called cryosurgery, or crymotherapy, or – more facetiously – cold-blooded surgery, this hinges upon the fact that cold tissues require less oxygen; but first a word about natural hibernation.

Hibernation The hedgehog in summer has a normal temperature of 92°F–95·5°F (33°C–35°C) – a few degrees lower than humans. The creature becomes comatose, with its own temperature fluctuating between 86°F (30°C) and 60°F (16°C), when the outside temperature falls below 62°F (17°C). This intermediate stage becomes true hibernation when the outside temperature drops below 57°F (14°C). The animal then remains about one degree (Fahrenheit) above the outside temperature and, save for that one degree, is apparently cooling off like a block of wood, or anything similarly inert. However, unlike wood, its temperature never falls (except in death) below 43°F (6°C). The world may continue to freeze, but the hedgehog stays at 43°F and can even wake up. According to Nathaniel Kleitman, the American expert on sleep, the hedgehog may be said to be cold-blooded when it can economise fuel by being so, but not when its life is endangered. Heart rate always falls with the falling temperature. In the squirrel it can go from 300 beats to 2 beats per minute, in the hamster from 500 to 4 beats per minute and in the marmot it can beat only 23 times per hour – slower than a funereal minute gun.

None of this is possible with human beings. The regulation system goes rapidly awry. It certainly cannot keep a watchful eye on the situation or prevent undue cooling as with the hedgehog. The human heart starts to fibrillate (or suffer tremor) at 64°F (18°C), and the nervous system stops transmitting impulses at 50°F (10°C). Mankind is definitely not a hibernating creature.

At least, not so under natural conditions. Since the last war progress has been abundant in unnatural imitation of the cold state. Here is Ronald Woolmer describing (in *The Conquest of Pain*) an operation using deep hypothermia: 'The patient was a woman in her late thirties. Her face had the paleness of death. Her skin had the coldness of death – and more. Her eyes were glazed, and her pupils widely dilated. There was no hint of respiration. The heart was not beating, and the pressure in the blood vessels could not be measured. This state of affairs persisted for forty minutes, and during that time by all the usual tests the patient was dead.'

Two hours later she was chatting away in bed. Obviously there are great surgical benefits in not having a vibrant heart and in not having tissues urgently demanding oxygen. The maximum permissible time for stopping circulation at the body's normal temperature is three to five minutes. With a general cooling of 14°F (8°C) (and down to 84°F/29°C) the time of stoppage can be ten minutes. Nowadays cooling a body to 60°F (16°C), and stopping circulation for an hour, is a standard procedure. Plainly lower temperatures and longer times will gradually occur. A textbook now refers to 48°F (9°C) as being the minimum yet tolerated 'without apparent alteration in neurological or intellectual functions'.

The procedure is either a matter of cooling the blood or cooling the body. Extracting blood, cooling it, and pumping it back again, can cool a patient from 98°F (37°C) to 60°F (16°C) in less than an hour. Packing ice around a patient, as if he were champagne in happier circumstances, takes longer. One hour would bring the temperature to 86°F (30°C). Selective cooling, as against whole body treatment, can be simpler. Ice can be packed around the heart; or the brain, that most demanding tissue of all for oxygen, can have its own circulation blocked.

At a Copenhagen Conference in 1965 a Japanese neurosurgeon, Dr Tatsuyuki Kudo, described the cooling of brains to an unprecedented 43°F (6°C). His patients were first surface cooled with ice to 90°F (32°C). Then, by suitable exposure of arteries and veins, he drained the brain of blood, and perfused that organ instead with a cold blood-substitute. Within twenty-five minutes the temperature of the brain was down to 43°F (6°C), and the rest of the body remained at 90°F (32°C). The surgeon then started work on the cold, white and bloodless brain. Subsequent warming up was also, according to Dr Kudo, a relatively fast business. Some three to five pints of new blood had to be used and, if surface warming of the whole body accompanied this transfusion, 'revitalisation' of the brain took about an hour. There were, he concluded, neither cardiac complications, nor any other major side effects in the six patients who had experienced this particular trial procedure.

Is human hibernation possible? At a symposium held in Toronto 'On Natural Mammalian Hibernation' in 1965 Dr A. W. Dawe, of the United States Office of Naval Research, said it 'might' be achieved in the 'foreseeable' future. Well, it still might be achieved in the foreseeable future. Much impetus behind such work is associated with the space programme. Flights in space, at least beyond our immediate planetary neighbours, will be long and tedious affairs. Suspended animation could not only relieve the tedium but could also ensure, for the lengthier trips, that old age did not kill off the crew before landfall.

Apart from such space advantages, and present-day earth surgery, the science of hypothermia will inevitably have more to offer. Cold can kill tissue, and can kill it selectively. Undesired nerves, which die after five minutes at 5°F (−15°C), can be put out of action by cold-tipped probes. These devices have already been used in localised tissue destruction, and have helped with Parkinson's disease (by causing brain damage in just the right place), breast cancer (by destroying the associated pituitary gland), and with operations on the eye. Plunging new babies into iced water instead of warm blankets is a bizarre aspect of hypothermia, but one sometimes recommended for new-born infants who do not begin to breathe as they should. The theory is that, as such infants often suffer thenceforth from irreversible brain damage and cerebral palsy due to temporary oxygen lack, it is sensible to cut down their brain's stringent demand for oxygen during the respiratory difficulties following birth.

A report from St Bartholomew's Hospital, London, described how a small girl on the point of death from meningitis, recovered after ice packs had kept her 5°F cooler for a whole week. Normally the swellings of meningitis restrict the brain's blood flow, and therefore its supply of oxygen. The suggestion in this case is that the reduced demand for oxygen (caused by the cooling) meant that the brain could survive until the meningitis swellings had subsided. Without doubt induced hypothermia, with its countless uses, will enforce many of us to spend many more hours on the chilly side of 'normal' in future years.

Traditional temperature taking, now over a century old, may also change. The mouth, the armpit and the rectum, time-honoured sites for the ordinary thermometers, may yield their importance. Not only are other sites being constantly recommended, such as the navel and even the ear, but thermography is gaining ground. There are several techniques for recording the whole pattern of a person's temperature, rather than the heat at one point. Patterns can be more informative. They can show regions of abnormality, such as cancers and arterial disease. Carl Wunderlich's régime of temperature taking may yet prove to be indispensable, but the new heat cameras are all-embracing. They have much to say that no temperature chart, however doggedly kept, could ever indicate.

Cryosurgery, crymotherapy, induced hypothermia and thermography are all fairly novel, but cryo-burial is, as they say, something else. Dr James H. Bedford of California, paid £1,500 before his death in January 1967 to ensure permanent freezing of his body. He also left £70,000 for cryo-biological research. Whereas the money will soon be consumed his body will remain for ever inside its metal container at liquid nitrogen's temperature of $-321°F$ ($-196°C$). In theory he will be revived as soon as science is capable of such resurrection. In practice, according to present-day science, the freezing will have caused ice crystals to play havoc with his cellular structure. Revival of such a shattered body is likely to be impossible even for tomorrow's resurrectionists.

Nevertheless, today's cryonics societies, with their motto of 'Never say die', are expanding. Within two years of Dr Bedford's death from lung cancer and his interment for the future, ten more Americans had followed him and a thousand more have joined the societies. An underground 'cryotorium' has even been built at a cost of $100,000, and it will sleep 40 – until executors form themselves into another society having 'Let the dead die' for its motto. Dr Bedford is still not a-mouldering in his grave, along with several hundred other Americans, but this particular Californian cult has not caught on as some feared it might. The dead, it seems, are still (almost entirely) dying in standard fashion, with dust to dust and ashes to ashes rather than dust to icicle.

30. The Senses

*'What we perceive comes as much from inside our heads as from
the world outside.'*

WILLIAM JAMES

Background   It is all very well accepting the general principles of
natural selection, that the pressure of this selection favours the
possession of certain characteristics. It is harder accepting the
advantages of many of these characteristics when they are first
being developed. To have a slightly longer neck, or a slight ability
to fly, or a slightly stronger skin, is plainly a modest advantage
right from the start; but to have the primitive rudiment of an eye is
less immediately beneficial. Originally it must have been an area
of skin faintly sensitive to light. It could just have discerned, one
assumes, a difference between bright sunlight and everything else.
Was this ability so beneficial that natural selection could act upon
it? Even the later stages, the development of lens and orbit, are
nothing like so straightforward to contemplate as the giraffe's
steadily lengthening neck, the cheetah's gathering speed. The
world of botany, entirely dependent upon light, has formed no
eyes: in the world of zoology animals must have originally used
their eyes much as a plant uses phototropism to manoeuvre either
towards or away from light, and yet eyes are now paramount.
Nearly every vertebrate has a pair, and there are innumerable
forms of eye in the invertebrates right down to the 'light spots' of
many single-celled organisms.

Mankind has the typical pair of vertebrate eyes, although they
have moved round to the front. The majority of vertebrates have
lateral eyes; so does man in the first instance, when the embryonic
eyes grow as two lateral stalks and bulbs from the brain. This fish-
like appearance persists for many weeks, but the final result is two
eyes at the front capable of binocular vision to a large degree. Both
eyes can see the same object at the same time in most of the visual
field; hence an improved ability to determine distance and assess
speed.

Useful as a third eye could have been to the vertebrates in general, and to those wishing to see upwards in particular, it never materialised in any proficient form. The lampreys (and others in this cyclostome group) possess such a third eye, and it contains some retinal material; but it is nothing like so definite an organ as the lamprey's two other eyes. The cyclostomes are primitive fish, and in more advanced forms the third eye ceases being an eye at all: instead the organ develops into a gland. The reptiles developed after the fishes but certain reptiles, including the archaic Sphenodon of New Zealand which has somehow survived out of its time until the present day, also possess a kind of third eye. However it is as a gland, although of uncertain function, that the extra eye persists in animals of later evolution.

Probably the eye began as a pair of eyes, and evidence from the fossil forms of the world's earliest fishes suggests that this was so. Also, although the third eye is customarily spoken of as if it is single and in the mid-line, it is developed from one or both of two distinct outgrowths from the brain. They are the parietal and pineal organs. In man the pineal outgrowth is a small, flat, oval portion of tissue attached to the brain by a short stalk: it is this fragment from the past that is the most positive human remnant of the vertebrate attempt to possess more than two eyes. George Orwell's creatures of his *Animal Farm* originally recommended four legs, but later settled for two: the animal kingdom did likewise as two primaeval pairs of eyes, one dorsal, one lateral, were dropped in favour of a single lateral pair which, to a varying degree, moved round to the front. The pineal body was left for René Descartes to call the seat of the soul. (Once again this is a subject more liberally explained within *The Mind*.)

The eye   Each human eyeball weighs one-quarter of an ounce (7 gms.), and has a diameter of slightly less than one inch (2·5 cms.). A male eye is bigger – by about a fiftieth of an inch – than the female eye, and an adult eye is less round than the new-born eye. At birth the eye is much nearer adult proportions even than the brain: whereas the body increases its volume twenty times or more after birth, the brain increases 3·75 times and the eye only 3·25 times. One 400th of a baby's weight is eye: 1/4,000th of an adult's weight is eye. The human eye is not a close fit within its orbital cavity: eye volume is 6·5 ccs. but orbital volume is 29 ccs. Fishes have large eyes, amphibians and reptiles have relatively small ones, but bird eyes are large. An eagle's eye is larger than a man's, and the owl's eyes occupy one-third of its head.

The sixteenth-century inventor of the camera obscura had to defy current opinion and state that the eye was always a receiver, never a transmitter, of rays. Ever since then the science of optics and an understanding of the eye have run parallel. Light waves entering the human eye first pass through the thick, transparent and bloodless cornea, then through the anterior chamber, then through the pupil gap between the iris diaphragm, then through the lens, then through the large posterior chamber (both chambers are filled with fluid) until it eventually lands on the sensitive retina.

It is the iris that gives colour to the eye. Named after the Greek for rainbow the iris tends to be without much pigment at birth and appears blue (except in the dark races when it is brown even at birth). At the other end of life the pigment changes, and some time after death all European eyes look a greenish-brown. Although blue eyes may seem to possess more colour than brown eyes the effect is an illusion: brown eyes have most pigment. Irises can be piebald, or different colours for each eye. The patterns on an iris, of rays, rings and spots, are highly individualistic, and have been suggested as alternatives to finger-prints for identification.

The circular opening slightly to the nose side of the iris is called the pupil (after the Latin pupilla, allegedly because the minute image of an observer within someone else's pupil looks like a little girl). In humans the pupil is circular both when dilated and contracted, and it is customarily circular in animals when dilated. In the contracted state animals show much pupil variation, changing the dilated circle into a slit or even a dumb-bell form. Many animals, such as the lower mammals and birds, can contract and dilate one eye at a time. In man both behave equally even if one lid is closed. Normally the human pupil is between 2·5 and 4 mms. in diameter; but extreme and abnormal dimensions of 1·5 mms. and 8 mms. have been measured. Pupils are affected primarily by light intensity and the distance of the object; but fear, interest and numerous emotions can dilate the pupil. A pretty girl has been shown to open up a man's pupils by some 30%; conversely, men have unwittingly preferred female photographs in which the pupils have been touched up to look larger. (But then nothing is new. Belladonna, or beautiful lady, was traditionally used to give this effect.)

The lens is flatter at the front than the back, it is slightly yellow and it is soft. With age it grows flatter, yellower and harder. There is no blood in the lens, and it is sustained by the liquid on both sides of it. Cataract is the principal disease of the lens, with its first symptom being the appearance of stationary motes. Normally all motes seen by the healthy eye float and jerk their spasmodic way in a person's vision.

The retina is at the back of the eye, and is the destination of the light waves. By then they should have been correctly focussed mainly by the cornea and partly by the lens, and the resulting image is both upside down and smaller. Two 6-foot men standing 10 feet from each other are shrunk down to inverted images only 0·4 ins. (10 mms.) high in each other's retinas. The 20:20 vision of Americans or the 6:6 vision of the British, which are essentially the same, one being expressed in feet and the other in metres, refer to letters on a card 20 feet or 6 metres away from the observer. The letters in question are 0·35 ins. or 8·75 mms. tall. Such letters subtend an angle of five minutes, or one-twelfth of a degree, from the observer's point of view, and such a letter is shrunk down to an image only 0·025 mms. high, or a thousandth of an inch. It is this size of letter image which the retina and the brain have to decipher if the observer is to be adjudged visually competent.

The light-sensitive layer of the retina responsible for this competence consists of a multitude of cells called rods and cones. In the human eye there are approximately 125,000,000 rods and 7,000,000 cones, but information from these millions of receptor cells has to pass to the brain via the optic nerve, a relatively modest bundle of 800,000 nerve fibres. (Some say there are a lot more, but no one says a number which approaches that of the rods and cones.) Essentially the retina is just an extension of the brain carried on this thin nervous stalk. The exit of the optic nerve causes the blind spot, an area devoid of receptors. This phenomenon was first discovered in 1668, and can be detected by anyone with patience, two dots 3–4 ins. (7·5–10 cms.) apart, and one eye closed. The precise working of rods and cones is still largely unsolved; but, briefly, rods are for night vision and can only detect shades of grey, while cones are for day vision and can sort out colours. Birds active in day-time have a greater proportion of cones, and night-time birds have more rods. Nocturnal mammals, such as the bat and the rat, also have more rods but so, confusingly, does day-time man.

The fact that so many animals' eyes shine in the dark is due to the nature of the tapetum, part of the pigmented layer of the retina. There is no creation of light, as was formerly supposed; only an ability to reflect visible light. Human eyes will also shine if the observer is sufficiently near to the beam of light being shone at them; it is best to look directly along the beam, preferably by using a pin hole in the centre of a reflecting mirror aiming the beam. I do not know the reasoning or intent behind the experiment, but in 1704 it was observed that a cat's eyes no longer shine if the animal is immersed in water: abolition of the customary refraction from the cornea was responsible for the change.

There are ancillary anatomical features associated with each human eye. The epicanthic fold is a doubling of skin on the upper lid which gives the Mongolian race its characteristic appearance. Other races possess this fold in foetal life but it then disappears. Eyebrows are frequently alleged to keep forehead sweat from the eyes, but it is probable they are primarily associated with expression: some monkeys have white eyebrows arched in the human manner but surrounded by black hair – the effect upon sweat must be small but upon the appearance is startling. Cilia, or eye-lashes, are longer and more plentiful on the upper lid; they are continually falling out and each one of the total complement of about 200 per eye lasts from three to five months. They are usually as dark or darker than the hair, and do not whiten with age. The human eye does not possess a nictitating membrane, or third eye-lid. This useful extra, well developed in most mammals and other vertebrates and extremely blatant in the blinking of birds, could have been a protective asset: instead it is only a vestigial and immobile fragment of tissue in man.

Tears are essential to bathe the eye, and they are sterile. Each human tear gland is in the upper and outer part of each eye, and it produces tears continuously. Even though the eye is warm, constantly exposed to air throughout at least sixteen hours of the day, blinked over and generally susceptible to evaporation, it has been calculated that tear production is only from a half to two-thirds of a gram a day. Laughing, yawning, coughing, vomiting, cold wind, foreign objects, certain chemicals and crying all step up tear production, and the traditionally 'good cry' will certainly produce in excess of half a gram. All vertebrates living in air produce tears, and all humans over a few weeks old do so, but no animal other than man has yet produced convincing evidence that it can weep in response to emotional stress.

Epiphora, as scientific man has named this unique attribute (as well as a disease of excessive tears), is responsible not only for a considerable rise in lachrymal production but a considerable drop in the bactericidal property of tears. It was Alexander Fleming who investigated the property several years before the famous penicillium spore dropped in at his Paddington window. On the earlier occasion, somewhat similar in its casual serendipity, the object dropped in was a nose drip, the obscene and leaking accompaniment of a heavy cold. Where the drips dropped the bacteria on a culture plate soon died.

Soon lysozyme was isolated, the bactericidal agent responsible for their destruction that is present to a large degree in tears, and to a lesser degree in saliva and nasal mucus. Without some killing agent like lysozyme the eyes, nose and mouth would be unguarded entrances for the steady infiltration of unwelcome bacteria. The molecular structure of this natural antiseptic was worked out in 1965 by a team at the Royal Institution in London, the first enzyme ever to be so described, thereby providing an excellent opportunity for further understanding of the all-powerful enzyme world. Unfortunately, although Fleming squeezed lemon-juice into the eyes of his students, and collected the lysozyme which then rolled down their cheeks, it proved a poor remedy against disease. Generally speaking, the diseases we acquire know how to get round lysozyme. Fleming had to wait for the penicillium spore before a disease-killer was in his hands.

No invertebrates blink, but most vertebrates do so. Exceptions are some reptiles and animals living in water. According to Sir W. Stewart Duke-Elder, the ophthalmologist, true blinking does not begin in human infants until they are six months old. Whereas rodents and cats blink slowly the primates are much quicker. A human blink lasts 0·3 to 0·4 seconds, and occurs once every 2 to 10 seconds. As the eye is actually closed for 0·15 seconds of each blink, an average man, assuming a 6-second frequency between blinking, has his eyes definitely shut for nearly half an hour of each day spent awake. The blinking reflex can be halted for a time with conscious effort, and for still longer under water, but few people can resist blinking even for as long as a minute. In water the interval may be extended to five minutes.

The eyes themselves are never stationary. Not only is the head always moving slightly, causing ceaseless adjustment by the eyes, but the eyes themselves both jerk about and have a restless vibration to them. Only a very few seconds of arc are involved, but the constant activity enforces a constantly changing pattern of light upon the retina. Were it possible to fix the eye, and clamp it in one position, vision would probably cease altogether. Each fragment of retina would then be receiving an identical and fatiguing stimulation.

Reading is also a jerky business, with the fixation pauses lasting from a fifth to a third of a second before the eye hurries on to the next stopping point. Nothing can be read during the periods of movement as the speed is too rapid. A person reading out loud has his eyes ahead of his voice, and usually he can continue for some five more words if the light goes out.

Convergence is the ability to turn the eyes inwards, an ability which deteriorates with age. A child can bring both eyes to bear upon objects less than three inches from his eyes; anything nearer is seen as a double image. Duke-Elder says that such accommodation is about $5\frac{1}{2}$ ins. (14 cms.) for someone in his or her mid-thirties, and there may be rapid periods of deterioration both in the early forties and then again when he or she has really encountered old age.

Vision is subject to many defects. Night-blindness (nyctalopia) can be caused by disease, inheritance, lack of food, over-exposure to light or a deficiency of vitamin A. Day-blindness (hemeralopia) can be caused either by disease or inheritance. The former is a marked inability to adapt to conditions of modest light: a normal eye can increase its sensitivity to light by about 75,000 times within an hour of being placed in darkness. Day-blindness is an ability to see normally when light is modest, but poorly when light is good. There are also other failings, but the big three weaknesses in vision are colour-blindness, the errors of focussing, and blindness itself.

Colour-blindness Essentially there are six types of defective colour vision, of which four are common. Many forms of classification exist, and they are all based upon the fact that normal colour vision is a function of three variables (red, green and blue) and these three enable a normal person to distinguish 150 to 200 colours in the spectrum. A colour-blind person is either distinctly weak on one of the three variables, or he is having to make do with only two of the variables, or he sees every colour as a confusion of only one variable. These three types, based upon their three, two or one variables, are classified as trichromatic, dichromatic and monochromatic. Within each group, again bearing in mind that there are three possible types of failing, there are three sub-groups: a colour-blind person is either weak on one of the three variables, or lacking in one of the three variables, or having to make do with no variables at all. Hence the classification, and the words coined to describe the deficiencies. It may seem a moot point to distinguish between someone who cannot tell yellow from orange and someone else who cannot tell orange from yellow, but the differences between them are real and valid.

Men everywhere are more colour-blind than women, and people who are leading or have recently been leading primitive lives are less colour-blind than urbanised groups. The sex difference is due to the fact that most of colour-blindness is inherited in association with the X chromosome. As males possess only one X chromosome any genetic defect upon it is likely to manifest itself. Females have two X chromosomes, and a defect on one can be masked by the normality of the other. (See the chapter on inheritance for a fuller explanation.)

	Common name	Colours confused	Frequency
Trichromatic (3 variables)			
(a) Protanomalous	Red-weak	Yellow and orange	Fairly common
(b) Deuteranomalous	Green-weak	Yellow and orange	Fairly common
(c) Tritanomalous	Blue-weak	Blue and green	Rarest (possibly non-existent)
Dichromatic (2 variables)			
(a) Protanopic	Red-blind	Red, yellow and green	Fairly common
(b) Deuteranopic	Green-blind	Red, yellow and green	Commonest
(c) Tritanopic	Blue-blind	Blue and green	Very rare
Monochromatic (all one colour)			
	Total colour blindness	All colours	Very rare

The difference between primitive and other people may have been brought about by the relaxation of natural selection previously working against colour-blindness. Whatever the cause the frequency of this defect among Europeans and North American whites is greater than among Asians. Its frequency among Asians is greater than among blacks, American Indians and Australian aborigines. Eskimos have the lowest incidence in the world. In England the frequency is greatest in the south and east, in France nearest the sea and in India among Brahmins. Perhaps all these high incidence groups are furthest removed in time from direct food gathering and from the disadvantages of colour-blindness. Finally, for some unknown reason, there is a link between colour-blindness and alcoholism: more alcohol addicts are found to be colour-blind than similar but non-drinking people.

European and white American frequency of colour-blindness is about 8%. A large survey in Norway and Switzerland reported that males afflicted with red-blind vision formed 1% of the male population, with red-weak vision also 1%, with green-blind vision 5% and with green-weak vision 1·5%. The total frequency among European women is 0·5%. Among Asiatic males it is about 5%, among blacks and American Indian males about 3%, and among Eskimo men about 1%. Considering the extent of the affliction among motorised communities, and the global convention of red and green stop lights, it is curious that most licensing authorities totally disregard a driver's ability to distinguish the colours for stop and go. Even curiouser is the fact that red and green were chosen as the crucial colours either for entering an area or for not doing so, a vital distinction on roads and railways, and equally important at sea where they denote port and starboard. Perhaps natural selection is still operating and killing off a greater proportion of red and green blind travellers?

To be colour-blind is not always a disadvantage: camouflage of any kind is easier to detect, whether military or natural, and the colour-blind entomologist can often succeed in finding some concealed insect where others have failed. Everyone is colour-blind when the light is weak. In the evening an ability to discriminate red is the first to go and the other end of the spectrum is the last to go. Blue is the first colour to become perceptible in the morning.

Colour vision in animals is thought to be poorly developed. It is hard devising tests to assess colour discrimination as against luminosity differences, but it is thought that fish and amphibia cannot appreciate colour, although they often prefer an area lit with a particular hue. Some birds can perceive some colours, and many insects undoubtedly can. Mammals are generally poor at discrimination; dogs may have a slight sense, but sheep, mice and rabbits have all proved themselves disinterested in colour. Wave a green rag at a bull, and he will still behave in the traditional fashion. Wave a green rag at a monkey and he will know it from a red one. The higher primates seem to be the only animals with a colour appreciation either approaching or equalling that of mankind.

Focussing errors   The three main troubles are:

Hypermetropia. Too short an eye, or too weak a lens. Focussing is therefore behind the retina. Called long-sightedness.
Myopia. Too long an eye, or too powerful a lens. Focussing is in front of the retina. Called short-sightedness..
Astigmatism. Unevenness. The focus point for horizontal rays is not the same as for vertical rays. Usually caused by a warped cornea.

Nero is said to have looked through an emerald to magnify objects, but spectacles in the modern sense did not appear in pictures until the fourteenth century, having allegedly been invented by a monk of Pisa in 1299. Ever since then the victims of the three principal errors of focussing have had a chance of seeing more clearly through spectacles. Nevertheless these rarely help to preserve vision, as is often believed. In fact, notably in long-sighted people, they may commit the patient to an increasing reliance upon them.

Myopia is named after two Greek words meaning to close the eye because short-sighted people have a habit of half-closing their eyes. This causes a smaller opening available to the light waves, thus better vision. Contact lenses, which do away with all of the spectacles' external attachment, leaving only the vital refractive material, have by no means proved as popular an alternative to spectacles as might have been supposed when they were new. Much more recently the idea has been put into practice of changing the focussing power of the cornea by grinding away some of its substance. For short-sighted people in particular, whose corneas are too greatly curved causing too great a bending of the light waves, the corneas have been effectively changed: they have been flattened to some degree and normal vision has resulted. For long-sighted people, traditionally found among the older section of the community, such an operation is more complex: it necessitates increasing the curvature of the cornea. However much the cornea may be changed it is a static change: although the cornea does 60% of the focussing it is the lens which has to adjust itself to alter the combined focussing power of both cornea and lens.

The need to wear glasses (or contact lenses) rises with age. In one British survey of over 20,000 people the 16–24 age group possessed 75% without such a need, the 25–34 group 64%, the 35–44's 57%, the 45–64's 14%, and the 65 and overs 4%. In other words faulty vision tends to rise along with all the other disadvantages of reaching middle age. In this 1982 survey the percentages wearing contact lenses for the same age groups (and in the same order) were 3, 5, 3, 1 and 0. In short they formed 13% of those with a need for some sort of correction in the youngest age-group, 16% in the next group, 7% in the 35–44's, 1% in the 45–64's, and 0% in the elderly. The wish to be without glasses, either for appearance or for convenience, therefore wanes with age. More women than men wear glasses in every age-group, and roughly twice as many women wear contact lenses as men do. However, their enthusiasm for doing so also diminishes with advancing years.

Dark glasses are predominantly a feature of the twentieth century. Practically no one needs them in ordinary conditions, but sales are phenomenal. In Britain over 3,000,000 pairs are bought annually, and in the United States enough dark spectacles are sold to give everyone a new pair every other year. Presumably most pairs are bought without any intention of protecting the eyes, but of enlarging or concealing the ego. For those who wear very dark glasses, some being far darker than others, and who wear them steadily, whether at night or indoors, the eyes could well become too sensitive to ordinary light. In which case the wearer will have provided himself with one of the few valid reasons for wearing dark glasses.

Blindness In Britain the National Assistance Act of 1948 defines a blind person as someone 'unable to perform any work for which eyesight is essential'. The Royal National Institute for the Blind works on a figure of 130,000 blind people in the United Kingdom, or one person in every 420. The figure has been rising in recent years, partly (or mainly) because of the increasing number of old people. Over half of the registered blind have some degree of useful vision; the actual number of people without any useful vision is nearer one in a thousand of the population. Total blindness, with no perception of light whatsoever, only affects less than 4% of those registered as blind, namely 5,000 people. Most of those on the blind register are elderly; only a quarter are less than sixty years old.

At the other end of life only ten of the totally blind are less than one year old, and only 2,000 are less than fifteen. Females outnumber males by 3:2, with a contributory cause to this sex difference being the fact that women live longer than men and a gross deterioration of vision is a frequent companion of old age. Unfortunately, although cataract, for example, is less a cause of blindness than it used to be, the numbers of blind people are not being reduced. Medical science is advancing, but it is also permitting more people to live to an age likely to be accompanied by poor eyesight. Unfortunately no one predicts any marked decline in the length of the blind register in the near future.

In the poorer and tropical parts of the world the situation is far worse. There is no register of the world's blind population, and estimates fluctuate much like estimates of global malnutrition, but onchocerciasis (caused by a fly-borne worm parasite), trachoma (a form of conjunctivitis, discovered by Peking scientists in 1957 to be caused by a virus), smallpox and malnutrition in the young all add millions of blind people to the current total. The World Health Organisation has produced estimates that the nematode worm of onchocerciasis has affected 200 million to 300 million people, with perhaps 10% blinded as a result of the affliction. A recent report from India gave a total of 5,000,000 blind people in India alone, and added that this number was about 25% of the world's total. So, whether there are 20,000,000 blind people in the world or even more, as the W.H.O. figures imply, this means that at least one in 200 of the world's citizens is without competent sight.

Although there is no immediate hope of reducing Britain's blind population markedly, due to so many of its causes being linked with the general problem of old age, the worm of onchocerciasis and the virus of trachoma are far less intractable. Also, quite apart from the blindness resulting from these two, they cause plenty of misery long before they cause loss of sight. Each has been accused of affecting 300 million people, making a joint total of nearly a seventh of the world.

Hearing   The prime purpose of the ear is not that of hearing. Without the ability to detect sound waves a man is undoubtedly in a silent world of his own. Without the ability either to assess the position of gravity or the rotational movements of his head a man would be without equilibrium, and in a disastrously topsy-turvy world. To some extent, as in all higher animals, the eyes could compensate for the loss of balance and orientation, and they always assist anyone's appreciation of his situation; but such a full-time addition to their major role of vision would be entirely abnormal. A man can walk fairly well in total darkness; he would walk far less well in broad daylight if his ears were destroyed. The so-called auditory organs of the earliest vertebrates were probably quite unconcerned with hearing; they were more probably totally concerned with the detection of changes in spatial orientation. The fish ear is without an external ear, such as the flattened pinna of human beings, and without a middle ear with its membrane and bony ossicles: only the inner ear exists. It is this basic ear structure of all vertebrates, including man, that has more than one job to do.

The inner ear consists of three quite distinct sensory systems. There is a relationship between them in that they are all dependent upon small mechanical forces, but this does not alter their entirely separate functions. Firstly there are the three semicircular canals, each one set in a different plane, and all three acting as a receptor system for detecting rotational movements of the head. The acceleration and deceleration of fluid within these canals gives the ability to record head movement. Secondly, there are the saccule and the utricle. A combination of hairs and chalky particles called otoliths inside these containers enables the position of gravity to be detected: a person, or a fish, can therefore tell which way up he is in relation to the Earth. Both these two systems, the fluid and the chalk, the canals and the containers, are termed the vestibular apparatus, and without them any vertebrate would be far more incapable than one merely deaf to sound waves. Tortoises are without hearing and have done pretty well.

The third function of the inner ear is that of hearing. The organ of Corti, which lies within the $2\frac{3}{4}$ turns of the snail-shaped cochlea, is the receptor system for sound waves. It can measure both the intensity and the frequency of sound waves oscillating between 20 cycles per second and 20,000 cycles per second. Unlike the two organs of the vestibular apparatus, which are self-sufficient, the hearing part of the inner ear has to work in conjunction with devices for the amplification of sound waves; hence the mechanical complexities of the middle ear, and also the larger but less crucial shapes of the outer ear.

The squashed-up shape of the human outer ear is a poor remnant of the mobile catchment cups of many other mammals. Probably the outer ear has more to do with the location of sounds than with mere amplification. Even a human being can tell to within three degrees the direction from which a sound is coming. An owl, blessed with the possession of one ear slightly forward of the other, can locate a sound to within one degree while still looking straight at it. A human being tries to achieve the same accuracy by cocking his head on one side, and a blind man can demonstrate many human abilities previously only latent. For instance, if placed a few feet away from a disc, blind men have shown their skill at assessing any changes in its area: the minimum changes they could detect, just by making noises and listening like sonar for alteration in echo, varied from only 7% to 23%.

The middle ear is intriguing partly because of the straightforward way in which it achieves its amplification, and partly because of its curious origins. Throughout evolution the process of making do with inherited characteristics and transforming them into novel requirements has been continuous. Stabiliser fins became legs, and two of these legs then became arms. Similarly the air bladder of fishes, a buoyancy device, became the lungs of terrestrial animals, a respiratory device. In just the same fashion some gill bars of the fishes (used in breathing) became jaw articulation elements in the reptiles (used in eating) and then became the auditory ossicles of the middle ear (used in hearing).

Hearing occurs when sound waves, gathered in to a very slight extent by the external ear, enter the ear and cause the tympanic membrane to vibrate. (Anyone unfortunate enough to have some modest and scarcely visible insect invade his ear through the protective wax and hairs will be astounded by the deafening heavy-footed tread as it steps lightly on to the membrane. Sounding much like an elephant on corrugated iron it gives a deafening idea of the delicacy of the system.) The membrane's oscillations are directly transmitted to three bones, lying one after the other, named the malleus, the incus and the stapes because their shapes bear fleeting resemblances to a hammer, an anvil (both particularly fleeting) and a stirrup. They are the principal contents of the middle ear cavity, an air-filled hole about a third of an inch (8 mms.) wide and a sixth of an inch (4 mms.) deep.

It is this pocket of air that can cause trouble in flying; its pressure has to be constantly adjusted to the outside air pressure. Normally this is done effortlessly via the $1\frac{1}{2}$ ins. (38 mms.) Eustachian tube which leads from the cavity to the upper part of the throat behind the nose; but, when the nose is congealed with the mucus of a heavy cold, or when an aircraft changes height too rapidly for swift personal adjustment, the result can be either curious squeaks as pressure levels are sorted out or temporary deafness. (Babies are particularly bad at this pressure adjustment and, having slept soundly to the relief of all throughout the flight, will awake and yell the moment descent begins.) Within this cavity the middle ear's three ossicles, by their own jostling, pass on the tympanic membrane's vibrations to a far smaller membrane which effectively seals off the liquid-filled compartment of the inner ear. It is the organ of Corti within this inner ear that transforms the vibrations into nerve impulses, and these are interpreted by the brain as sound.

The human ear's ability to detect frequencies from 20 to 20,000 cycles per second is beaten at the lower end by, for example, the grasshopper and at the upper end by, for example, the dog. Whereas middle C is 256 cycles per second, and top C on a piano is four octaves higher, therefore 4,096 cycles per second, the sound at 20,000 cycles per second is heard as little more than a hiss. The fundamental tone of the human larynx is of course much less: about 100 cycles per second with men, and 150 cycles per second with women, but speech makes use of high-pitched hisses as well; the lack of these, as in all telephone conversations, can make speech less intelligible. (Try saying 't' and 'd' to the person at the other end, and he or she will have difficulty in detecting the difference, apart from wondering what on earth you are on about.)

The sensitivity of a good ear is remarkable. It can pick up sounds which only deflect the tympanic membrane by 0·00000001 millimetres. (Small wonder an insect is so audible when treating that membrane as an entomological trampoline.) Also a good ear can assess about 1,600 different frequencies between the highest and the lowest, and about 350 different intensities between the quietest and the loudest. Senile deafness affects the highest tones first, and bird song – particularly from the smaller species – can be quickly lost after the age of sixty.

Just as the organ for seeing light can be damaged by too intense a light, the human ear can be damaged by noise. The middle ear possesses muscles which dampen the membrane's excessive vibrations, but either short bursts of very loud noise or prolonged exposure to noise well below the painful level can cause permanent loss of hearing. A survey before and after New Year's eve in Denmark gave audiologists reason to believe that the impaired hearing of half of all Danish children can be traced to the firework festivities of December 31st. Riveting, drop forges, pneumatic drills, and big guns have long been known to be capable of causing deafness and the jet engine is now the loudest industrial noise; but less fearsome noises are perfectly capable of causing damage. Some visiting ear specialists in the Sudan noted that the ears of remote Sudanese groups, accustomed to nothing louder than human speech, suffered if they were subjected to sounds which many of us tolerate throughout our lives.

Although noise can also be irritating, and painfully aggravating to the system as a whole, the only damage which can definitely be confirmed as a result of it (according to F. I. Catlin who reviewed the literature) is hearing loss. Intensity of noise is measured in decibels. The bel is named after Alexander Graham Bell, who gave the world the telephone but unaccountably only one 'l' to the unit of loudness, and the threshold of sound when it is just audible is called 0 bels. A whisper is about 1,000 times more powerful: therefore, as 1,000 is 10^3 and as bels are measured logarithmically, a whisper has the loudness of 3 bels. Conversation is 1,000 times louder than a whisper, namely 6 bels. A loud shout is 9 bels, and noise becomes unpleasant at 12 bels, causing pain when of greater intensity. As bels are too large a unit of measurement the decibel is used instead, with each decibel equal to a tenth of a bel. Therefore a whisper is 30 decibels, and conversation is 60.

Deafness There are two main types: conductive deafness, caused by any trouble to the conducting apparatus (as, for example, a punctured ear-drum) and perceptive deafness, caused by any trouble to the cochlea or its nerve. Hearing is made adequate with the former if sounds are made louder. With perceptive deafness this may be true up to a point, but beyond that point the greater the amplification the greater the confusion – notably with speech. It was Thomas Willis, the sixteenth-century physician, who drew attention to the advantage of loudness for conductive deafness. He told of a deaf woman who could only hear speech if a drum was beaten simultaneously, 'wherefore her husband kept a drummer on purpose for his servant, that by that means he might have converse with his wife'.

The first hearing aids were ear trumpets, giving a gain of 20 decibels. Electrical aids arrived in the last century, and they were radically improved first by the thermionic valve and later by the transistor. All such aids fail to achieve the selective ability of the human ear which tends to hear what it wishes to hear: without selection the environment is incredibly noisy and raucous.

Deafness is not a notifiable affliction in Britain. Therefore no register exists which accurately lists the numbers of all deaf people. A Survey of Sickness was carried out after the last war, and before the National Health Service was introduced, to assess the extent of numerous problems, including deafness. Almost two million people were thought to be deaf, but in markedly different degrees. The majority, 1,650,000, were classified as hard of hearing, although half of this majority had difficulty in hearing normal speech unless they had a hearing aid (which is frequently referred to as a deaf-aid despite, as the deaf speedily assert, their deafness flourishing without need of assistance). The hard core minority included 70,000 deaf to all ordinary speech (although some could hear abnormal speech designed to assist them, such as the exaggerated movements helpful to the lip reader), plus 30,000 who were totally deaf, and 15,000 who were both deaf and dumb. Therefore a total of 115,000 were grossly deaf in Britain at that time, and another fourteen times as many suffered from deafness, but less severely. Today's numbers are similar, with about 2 million British people (or 3·6%) having a 'moderate' hearing loss of 45 decibels or over. The figure customarily quoted for the United States is of 4,000,000 Americans suffering from hearing loss in both ears plus another 2,500,000 with loss in one ear.

In 1965 a new machine was announced in Canada capable of testing children for deafness within a week of birth. The development of such a device is the culmination of mounting pressure to take note of deafness as early as possible. After all, a baby starts taking note of sounds from the day it is born, and there is no point in delay. A deaf child has to be fitted with a hearing aid before he is a year old if he is ever to speak properly, and children only three months old have been fitted with them.

The custom in former days was to disregard the deaf child's right to the power of speech, and permit him or her to become a deaf-mute. Suzanne de Parrel, the French expert on speech disorders, has written that 'the use today of the term deaf-mute revolts the conscience. It should be banished from the vocabularies of the medical and teaching professions. True muteness is a symptom of aphasia, damage to the brain. ... The deaf are not mute. Also very few deaf people are totally devoid of residual hearing'. She has recommended that exercises in remedial education for the deaf child should begin at eight months. Children in general begin to speak later, but they have been listening, and making noises, and preparing for speech long beforehand. A generalisation, with exceptions (Einstein, who did not speak until aged four, being one of them), is that a child not speaking by the age of two is either backward or deaf. Most certainly a child with late speech should be tested for deafness, as high frequency deafness on its own is a very important cause of late speech. The ability to hear bangs or the radio or footsteps does not preclude forms of deafness crucial to the acquisition of proper speech.

The Canadian machine discovered that new-born infants initially have a hearing loss of 40 decibels (or roughly the same as many old people experience at the other end of life). As with swimmers who can suffer temporary deafness from the presence of water in their ears, the formerly aquatic neonate has fluid in his auditory meatus. By ten or twelve days this has all gone. Young babies can react alarmingly to quite modest sounds, behaving as if hand grenades rather than milk bottles are falling nearby. Gradually this reflex hearing becomes inhibited, and by five months an infant may not respond to quite a loud bang, particularly if it is familiar. Later he will start smiling when he does hear a bang or a loud noise if he can associate that sound with pleasures like food or company.

It took Britain a long time before it cared for the deaf. The first school for the deaf and dumb, many of whom had been traditionally assumed to be idiots as well, was established in Scotland in 1760, and London followed suit in 1792. Only in 1884 was a College for Teachers of the Deaf founded. Some causes of congenital deafness are known, such as German measles, syphilis, and Rhesus incompatibility; most of these children born deaf have a failure in their perceiving rather than their conducting apparatus. Some causes of deafness after birth, apart from the human and environmental cuffs to the ear, are diseases like meningitis, mumps, syphilis, and scarlet fever. Senile deafness, which is responsible for such a rise of deafness with age, is due to an atrophy of the nerve fibres.

Since the early 1970's considerable attention has been paid to the idea of a bionic ear. The House Ear Institute in Los Angeles, for instance, has inserted (or implanted) over 300 devices within the cochleas of deaf patients during the past ten years. Essentially these transmit current to the auditory nerve and bypass each (damaged or malformed) cochlea. The implants have steadily become more sophisticated, and the operation costs up to $10,000. Claims are being made that modern methods should be able to restore the hearing of 70% of the 500,000 Americans currently unable to benefit from hearing aids. Many experts are less sanguine. Donald Eddington, of the Massachusetts Institute of Technology, said in 1984 that even the most advanced multi-channel devices implanted within the most suitable patients are only an aid to visual communication, such as lip-reading. Nevertheless, even that is a blessing. Implants can reduce the feeling of isolation that many deaf people experience.

Finally there is malingering deafness. The art of pretending not to hear can be practised with considerable success, but it can also be confounded by an equally cunning procedure. The malingerer is asked to read a piece of prose. His words are picked up by a microphone and then fed back at him through a loudspeaker one-tenth of a second later. Much like some rapid echo this near-instant repetition of one's own words will transform the normal person into a confused and stammering dunderhead. The deaf person, on the other hand, unhearing and unconfused, will be able to read happily on.

Smell The human body possesses strong powers for evoking phrases of wonderment and awe, as if it had no visible faults, as if no aged, acne-ridden, asthmatic with myopia and a high blood-pressure had ever been watched running for a bus. Clearly it does have imperfections, and it does break down; but some of its systems are outstandingly skilful. Such a system is smell. No other receptor has the sensitivity of the olfactory sense. It is exceptionally precise, and accurate; and a smell is extremely memorable. Although they are so readily identifiable, and each one of us can distinguish thousands of different odours, the problem of passing on information about any single smell to someone ignorant of that particular smell is insuperable. Burning rubber, wet dog, new earth, old books, ripe cheese, fresh rain, cut grass – they are all unique, entirely distinctive and quite indescribable except with vague terms like acrid and pungent which also embrace a few hundred other odours. Describing colour to a blind man is one impossibility: describing a new odour, say whale breath or palm wine, even to someone with a perfect sense of smell is another.

Amazingly, the ability to smell is conveyed by very small organs that are not even in the mainstream of inspired air. They are in a blind alley, and as soon as they detect the whisper of an interesting smell the nose is wrinkled into a sniff. These olfactory receptors are cells with long hair-like cilia attached to them. The cilia do not wave about in the air like a field of corn; instead they are embedded in a mucus layer which rests on the cells, and so they are more like reeds beneath the surface of a lake. No one knows how the cilia detect smells, but all odours have to be soluble in the mucus before they can be detected. Somehow these submerged cilia have the power of sending differing impulses to the brain so that a certain assortment of impulses is instantly (i.e. within a second) recognisable as rubber or dog, grass or mouldy cheese.

As a species, mankind is not good at detecting smells. This is partly because a dog, for example, has its olfactory receptors more sensibly placed with regard to the incoming air flow, partly because its receptors can take note of a wide variety of smells, and partly because its brain is better equipped to record and interpret olfactory stimuli. Much of mankind's success in the world can be attributed, loosely speaking, to his disregard for smells and his development of that part of the brain for other purposes. Consequently, from our less efficient point of view, a dog's powers of smell are near miraculous. With some odorous sources a dog's nose is over one million times more sensitive than a man's. Acetic acid, for instance, can be detected by man if there are five hundred thousand million molecules of it in a cubic metre of air. A dog needs only 200,000 molecules in a cubic metre for the smell to be detectable. Man's ability is incredible; a dog's is over a million times more so.

Canine abilities at following their human masters have been scientifically investigated. Two minutes are sufficient time for the finger tips alone to touch a stick, or a few seconds of a whole grasp, if that stick is to be positively identified by a dog with a particular person. It does not matter how many other people have handled the stick before or after the dog's chosen master.

There have been other experiments. If eleven men set off in single file, with the master leading and everyone else treading in his footsteps, and if that single file divides to form two files with the master still leading one of them, a good dog will have no difficulty in following the right group, whether the trail is over grass or earth, asphalt or rock. Nevertheless a dog can be muddled by identical twins. Even if the twins are married, living apart, eating differently and doing different jobs, a good dog can confuse the two scents. It will tend to pick up one glove, for example, out of a whole pile of gloves provided that it belongs to one twin even though the other twin may be its master. New Scotland Yard reported that police dogs had never been concerned in the tracking or arrest of identical twins, but research had proved that the odour of twins could be distinguished by a well trained dog. If the experiment is so arranged that the dog knows it has definitely to distinguish between two gloves from the twins it will pick out the correct one.

A science which cannot be neatly classified is always vexing, and this encourages many to try defining its categories. No one has been able to delineate a few primary odours equivalent to the three primary colours, but H. Zwaardemaker drew up a list of smells which did not appear to influence each other confusingly. He felt there were nine distinct groups, and he therefore postulated the presence of nine types of receptor. No one accepts the list as being much more than an intelligent attack upon the problem, but it is still quoted even though it was first published in the nineteenth century. Its nine divisions are:

Ethereal (such as fruits)	Empyreumatic (burnt odours)
Aromatic (camphor, almonds)	Caprillic (cheese, fat, sweat)
Fragrant (flowers)	Repulsive (bed-bug, deadly nightshade)
Ambrosial (musk)	Nauseating (rotting meat or plants, faeces)
Alliaceous (garlic, sulphur, chlorine)	

Nowadays, although the existence of primary odours is still elusive, and no definitive classification can be made, it is suspected that each olfactory organ has more than nine kinds of receptor. The human nose is thought to have at least fourteen kinds of receptor, and the dog is thought to have nearer thirty. The reason for these estimates is that the nose is believed to be working (much like a computer) on a yes–no basis: a particular smell will either trigger or fail to trigger an impulse in a particular receptor. Therefore a group of, say, five receptors will perhaps record yes-yes-no-yes-no in the presence of a certain odour, and this batch of stimuli will then be passed to the brain. Another odour will cause another message to the brain, such as no-no-no-no-yes. As mankind can detect thousands of odours, and more than 10,000 with practice, and as each nerve fibre is limited in the number of impulses it can transmit per second, and as a smell is discernible in less than a second, the problem of assessing the actual number of receptor types in the human nose is considerable. Perhaps fourteen is the answer. Probably it is higher. At present no one knows. It is also unknown why some substances smell, and others do not. So many things smell: why not the rest?

The act of smelling is carried out at the molecular level. It has been calculated that eight molecules can trigger off an impulse in a nerve ending, and about forty nerve endings have to be triggered before a smell can be detected. When observed from this molecular standpoint the procedure of smelling, with its remarkably low threshold levels, does become more comprehensible. For instance, take the ability of a dog at tracking a man's invisible footprints. A man produces over a pint of sweat a day, but the bulk of it is ordinary water, and certainly only a small fraction of sweat passes through the sole of each shoe; yet, at the molecular level, the amount of odorous substance passing through each sole at each step is enormous. Butyric acid is a smelly constituent of sweat and, following suitable assumptions, it has been estimated that 250,000,000,000 molecules of this one substance are deposited in each footprint. For a dog a millionth of that quantity would be detectable; hence the dog's ability to follow an old scent when so many of the original molecules have evaporated. Even a man can follow a fresh human foot trail if he is prepared to get on his knees, and if the trail has been made – invisibly – over blotting paper on a firm floor.

An animal's ability to smell other animals is much enhanced by the presence of smell-creating, or scent-producing, glands. The elephant has a scent gland on its forehead, the rhinoceros on its feet, the rabbit near its anus, a lemur on its arms, the marsupials (generally) on their necks. Mankind has no such glands, but modified sweat glands produce characteristic smells both in the armpit and in the genital area.

It could be that the racial groups of mankind also have specific group odours, as well as the individual. A Japanese anatomist visiting Europe said he found the body odour 'objectionable, strong, rancid, sweetish, sometimes bitter'. Later on, happily, he found it sexually stimulating. Food and mode of living rather than racial inheritance may have caused this European smell, but both are probably important. The smell of an animal is important at the social level, at the sexual level, and at the individual level. Not only can dogs recognise their masters solely by smell, but mice will fail to maintain their former pregnancy if re-mated by mice of a different odour. The experiments with human twins show that everyone has a unique smell, genetically determined. Presumably the ability to smell others, and other things, is also genetically determined. A test conducted at the University of Pennsylvania showed that both male and female judges could guess the sex of donors (via a tube of exhaled breath) on 95% of occasions. Both sexes found it easier to identify the opposite sex, and male odours were generally thought to be more intense and less pleasant than female odours. Albinos, already with more than a fair share of disadvantages, have a weak sense of smell. And so do smokers.

Taste   Taste and smell are often spoken of in the same breath; but incorrectly, for there are few similarities. However, to twist the phrase, an actual breath may stimulate both types of sense organ because smell and taste are associated in most minds. A common cold is said to destroy the taste of things, but it only blocks off much of the attendant smell; the taste remains. The true seeker after taste truth should therefore consume everything with a heavy cold and his nose quite closed.

There are four fundamental or primary sensations of taste, namely sweet, bitter, sour (or acid) and salt. Sometimes alkaline and metallic are added, but it is generally felt that four types only are involved. (In recent years this ancient fact has been disputed, and it is currently being suggested that all tastes are nothing but a variation of just two sensations, bitter and sweet.) All tastes are either a combination of effects upon these four (or two) types or upon these four (or two) plus the various ordinary nerve endings in the mouth. A curry has a taste, but it also has a burning effect upon the normal nerve endings; it would not taste like curry but it would burn like curry if placed elsewhere, for instance in the sensitive areas around each eye.

The four types of taste receptor are located in taste buds, and these are mainly on the front, back and sides of the tongue's upper surface. Sweetness and saltness predominate at the tongue's tip, sourness at the sides, and bitterness at the back. The centre of the tongue is virtually without the ability to taste. Not all taste buds are on the tongue: some are on the palate, pharynx and tonsils. Nothing solid can be tasted unless part of it is dissolved; and without saliva on the tongue to do the dissolving a solid substance is wholly without taste.

As with smell no one knows how the taste receptors collect their information. And as with smell different substances have quite different thresholds, with the general threshold for taste being much less sensitive than for smell. Whereas bitterness at the back of the tongue, with some of its taste-buds raised on quite long nodules, can be detected at a dilution of one in 2,000,000, sourness needs a dilution of one in 130,000, saltness of one in 400, and sweetness of one in 200. Odours, by comparison, need dilutions only of one part in a thousand million or more.

Mankind has about 3,000 taste-buds. The crude and insensitive pig has about twice as many. The teleost fishes have their taste-buds extending the length of their bodies as well as in their mouths. Taste and smell sensory organs, or an indistinguishable blending of the two, certainly arose in evolution earlier than hearing or sight, but the present purpose of taste is much less positive than that of smell, and much, much less of the brain is concerned with it. Some harmful plants and foods can be detected by their taste, and therefore some selective advantage exists; but taste as an ability is completely outclassed by the sense of smell.

Touch Sometimes called the fifth sense the sensation of touch is five senses on its own – touch, pressure, pain, heat and cold. Unfortunately, having said that and having delineated five functions, the five can then become blurred and cease to be five neat, separate, distinct types of receptor. Sometimes they seem to be, and the different receptors have been described – corpuscles of Ruffini for detecting warmth, end-bulbs of Krause for detecting cold, Meissner's corpuscles for touch, and Pacinian corpuscles for pressure. At other times they seem to be far less definite as the external ear contains no such receptor endings for the nerve fibres; yet the ear is perfectly capable of detecting temperature, touch and pressure. Presumably, as in most discussions, both sides are partly correct – sometimes the nerve fibres end in special end-organs, sometimes they do not; and sometimes a nerve ending is sensitive to more than one type of sensation and sometimes it is not. By no means, as with the entire story of sense organs, has anything approaching a full understanding of the mechanism of the sense of touch yet been worked out.

Taking the body as a whole there are many pain endings or receptors, there are fewer touch endings, far fewer heat receptors, and fewest cold receptors. Proportional to these numbers is the ability to describe the locality of a particular sensation; the feeling of cold or warmth is much less precise in any area than either touch or pain. As is well known the various parts of the body are also unequal for any given sensation. The tip of the penis, the clitoris, the tongue, lips and finger-tips are all highly sensitive to touch. The hairy parts of the body, which include almost everywhere except the soles and the palms, are made more sensitive by the leverage of a touched hair. The touch receptors grouped around the base of each hair are efficient at detecting hair movement, and hairs are moved by the gentlest of forces. In animals the whiskery vibrissae attached to most mammalian snouts are extensions of this ability.

The simplest way of proving the closeness of touch receptors in different areas is to measure two-point discrimination: how far apart do the two stimuli (say from pencil points) have to be before two separate stimuli are in fact felt? In the middle of the back the distance is $2\frac{1}{2}$ ins. (63 mms.), on the forearm it is $1\frac{1}{2}$ ins. (38 mms.), on the hand's palm it is $\frac{1}{2}$ in. (13 mms.), on the tip of the nose $\frac{1}{4}$ in. (6 mms.), on the finger tip $\frac{1}{10}$ in. (2·5 mms.), and on the tongue's tip 0·025 of an inch (0·6 mms.). Similarly any minute slit-shaped wound to the tongue seems to possess the dimensions of a canyon.

Pain, the sense of injury, the warning system that damage is in progress, is a mixed blessing. Not only does it sometimes arise when damage is not in progress, but it can entirely fail to give warning of, for instance, a lethal tumour. There is also the very rare congenital absence of pain which gives to a few the mixed blessing of having no toothache but no warning of decaying teeth. The usefulness of pain is also limited by the fact that it does not correctly indicate the severity or amount of damage: a pin-prick gives a ludicrously impressive quantity of pain. A headache may either be associated with physical damage or not.

There are, according to C. A. Keele and others, three layers in the skin from which quite different pain sensations can be evoked: the epidermis produces itch, the dermis a sharp pain, and the deep dermis, or lowest layer, an aching pain. The last two can be distinguished by personal efforts with a pin. Provided some masochistic force is put into the experiment there will be a sharp stab of pain followed by a deeper and prolonged throbbing. Itching is quite different from both. The successful itching powder of schooldays is made up of the spicules of cowhage, *Mucuna pruriens*, and they are extremely adept at inducing the curious and apparently valueless tingling. Whereas pain is not always beneficial, the 'unpleasant cutaneous sensation which provokes the desire to scratch' – as S. Rothman defined an itch, seems markedly less so.

A girl called Rosa Kuleshova achieved considerable distinction in the early 1960's for her reported ability to see with her finger-tips. For a long time, first in her home town of Nizhni-Tagil and then at Moscow, she astounded groups of investigators by her ability not only to see but to read when blindfold and with her finger-tips running over the type. Here, according to Dr Isaac

Goldberg, a Russian psychologist, was a modern exponent of a feat which had repeatedly been achieved since first reported in the eighteenth century. Many other girls then did as well or better than Rosa: eyeless vision was all the rage. Some girls said they could read in the dark, or see a picture in the middle of a book.

Then, suddenly and sadly, the whole tower came toppling down. *Literaturnaya Gazeta* reported in 1965 that the skilful readers had merely been doubly clever with their own two eyes either before or after being blindfolded. It was a shame, for undoubtedly there are many latent and unused abilities in finger-tips. They could, it is frequently assumed, be trained to detect colours by the differing warmth they radiate, and various claims at doing this have been made, or take more subtle note of textures, or read big print as if it were braille; but it is asking too much to consider them capable of reading something in the middle of a book.

Sixth sense It is remarkably easy to understand how it was first postulated that a sixth sense might exist because so many animals demonstrate abilities which seem blatantly beyond the range of the five ordinary senses. A hawk, scarcely visible to our eyes, is looking for and seeing brown mice on the brown earth. An owl does the same trick at night-time. Bats fly listening for their squeaks, and miss cotton threads hung in their paths. Seals return from a 6,000-mile Pacific migration, never once in sight of land, to the small speck of, for example, the Pribilof islands. Birds fly to Africa and return to the same eave. Huge flocks of birds manoeuvre with guard-like precision. Salmon find the same river. A dangerous scent, days old, can cause animals to stop as if they have run into a brick wall. Birds navigate over the ocean by night. Bees hurry off to a new source of honey.

In general, as is reasonable, mankind is most astounded at abilities demonstrably beyond him. He is also amazed when the long arm of coincidence strikes home, and when some talked of person chooses that very moment to enter the room. His traditional beliefs in the supernatural foster a desire also to believe in inexplicable events, in telepathy, mysticism, the paranormal, the psychic. Moreover the animals have demonstrated not only a brilliant exploitation of the five senses but of abilities quite untapped, or seemingly so, by us and our five. The horns of snails retract from ionising radiation. Rats seem to smell it. Both snails and worms are affected by magnetic fields. Paramecium swim along mild electric currents. Other single-celled animals line themselves up at right angles to electromagnetic fields. And some puzzling things happen to man, frequently linked to the supposition that minds can influence other minds without going through the normal channels.

How telepathy could possibly have originated among neurones customarily dedicated to transmitting impulses along nervous pathways is entirely puzzling, but possibly entirely irrelevant. The first problem of extra-sensory perception is the need to prove both that perception is occurring (as against chance, wishful thinking, etc.) and that it is extra-sensory. The present state of ignorance about so many aspects of the body in general, which include countless unknowns about the senses in particular, make it hard for anyone to be totally dogmatic that all human perception must come through the five senses. Such evidence as does exist for other forms of sensation weakens the dogmatic case still further. In the meantime E.S.P. workers gain and lose ground steadily, with the entrenched prejudices on either side swaying much as the front line swayed in the first world war. The biologist W. H. Thorpe made a cool summing up when he wrote: 'It does seem to me that [the evidence from psychical research] has shown that it is possible for separate minds to influence one another in a manner which is totally inexplicable for any present-day theory of the physiology of mind, and is perhaps in principle inexplicable by science as we know it.'

31. The Skin

Dermis and epidermis/Skin patterns and prints/Sunbathing/
Tattooing/Beauty/Hair/Lanugo and vellus/Bristles and soap/
Teeth, deciduous and permanent/Decay/Cleaning and fluoride

The only unprotected tissue which has the living body on one side
and the outside world on the other is the skin. Taken as a whole it
is the human body's largest organ; it is enormously versatile; it
keeps out foreign agents; it keeps in body fluids; it shields against
harmful rays, and takes steps to improve its shielding should the
rays increase – as in sunbathing; it is associated with body cooling
and with heat conservation; it helps to regulate blood pressure; it
contains the sense of touch; it forms an individual's shape; it is the
main organ of sexual attraction, linked as it is with shape and
touch; it is constantly and efficiently renewing itself; it is
occasionally and less efficiently having to repair itself when
wounds occur; it can differentiate itself into the structures of hair,
spines, horns, hooves, claws, scales, feathers; and it lasts fairly
well throughout life, although it loses elasticity to a marked
degree in later years.

Essentially it is a two-layered system made up of the dermis and
epidermis. The outer layer is epidermis, and all the outer cells of
this layer are dead. Living cells cannot survive exposure to air;
and, however beautiful a living mate may appear to be, he or she
is externally a vista of quite dead cells. The death of each
epidermal cell is not caused by the external air but is a deliberate
act of self-immolation by each cell. As it is forced nearer and
nearer the surface by steady replacement from beneath, each cell
produces keratin within itself, a horny substance that is also the
principal constituent of nail and hair. By the time the cell reaches
the surface it is dead, packed with fibrous keratin, and ready for its
share of defensive action. It will soon flake off to be lost as yet
another fragment of dried skin.

The dermis lies beneath the self-sacrificing epidermis, and is
only seen when the skin is cut. As the dermis contains both nerves
and blood-vessels any small cut produces pain and blood. The
dermis also contains fat cells, the roots of hair, erector muscles for
each hair, sweat glands, sebaceous glands and much loose
connective tissue. A woman has just as many hair follicles as a
man, although less conspicuously, and each human being has

about as many as his more visibly hirsute primate relatives, such as the chimpanzee and gorilla. Melanocytes, the pigment-producing cells, are equally scattered in the skin of both white and black people, with the difference in colour being the productive capacities of each melanocyte rather than their number.

There is a predictability about the amount of skin which can be lost, as in burns. Above a certain percentage of the total body area, and everyone will die; below a certain percentage, and no one will die. Age is highly relevant to these predictions. Until the age of thirty it is generally reckoned that death is inevitable if over 75% of the skin is burnt. Below 22% death is improbable, with the chances of survival gradually deteriorating as the area increases from 22% up to 75%. For the middle-aged (45–49) the two percentages, probable survival and probable death, have dropped to 12% and 58%. For the aged (80–84) they have dropped even further, with death being probable above 23% and survival never a certainty: even a 2% burn has a one in five chance of causing death to the very old. The skin area injured can be rapidly assessed by the law of 9's: each leg is 18% of the whole (9% for the leg's front, 9% for its back), each arm is 9%, the head is 9%, the front torso and back torso are each 18%, and the final 1% is, for convenience, attributed to the genitals.

Prints   It is the dermis and its fibres that principally determine the pattern of skin, the entirely distinctive assortment of shapes, ridges, loops and whorls laid down for life by the third or fourth month of foetal existence. The uniqueness of finger-prints was first shown by the anatomist J. E. Purkinje in 1823 (the same individual whose name is immortalised in the Purkinje cells of the cerebellum, the Purkinje phenomenon concerned with low light intensities, and the Purkinje tissue of the heart, etc.), and no authenticated case of indistinguishable finger patterns in different people has yet been found, although identical twins come very close. The Chinese, who have a habit of beating Western records by ludicrous margins of a millennium or two, were using finger-prints as legal identification in AD 700.

The four main types of finger pattern are the arch (which resembles strata beneath a hill), the loop (which normally has the rounded or sealed part of the loop towards the thumb), the whorl (which is rarer in women than men; women have more arches), and the composite (which is a muddle of loops and forks). It is not just finger patterns which are individually unique, but palm prints, sole prints, toe prints, and even the pattern of the palate ridges; it is the convenience of fingerprints that has made them most important. The United States Federal Bureau of Investigation has a stack of finger-prints almost as numerous as the current American population, and one day a computer will be competent to sift through this collection to make the necessary comparisons. The present law for England is that a person's prints may not be taken before conviction without a magistrate's permission; in Scotland they can be taken from anyone under arrest.

The habit of fortune-telling via the palms is venerable only for the length of time it has happily engaged both teller and told; but the science of dermatoglyphics has recently become useful. About fifty years ago it was first suspected that palm-prints might be associated with disease, and by 1965 some two score disorders had been linked with abnormal palm prints. As palm prints are indelibly engraved five or six months before birth the disorders are congenital and predominantly decided at the moment of conception; but in 1966 abnormal palm-prints were linked for the first time with virus infection. The catching of German measles by mothers early on in their pregnancies had been known for a quarter of a century to cause large numbers of congenital malformations; but its infection was suddenly linked with abnormal prints. Three New York paediatricians, having first devised a better system for recording the prints of new, tight-fisted wriggling babies than the policeman's inky pad, found that about half of the babies affected by their mother's infection also had abnormal prints. Normally, and in babies without noticeable malformations, the proportion of strange prints is 14%. A typical palm abnormality is the presence of a simian crease, namely one fold traversing the palm from the area at the base of the forefinger instead of the conventional two. The mere existence of a palm abnormality is certainly not diagnosis of a larger congenital abnormality elsewhere, but it is a striking hint that one is more likely to exist than in a baby whose podgy palm is normal.

Melanophilia Ultra-violet radiation, present in sunlight, was discovered in 1801. It is the presence of ultra-violet which makes so many people go down to the sea, for it is the causative agent both of sunburn and of suntan. The former, known as erythema, results from the increased blood flow to the small blood vessels in the dermis. The latter, upon which many of the Mediterranean and other coastal economies are based, results from the change in position and the increase in quantity of melanin pigment in the epidermis. Pigment darkening appears within a few minutes of exposure to sunlight; pigment production takes a day or two to show itself.

Current addiction to suntan, presumably enhanced by long winters spent largely under light bulbs, is scarcely influenced by the possible harmfulness of solar grilling. L. Dudley Stamp, in *Some Aspects of Medical Geography*, stated that skin cancer is four times as prevalent on the Black Sea coast as on the Baltic coast and, concluded that 'voluntary exposure to the sun's rays may be more dangerous than fall-out'. As fall-out has diminished considerably in the two decades since that book was written his remark has become even more true with each passing year. Certainly skin cancer in white races occurs predominantly on the face, with another important area being the back of the hand. The *British Medical Journal*'s anonymous expert wrote in 1964 that 'sunburn and sun-tan are the early manifestations of injury to the skin by noxious radiation that is eventually carcinogenic. The only significant beneficial effect of sunbathing is the production of vitamin D'. This vitamin is amply supplied in a normal diet, and the expert added that sunbathing and its attendant lubrication 'can be regarded as a harmless manifestation of narcissism that one day may interest sociologists, anthropologists and the like'.

Harold F. Blum, the radiation biologist who has written an excellent review of the subject, concluded that it is undesirable to instil fear of exposure to sunlight (partly because skin cancer is still a rare cancer even under conditions favourable for it, such as the southern United States) although good to recommend certain prophylactic measures, such as moderation. Anyone possessing skin changes recognised by dermatologists as pre-cancerous, he adds, should avoid the midday sun and know that ordinary window-glass offers good protection against carcinogenic radiation. Finally, everyone, except albinos, tans on exposure to sunlight, and even those born black may become twice as black – or at least possess double the amount of melanin – after prolonged exposure.

Tattooing is an ancient art, with the word itself being brought from Indonesia by Captain James Cook. China (yes, China again) practised it four or five thousand years ago, and it still thrives throughout much of the world despite the Koran forbidding it, Leviticus condemning it, and Pope Hadrian I calling it barbaric. Many states have found it convenient. The Greeks and Romans tattooed slaves and criminals; the British and French tattooed their convicts even in the nineteenth century; the British Army marked BC on their bad conduct soldiers until 1879; and the

Germans indelibly numbered many concentration camp inmates in the last war.

Despite a long association with punishment most tattooing has been for entirely voluntary reasons. The young or the newly enlisted have been particularly prone to the custom. Whereas doctors in Britain have to obtain parental consent for some entirely beneficial injections, and can be accused of assault if such consent has not been obtained, the tattooist can inject his nearly immortal particles into any skin, old or young, without fear of subsequent action. Tattooing is easy, being no more than the injection of an insoluble dye such as carbon black into the dermis; about a millimetre is deep enough. Carbon lasts a long time. The coloured pigments are more likely to fade, but the evidence of tattooing often lasts longest in the lymph glands of the armpits. However much dermatologists may deplore the practice it is unlikely to die out, but at least it could be a specific offence to operate on children.

Beauty   Allegedly skin deep, or a matter of bone structure, but obviously neither one nor the other, what on earth is beauty? *New Scientist* once had a lengthy correspondence on the subject. This ended only with a suggested yardstick: the unit of beauty should be a millihelen, or enough beauty to launch one ship. Presumably a microhelen would launch just the one plank. Science has been notoriously deficient in this subject. Even its language cannot be used. How can the eye, the window of the soul, be described fully in its anatomical terms. Externally its folds of skin are entirely beautiful: anatomically these are the orbital fold, the superior and inferior orbito-palpebral sulci, herniation of orbital fat, and the malar and nasojugal folds. 'For where is any author in the world teaches such beauty as a woman's eye?' asked Biron in *Love's Labour's Lost*. Plainly any author would use other language than science provides.

Is beauty a matter primarily of the face? The reversal of the traditional animal mating position, which accompanied man's increasingly upright stance, changed his viewpoint. This new attitude, to quote C. D. Darlington, 'directed man's interest to the face of his mate, and led to the change from an older to a newer version of Venus'. Conversely it is reported that some African tribes still make their choice of wives from those who project furthest behind. Without doubt beauty is no one thing among the peoples of the world, and without equal doubt these concepts change with time. Is it rarity? In a country of the beautiful does the ugliest reign as queen? After a while, as Charles Darwin pointed out, the preferred choice of female influences the characteristics of the male. Choose long-legged women, mate with them, and second generation males will be longer in the leg. Darwin ascribed human hair loss to man's steady preference for women without bodily hair; their own woolliness withered as a result.

Is beauty a breaking of the conventional rules? Francis Bacon said 'there is no excellent beauty that hath not some strangeness in the proportion'. Greta Garbo used to be regarded as the beauty of the western world; and she showed strangeness. *Nova* reported that 'her looks were not pretty. Her expression was of an aching sadness. She moved awkwardly. Her feet were big. Her mouth, downturned and pessimistic. She rarely laughed'. Perhaps it is entirely right that beauty should be inexplicable. It needs mystery. It would be destroyed by delineation. It is haunting, eternal, most temporal, and can suffer no yardstick. 'Real is what can be measured,' said Max Planck. 'That which cannot be grasped quantitatively is scientifically invalid,' added J. Schorstein. Better so.

Hair   The hairs develop from the epidermis, but then penetrate through the dermis to the tissue beneath the skin. The root of the hair is a knob called the bulb, and the socket containing each hair shaft is called the hair follicle. To each hair is attached a muscle, and with goose-flesh each muscle makes its power visible. Hair follicles are gradually formed between the second and fifth months of foetal life, and all races and both sexes have a similar number. Each individual produces three types of hair in sequence: lanugo, vellus and terminal hair.

Lanugo, seen on premature babies, is normally pre-natal. It can be quite long, grows a lot on the face, is without pigment, is very soft and silky, and is usually lost by the seventh or eighth month of pregnancy. Its place is taken by vellus, which is traditionally the first post-natal crop of hair. This is also soft, occasionally with pigment, and hardly ever longer than an inch. The change-over from lanugo can sometimes give babies a two-toned appearance at birth.

Vellus gives rise, sometimes gradually, sometimes blotchily, to terminal hair, the coarser, longer, pigmented hair of later life. The changes are not necessarily immediate; the fineness of a baby's hair can slowly grow coarser, and equally slowly change colour. The change can be quite sudden in, for example, the rapid growth of coarse terminal pubic hair. On the face the change from vellus to terminal hair occurs first on the upper lip, then on the chin, then on the cheeks and then elsewhere.

At the other end of life vellus is liable to occur again. Even lanugo can occur again, and obscure a person's face with hair, but this change is entirely pathological. Terminal hair is lost, not only from the head but from all other parts of the body, with age. It tends to depart in the reverse order of its arrival, and so goes from the shoulders, arms, chest, abdomen, forearm, thighs and armpits – roughly in that order. Pubic hair also goes, more so in women than men.

An average scalp contains 100,000 hairs. Whereas new hair follicles do not in general arise where they did not exist before it can seem as if they do because all hair follicles follow a cycle: they are active, they then wither and then they rest. It is after the resting phase, which lasts some three to four months, that the hair grows again, and continues to do so for a few years. The period of withering is a matter of a couple of weeks, and during this period each hair shaft falls out. Daily loss from a full scalp of hair is about thirty to sixty individual hairs, but the follicles which grew them are not lost. Only with baldness do the follicles not spring to life again.

The speed of hair growth during its active period varies according to its position on the body. On the scalp and beard it is fast; on the limbs and back it is slow, although the occasional rogue hair, especially with males, can suddenly grow out of all proportion to its sluggardly neighbours. Scalp hair growth, according to Dr Arthur Rook of Cambridge, proceeds at about 0·33 mms. daily, beard growth at 0·38 mms., and limb hair growth more slowly. Therefore barbers make their living out of a steady increment of 4·75 ins. (12 cms.) of hair per year. Assuming a similar rate of growth, and men and women are equal in this regard, it would take a girl (or a man these days) some six years before their hair is long enough to sit on. At this rate, even though the speed slows down in later life, a person produces about 25 ft. (7·6 m.) of hair in a lifetime, but the cycle of growth and rest prevents anyone possessing such lengthy bird-of-paradise locks. The follicle cycle, and the regular moulting of each shaft every few years, means that no one shaft produces anything like 25 feet of hair continuously. As the cycle lasts for one to six years a maximum shaft length is likely to be nearer $2\frac{1}{2}$–3 ft. (76–90 cms.). Some women can grow their hair longer, and can even stand on their hair, but the regular moulting prevents lengthier ambitions.

Baldness is customarily a male attribute, and appears to be genetically determined. Girls unhappy at the thought of husbands going bald in later years should study the family album (and see the earlier chapter on inheritance). Bald fathers and grandfathers are heavy-handed pointers towards baldness both for spouses and male offspring. Women can grow bald with just as much severity as many men, but it is far rarer and usually starts later. Cures for baldness are legion, which in itself is a kind of indictment, considering their high sales and the persistence of so much baldness. Without doubt there is one treatment which either prevents it or arrests it if it has already begun: castration removes the predisposing male hormones, and almost always is a cure.

The *Journal of the American Medical Association*, in its questions and answers column, once gave an intriguing reply to a query about beard softening, and how best to prepare the bristles. 'The most effective beard softener in preparation for shaving is water' was the reply which must have cut across an entire nation's early morning prejudices and preferences. Two minutes' contact with warm water is all that is necessary for bristles to become hydrated and softened, and the warmer the water the shorter the time. Soap and cream merely maintain the softness achieved by the water. Reputedly, Albert Einstein used only water. So, reputedly, he was right once again.

A last skin custom, totally different but equally engrained and therefore probably impossible to change, also fails to have the whole-hearted support of the medical profession: it concerns a baby's skin and how to wash it. Not only are there many who discourage the traditional soap and water, but it is frequently advocated that newborn babies should not be bathed at all. Soaping has many disadvantages: it takes time; it makes the baby slippery, thereby increasing nervousness all round; it can get into the baby's eyes, thereby causing yelling and an early resentment of the whole wet business; and soap can readily form unwelcome scum. The substance hexachlorophene is gaining ground as a soap replacement in the wards, but soap has a strong hold on the home however much it causes a chancy hold on the baby.

Teeth Mammalian teeth are formed in two series. The first milk or deciduous teeth are lost, and their place is taken by the permanent teeth. Many primitive vertebrates possess a different system with marked advantages: they produce series after series of new teeth. The sharks are particularly adept at this. The mammal man, envious of sharks while so-called permanent teeth fall out and decay with extreme impermanency and without hope of replacement, could well do with a third set in middle age. Such third sets have been recorded (for seemingly every possible aberration is written somewhere in the literature), but the condition is extraordinarily rare. So too the complete absence of teeth although congenital absence of individual teeth has been amply recorded for every tooth position. The most curious anomaly associated with teeth, painfully so, is that no other structure in the human body is as likely to perish during life and yet no other structure is as resistant to decay after death. As castration is a remedy for baldness, death is a cure for caries.

Most human babies are born without visible teeth. Julius Caesar, Louis XIV, Napoleon and Richard III, together with one in 2,000 of the general population, possessed a tooth at birth. Usually this early dentition is loose, and is one or both of the central lower incisors. First teeth do not normally appear until $7\frac{1}{2}$ months, with the customary range being $5\frac{1}{4}$ to 10 months. For the next twenty months or more, as there is great variance in the timing of teeth, the child produces not only a ready made excuse for its mother of any wayward behaviour in public, but its full complement of twenty deciduous teeth: four incisors, two canines and four pre-molars on each jaw. The final order, starting from the centre of each half of each jaw, is central incisor, lateral incisor, canine, first pre-molar, second pre-molar. According to Harold Stuart the customary order and age of arrival for the milk teeth are:

Order of arrival	Average date of arrival
Lower central incisors	$7\frac{1}{2}$ months
Upper central incisors	$9\frac{1}{2}$ months
Upper lateral incisors	$11\frac{1}{2}$ months
Lower lateral incisors	13 months
Upper first pre-molars	$15\frac{1}{2}$ months
Lower first pre-molars	16 months
Upper canines	19 months
Lower canines	19 months
Lower second pre-molars	26 months
Upper second pre-molars	27 months

Arrival times many months earlier or later than these dates can be entirely normal. Only two-thirds of children produce their upper second pre-molars within five months of that average date of twenty-seven months. And only two-thirds of children produce their first tooth between $5\frac{1}{2}$ and ten months after birth. The teeth of girls appear earlier than those of boys, but boys tend to lose their deciduous teeth earlier than girls. Whereas teething is now an excuse for mothers, and a presumed unpleasantness for children, it used to be listed as a major cause of death: the Registrar-General reported in 1842 that 4·8% of all London babies dying before their first birthday had been killed that year by teething.

The permanent teeth by no means mimic the deciduous order of appearance. Although permanent teeth do not appear for six years or so after birth the buds for them were formed six months before birth. These small protuberances, developed on the tongue side of the deciduous teeth, lie dormant until their sudden growth and activity help to push out the relatively short-lived deciduous teeth. (An alternative plan to three series of teeth would be to have two series of equal duration, say thirty-five years and thirty-five years rather than half a dozen years followed by half a dozen decades.) Children acquire their first permanent teeth, the first molars (named after the Latin for mill-stone), before losing any of their deciduous teeth. Girls get them at 70–72 months, boys at 73–74 months. These permanent molars are not replacements for the deciduous pre-molars, but the first of the extra teeth, the many-cusped molars possessed by an adult. The deciduous pre-molars are replaced by two-cusped teeth, the bicuspids or pre-molars, and these erupt during the child's ninth or tenth year. (Again there is wide, entirely normal variation from the average.)

The first deciduous teeth to be replaced are the incisors during the seventh or eighth year, followed by the pre-molars, and then by the canines in the twelfth year. Whereas the first permanent molar appears in the sixth year, the second one does not arrive until the thirteenth year. The third – or wisdom tooth – is most casual, arriving perhaps at seventeen, perhaps at twenty-five, perhaps never. The permanent teeth are entirely leisurely about their onset: the first series of twenty deciduous teeth takes less than two years to be completed, but the second series of thirty-two permanent teeth takes at least six times as long from first to last appearance.

A mature tooth consists predominantly of enamel and dentine. The surface enamel is the hardest bodily tissue, and is 96% mineral. The inner dentine – also called ivory – is much like bone, but harder; 70% of it is of inorganic material. Not only are human permanent teeth not replaced, but they do not grow permanently. The incisors of a rabbit, for example, grow steadily, and wear is continually replaced, while the tusks of an elephant grow more and more formidable year by year. Mankind's set of teeth, remorselessly fixed and limited, can only deteriorate with time. This they do but never so determinedly as in this twentieth century.

Decay Modern food has been described as overcooked, soft, sticky and apparently designed for a nation with ill-fitting dentures. Without doubt the quality of today's diet has much to do with today's caries; so have the refined carbohydrates and the frequent consumption of sweet things between meals. Neolithic and Iron Age man would be astounded to hear modern dental advice, that food should both be chewed and eaten only at meal-times, but even Palaeolithic man suffered from tooth decay. Over 1% of South African Australopithecine teeth had been affected by caries. A survey of 12,000 Stone and Iron Age teeth from France and Belgium showed that decay frequency was sometimes as high as 8%, although rotting teeth were normally an affliction of adults. Teeth from the more recent Middle Ages cannot be dug up and examined with such impunity, but a Norwich survey found that a quarter of all mediaeval teeth had been lost before death and 18% of the remainder were victims of caries. Poor Anne Mowbray, unearthed by an excavator in 1965, a Plantagenet who had not reached her ninth birthday, had four milk teeth rotten and even two permanent teeth already bad. Perhaps a fifteenth-century royal diet had much in common with twentieth-century commoners' food, although she could not have encountered much sugar and certainly no refined white flour.

Today Anne would not be so remarkable, but the situation is improving. During the 1960's and in Britain 84% of five-year-olds had at least one decayed tooth, and eight years later a third of British juvenile teeth had been damaged by caries. There had been an improvement in World War Two, but this had been lost with the peace. Some 5,000 children under twenty were fitted with complete upper and lower dentures annually during the 1960's. (A letter to the *British Medical Journal* erroneously reported in 1964 that 90,000 children a year left British schools with false teeth: the report was reprinted in the newspapers, but even its shockingly false total did not cause national alarm.)

Improvement in Britain began only after those allegedly swinging sixties. In 1969, and in England alone, 27 million fillings were inserted into permanent teeth. In 1970 the figure leaped to 30 million, but that was the peak. By 1979 the total was 28 million, and there it has stayed, give or take a few hundred thousand annual drillings of decay. Peak year in England for filling milk/deciduous/baby teeth was 1969, when $2\frac{1}{2}$ million were treated. By 1977 this figure had dropped to 2·4 million, and by 1981 to 1·8 million. This total is still going down. Permanent teeth extractions have also been less necessary in recent years. Seven million such teeth were removed in 1969, but less than 4 million in 1981. The same sort of figures have been reported from all over Europe, and the dental profession is pleased at the trend but also selfishly concerned by the lessening work for its (steadily increasing) members.

No one knows why there should be this good news for us and bad news for the dentists, but we are certainly being more sensible about sugar, about fluorides, about making visits to a dentist, about brushing teeth, keeping them cleaner, and maybe the dentists themselves are doing a better job, not just in propaganda but with their instruments. Unhappily the good news is not universal. It used to be stated that simple citizens of some undeveloped land generally had excellent teeth, but the World Health Organisation has recently produced figures to belie that belief. It uses an index which registers the number of teeth that are decayed, missing or filled in twelve-year-olds. During the 1960's the average index for developed countries was ten as against one for the undeveloped nations. In the 1980's the figure is 4·1 for the third world in general (and 5 in many of its urban areas) while it is 3·3 in the industrial world. In short the wealthier nations have improved from 10 to 3·3 while the poorer ones have worsened from one to 4 or 5. The third world, it would seem, has inherited many less savoury practices (bad diet, sugary concoctions, less fibre) while the more educated world is at last learning what not to do to encourage caries.

George Washington himself helped to found the national American tradition of bad teeth. He needed dentures, and one of his ill-fitting sets was carved from hippopotamus ivory. Animals and other donors, such as slaves, helped to fill the gaps in wealthy jaws, and Waterloo teeth were in plentiful supply for many years; but twenty-four years after that battle Charles Goodyear discovered a process for vulcanising rubber. Not only could dentures then be made to fit accurately, but they could be kept in place by suction. Nowadays dentures may look well and feel well but they can still not permit the pressures of chewing and biting exerted in a normal mouth: the 150/250 lbs. pressure of biting is reduced to 10/30 lbs. because, although jaws are just as strong, the tissues beneath the dentures are less capable of withstanding the load. On biting, Albert Schweitzer told John Gunther that a human bite is the most dangerous, then serpents, then monkeys.

How to clean teeth? At a dental conference held in London Professor R. D. Emslie described an experiment in the Sudan which assessed various methods. The results, in order of their cleansing abilities, were:

1. Chewing stick (dominant in Africa, referred to by Mohammet).

2. Towel on finger. (Part of Moslem code involves washing out mouth with finger and thumb. Not good for mouth areas inaccessible to the towel.)
3. Tooth brush and paste (dominant in industrialised western world; various forms of brush and paste have been proved highly abrasive).
4. Chewing luban (a natural resinous chewing gum).
5. Eating dom (a fibrous fruit).
6. Eating sugar cane.
7. Eating orange.

The reduction in debris varied from 80% with the chewing stick to 40% with the orange. This debris (or plaque) is important; without it tooth decay would be less rampant. No one is entirely positive about all the reasons for caries, but there must be acid in adequate concentration and there must be time. The starches are generally swallowed before they are broken down; the simple sugars not only stick around but are easier to break down into acid. Saliva would be able to buffer this acid were it not for the plaque, which tends to be neutral on the surface where the saliva can get at it, but more acid with depth. Given sufficient acid for sufficient time even enamel, the hardest tissue of the body, capable of lying in the earth for thousands of years, is eaten away voraciously.

For some reason the addition of fluoride to the diet prevents decay. Decay is not stopped, but fewer cavities develop – perhaps one-third fewer, perhaps two-thirds – if fluoride is present in drinking water to the minuscule tune of one part in a million. The fact was discovered by an American dentist, Frederick S. McKay, in the early years of this century, and it has frequently been confirmed since then. Some towns have a natural abundance of fluoride in their water – and less caries; others have added it to the supply, changed their minds, and then added it again – with the incidence of caries going down, then up, then down again in consequence. Antigo, Wisconsin, see-sawed in this fashion, being worried initially about tooth decay, then worried about sodium fluoride – which is a poison if taken in large doses, and then worried again after the decay rate had shot up. Other arguments against the addition of fluoride to the water, called contamination by the opponents, are levelled against the allegedly unproven beneficial effects, against the proven harmful effects (modest over-doses cause teeth staining), and against tinkering by the state with the freedom of the individual. Not until January 1, 1967, did a state – Connecticut – have a mandatory law making fluoridation compulsory within its borders.

So far about 90,000,000 Americans drink water to which fluoride has been added by man, and a further 7,000,000 to which it has been added naturally. In Britain only 5½ million people drink fluoride, with Birmingham, Anglesey and Watford being the first to add the tasteless, colourless, odourless and inexpensive substance to their water supplies. Dr Luther L. Terry, former Surgeon General of the United States, has called fluoridation the most recent of the 'great mass preventive health measures of history': the three others were the pasteurisation of milk, the purification of water and immunisation against disease. By 1983 it was reckoned that 217 million people around the world were using water supplies whose natural fluoride content had been supplemented artificially for dental health purposes. That leaves, of course, some 4,000 million other people whose supplies have not been treated.

Even in the developed countries, despite fluoride and other benefits, the battle against caries has hardly been won. In Britain, which is favoured with a national health service, the average fifteen-year-old has ten permanent teeth either decayed, missing or filled. Almost one in three of all adults has no natural teeth, and about 7,000 sets of dentures are supplied annually to schoolchildren. Dentists need not worry unduly about their unemployment. Besides, they can always go to the Third World if work is ever excessively slack in their developed nations. It looks like being plentiful in those poorer regions for quite a time to come.

32. Digestion and Nutrition

'Fate cannot harm me – I have dined today.'

REV. SYDNEY SMITH

On the one hand the process of digestion is entirely elementary; the hollow tube of the digestive system stores food temporarily, prepares it for absorption, absorbs what is absorbable, and rejects the rest as faeces. On the other hand such simplicity has challenging riddles within it. How is it that the gut can digest meat without digesting itself? How does it consume tripe, or the gut wall used as casing for some sausages, or even bone, without the cannibalistic self-destruction of its own tissues? The stomach contains hydrochloric acid, and in quite a concentration (gastric juice is about 0·6% HCl). The popular appraisal of its strength is that a stomach's contents could burn a hole in the carpet. So why does the stomach not burn a hole within itself? And how can glands of living tissue actually manufacture such a corrosive substance as hydrochloric acid?

There are further problems. The human being, traditionally called an omnivore, can indeed eat a bewilderment of foods; yet he can starve to death with remarkable ease. Basically the human system cannot cope with plants, trees, mosses and most of the botanical world; it can cope best with the botanical end-products, such as fruits, seeds, nuts. The few plants known as vegetables are exceptions. The plant world, so suitable for all the herbivores and consumed so avidly by well over half of the world's animal species, is virtually forbidden to man. Forbidden too, despite our digestive ability to break down so many chemical compounds, are a few substances that have the power to poison us, to destroy our lives or merely to affect our well-being disastrously. Eating earth or leaves or old newspapers will not do us good, and may do us harm, but poisons are in a different category. A substance is said to be a poison if less than 2 ozs. of it – about 50 gms. – will either kill us or be seriously harmful. (Everything is harmful if consumed to excess, even bread and water. Poisons are harmful if consumed minutely.)

Proteins are a further complexity. The body must have these within the diet and certainly absorbs them, but the body is normally resentful of foreign proteins, and brings its powers of immunity to bear upon any such invasions by amassing anti-bodies to counter the antigens. However, foreign proteins taken into the gut are customarily absorbed, transformed and utilised without any disturbance whatsoever.

Also what is hunger and what is thirst? In general we eat what we need in that we stay reasonably constant in weight. We drink without too much thought of the need for liquid, and yet do not dehydrate ourselves. We say 'enough', and scarcely pause to marvel at such precise comprehension of our various requirements.

A final conundrum, more perplexing to the chemist than the ordinary consumer, is that the body performs with speed and precision large numbers of chemical reactions that would normally take far, far longer if carried out in a laboratory at the same temperature and pressure. Anyone who has ever wielded a test-tube or a frying pan will know that chemistry happens faster when things are hotter; yet the body breaks down molecules, combusts them with oxygen, and builds up the molecules at the modest 98°F (36·7°C) temperature of the human frame. This is less than the temperature of bath-water, and the chemistry both of cooking and of the test-tube would be immeasurably slow if confined to such a heat. A partial answer to the body's abilities is the profusion of enzymes, the natural catalysts which assist and promote biochemical reactions and are not used up in the process; but to define a catalytic enzyme is one thing. To explain how it achieves its remarkable role of initiating or accelerating any reaction, without being unduly involved, is quite another.

Digestion summary　Unfortunately the language of digestion is largely indigestible but, as this barrier does exist, it might as well be encountered right at the start. Food consists of proteins, carbohydrates, fats, salts, vitamins and water. Some of all are essential, but only the first three have to be altered by digestion.

Proteins (after the Greek word for primary or fundamental) are large molecules made up of chains of amino-acids. These amino-acids are joined together by what is called the peptide linkage, whereby each amino group (NH_2) is attached to a carboxyl group (COOH). Some enzymes can break the linkage, and can add a molecule of water at the same time – hence the action called hydrolysis. Some links, as might be expected, are easier to break than others; so the long protein molecules become broken, firstly, into shorter lengths (polypeptides), then into very short lengths consisting of three amino-acids (tripeptides) or two amino-acids (dipeptides). Finally there will just be the single amino-acids. Traditionally called the body's building blocks the twenty different kinds of amino-acid which go to build up protein are indeed, when fully digested and broken down into single units, the raw material for the manufacture of the body's tissues. The human body, apart from its bone and its fat, is rich in protein, a fact appreciated by the occasional carnivore; but all proteins have to be broken down into their amino-acids before they can be built up again into useful protein. Digestion does the breaking down.

Carbohydrates (after the Latin for coal and the Greek for water) are compounds of carbon, hydrogen and oxygen (but they lack the nitrogen crucial to protein). These three are joined together to form the three kinds of monosaccharide, namely glucose, fructose and galactose, and all carbohydrates are built up of monosaccharide units. Once again the problem of digestion is to break down big molecules into smaller ones. The big poly-saccharides, often called starches, have to be broken down. They too have linkages, called glucosidic linkages, and the big poly-saccharides are attached by enzymes at their linkages until they form disaccharides (two units) and monosaccharides (one unit). Unlike the proteins, which are all big and which all have to be broken down, some of the carbohydrates are small. Glucose is just composed of single monosaccharide units, but glucose has to be made artificially. There are two disaccharides present in food: sucrose – found in cane sugar, and lactose – found in milk. Whether artificial or natural such carbohydrates need little or no breaking down, and therefore next to no time is necessary for their digestion; hence their use for those with disrupted digestions or the need for instant energy.

Fats (an Anglo-Saxon word) are, like proteins and carbo-hydrates, combinations of simpler units. These units are glycerol (or glycerin) and the fatty acids, such as stearic acid, palmitic acid and oleic acid. All such fats are, like the carbohydrates, made up of carbon, hydrogen and oxygen, but in different proportions – the fats have very little oxygen. A typical carbohydrate has equal amounts of carbon and oxygen and twice as much hydrogen. A typical fat has twice as much hydrogen as carbon, being made up principally of CH_2 units, but only an atom or two of oxygen at one end of the molecule. Another major difference between fats,

418

carbohydrates and proteins is that fats can easily be stored. The body's available stores both of protein and carbohydrate are very limited, but the body's willingness to store up supplies of fat bedevils large portions of the population. The fact that honey (a carbohydrate) lasts and so does biltong (a protein), while butter (a fat) goes rancid does not destroy the generalisation that fat is easier and better to store. Weight for weight, fat has twice the fuel value of either protein or carbohydrate.

The all-important and all-skilful digestive enzymes, which act so effectively upon the casual assortment of foods consumed by the average human, make a long list. Their names and their products also tend to be lengthy. Nevertheless the most crucial enzymes in digestion ought to be mentioned. Their products follow in parentheses.

From the *salivary glands:*	Salivary amylase (maltose – a disaccharide
	Maltase (glucose)
From the *stomach*:	Pepsin (peptides)
	Rennin (casein – a milk protein)
from the *pancreas*:	Trypsin (peptides)
	Lipase (glycerol and fatty acids)
	Amylase (maltose)
	Ribonuclease (nucleotides – proteins of cell nucleus)
	Deoxyribonuclease (nucleotides)
From the *intestine*:	Carboxypeptidase (amino-acids)
	Aminopeptidase (amino-acids)
	Enterokinase (trypsin – same as the pancreatic enzyme)
	Maltase (glucose)
	Sucrase (glucose, fructose)
	Lactase (glucose, galactose)

Those from the salivary glands work best in neutral conditions, those from the stomach work best in acid, and those from the intestinal glands work best in neutral or alkaline conditions. The stomach's acidity is partly neutralised by an intestinal secretion of sodium bicarbonate, the same chemical which so many people pour, with such enthusiasm, into their stomachs.

Food Perhaps there was a time when humanity existed in luxuriant gardens of Eden, plucking delicious fruit here and there; but one wonders. The human digestive system can never have had an easy time. Instead, subjected to trial and error, it must have been the steady recipient of good and bad, of beneficial and disastrous. Countless unsung heroes of the past must have noised it abroad, painfully, that deadly nightshade berries, henbane, wild hyacinth, bluebell, deathcap toadstool, monkshood (wolf's bane), water dropwort roots, cowbane roots, yew berries, ivy leaves and fresh anemone were not for eating. Other heroes, having gobbled gristle greedily from some freshly hunted creature, must have come up with the idea of roasting or boiling food to render it more digestible. Still others must have realised that fat, by boiling at a higher temperature than water, can break down food more effectively.

Even had they been presented with a Garden of Eden stripped of all poisons, one suspects that our human ancestors would not have been content with the dull diurnal round of fruit-gathering. For one thing, virtually every tribal group in the world has known how to make alcohol. For another, few people just pluck fruit and harvest nuts; they do things to the food, chop it up, store it, preserve it, flavour and pickle it, and let it rot to the right degree of putrefaction. They want variety, they want the exotic and the rare, the strong-tasting and the rich. The human digestive system, ably equipped with enzymes, has just had to cope with the very mixed assortment of commodities sent down to the stomach in each oesophageal bolus.

Nowadays, such naturalness as ever existed is being diminished still further. Take this declared analysis of a particular biscuit: 'Wheat flour, processed Cheddar cheese solids, cotton seed and soya oil, non-fat milk solids, cornflour, cheese flavour, artificial flavour, salt, sugar, mono- and di-glycerides, egg yolk, baking soda, mono-sodium glutamate, butylated hydroxyanisole, butylated hydroxytoluene, certified colour, propyl gallate.' Standard flavourings, although kept under constant vigilance by the Food Standards Committee, still sound unpleasant: popular ones include Tartrazine, Ponceau MX, Ponceau 4R, Red 10B, and Amaranth. Ponceau 3R, found to act as a carcinogen in rats, was recommended for withdrawal by the Committee not so long ago. However, neither food adulteration nor colouring is new. Think of saffron, turmeric, cochineal and carmine.

Today's average Britons spend about a quarter of their disposable income on foods (plus another £100 million on pet foods, a sum and a quantity of nourishment that would keep large sections of Africa's Sahel amply supplied). Not only are they spending more on food than a few years ago, but they are spending a greater proportion of their actual income on it. A typical Englishman eats 3 lbs. 5 ozs. (1·5 kgs.) of potatoes a week,

420

18½ ozs. (524 gms.) of sugar (plus jams and other preserves), 4 eggs, 2 ozs. (56 gms.) of sausages, 1½ ozs. (42 gms.) of fresh fish, 2 lbs. 11 ozs. (1·2 kgs.) of bread (it was 4 lbs. (1·8 kgs.) in 1950), 5 ozs. (142 gms.) of biscuits, 3 ozs. (85 gms.) of tea and much else besides. The average diet more than meets the nutritional standards set by the British Medical Association, although by no means does every individual eat adequately.

A major victory of World War Two was the satisfactory nourishment of the British people, and the distribution of a food supply which had much to be said for it medically over today's excesses. Food rationing began in January 1940; and eighteen months later, when at its most severe, the weekly allowance for each civilian adult was 4 ozs. (113 gms.) bacon or ham, 8 ozs. (227 gms.) sugar, 2 ozs. (57 gms.) tea, 8 ozs. (227 gms.) fat (of which only 2 ozs. (57 gms.) were butter), 2 ozs. (57 gms.) jam, 1 oz. (28 gms.) cheese, and a shillingsworth of meat. At least the food bills were also low: that weekly supply, including the meat, cost 2s. 7d. (or 13 pence in today's decimal currency. Even with the rise in food prices since then the wartime bill would, in 1984, still be cheap and amount to £1.03). In addition, there was a modest supply of milk (with priority for children and expectant mothers), roughly one egg a week, and tinned meats and fish, although such desirable extras were included in a 'points' system of rationing which enabled everyone to select his or her preferences, provided these were available and the ration-book still had sufficient unused points. Nevertheless, many a dietitian would long for Britain to be restricted to war-time food supplies were such an enforced method of national abstinence feasible.

Globally today's food picture is both unsatisfactory and obscure. Whereas many Europeans eat too much, and the Americans in particular carry so much of their food surplus around with them (Hugh Sinclair has calculated this transported excess to be 2 million tons of fat), there is obviously great undernourishment in the world; but there is also great uncertainty about the extent of this lack. The traditional figure, still quoted, is that two-thirds of the world are suffering from malnutrition or hunger. In 1964, Colin Clark, director of the Agricultural Economics Research Institute at Oxford said, 'This extraordinary mis-statement . . . is believed by almost everyone, because they have heard it so often'. It was first made in 1950 by Lord Boyd Orr, formerly director-general of the Food and Agricultural Organisation, a body once described by the *Economist* as 'a permanent institution devoted to proving that there is not enough food in the world'. Colin Clark drew attention – at a Ciba symposium on the subject – to the estimate given in May 1961 by Dr Sukhatme, F.A.O.'s own director of statistics, that 10% to 15% of the world's people were hungry.

At any one time a proportion of that percentage must be starving, in that hunger must mean inadequate intake. Such a life-style will inevitably lead to starvation if excessively prolonged. And, if starvation is itself prolonged, death is equally inevitable, mainly among the youngest and oldest who are least able to cope with lack. How many people are actually on this downhill path at any one time is difficult to assess, even in the 1980's when the world is more acutely aware of this problem and when the 'aid' agencies are more active than ever before. Some people starve on an annual basis, always being hungry for a season of the year. Others starve on a ten-year or so basis, suffering whenever rainfall is slightly (or not so slightly) less than average. The greatest starvations, as in the Sahel in 1983 and Ethopia in 1984, tend to acquire publicity, and the rest of the world then hurries (usually) to make amends, as if starvation has come suddenly, like earthquakes or a tidal wave. An important factor is that many people live (having nowhere else to go) where people should not live, and where the smallest shift in climate brings disaster. Worse still, according to Professor Jack Mabbutt of New South Wales, the whole of the earth's sub-humid tropical region is liable to become desert. This area is currently the home of 850 million people, or 20% of the total population.

Of course people can be eating substantially but be malnourished at the same time. Accurate estimates of the extent of poor nourishment are hard to make, for adequate nourishment depends on ambient temperature, sex, body weight, pregnancy, lactation, exercise and work.

Three generalisations are possible. Firstly, scientific estimates of human calorific requirements have been falling in recent decades; the original calculations put our needs too high. Secondly, world food production is keeping pace with population growth, and looks like doing so – or doing even better – for quite a time. Thirdly, the existence of hungry or even starving people in the world is poor distribution rather than any general planetary lack. ('On a clear day,' said a cartoon figure in *The Times* who had just climbed to the top of Europe's butter mountain, 'you can see Ethiopia.') And fourthly, all manner of people will inevitably continue to reiterate that two-thirds of the world are suffering from malnutrition or hunger.

Whatever the actual fraction of people who would like to eat more, there is a big fraction constantly trying to eat less. Diets abound, and for every scheme, however ridiculous, there are loyal devotees: nuts at night-time, no fats, lots of nibbles, no nibbles. Prescription slimming drugs sold in the United States doubled between 1960 and 1965 alone. Stipulated low-calorie foods quintupled their sales in the same five-year period. Low calorie foods are still marching forward in their sales. Items proudly proclaim either one calorie or no calories whatsoever. It is bizarre that some of mankind now grabs at tins asserting their nutritional pointlessness, while so many are dying from a lack of calories; but at least the eating nations are trying to trim their appetites. In the United States the annual per capita consumption of all foods is down 197 lbs. (89·4 kgs.) from the 1909 figure, the year totals of this kind were begun.

Actual requirements   For every living creature food has a two-fold purpose. It must supply the raw materials for the construction or replacement of human tissue, and it must act as fuel for supplying energy to the body. Generally speaking the first purpose requires the intake of quite a small number of basic substances, and the second requires the intake of sufficient calories. It is possible to eat more than enough calories, and still die owing to the lack of some essential substance. It is equally possible to ingest a fully representative tally of essential nutrients, but to die owing to an insufficiency in the total bulk and a lack of calories necessary for the maintenance of life.

Nutritional needs are also a compromise, for any given species, of what is available, what can be digested and what can be manufactured against what is actually required. Chemically, a zebra and a lion and a vulture would yield similar analyses, but the zebra lives on grass and the lion can live on zebras, and vultures can live on the decaying corpses of both of them. They all have similar requirements, but they attain them differently. The zebra is making grass fit for lions to eat, and the lion is producing meat fit for vultures to eat, and all three are producing flesh fit for the bacteria to consume eventually.

The human being, like any other creature, has a limited range of food-stuffs and a limited power of converting digested materials into actual needs. He or she does not have to eat meat to make the meat of muscle tissue, but he or she does have to eat the right vitamins, for example, and the right kinds of protein to stay alive. Food is the supply of raw materials, but when broken down by digestion these are still partially constructed for our needs. A house-builder requires neither trees nor sawdust to make a house, but needs the intermediate plank form. The body requires neither meat nor its elemental constituents of nitrogen, hydrogen, carbon and oxygen, but the intermediate amino-acid form. As the house-builder needs more than one type of plank to make his house, so the body requires some thirty to forty essential nutrients.

These essentials consist of about a dozen vital minerals, about a dozen vitamins, some ten amino-acids, a lot of water and a sufficiency of fats, carbohydrates and proteins. From these the countless profusion of different bodily constituents will be manufactured, such as the haemoglobin of blood, with each molecule possessing 64,000 atoms, and the nucleic acids possessing even more, and the digestive enzymes that can break down still larger molecules. Yet all this profusion also consists of just a very few elements. If a human body (of 156 lbs./70 kgs.) were analysed completely in a test-tube, and all its complexities were rendered into their constituent elements, it would be found to consist of oxygen – 100 lbs. (45·4 kgs.), carbon – 28 lbs. (12·7 kgs.), hydrogen – 15 lbs. (6·8 kgs.), nitrogen – 4·6 lbs. (2 kgs.), calcium – 2·3 lbs. (1 kg.), phosphorus – 1·6 lbs. (0·7 kgs.), potassium – 8·5 ozs. (241 gms.), sulphur – 6 ozs. (170 gms.), sodium – 3·7 ozs. (105 gms.), chlorine – 3·7 ozs. (105 gms.), magnesium – 1·25 ozs. (35 gms.), iron – 0·15 ozs. (4·2 gms.), zinc – 1·9 grams, copper 0·2 grams, manganese 0·02 grams, molybdenum 0·015 grams, some cobalt, some selenium and some still smaller fractions of other elements.

Schoolmasters sometimes comment facetiously that, chemically, the body is worth about £1, and certainly there is nothing exotic about its main constituents, although some of the trace elements might fetch more money these days. Nevertheless, no body could survive were it to be presented, like a chemical Pygmalion, with a ready supply of its constituent elements in elemental form: it has to have them as protein, as fat, as carbohydrate, as vitamins, as water, and only to a limited extent can it have them as salts and simple inorganic chemicals.

Protein   Every cell contains proteins, and about 18% of body weight consists of protein. They are the most complex compounds found in nature, and consequently have large molecular weights varying from a few thousand to several million. (A molecular weight is relative to that of an atom to oxygen which is taken to be 16.) However big, the structure is always built up of amino-acids, the small constituent units of every protein. Egg albumin, for example, the ordinary 'white' of egg surrounding the yolk, is a small sized protein of molecular weight 45,000. It consists of 418 amino-acid units bonded together to form each molecule of albumin. Each amino-acid unit has of course a smaller molecular weight than the large built-up protein, and amino-acid weights vary from 75 (glycine) to 240 (cystine).

The list of amino-acids makes unexciting reading, but their totally vital role demands that they should be heard. Some are more vital than others, and the ten which have been found to be indispensable in the diet of a young growing rat have been under-lined. Without every one of these ten essential amino-acids the laboratory rats suffered. With sufficient supplies of all ten the young rats were able to make the remaining amino-acids necessary to the build-up of their own proteins. There are several variants in the list of amino-acids, but the basic types include:

Glycine $C_2H_5NO_2$	(19)	Glutamic acid $C_5H_9NO_4$	(52)
Alanine $C_3H_7NO_2$	(35)	Hydroxyglutamic acid	
Serine $C_3H_7NO_3$	(36)	$C_5H_9NO_5$	
Threonine $C_4H_9NO_3$	(16)	Arginine $C_6H_{14}N_4O_2$	(15)
Valine $C_5H_{11}NO_2$	(28)	Lysine $C_6H_{14}N_2O_2$	(20)
Norleucine $C_6H_{13}NO_2$		Phenylalanine	
Leucine $C_6H_{13}NO_2$	(32)	$C_9H_{11}NO_2$	(21)
Isoleucine $C_6H_{13}NO_2$	(25)	Tyrosine $C_9H_{11}NO_3$	(9)
Cystine $C_6H_{12}N_2S_2O_4$	(6)	Tryptophan $C_{11}H_{12}N_2O_2$	(3)
Methionine $C_5H_{11}SNO_2$	(16)	Histidine $C_6H_9N_3O_2$	(7)
Aspartic acid $C_4H_7NO_4$	(32)	Proline $C_5H_9NO_2$	(14)
		Hydroxyproline $C_5H_9NO_3$	

Like carbohydrates and fats, the proteins consist largely of carbon, hydrogen and oxygen; but, unlike carbohydrates and unlike fats, the proteins are distinguished by the presence of nitrogen. Every single one of the amino-acids contains some nitrogen, and a couple contain sulphur as well. The proteins in human diet are practically the only source of new nitrogen. Constant replenishment of nitrogen is necessary because irreversibly damaged proteins lead to the loss of nitrogen; it passes out of the body in urine as urea. Proteins are also the chief source of sulphur. The figures listed in parentheses after the chemical formulae are the numbers of each of these amino-acids units present in egg albumin.

This one example of the constituents of one fairly simple protein should give some idea of the hideous complexity of organic chemistry, particularly with regard to proteins. Ordinary inorganic chemistry seems infantile when set beside the interactions of such large molecules. Every schoolboy is taught basic inorganic reactions, like sulphuric acid acting upon zinc when $H_2SO_4 + Zn = ZnSO_4 + H_2$. The production of zinc sulphate and hydrogen is straightforward and entirely elementary compared with the changes that must take place when, for example, egg albumin is merely heated. Everyone knows it changes irreversibly with heat from being transparent and fluid to opaque and stiff (in four minutes if you like your egg done that way), but the chemistry of that change and the manner in which heat affects those 418 amino-acid units is formidable. Even writing down the 418 constituent formula units of just one albumin molecule is a major endeavour.

Proteins, in short, are complex. However it is not their complexity that is vital to our diet, but their amino-acids (plus their nitrogen and sulphur). Digestion breaks down the large molecules into these amino-acids, and the body then builds up its own vast protein molecules, its albumin, its haemoglobin, its nucleoproteins, its enzymes, collagens and keratins. As might be expected animal proteins are, when broken down, nearer to human requirements than plant proteins, but our normal human diet contains a bit of both.

In Britain the average consumption per person per day is 3 ozs. (85 gms.) of protein, almost 2 ozs. (57 gms.) of which are animal protein. In New Zealand, Canada, United States and Australia the consumption is nearer 4 ozs. (113 gms.) daily, with nearly 3 ozs. (85 gms.) of animal protein. Conversely, India eats less than 2 ozs. (57 gms.) of protein a day, of which only a fifth of an ounce (6 gms.) is of animal protein. If all these figures, even for North America, seem low it should not be forgotten that water is the main weight of most foods. There is only a quarter of an ounce (7 gms.) of protein in a pint of human milk and 3 ozs. (85 gms.) or less of protein in every pound of beef.

Anyone's intake of either total or animal protein is a very fair guide, and often an extremely precise one, to his or her income in the world. The United States Food and Nutrition Board recommends 1 gm. of protein daily for adults for each kilogram of their body weight (and therefore $2\frac{1}{2}$ ozs. a day for a 156 lbs. man). This is less than the average American eats, but probably more than he needs, and decidedly more than most people in the world achieve.

Vegans, the vegetarians who eat no animal products, not even the eggs or cheeses that are entirely acceptable to other groups, have to subsist wholly on plant protein. This can lead to dietary difficulties, but plant proteins can be mixed in such a way that the amino-acid content of the mixture is adequate. Many strict vegetarians have adopted the diet for ideological reasons, but tend to look upon the whole business of eating as a slightly sordid necessity. Consequently, they can be lax about diet, and there can be deficiencies in it. Vitamin B_{12} is an additional complexity for it is short in a vegan's diet. It can easily be supplemented, but its lack may also encounter a general lack of enthusiasm for caring about the unpleasant demands of mere food.

Where to find proteins? All flesh, whether of fish, fowl or mammal, is rich in protein. Cow's milk can have more than twice as much protein as human milk, the dried or condensed milks have a still greater proportion by weight and the cheeses have even more. Cereals contain some protein, roughly 5%–10% by weight, with rice and rye being less good in this respect than wheat and maize. The sugars have no protein, but nuts, beans, lentils and all such firm botanical end-products are rich with it. Fruits also contain protein, but only 1% or so by weight as so much of their bulk is water. The same also applies to vegetables although, with their water content usually less predominant, the protein proportion is greater.

Where to acquire more protein? The world is short of it, and this lack is the greatest single cause of malnutrition. Kwashiorkor (which in South Africa means the 'disease of a child when another is born') is protein-deficiency, and generally occurs after weaning. Even today, with so much chemistry applied to the food industry, no protein is produced synthetically for food on any major scale. Attempts to short-circuit traditional gastronomic procedures have generally been frustrated by human conservatism. 'The force of habit of millions of people is a terrible force' said Lenin. Bill Pirie, of the Rothamsted Experimental Station, has constantly produced new schemes for stepping up protein supplies. His 'mechanical cows' have extracted it from leaves, from grass, from cereal waste; and yet people are still suffering from lack of it. The wealthier countries, with enough protein to eat, are more tolerant of new foods than the poorer countries who are in greater need of supplementary – and novel – foods.

Most traditional methods of protein production are wasteful and lengthy. In twenty-four hours half a ton of bullock will make 1 lb. (0·45 kgs) of protein; in twenty-four hours half a ton of yeast will make fifty tons of protein. One wonders how long the world will permit itself to be semi-carnivorous. It is so much more economical not to process available protein through creatures like the cow, the sheep and the pig before eating it. The vegetarians have long advocated their policy on the grounds that the raising and killing of animals for food is thoroughly distasteful. Other arguments are equally strong, such as the inefficiency of the process and the great quantities of land consumed not just by the animals themselves but by food grown specifically for them. England, for example, could easily become self-sufficient in food if it became vegetarian, and have lots left over for the hungrier regions of the world.

Fats   Broadly speaking, fats have two roles in the body. For the first they are essential ingredients of every cell, being vital to innumerable cell mechanisms. As such they weigh about 1% of the body's total, and are not appreciably consumed during periods of starvation. That kind of consumption is limited to the second role of fats; their use as a store of energy. Weight for weight 1 lb. (0·45 kgs.) of depot fat – as it is called – will yield over twice as much energy as either 1 lb. of protein or 1 lb. of carbohydrate. The ratio, roughly 9:4, is due to the straightforward nature of the typical fat molecule. Essentially it is just a chain of carbon atoms tied, except at one end, to an unvarying series of hydrogen atoms. Except for the occasional oxygen atom at the end there are no more; hence no water is locked up within the molecules, hence its greater capacity to act as a fuel. Water cannot serve as a fuel, and molecules rich in H_2O are less effective as energy liberators when oxidised. Fortunately for creatures like the camel not only is fat such a well-compressed store of energy, but when it is oxidised it also liberates water. Nature is frequently adept at gaining on both swings and roundabouts, and the camel rides on both. (The animal can also allow its body temperature to rise; thus obviating some water loss.)

Depot fat – and more about it under starvation and obesity – is not a fixed entity, like a spare tin of lard in the kitchen. However constant an abdominal outline may in fact be, the fat within is being perpetually removed and replaced. Only about half of depot fat is stored beneath the skin. A lot is attached to the mesentery, the membrane supporting the small intestine, and a lot more is around the kidneys. Some depot fat is entirely normal; obesity is often defined as the condition where over 30% of the body weight is fat. This depot fat also acts as a good insulator due to the relative lack of blood vessels ramifying through this kind of tissue. It consists essentially of cells, which are just droplets of fat each surrounded by a thin membranous shell of protoplasm.

Unlike beef fat, and various other animal fats, human fat melts at quite low temperatures: its melting point is 63°F (17°C), or normal room temperature and much less than body heat, as against 121°F (49·5°C) for beef fat. Although many fats that are liquid at normal temperatures are called oils, there is no distinguishing difference; olive oil is strictly olive fat. Fats also have a low specific gravity. Consequently, fat people float in water more readily than thin people. This is one advantage of growing old, a time of greater fatness, for one's specific gravity decreases as the proportion of fat increases. Even thin old people have more fat in them than in their youth, and therefore float more readily without the exertion of swimming.

Fat needs are very obscure. They are mainly hidden by the organic changes going on within the body. It is all very well to talk of proteins, fats and carbohydrates as if they were entirely separate from each other, but to some extent they can be made from each other. Potatoes only have 0·1% fat, but they are undeniably fattening: much of their carbohydrate is turned into fat by the body. Similarly, some fat can be turned into carbohydrate and many proteins can be turned into sugar. However, although both carbohydrate and fat can go some way towards forming proteins, the essential amino-acids already referred to *are* essential to the formation of new proteins; the fats and carbohydrates cannot do all the conversion by themselves.

Such interchangeability complicates the issue of dietary fat needs. Certainly some mammals survive on outstandingly little fat, and the human species as a whole is remarkable in its ability to eat either a little or a lot of it. In Britain the average fat intake is 4 ozs. (113 gms.) per person per day. (One should perhaps remember again that wartime ration of 8 ozs. (227 gms.) of fat a week, and remember too the enthusiasm of the medical profession for war-time diet to be feasible once again.) A Kikuyu of Eastern Africa eats less than a quarter of this, while an Eskimo eats on average double the British amount. Extra fat is not always beneficial if the diet is too one-sided: too great an intake of fat with insufficient carbohydrate in the diet can lead to ketosis, a dangerous affliction which is a kind of internal poisoning. The general opinion on fat intake is that, whereas a high carbohydrate intake can do much to maintain high fat stores, it is probably necessary to eat at least a small amount of fat, particularly if it contains what are called unsaturated fats. (They are called unsaturated because not every carbon atom has its full potential of hydrogen atoms around it; instead there are some double bonds in the carbon chain.) Quite apart from actual fat needs, but integral to them, is the need for vitamins. Animal fats to a large degree, and vegetable oils to a smaller degree, are rich in vitamins A, D and E.

There is next to no difference between animal and vegetable fats in their energy values, but there is eternal dispute over their respective merits. The principal villain is alleged to be cholesterol, a sterol (after the Greek for solid. As chole is Greek for bile, the word cholesterol means solid bile – or gall stone) which is found in animal fats and which can also be manufactured by the body. Each of us has about 6 ozs. (170 gms.) of cholesterol within us at the best of times. Gall stones are largely cholesterol (up to 97%) and evidence of an association has sometimes been found between high cholesterol levels in the blood and the incidence of heart disease and arterial degeneration. However, an association between two factors does not indicate a causal relationship; the cholesterol may well be in the blood because of some breakdown in fat metabolism which is also affecting the arteries. Whether cause or effect the proportion of cholesterol is at the centre of the storm – and of countless contradictory statements about animal and vegetable fats.

In 1961 the American Heart Association said: 'There is no final proof that changing the fat content of the diet will prevent cardiovascular disease. . . .' In 1965 the same Association turned round and said 'atherosclerosis could originate as a result of high-fat diets': it recommended an increased consumption of vegetable oils, and a decrease in foods and fats rich in cholesterol. Conversely, only a few months beforehand, a professor of food science from California had said that too great a reliance on vegetable oils would greatly increase ageing and decay. Conversely again, the National Academy of Sciences produced a report in 1966 recommending no general decrease in fat consumption: 'Any drastic reduction . . . would alter the body metabolism in unpredictable, possibly deleterious ways.' With such a complex and little understood subject as the ingestion of polyunsaturated fatty acids, triglycerides and all the rest, and with the equally tricky problem of arterial decay, it is small wonder that no hard and fast answer can be given. In summary, therefore, cholesterol had a bad press in the late 1950's, but its reputation is now better; vegetable oils boomed in the 1950's and have done well ever since. However there is a see-saw character to the perpetual skirmishing between the animal fat brigades and the contingents supporting vegetable oils. Sometimes butter forges ahead, following a medical report in its favour or just the whim of the buying public. Sometimes margarines do better for similar reasons (or the lack of them) and no doubt this mercurial warfare will continue.

Where are fats found? Roughly speaking, meats have a comparable weight percentage of fat and protein, although something like lean pork has rather more protein and something like rump steak has rather more fat. Rabbit, duck and chicken all have extremely modest amounts of fat in their meat. Fish vary considerably, with eel, salmon and herring having quite a bit, with cod and haddock having virtually none at all. Human and cow milks have a little, and all the condensed milks and cheeses have far more. Egg white has scarcely any (about 1%), egg yolk has a lot (over 30%). Substances like lard, dripping, olive oil and fish liver oil have 100%; the butters and margarines have less as they contain some water. Most of the cereals are extremely low in fat, and the sugars, jams, treacles and honeys have none. Nuts have a lot, Brazil (70%), hazel (62%), walnut (60%), and peanuts (48%). Fruit and vegetables have less than 1% fat.

Carbohydrates   These include all sugars and starches. Chemically they are built up of carbon, hydrogen and oxygen, and they follow a general ruling: there are two hydrogen atoms for each atom of oxygen – as in water. The general formula can be written as $C_x(H_2O)_y$, and this is called a hydrated carbon or carbohydrate.

British people eat carbohydrates, fats and proteins roughly in the proportion of 4:1:1. This means, bearing in mind that a given weight of fat yields over twice as much energy as a given weight of either protein or carbohydrate, that 55% of British energy comes from carbohydrate. Many other societies, because of the universal cheapness of carbohydrates over fats and proteins, eat an even greater proportion of sugars and starches. It is customarily recommended that carbohydrate should not supply more than 66% of anyone's energy; but it frequently does so.

The total weight of carbohydrate within a 156 lb. (70 kg.) man is about 13 ozs. (368 gms.), with some two-thirds of this being in the muscles. Carbohydrate is steadily being used by the body, but can only be stored to a very limited degree. Within thirteen hours of the last replacement, even assuming a sedentary occupation, all available supplies have been consumed, and fat stores will be called upon to bridge the gap. Basically, the body's method of dealing with the various forms of ingested carbohydrates is to try and turn them all into glucose. No starches can be absorbed into the blood: the enzymes amylase and maltase must work on them to produce glucose. The sucrose of cane or beet sugar is turned by sucrase into glucose and fructose, and the fructose is turned into glucose both by the liver cells and other cells. The lactose of milk is turned by lactase into glucose and galactose, but the liver soon turns the galactose into glucose. There is a stubborn persistence about the body's attitude towards carbohydrates. Glucose is always the end result.

Unfortunately ability does not always match this persistence. A

431

notable exception is cellulose. Were the human body able to absorb this sugar the world would not go short of carbohydrates because cellulose is omnipresent in the vegetation all around us. We do eat cellulose in large quantities with leafy vegetables and most forms of plant food; but nothing happens to it and the undigested raw material is expelled in the faeces. It may have suffered somewhat from decomposition by bacteria, but they have not had time to break it down into a form that human enzymes can work upon. The greater length or capacity or complexity of the herbivore's alimentary canal permits greater opportunity for the bacteria, and therefore an absorption of the valuable glucose locked within cellulose. Some human beings claim to have the key and eat a lot of grass, and possibly they do extract some nourishment from their lawn mowings; but a human being is, in general, incapable of benefiting from the universality of cellulose. He can starve in a forest of vegetative profusion.

There is an argument that cellulose supplies beneficial roughage. No such argument can be applied to the refined sugars which form an increasingly large lump of our diet. In fact, sugars have been receiving bad notices for quite a while. A writer to *The Lancet* in 1964 said 'the refining of sugar may yet prove to have been a greater tragedy for civilised man than the discovery of tobacco'. Refined sugar used to be listed with spices as a rare and expensive delicacy. Colonisation of the new world with cheap labour from the old led to huge sugar cane plantations, and by 1837 the average consumption of refined sugar in Britain had risen to 20 lbs. (9 kgs.) per man per year (or about an ounce a day). By 1850 it was 30 lbs. (13·6 kgs.) per man per year, by 1900 82 lbs. (37·2 kgs.), by 1936 100 lbs. (45·4 kgs.) – and then came the war with its enforced ration of only 26 lbs. (11·8 kgs.) per man per year, but by 1961 it was up to 118 lbs. (53·5 kgs.) (over 5 ozs. (142 gms.) a day) and it is now even higher. In the United States the recent rise has been steeper still, going up 120% in the past seventy years.

In 1964, Professor John Yudkin and Mrs Janet Roddy published an exceptionally important paper entitled '*Levels of Dietary Sucrose in Patients with Occlusive Arteriosclerotic Disease*'. For years the fats had been harangued for their suspected role in arterial disease, and for years the sugars had been relatively free of such insinuations. The joint paper presented evidence that 'sugar rather than fat is responsible', and it added that sugar was the more likely culprit on both evolutionary and historical grounds Not only had the authors discovered evidence of association between high sugar intake and arterial and heart disease, but also

432

the increasing consumption of sugar mirrored the increasing incidence of cardiovascular diseases. The rise in fat consumption – where such a rise even exists – is much less positive than the meteoric rise in sugar intake.

Many authors supported Yudkin and Roddy with similar findings of their own. Various others (for such is the meat and drink of scientific argument) took issue either with the paper's conclusions or with the manner in which the evidence had been collected. All in all, with sugars now indicted, with fats still criticised, with protein always expensive, and with any form of gluttony deplored, the latest revelations add weight to a remark made by Alistair Cooke about the United States Food and Nutrition Board. He pictured them 'haunted by the fear that someone, somewhere, may be happy eating'.

Where to find carbohydrates? Meats, in general, do not possess any, but there is some in liver, and quite a bit in such artefacts as pork pies, sausages and fish paste. Milk is about 5% carbohydrate; therefore, by weight, the condensed and dried milks possess much greater percentages of carbohydrate, although the cheeses either contain the same amount as milk or less. Egg contains a little (less than 5%) and chocolate a lot (some 55% on average). Of course the cereals contain massive amounts, with white flour being 75% carbohydrate, rice even more, sago and tapioca more still. Sugar is 100% carbohydrate, but brown sugars contain 2% or so of impurities, such as a few minerals and possibly some vitamins which do not contribute any significant advantages to normal diet. Nuts all contain carbohydrates. So do all fruits and vegetables – it is their greatest constituent after water. As a generalisation, albeit with exceptions, starvation happens in the world when carbohydrate supplies are exhausted, and malnutrition happens when carbohydrate is available but either protein or fat is not.

Minerals Carbon, hydrogen, oxygen, nitrogen and sulphur are necessary to life, but are always liberally contained in the carbohydrate, fat and protein of the diet. The remaining elements necessary to life are customarily grouped together under one heading – minerals. Unlike the first five, which *have* to be eaten as part of a molecule, the remainder can be eaten in their ionic form. Most elements are eaten along with the food without there being a chance of inadequacy; but in certain areas certain elements can be in short supply, notably calcium, iron and iodine. Minerals are not used in the production of energy, but are eaten solely – except when the body is actually growing – to replace the minerals that are lost, principally by excretion. Even the minerals such as calcium and phosphorus needed in greatest quantity by the body are required in very small amounts; while others, just as vital, are needed in microscopic fractions. No one can be deprived of iodine in their diet, but needs are measured in millionths of an ounce per day. (The extreme modesty of mineral requirements demands their measurement in grams rather than ounces – 28·35 grams equal 1 oz.).

Calcium Needs are roughly 1 gm. a day for children, 1·6 gms. for pregnant or lactating women, and 0·68 gms. a day for adult men. Most calcium eaten is not even absorbed through the intestine, but passes straight out with the faeces. The chief roles of calcium are associated with bones and teeth, blood clotting, muscle contraction, and nerve impulses; and there must always be a certain level of calcium in the blood. Calcium is plentiful in food. Milk has about a gram a quart. There is little in meat, quite a lot in egg yolk and it is present in most cereals and some vegetables, although generally in a form less suitable for absorption. Countries with poor milk or cheese supplies are most likely to suffer from calcium deficiencies – small stature, poor teeth. One per cent of the edible part of cheese is calcium, but less than 0·1% of wheat and rice.

Phosphorus Needs are 1 gm. a day for children, 1·9 gms. for pregnant or lactating women, and 1·3 gms. a day for adult men. This element can be used either in an organic form, such as part of a protein, or in an inorganic form, such as part of a salt. Phosphorus is used in the body in association with bones and teeth, with the fat content of every cell, with enzymes, and with many intermediates in metabolism. It is found in most foods. Good sources are meat, fish, milk, cheese, eggs, many vegetables and many cereals. Normally, provided someone is getting enough to eat, he is getting enough phosphorus.

Iron Daily needs, although vital, are only about 0·012 of a gram. This means the loss and the replacement of ¹⁄₂₅₀ of the body's iron content every day. Iron's chief role is in the formation of blood (about 60% of the body's iron is in the haem portion of haemoglobin), but there is some in enzymes and in the liver. A normal diet has more than enough iron in it to satisfy the body's milligram requirements, and good sources are liver, kidney, egg, many vegetables and cereals. Unfortunately there is little iron in human milk and even less in cow's milk. Therefore babies fed lengthily and solely at the breast or the bottle are liable to suffer from an iron deficiency anaemia. This is no problem for six months or more because the baby is born with a modest store of iron and does receive a little in its milk. There is varied enthusiasm and considerable debate about giving additional iron to pregnant and nursing women.

Iodine   Daily needs are 0·00005 grams (or slightly over a millionth of an ounce). Most areas have more than sufficient iodine in the water and vegetables to provide this modest intake, but there are iodine-deficient regions, such as Derbyshire, Michigan and some Swiss valleys. Goitre, or neck swelling due to over-activity of the thyroid gland, can result, and disruption of the thyroid can lead to stunted growth. Women are far more affected than men, and treatment generally consists of the addition of minute fractions of iodine to the table salt. The thyroid gland contains over half of the 0·025 grams of iodine normally in the body. As Iodine 131 is a radioactive fission product, happily short-lived with a half-life of eight days, nuclear explosions or accidents can lead to a sudden rise in iodine contamination of local water, milk and vegetables. Conveniently most of this radioactive iodine will, as with ordinary iodine, be absorbed by the thyroid gland, thus localising the radiation and helping geiger counters to measure the dose which has been received, although it is unsatisfactory medically that the dose should concentrate in just one area.

Sodium   This element can hardly be mentioned without its principal salt, sodium chloride. Man cannot exist without sodium and without sodium chloride, the dominant salt of common salt; yet he cannot consume too much salt, as everyone knows, thirsty ancient mariner or not. Normal daily intake of sodium chloride is 20 gms. per adult, well above actual needs. Although bulk salt is not found everywhere (the sea, saline springs and salt mines are the principal sources), the normal diet contains quite enough salt without any extra sprinkling either at the cooking stage or at the table. Such additions are for taste; but in areas accustomed to food preserved in salt the acquired craving for it can be considerable. (The bad news, actively promoted in the early 1980's, is that salt is one more substance, along with all the others, that can do harm. As the droll individual stated on leaving his bed in a *New Yorker* drawing: I wonder what science will discover today to be acting harmfully upon my health.)

When abnormal quantities of salt are lost in sweating the losses have to be made good, but one aspect of acclimatisation to hot places is a reduction in the amount of salt lost both by sweat and urine. Salt intake does not need to be much higher in the tropics, particularly as most individuals normally consume more than enough either for temperate or tropical conditions. Of the 60 gms. of sodium customarily found in the body one-third is in the bones, and two-thirds are in the extra-cellular fluids.

Potassium   This element is chemically very similar to sodium, and its lack can be just as fatal to a body totally deprived of it; but, whereas most of the body's sodium is outside the cells, 99% of the body's potassium is inside them. There are about 150 gms. of potassium in the body, and daily turnover is about 3 gms. Potassium deficiency is unlikely, but possible.

Magnesium   It plays a similar role to calcium, but is scarcely ever deficient in diet. Most of the body's magnesium is in bone.

Chlorine, bromine and fluorine.   These are grouped together chemically, but they have few similarities for the body. Chlorine is mainly present as the chloride of sodium chloride, and therefore plays a vital part in the osmotic regulation of the body's water content. Normally about 15 gms. of chloride are lost and replaced every day. Bromine is a mystery. It is distributed throughout the tissues, there is more bromine than iodine, and its importance is quite unknown. Fluorine has achieved distinction because of its association with tooth decay. There is convincing evidence that fluoride in the drinking water inhibits dental caries. There is doubt about the actual role of fluoride in the body, as the lack of it causes no detectable effect beyond increased tooth decay, but there is absolute certainty that it can be harmful in excess. A mere 2·5 gms. of fluoride are lethal, and the compound sodium fluosilicate can be fatal if only 0·2 gms. are eaten.

Copper, cobalt, zinc, manganese and molybdenum   These five are the small five, the essential trace minerals. An average man possesses 0·96 gms. of zinc, 0·15 gms. of copper, 0·018 gms. of molybdenum, 0·012 gms. of cobalt, and 0·011 gms. of manganese – a total weight of scarcely more than a gram, but he *has* to have them. They play vital parts in the creation of vital enzymes. Strangely there is more copper in the brain and liver of a baby than an adult.

The rest   Twenty elements have been mentioned. Of the

remaining seventy-two naturally occurring elements, a good many are both eaten and customarily present in the body, but what they do, or whether they have to be present, is unknown. Take aluminium. Ever since the manufacture of aluminium saucepans there has been controversy whether the extra quantities of this element now inevitably added to human diet are harmful, beneficial or totally irrelevant. According to existing knowledge, the human body manages with about a fifth of the Earth's natural elements; and the remaining four-fifths are considered – at present – to be quite unnecessary.

Vitamins Give a man only the correct amounts of fat, carbohydrate, protein, minerals and water, and he will surely die, probably within a few months, certainly before long. The understanding of vitamins has a long or short history, according to your outlook. The actual word is younger than the current century, having been coined by the Polish Jew Casimir Funk when working in England in 1912. His word vitamine, used to describe what had been called accessory food factors, was a union of vital and amine. They are still vital, but between the two world wars it was realised that few were amines. Therefore the spelling was adjusted to something less emphatically inaccurate.

The idea of vital substances existing surreptitiously in foodstuffs is much, much older. Hippocrates was recommending Vitamin A when he suggested liver for night-blindness. The Indians were suggesting Vitamin C when they produced a leafy brew in Quebec for Jacques Cartier's scurvy-ridden men. Two centuries later, when Captain James Cook received accolades owed more accurately to various nautical predecessors for insisting upon lime juice for his crew, he was only recommending Vitamin C. In that same eighteenth century cod-liver oil, rich in A and D, achieved fame in industrial England as a cure-all primarily because it was making good dietary inadequacies. A century later the Japanese were fighting beri-beri in their sailors by giving then barley. By the turn of the century the scientists were doing important experiments proving the same sort of point, and in the twentieth century's first decade the subject of accessory food factors started to open up. The food deficiences of the 1914–18 War were a great spur to research, but the work took time: in 1921 the magazine *Discovery* reported 'it is at present generally accepted that there are three distinct vitamines...'.

Now there is evidence that some forty vitamins exist, at least a dozen of which are essential to the human diet. Unfortunately, despite the similarity of the various roles of vitamins, in that we cannot get by without their presence, there is no similarity between the vitamins themselves. They are a hotch-potch of chemicals which the body must have but cannot manufacture. Professor A. S. Romer has called them 'the odds and ends of vitally needed materials'. Others have called them 'hormones produced outside the body'.Unlike the vital amino-acids, which also cannot be made by the body, the dominant characteristic of vitamins is the pathetically small amounts needed to keep a person healthy. A man, for example, needs 0·003 grams of Vitamin A per day (or a ten-thousandth of an ounce). If totally deprived of this vital speck in his diet he will suffer visually, he will be more easily infected, he will suffer internal lesions, and – if still growing – his growth will be disturbed. In his whole lifetime he will need to consume only 3 ozs. (85 gms.) of this entirely necessary substance. He needs even less of some others, 1 oz. (28 gms.) of thiamin, distributed over his seventy years, will be ample.

To cope with such pathetic requirements the League of Nations, in the 1930s, defined International Units of vitamins in a suitably small fashion. For example the I.U. of Vitamin B_1 is defined as 'the activity of 3 micro-grams [millionths of a gram] of the International Standard Preparation of pure vitamin B_1'. Consequently, due to such minute definitions, any vitamin dose seems fairly hefty when it refers to thousands of units being encased within a single vitamin pill, but the scale of things is again shrunk to its proper size when it is remembered that a few thousand millionths of a gram are still only a few thousandths of a gram, i.e. very little.

All animals and not just man require vitamins, or so it would seem because even the smallest micro-organisms require fairly similar groups of vitamins to our own. Naturally, not too much work has been done on the dietary refinements of 99% of the animal kingdom; but, from the work which has been done, it is probable that all animals require vitamins, that requirements vary widely, and that some animals are capable of making common vitamins for themselves. Of course, as soon as a vitamin can be manufactured by an organism it is no longer a vitamin, but just one more complex organic molecule being created as part of an organism's biochemistry. And if some animals and plants were incapable of making vitamins for others to consume, the natural supply would soon expire. Vitamins are what each creature cannot make but can normally find in its diet. Only when diets become grossly abnormal, as in forced labour camps, or on long sea voyages, or when rice grains are polished, does avitaminosis manifest itself in its various unpleasant ways.

The haphazard classification of vitamins reflects to some extent the manner of their discovery. The alphabetic listing of each new discovery was upset by subsequent discoveries (J. B. S. Haldane made use of the word 'covery' to describe work which upset or corrected an earlier 'discovery'). The single vitamin B was subsequently found to contain more than one vitamin, so B1, B2 and so forth were coined. Later, various 'coveries' made it clear that some sub-divisions were one and the same. Today the numbered B vitamins are B_1, B_2, B_6 and B_{12}, and today these four – like most vitamins – have proper names as well.

Briefly the characteristics, names, functions, and peculiarities of the leading vitamins are:

Vitamin A   Certainly all mammals, probably all vertebrates, and possibly large numbers of invertebrates need either Vitamin A in their diet or some similar substance enabling them to manufacture it. Lack of A causes blindness in rats. Deficiency of A in humans causes night-blindness, a dry and inflamed eye, lack of resistance to infection and bad skin. The major source of A in human diet is liver, but there is huge disparity of A content in food. International units per ounce are sardines 25, milk 75, peas 200, carrots 550, butter 725, rose-hips 2,100, ox-liver 4,600, calf's liver 13,000, and cod liver oil 14,000. Excessive intake of A can be poisonous, causing painful swellings, rashes and loss of hair.

Vitamin B   Every animal needs vitamins of this group to carry out vital reactions within the cell. (One wonders why such crucial components of living matter have to come from outside.) The four numbered B vitamins are also called Thiamin (B_1), Riboflavin (B_2), Pyridoxin (B_6) and Cobalamin (B_{12}). Various other vitamins, all water soluble, are often classified in the B complex: important ones are Nicotinic acid (or Niacin or P-P), Biotin (or H), Pantothenic acid and Folic acid.

B_1 was the first of the group to be isolated – by the American Robert Williams who in 1910 forced some rice-bran syrup containing B_1 into a victim of beri-beri (after the Sinhalese word for weakness). Victims of alcoholism often suffer from a form of beri-beri – called alcoholic polyneuritis – due to poor diet. The tales about an alcoholic's suffering liver should often be applied, more correctly, to the alcoholic's suffering body as a whole. All B (and C) vitamins are likely to suffer during boiling because they

439

are soluble in water and therefore can diffuse into the cooking water.

B_{12} is the most spectacular being the largest vitamin and the only one to contain a metal atom (cobalt). It can only, so far as is known, be manufactured in nature by micro-organisms, and it is the most potent of known vitamins – the daily human requirement is a few millionths of a gram. B_{12} is deficient in the strict vegan's diet, and needs to be supplemented. The source of dietary B_1 is largely the germ of cereal grain; hence the rash of beri-beri when this was polished away, but the richest source is yeast.

The other B vitamins are obtained in a normal diet from a great variety of foods including the meats, cereals and vegetables. During World War Two national flour was introduced in Britain. This had to have a certain percentage of the wheat germ, and therefore more B_1. White flour is now again permitted in Britain, but is still subjected to beneficial additions, such as B_1, nicotinic acid, iron and calcium carbonate. Pellagra, a disease suffered occasionally by American and European poor, is caused by a lack of nicotinic acid: 7,000 Americans died from it in 1930, the peak year for this fatality.

Vitamin C, or Ascorbic acid, prevents scurvy. It is amply supplied in fruit and vegetables and only a couple of ounces daily of blackcurrants, Brussels sprouts, strawberries or peppers, for example, would be sufficient for each human being. Fresh fruit and uncooked vegetables are much richer than old fruit or cooked vegetables. The famous lime juice may either be rich in vitamin C or totally devoid of it: a polar expedition of 1875 relied upon lime juice and suffered scurvy. The disease has a venerable history on expeditions, and a Spanish galleon was once found with every man on board dead of it, but it now survives in a less adventurous fashion; the old and the lonely are its current victims. Lazy bachelors, widows, widowers and the poor are likely sufferers.

Many old people eat a lot of potatoes, but stored potatoes have lost a lot of their vitamin C and boiling will destroy much of the remainder. There may be hundreds of thousands of scurvy sufferers among the elderly in Britain, reported Dr Geoffrey Taylor in 1965, with most of these cases being undiagnosed. Babies fed entirely on cow's milk have also suffered, and need additional C; the boiling or even heating of the milk will have destroyed much of the vitamin C that did exist. Human milk contains enough C provided the mother is getting enough, but even such breast-fed babies are generally offered supplementary spoonfuls of orange juice these days. In many households the elderly could well imbibe a sip or two of that C-rich orange juice.

Vitamin D   This protects against rickets, and is called the anti-rachitic vitamin. Not strictly a vitamin, as it can be made in the body, its existence was first proved in the 1920's, although an association between the bandy, stunted legs of rickets and bad nourishment had been known only too well. Good sources of D are the fish liver oils, but there is colossal variance: there are 40,000 international units per liver oil gram from bluefin tuna, 1,000 per gram from halibut and mackerel, 100 from cod, and none from sturgeon. There is some D in egg-yolk and some in milk, but the cheapest source by far – although still unavailable for many – is sunshine. Solar radiation, although with a great capacity for doing harm, can also turn sterols (including the cholesterol linked with heart disease) into vitamin D. Eskimos, seeing little sunshine, get their D from fish oil; equatorial Africans are the other way around. Unfortunately some slum children in Britain have experienced neither sufficient sunlight nor sufficient D in their diet. Even in the 1960's, with cod-liver oil supplies available at the welfare clinics and with cheap foods, like dried milk, fortified with vitamin D, there were forty cases of rickets in five years from just one Glasgow slum area.

Conversely there have been cases of hypercalcaemia, an unpleasant assortment of disorders including mental retardation of babies, which has been blamed – by some – upon excessive vitamin D intake. In 1957, due to these suspicions of guilt, the D-content of British welfare foods was halved. No one blames this reduction for the cases of rickets which can appear in welfare Britain. The United States, alarmed at the huge intake of vitamin D by many pregnant women and its possible effect upon their babies, has sought to cut the quantities of the vitamin in non-prescription foods. The Food and Drug Administration proposed, in 1965, that the normal daily dose should be limited to 400 units from all sources. Some pregnant women had been taking 2,000 to 3,000 units a day causing, according to some of their doctors, the greater likelihood of disastrous results to their babies.

Vitamin E (the tocopherols)   Like some magic elixir the claims made for this vitamin range widely, including cures and treatment for infertility, menopause, heart disease, lupus,

441

diabetes and muscular dystrophy. Unfortunately, although it is generally agreed that vitamin E does none of these things, there is much uncertainty about what it does do. No optimum human requirement can be given, since a human lack of the vitamin is extremely rare and the lack does not manifest itself in any clear-cut fashion. This is not so for animals; without it chicks die from cerebellar degeneration, pigs develop liver disease, rats fail to multiply. The rat story helped to found the tale of E being the sex vitamin, for without it rats suffer progressive reproductive deterioration: first, gestation is prolonged, then stillbirths occur, finally the unborn young perish.

The vitamin is found in a wide range of vegetable foods, and some is even found in animal foods although there is no evidence that the animal has actually manufactured the vitamin itself. In the sixty years since its discovery, due one suspects to the very positive effects upon animals, the search for the vitamin's role in humans has never been allowed to fade. In the meantime vitamin E, called everything from a will-o'-the-wisp to the 'shady lady of nutrition', will have to suffer such nomenclature, as well as its alleged association with fertility in particular and sex in general.

Vitamin K   A human baby is born with virtually no bacteria in its intestine. This fact can be highly relevant both to decay and vitamin lack. A stillborn child is likely to mummify because decomposition does not start from within as with longer-lived bodies, and a new-born baby can die from lack of vitamin K. This vitamin is divided into K_1 and K_2; the former is abundant in any normal diet, and the latter is amply produced by the intestinal bacteria. Unfortunately some babies receive inadequate supplies of vitamin K from their mothers and are born with a major need of it: they can then suffer haemorrhagic disease of the new-born. If blood loss or haemorrhage does not kill them within the first two or three days, the disease will rapidly recede because the bacteria invading the intestine start their production of K_2 within a few days of birth. The critical first three days can be made safe by daily doses of vitamin K in the sort of quantity that only makes sense in the world of vitamins: one thousandth of a gram every twenty-four hours.

Lack of vitamin K in older children and adults is unlikely, but it may occur during starvation or when antibacterial drugs, such as the sulphonamides, have killed off the bulk of the intestinal bacteria or, as in obstructive jaundice, when bile is not getting into the intestines. (No K can be absorbed without the help of

442

bile.) The vitamin is extremely important in birds and the poultry industry; without K in their diets they die rapidly.

The novelty of vitamins and their fearful revenge when they are inadequately absorbed has led to great respect for their powers. Sometimes this respect has overflowed into a belief that, while some vitamins are good, more must be even more good. They are not. Each vitamin is required to a degree, but no more, and much more can be harmful. One also suspects that the full vitamin story has yet to be told, and not every essential organic chemical in the human diet has yet been identified.

Water   There is nothing quite like water. It has no equal in the number of functions it carries out in living tissue. It is the best solvent in existence; it is extremely stable, carrying many chemicals either in suspension or solution without itself being changed; it is the most efficient cooling agent because it needs so much heat to evaporate each gram; it vaporises easily at human temperatures; it is by far the most abundant chemical in the human body, and it is also the most abundant chemical in the human diet. It dominated mankind's evolution; it is still dominating progress. Its own chemistry is simple, being easily made by the meeting of hydrogen and oxygen and less easily parted. Its absence from the diet will cause a quicker death than if every other dietary need is withheld. There is, in short, nothing like it.

An average human being, doing light work in a temperate climate, loses nearly five pints of water a day. Therefore he has to replace it. As about half of ordinary food is water he can take in three-fifths of a pint (just under $\frac{1}{2}$ litre) at each meal without drinking at all. He also creates about three-quarters of a pint of water a day within himself in the oxidising of his food; both energy and water are by-products of this combustion, notably of the combustion of fat. The remaining $2\frac{1}{2}$ pints ($1\frac{1}{2}$ litres) or so which a man needs to maintain the balance are drunk as liquids. The daily 5 pint (3 litre) loss of this average man is made up of one-fifth of a pint (0·1 litre) in his faeces, nearly a pint ($\frac{1}{2}$ litre) from his lungs, some $2\frac{1}{2}$ pints ($1\frac{1}{2}$ litres) of urine, and about $1\frac{1}{4}$ pints (0·7 litre) from his skin. Sweating is often spoken of as if it only begins during strenuous exercise; instead it is customarily only visible during such exercise, when evaporation fails to keep pace with perspiration, but even a man at rest is losing over half an ounce (14 gms.) of water through his skin every restful hour.

Intake and output are both highly variable. Diarrhoea can step

443

up the faecal loss greatly. Whereas a sedentary European is only consuming some 5 pints (3 litres) a day, a miner may lose 13 pints (7½ litres) in a shift, and an Indian working in the sun may have to replace over 20 pints (11¼ litres) every normal day. Dehydration and sweating can each cause a great drop in the excretion of urine. Tropical conditions can cause urine loss to be less than a pint a day, often much less. On occasion sweat loss in the heat can even be 3½ pints (2 litres) an hour, and 1½ pints (just under 1 litre) an hour can be lost for several hours in succession provided that the fluid is being replaced in drinks. The maximum possible daily loss – and replacement – ever recorded is about 50 pints (28½ litres). Even under normal sedentary conditions in a temperate climate there is a rapid turn-over in the body's water supply: a 156 lb. (70 kgs.) man possesses 70-80 pints (about 40-45½ litres) of water within his frame, and half of these pints will have been lost and replaced every ten days. Incidentally, a woman has less water within her than a man of equal weight. The reason is her greater abundance of fat, and the low water content of all fatty tissue. (Female water content also oscillates with the menstrual cycle.)

Man is not a good desert animal. His organic system squanders its precious liquid. The kangaroo-rat, which never needs to drink water, is far more careful. It has, for example, a cunning by-pass system within its nose for cooling down expired air so that much of the water this inevitably contains is condensed as a kind of nasal dew; therefore it is not lost to the system. Man just breathes out, and wastes nearly a pint a day by doing so. The kangaroo-rat may still lose 13% of its modest 3½ oz. (99 gms.) weight in an hour by water loss; a human being is dead or dying after such a loss, whether after hours or days. Given cool conditions, a totally inert existence and neither food nor water, a man is dead when he loses 15% of his body weight, and such a loss has usually occurred within ten days. (A camel can last for two weeks with no water at all, it can lose up to 30% of its body water with no ill-effects, and it can then drink 30 gallons of water in 10 minutes and return to normal.) Given water and no food a man can survive far longer, for two months or more, and given prolonged starvation he can still live having lost over half his weight. With water, therefore, a 12 stone man can drop to less than 6 stone; without water he is dead well before he has even reached 10 stone. Whereas acute hunger fades away after a few days acute thirst is continually rampant.

Thirsty mariners, unable to drink sea-water, may well have enviously noticed sea birds drinking it. The human system cannot absorb large quantities of sea-water because the salt it contains can only be excreted by the use of even greater quantities of water. The body's capacity for salt is strictly limited, and the circle is therefore as vicious as any. Most sea-birds possess nasal glands. These lie above each eye, and they secrete a fluid extremely rich in salt. Like thick tears these secretions are shaken off and the bird is therefore able to make use of the desalted water it has drunk. Those who stress the perfection of the human being should ponder for a moment upon the advantages of such a nasal gland.

How much food?   Most human food goes to produce energy; only a small fraction is used for growth, repair and replacement of tissues. Vitamins, water and minerals in the diet produce no energy. Instead it all comes from the burning of carbohydrate, protein and fat fuel in the ratio of 4:4:9 for any given weight of these three. It is entirely correct to refer to them as fuel. Just because something is eaten, digested and then acted upon during metabolism in the presence of oxygen does not make it basically any different from a lump of coal burning in the grate. In fact, 1 lb. ($\frac{1}{2}$ kg.) of coal produces about as much energy as a man needs from his food in one day. The energy potential of a piece of food, say a sandwich, can be discovered in a most physical fashion by placing it inside a calorimeter (essentially just a can), burning it in the presence of oxygen, and measuring the amount of heat produced by the combustion. Both sandwich and coal will be ash at the end of the experiment, and an eaten sandwich will have been just as emphatically consumed by the body.

The energy of coal and of sandwiches are both measured in heat units, namely calories. A single calorie is the amount of heat required to raise the temperature of one gram of water by 1°C. This is too small a unit of heat for most measurements of the energy of food; therefore the kilocalorie is used. This unit, one thousand times larger, can heat one kilogram of water by 1°C. Unfortunately the precise term kilocalorie is forgotten in nearly all discussion about food, and the word calorie is used. The daily human requirement of some 3,000 calories is truly 3,000 kilocalories, or enough to heat 3,000 kilograms of water through 1°C, or 30 kilograms of water through 100°C – or enough to maintain an active human being throughout twenty-four hours.

Human needs have a touch of economics about them; there seem to be as many opinions as there are experts. (Perhaps there are not enough experts. University chairs for animal nutrition tend to outnumber those for human nutrition.) Essentially there is the basic metabolic need, the actual cost of maintaining the warm, healthy and resting body. On top of this comes the cost of doing physical work, whether sewing or coal-mining.

The basic need has been calculated as 40 calories an hour for every square metre of a man's bodily surface area, and 37 calories for every female square metre (or roughly 3·7 and 3·5 calories per square foot). The difference in surface area between a lissom young girl and some blacksmith of a man is not as great as might be expected; and can be determined if both weight and height are known. Some examples are:

Weight	Height	Surface area of body Sq. metres	Sq. feet	Basic needs per hour Man	Woman
7 stone	4' 10"	1·355	14·5	54	50
8	5' 2"	1·505	16·2	60	56
9	5' 6"	1·65	17·7	66	61
10	5' 8"	1·765	18·9	70	65
11	6' 0"	1·915	20·5	76	71
12	6' 2"	2·03	21·7	81	75
13	6' 3"	2·115	22·7	84	78
14	6' 4"	2·195	23·6	87	81

Therefore, between the short, light girl and the 6 ft. 4 ins. (1·9m.) heavyweight who could carry her off with ease, there is double the weight but only 63% more surface area, a mere 9 square feet or 0·8 square metres, or the top of a card-table 3 ft. by 3 ft. The difference in basic energy needs between the two is greater owing to the more demanding metabolism of the male. She, if 7 stone (44·5 kgs.) and 4 ft. 10ins. (1·5m.), needs a basic 1,200 calories a day while he, if 14 stone (89 kgs.) and 6 ft. 4 ins. (1·9 m.), needs a basic 2,090, quite apart from the energy needed to lift her up and carry her off. Work done usually demands at least as much energy as the basic need.

Calories become liable to a more flexible interpretation in assessments of their extra supply according to work done. Coal mining, for example, is often said to demand an extra 120 calories an hour, or one and a half times the basic needs of a fair-sized individual; but there must be coal-miners and coal-miners, even assuming an equal task. Some people do a job with a minimum of energy expenditure, only bending down when they have realised the absolute necessity of doing so. They cut corners, lift up a shovel with a foot rather than reach for it, and never squander their resources. As was said at an obesity conference, not only do some take less exercise but some exercise less energetically when taking it. A few fat girls were once filmed playing tennis, and even during singles they were motionless for 60% of the time. A fat individual is often expert at such manoeuvres, thereby adding to his or her fatness.

Similarly, defying all the rules, some people eat like giants and remain like rakes; others eat like sparrows and expand like balloons. Also people with large frames are more likely to grow heavier. As the *British Medical Journal* once put it: 'To him who hath, it seems, weight shall be given.' Fat men do in fact need more food; not only do they have more to maintain, but they need more energy to transport themselves around, and yet they still tend to fatness more than the thin ones.

Plainly calories do count, although a book sold well by saying the opposite, but by no means is calorie theory reconciled to calorie practice. Suppose, for example, a man is eating the correct amount. Suppose he is then permitted some minor indulgences, such as a banana with his breakfast, one bottle of beer and one more slice of bread with his lunch, a cup of lemonade and one piece of cake for tea, and then a bar of chocolate on the way home. These humble additions to his fare, not everyone's taste admittedly, nonetheless add up to 600 calories in that one day. In theory, he would have to walk fast for over two hours, or row once from Putney to Mortlake, to burn up the extra calories. If he fails to do something of the sort, and continues to consume the additional items, he will have put on 1 lb. ($\frac{1}{2}$ kg.) of fat by the end of a week. And to get rid of 1 lb. of fat means walking thirty-four miles.

However, despite theory, people do often eat more than they should, they often take less exercise than they should, and yet they do stay reasonably constant in weight. There is no other way of acquiring calories except in food, and yet there do seem to be ways of disregarding calories without putting on fat or doing anything positive to counter their rapid intake. One wonders, for example, about the calorific content of faeces from different people at different times: there must be great variation. Nutritionists are in general agreed that the former straightforwardness of calories tends to grow more complex year by year. An American, for example, has listed twenty-seven different types of obesity. He would also, one presumes, agree that there is more than one reason why some people eat too many calories and yet remain eternally thin, while others are perversely the other way about.

The Food and Nutrition Board of the United States sometimes alters its estimates of calorie needs. Until 1964 the standard man and the standard woman were permitted 3,200 and 2,300 calories daily. Then these allowances were abruptly dropped to 2,900 and 2,100 respectively. The Board is still non-committal about the means of providing these calories. Apart from alcohol which is certainly a food (yielding 7 calories per gram, therefore more than protein, more than carbohydrate, less than fat), and which can certainly add to a drinker's fat, the American receives 47% of his energy from carbohydrate, 41% from fats and 12% from protein. British figures quote more energy coming from carbohydrate, and poor countries receive two-thirds of their energy or more from this one source, a source undeniably rich in energy but inadequate in so many other needs.

A baby and a young child both need much less food than an adult. However active the infant, he is having to move much less bulk about, and requires less energy to move his smaller frame from A to B. By the age of ten or twelve a boy starts needing as much food as his father, by fourteen he probably needs a third as much again, and by eighteen half as much again; the ages depend upon the timing of his growth spurt, and his actual size at any time, for it is his size and his work output which are most relevant. Growth itself is not very demanding of calories. The average growth rate of both boys and girls between eleven and sixteen is about 9 lbs. (4 kgs.) a year, requiring only 44 calories extra a day, allowing 4 calories per gram of protein.

Mankind is an increasingly idle being – his latest title is *Homo sedentarius*. Only a few centuries ago everything was done by muscle power. Now this is no longer so, but there is still great variance in activity and consequently in fuel needs. To swim demands 550 calories an hour, to run 500, to saw wood 400, to work stone 330, to walk fast 240, to dance 240, to bicycle 175, to walk slowly 140, to sweep 100, to drive 60, to dress 50, to stand 40, to write 30, and to think demands none whatsoever. These averages are wild averages, for some dance with the vigour of sawing wood, and some saw wood with the zest of a roadsweeper. The brain does indeed cause no measurable increase in energy output, although muscle tension accompanying the anxieties of cerebral activity consumes energy on its own separate account.

All in all no one runs, dances, bicycles or thinks all day long, and minimum requirements become more uniform. Comparative estimates for sedentary, light, moderate and heavy work have been assessed as requiring at least 2,500, 3,000, 3,500 and 4,000 calories a day. Should anyone in fact, run, dance or ride all day long his needs will be far greater; long bicycle races can burn up 10,000 calories in twenty-four hours, and even some lumbermen can regularly consume and use 8,000 calories a day. For the most part twentieth-century man in advanced countries is being increasingly transported to work, increasingly sat down when he gets there, and increasingly in need of less than 3,000 calories, although he probably eats more than that and so is fatter than he should be.

Obesity and overweight   Who is overweight? Or, rather, what is overweight, and might not average weight itself be overweight? Is it normal or correct for middle-aged spread to occur, and how thick should thick-set people permit themselves to be? The most telling definition takes as its premise that the right weight to be is the one associated with maximum longevity. No one is so single-minded about survival as the insurance companies, and no insurance company is as good as the Metropolitan Life Insurance Company (of the United States) at producing biologically intriguing statistics. Desirable weight, according to its figures, is within 10% of one's weight when twenty-five years old. To be 20% heavier means definite overweight, and anything greater is defined as obesity. Therefore three people weighing, 9, 11 and 13 stone (or 57, 70 and 82·5 kgs.) at twenty-five will have reached the 20% category when their respective weights have crept up to 10 stone 11 lbs. (68· 5 kgs.), 13 stone 3 lbs. (84 kgs.), and 15 stone 8 lbs. (99 kgs.). Similarly, to be within 10% of the twenty-five-year-old weight, means that the maximum permitted additions to the 9, 11 and 13 stone people are respectively 12½ lbs. (5·7 kgs.), 15½ lbs. (7 kg.) and 18 lbs. (8·2 kg.).

There are other more subjective definitions. One is to look in a mirror. Another is to stand nakedly before wives, husbands, lovers or mistresses and demand their genuine opinion. A third (for men) is to make certain that waist measurement is at least 2 ins. (5 cms.) less than deflated chest measurement. A fourth is to squeeze a fold of skin from such fat-potential areas as the side of the lower chest, the back of the upper arm, or just below the shoulder blade: if the fold is more than 1 in. (2·5 cms.) thick it is bluntly said that you are fat.

Finally, there are charts, which have usually been compiled with the aid of insurance morbidity data. The charts divide humanity into ectomorphs, mesomorphs and endomorphs (see page 301), or small, medium and large frames, and allow for the disparity between the three. The desirable weight, for example, of a 6 ft. (1·8 m.) man is permitted to vary between 10 stone 8 lbs. (67 kgs.) (light end of small frame) to 13 stone 2 lbs. (83·5 kgs.) (heavy end of large frame). What is not permitted is for either a small frame man or a medium man to regard himself suddenly as being in a larger frame category merely because he has put on weight. The frame will not change in life; it is the poundage hanging on to it that differs.

Women are overweight more frequently than men. Figures exist suggesting that one-third of all American women over thirty are at least 10% heavier than the standard weight, and 75% of women between forty-five and fifty-five are said to be overweight. Men tend to become heavier later than women, but between forty-five and fifty-five half of them are said to be overweight. There is thought to be less fatness in Britain, and the sex difference is less marked. Of course men in offices put on weight more decidedly, roughly 2 stone (12·7 kgs.) more decidedly than manual workers; but women are contrary, for it is the manual lady who is fatter than the professional one – on average. (As a digression I wonder if women really do see themselves as those very long, thin and elegant creatures dressed so beautifully in shop windows. Adel Rootstein Ltd, who make many of these models, produce a standard size – 5 ft. 8 ins. (1·7 m.) to the hairline, 34 ins. (86 cms.) bust and hips, 24 ins. (61 cms.) waist. There must be extraordinarily few real women with such long and thin dimensions.)

Fat children are more than likely to grow up into fat people; 80% of them do so. Similarly 30% of fat adults can remember being fat in childhood. A recent survey showed that, when both parents are overweight three-quarters of their children will be; when neither parent is overweight, 9% of their children will be. Once again both heredity and environment must be playing an interwoven part. Is this child fat because its genes promote fatness or because its household customarily fattens up the inhabitants? The problem is not simple; most children in Britain consume more calories and more carbohydrates than they require, yet only a few become obese. Mothers generally welcome rotund children, and deplore skinny ones. At the Institute of Child Health in London more visiting mothers express anxiety about possible underweight than possible overweight. Baby-show prize-givers tend to favour the rotund babies, not the slim ones, and the prize-winners then chuckle into their deep-set dimples.

Making young children thinner is a distinct problem, with a general feeling that any treatment before five is unprofitable. Even then diets are liable to be unsuccessful unless the whole family takes part. The best hope of success, according to a leader in the *British Medical Journal,* is to persuade the mother to reduce the calories in general, by giving meals richer in protein, poorer in carbohydrate. The older the fat child the greater the chance of persuading the child to take personal steps to reduce his or her intake.

The control of appetite, and the acquisition, retention and

disposal of food is, despite some obesity, remarkably well controlled. An individual is likely to process 30–40 tons of food through his system in his adult life-time, and yet remain within 5 or 10 lbs. (2·3 or 4·5 kgs.) of a certain figure on his bathroom scales. The so-called appestat, or appetite control centre, is presumed, with good reason, to exist in the hypothalamus, the small portion of brain which lies underneath the cerebrum. If a small central portion of a rat's hypothalamus is deliberately attacked with needles the animal recovers from the anaesthetic to start voracious eating. Sure enough, within a few weeks the rat is very fat. Should two lateral portions of the hypothalamus be destroyed instead, the animal will not eat at all. Occasionally a human being will suffer injury to his hypothalamus, notably when a tumour is growing nearby, and obesity may follow. However, most very fat people reveal no such detectable lesion at post-mortem, and a full understanding of the appestat's mechanism or method of control is a long way from being grasped. How can it possibly know when a meal has proved sufficient? What information can it possibly acquire that leads it to interpret so accurately one's needs of the day?

Medically, fatness is to be deplored. Not only is there an association between extra weight and shorter lives, but it is linked with a great incidence of high blood pressure, of arterial disease, of liver and kidney disease, of varicose veins and a small regiment of other ailments, including obvious associations such as tiredness and shortness of breath. ('Know the grave doth gape for thee thrice wider than for other men', wrote Shakespeare of the fat all those centuries ago.)

Journalistically, fatness is to be welcomed. Slimming articles are excellent value, and there is more than a battalion of slimming notions. Professor John Yudkin, the London nutritionist, has written that he is astounded at the way new slimming nonsense bobs up as soon as he believes he has scotched the previous batch. He has written pungently of those which slim you only where you wish to slim, which surround you with marine plankton, which allegedly work while you relax in the bath. He has himself summed up the whole dieting question by dividing it into three.

1. Low fat diet – likely to lead to hunger, irritability, bad skin, poor concentration, and a possible unhealthiness.
2. Low protein diet – average British intake of protein is 95 gms. daily. The authorities agree it should not be less than 70. Therefore a saving of 25 gms. a day – only 100 calories – is possible, but ineffective.

451

3. Low carbohydrate diet – the easiest kind. An average consumption is 400 gms. – or 1,600 calories – a day. The refined
sugars are responsible for much of the excess. Some slimming
schemes remove sugar, and then replace it by glucose, honey or
sorbitol. These three give, weight for weight, the same calories as
sugar, but are less sweet. Therefore more is taken, and therefore
more calories than before are absorbed.

Finally, many people say it is easy to slim. They have done it time
and time again!

The real slimmers are quite a different matter. These are the
huge people who try and get weight down to a level when they will
again require just one chair to sit on. Recently, instead of giving
them modest diets, some have been given nothing at all – just
vitamins, minerals and water. Sometimes the routine is only
marginally less severe. Take the case of William J. Cobb, born in
Georgia. In 1962 he weighed 802 lbs. (or 57 stone 4 lbs., or 364
kilos), and a struggling circulatory system was failing to supply
enough oxygen; every few steps he had to stop and rest awhile. In
one sense a 200 lbs. (90 kgs.) man was carrying an extra 600 lbs.
(272 kgs.) around with him; small wonder that the load proved
excessive.

He decided to slim. Manfully he reduced his weight to 644 lbs.
(292 kgs.), and then volunteered for obesity research in an
Augusta hospital. This meant a far stricter regimen. He was
permitted no exercise, and merely 1,000 calories a day. (Don't
forget he was four or fives times normal size, and therefore with a
far greater basic need of calories.) The diet varied between high
protein, high fat and high carbohydrate diets, and every eight
weeks he was shifted from one to another. After eighty-three
weeks he left the hospital, and was soon working as a shoe
repairer: his weight – a dainty 232 lbs. (106 kgs.), proving a weight
loss equivalent to the combined weights of three large men or five
slim women.

At least he left the hospital. A different story, from San
Francisco in 1965, told of a thirty-eight-year-old 5 ft. 2 ins. (1·6
m.) woman who was rushed to hospital. Twenty men helped to
carry her in, as no stretcher or lift would accommodate her, but
she was to die within twenty-four hours. Her weight, both on
arrival and on departure, was registered as 675 lbs. (48 stone 3
lbs., 307 kilos).

Starvation is accompanied, apart from loss of weight, by a
senile appearance of the skin, a slow heartbeat, a shortness of
breath if exercise is attempted, increasing urination, and a

general swelling, or oedema, of the tissues. This retention of liquid is a feature of most starved people, and occurs mainly in the legs, ankles, feet and abdomen. It often gives a reasonably well nourished look to the lower half of the body compared to the skinny chest and spindly arms. Diarrhoea may follow or partner the oedema. Normally, and starvation is so frequent in the world that average findings have been easily evaluated, a person can shrink to half of his or her weight, and still be perfectly capable of putting on all former weight if given food. When weight is less than 50% of its original amount death is much more likely. Its actual cause will probably be some infection, itself permitted by a lack of resistance to infection which, in its turn, is caused by a deficiency of protein.

C. J. Polson in *The Essentials of Forensic Medicine* has listed some of the post-mortem findings of starved people. Body weight is usually well below half normal weight (on average 38%). An 11 stone (70 kgs.) man and an 8 stone (51 kgs.) woman may therefore have been reduced to 4 stone (25·4 kgs.) and 3 stone (19 kgs.). They look old. All organs except the brain have shrunk both in size and weight. No fat is visible, and all muscles have atrophied. There has been de-mineralisation of bone.

Between the starved human being and the obese one lies a fantastic margin of viability. If a starved man of 8 stone (51 kgs.) can fall to 4 stone (25·4 kgs.) and live, and if William J. Cobb can swell up to 802 lbs. (364 kgs.) and also live, there is a gap of 746 lbs. (338·6 kgs.) between these two living adult human beings. One is over fourteen times the weight of the other. Between a highly emaciated woman and Mr Cobb the differential factor is nearer nineteen. There is a huge range to the potential weight of *Homo sapiens*.

Cooking The Hadza of Tanzania are a small tribe living near Lake Eyasi. All their food is either gathered or hunted. They cultivate nothing and store nothing. However they do cook, crudely, briefly and badly. They throw a piece of meat into the fire, remove it later, and chew away at it. As a tribe they are probably typical of countless tribal people who existed in the days before agriculture and animal husbandry, and who were our ancestors. They were primitive, nomadic, casual, and yet they did cook. The practice was initiated, it is believed, virtually as soon as fire was controlled.

Cooking performs various roles. It can improve flavour, as with meats. It can help to liberate substances which stimulate the secretion of digestive juices. It can break down and loosen connective fibres, as with the collagen fibres of meat and the cellulose framework of vegetables. It can also kill bacteria and parasites within the food; think of 'measly pork' and trichinosis. It does not always increase digestibility; raw meat is said to be most easily digested if kept raw but well disintegrated. Cooking helps to achieve that disintegration; but overcooking, causing a shrinkage of the coagulated protein, can decrease the digestibility. Some foods, like meats, have less water in them after cooking; some, like rice, have more. A general estimate is that 10% of the energy value of foods is lost by cooking, but many contend that the loss is much less. Undeniably some vitamins are lost in cooking, notably B_1 and C. All B and C vitamins are soluble in water, and so are likely to be lost in the surrounding water. All other vitamins are soluble in fat. Vitamin C, the fruit vitamin, is particularly liable to destruction by heat. Because vegetables possess enzymes which can destroy vitamins, and because these act more speedily if warmed up, the slow warming of a vegetable can lead to much vitamin loss. A better practice is to plunge the vegetable into boiling water (or hot fat), thereby destroying the enzymes.

The Hadza do indeed cook. However, throwing an impala's head on to an open fire, and then peeling off the charred meat a few minutes later, is about as different as could be from the near-religious devotion that the art of cooking evokes in some households. It has to start somewhere, and that burnt antelope head is clearly nearer to tradition for mankind's stomach than a delicate soufflé threatening to tremble into humiliating collapse.

Food poisoning has three sub-divisions. The food can be contaminated with a poison; it can be a poisonous food; it can have been contaminated by pathogenic bacteria. The first is customarily considered as wilful malevolence on someone's part. The second, with so much advisory folklore, is generally obviated these days. The third is almost always the fault of the staphylococci, the streptococci and the salmonellae – plus the person who permitted them to grow, or who failed to kill them off.

Their presence is insidious. They do not change the taste or smell of food (although other non-pathogenic forms may have simultaneously caused detectable putrefaction). They can all give rise to equally direct, emphatic and short-lived attacks of food poisoning. Mere decay, without a corresponding density of the three pathogenic forms, is generally harmless. Many foods are intolerable without some putrefaction; fruits have to ripen, so do cheeses, so does game which is best eaten high, and – for Eskimos – so does fish. Many inland communities also prefer their fish with a tang to it, accustomed as they are to well-travelled cod and herring.

Clostridium botulinum is a big exception to all the rules. For one thing it can often kill. Its virulent poison is always referred to in discussions of biological warfare (for instance, it has been stated that a one inch cube of botulinus toxin would be sufficient to kill everybody in North America). Some 65% of botulism cases are fatal. Fortunately, the bacillus cannot (usually) multiply without liberating a noxious smell and, equally fortunately, fifteen minutes of boiling will destroy all the toxin and the bacteria. The spores are made of sterner stuff, but even they yield to steam under pressure. Commercially botulism has been defeated, with the last industrial contamination in both Britain and the United States occurring in the 1920s, but the home canning and preserving business can still lead, on occasion, to a modest, but fatal, proliferation of *Clostridium botulinum.*

33. The Alimentary Canal

Beaumont's window/The stomach/Acidity/Ulcers/Vomit/
Intestines/Appendix/Liver/Pancreas/Spleen

'De temps en temps, il faut étonner l'estomac.'

MARCEL PROUST

Quite the most famous (and astonished) stomach of all time belonged to a French-Canadian. His name was Alexis St Martin, and the contributions made to digestion knowledge by his one stomach were enormous. Behaviour of his gastric juices is still quoted today even though the famous accident, which opened up the subject so dramatically, took place so long ago. It was on June 6, 1822, that the eighteen-year-old Alexis was only three feet away from a musket that accidentally went off. He received the shot and wadding in his chest on the left hand side; part of his sixth rib was destroyed, the fifth was fractured, his left lung was damaged, and his stomach was pierced. The other principal in this story, William Beaumont, surgeon in the United States army, was quickly on the spot. He found lung and stomach protruding from the wound, and he pushed them back.

Remarkably, in an age when a severely wounded man was usually just a dead man who had not yet died, after nearly two weeks of giving every indication that he would not do so, the young Alexis began to recover. One week later the wound was quite healthy and the patient was enthusiastically eating, but with one snag – everything he ate came out of the open wound. He was given nutriment via his anus, and Beaumont naturally made every effort to close the stomach wall. Alexis refused any kind of operation that would attempt to suture the wound and instead preferred bandages which served in place of a stomach wall. He ate regularly and well, his digestion was effective, his whole alimentary canal behaved as it should have done and, provided the bandage remained unmoved, all was well. By the following summer, when Beaumont finally gave up all hope of 'closing the orifice', it was $2\frac{1}{2}$ inches (6·3 cms.) in circumference, 2 inches (5 cms.) below the left nipple and apparently a permanently open reminder of the accident. All food and drink still poured out unless some suitable bandage plugged the gap.

Two years later Beaumont started his experiments. This army doctor was among the first Americans to make an important contribution to medical science, and he achieved fame by realising the splendid opportunities afforded by Alexis's stomach: the annoying hole was an excellent window into the physiology of a vital organ. It was little good just looking; so he devised experiments. He weighed morsels of food, tied them with silk, and then let the stomach do its work on them. At hourly intervals he removed them, noted the degree of digestion, and then replaced them. He also took specimens of gastric juice, and it was from such a specimen that hydrochloric acid was first positively identified. Beaumont also noted that a fasting stomach was empty and contracted, and that it become flushed with blood when its owner became angry. It also moved about with anger. (A woman of St Louis who, much later, had a stomach that could also be inspected was made similarly angry, but her stomach then grew pale and motionless. If she is characteristic, this fact may help to explain the greater proportion of peptic ulcers among males.)

Of course once Beaumont realised the remarkable value of his physiological window he wished to make full use of it. He therefore had to follow his patient – expensively – from place to place. Alexis is frequently called 'uncooperative' and 'difficult', and plainly the relationship between these two men had its ups and downs. When having either an up or a down one can imagine Beaumont longing to taste the accompanying gastric acidity (as was frequently his custom) in the interests of science. Anyway he published his *Experiments and Observations on Gastric Juice and the Physiology of Digestion* in 1833. Douglas Guthrie, the medical historian, has called it a 'fine piece of research in the face of unusual difficulties'. Meanwhile Alexis, the difficult Alexis, lived his life and was eventually left alone with his orifice, his perforated stomach and his well-renowned gastric juices. His burial place was kept secret, but in 1960 his granddaughter revealed it to be at St Thomas de Joliette, forty miles north-east of Montreal.

Serendipity in science is common, and gun-shots have frequently been revealing, but none so conveniently as that of Alexis St Martin. So much of what is known today has its origins in that one French-Canadian stomach, and many of those involved in today's perpetual debate on ulcers would dearly love to watch the rise and fall of peptic ulceration, or to have such an intimate acquaintance with practically any aspect of the organs of the digestive tracts.

The human stomach is named after the Greek for throat, the Greek word for stomach being responsible for words like gastro-enteritis, gastrectomy, etc. It is, apart from the mouth and oesophagus, the initial recipient of human food, the first impor-tant organ of the alimentary canal. Its capacity is about $2\frac{1}{2}$ pints ($1\frac{1}{2}$ litres), and a heavy meal takes some six hours to pass through it. Regularly referred to as being J-shaped – which I personally feel is a poor description for it is more like a boxing glove – the con-stricted wrist end is the pyloric region, and the bulbous end is the fundus, which has the oesophagus leading into it from above. Its walls secrete digestive juices, and waves of contraction, roughly three to the minute, cause the food to move into the duodenum, the next section of the alimentary canal. The stomach is storage container, digester and mixer; both its movements and its fluids render food more suitable for the duodenum and intestines than the casually chewed lumps and varieties of food and drink which pass down the oesophagus.

There are also many things which the human stomach is not. It is not vital – many ulcer victims have had two-thirds of it removed, and many others have experienced its total removal. It is not particularly important in digestion, in that the bulk of the digestive enzymes are mixed with the food after it has passed through the stomach. It does not absorb much through its walls; some water and some alcohol may go through, but nearly all absorption occurs later. Quite a few vertebrate species normally have no stomachs.

In fact the evolution of the stomach gives a clue to its current role in man. Almost certainly its primary function, when developed with the early jawed-fishes (earliest fishes had no jaws), was for food storage; the jaws could bolt food and this had to be retained. Such rapidly bolted and chunky food also had to be prepared so that the main alimentary tract could act upon it. The introduction of digestive enzymes to these priorities of storage and preparation was presumably, as A. S. Romer put it, 'a phylogenetic afterthought'. The human stomach, now some 500,000,000 years after those first storage stomachs of the early fish, is still dominantly a storer and a preparer. Consequently its total or partial absence can best be obviated by sending down pre-pared food frequently. Complete removal of the stomach has not only been successfully inflicted upon humans but on various laboratory animals, such as rats, dogs, pigs and monkeys. Of course life is not the same without it but all, with the possible exception of the rat, find the stomach dispensable.

Animals, such as the carnivores which always bolt their food, tend to have large stomachs relative to the alimentary tract as a whole. In dogs and cats, for example, the stomach's capacity is between 60% and 70% of the total digestive capacity. They also tend to empty their stomachs between meals. Herbivores, who tend to eat again before the stomach is empty, generally have long and complex digestive tracts, and the horse's stomach capacity is only 8% of the whole; but some herbivores have exceptionally elaborate stomachs possessing extra pouches that prepare the cud they chew. An ox's stomach is 70% of the whole digestive system. The actual capacities of some animal stomachs are: large dog 5 pints (almost 3 litres), pig $1\frac{1}{2}$–2 gallons (about 7 litres), horse 2–4 gallons ($13\frac{1}{2}$ litres), sheep 4 gallons (18 litres), cow 30–40 gallons (160 litres) (although, strictly speaking, only those parts of the ruminant stomach which secrete gastric juices ought to be called stomach). The human $2\frac{1}{4}$ pint ($1\frac{1}{4}$ litres) stomach is a relatively modest possession, as well as being dispensable if need be. (Both a yard of ale and most college sconces, or measurements of drink consumed manfully and rapidly on various exhibitionist occasions, happen to equal human stomach capacity.)

Some digestive juices are produced by the stomach. The salivary glands have already mixed the digestive enzymes maltase and salivary amylase into the food, and the stomach then adds the enzymes pepsin and rennin. (As with the stomach, the saliva plays other roles, such as helping in the food's preparation and making speech possible, and a man secretes over $2\frac{1}{4}$ pints ($1\frac{1}{2}$ litres) of saliva every twenty-four hours. As a horse secretes 9 gallons (41 litres), and a cow 12 gallons (54·5 litres) a day their occasional drooling seems entirely reasonable.) The stomach's pepsin is a protein-splitter, and creates so-called proteoses and peptones. However long the food stays in the stomach – and two to four hours is a likely time – these proteins will be broken down no further; so far as the stomach is concerned they are then indigestible. Rennin (familiar in the kitchen as rennet, the junket maker, which is extracted from calves' stomachs) has just one task; it converts the caseinogen of milk to casein. This solidification of milk is definitely caused in a baby's stomach by rennin but, as the acidity of a stomach changes with age, the rennin becomes progressively less useful and the pepsin can and does do its job. No one knows much about the role and presence of rennin in the adult stomach.

The stomach's acidity poses problems all of its own. Fairly strong hydrochloric acid (it is 0·17N) is produced by living tissue, and is then poured into the stomach. There is nothing particularly resistant about living tissue, for a snake's living meal starts to suffer digestion as soon as it has been ingested, but the hydrochloric acid is diluted by the food in the stomach, and the stomach walls are protected by alkaline juices. The actual manufacture of hydrochloric acid is still surrounded by theoretical notions, and the available evidence is inadequate: somehow or other, it seems, the acid glands have a membrane which is permeable only to water, hydrogen ions and chloride ions. Even so, this is far from explaining how an acid capable of dissolving iron is produced, collected and secreted by human tissue.

The stomach's acidity is a fact responsible for a great deal of personal medication. Americans spend over £100 million a year on substances to neutralise the acid. There are innumerable notions about foods that are thought either to be acid or which promote or discourage acidity. Sodium bicarbonate, being cheap and easy, has been poured repeatedly into suffering stomachs, and their acidity has frequently been instantly turned into alkalinity. Alkalosis (or alkalaemia), which can be fatal, is the result of excessive alkalinity in the blood. Unlike sodium bicarbonate, which can be absorbed into the bloodstream, there are other antacids which cannot be, and there is also food, often under-rated but an extremely effective neutraliser of stomach acid. Without doubt the stomach is guilty of imperfection, and without doubt it can be upset by the circumstances of life; it just so happens that acid is customarily indicted by possessors of upset stomachs as the guilty agent. Punishment with swift doses of antacids is therefore summarily meted out.

Ulcers Nevertheless there is positive correlation between the secretion of a lot of acid gastric juice and the origin of peptic ulcers. These open wounds are almost always single, usually half an inch to an inch in diameter, and about 60% of them occur in the duodenum. The remaining 40% are in the stomach but near its exit (near the wrist of the boxing glove). Contrarily few ulcers occur in the acid-producing area of the stomach, the bulky fundus. Men get ulcers more than women, 3:1 or 4:1, and those with blood group O are more likely to suffer from duodenal ulcers than the general population. There is a less marked correlation between the group O people and the formation of gastric ulcers. (Peptic ulcers include both gastric ulcers – in the stomach – and duodenal ulcers – in the duodenum.)

460

An excellent study of peptic ulcers in Kent recently investigated 265 sufferers in one part of that county, 212 with duodenal ulcers (83% men) and 53 with gastric ulcers (53% men). There were 4.2 new victims per 1,000 of the general population every year (making, if Kent is normal, 250,000 new peptic ulcer cases a year for Britain as a whole, or nearly one more a week for each general practitioner). Duodenal ulcers in the Kent study were more likely to appear first, in the 30-39 year decade for men and 40-49 decade for women. Gastric ulcers were more likely to be later, in the 50-59 decade for men and 60-69 for women.

Unfortunately ulcers are no will o' the wisp, here today and gone tomorrow. The degree of disablement they cause tends to increase as the years go by, but this reaches a peak after some 5-10 years and then recedes. Average peak times for duodenal men in Kent occurred 8·1 years from the start of symptoms, and for duodenal women 7·1 years. Gastric peaks were after 6·5 years for men and 6·8 years for women. After fifteen years only 2% of the victims were still severely disabled by their peptic ulcers. The two most usual complications of the ulcers were haemorrhage and perforation, and a fifth of the Kent sufferers suffered one or the other, but only one person out of the 265, a man of seventy, had a death attributable to his ulcer. Perforation normally accounts for two-thirds of all ulcer deaths. Only 16% of the 265 men and women of Kent were treated surgically for their ulcers.

Surgery for ulcers is a world of argument in itself, with individual and regional preferences for cutting the vagus nerve, and where to cut it, for cutting out stomach tissue, and how much, for enlarging the stomach's exit, and by how much, and even for freezing part of the stomach. Someone once said that the surest way to empty the room at a medical meeting was to announce that the next subject would be the surgical treatment of peptic ulceration. Someone else's ulcers also have the power to empty countless ordinary rooms if embarked upon as a subject with any determination by the ulcerous owner.

Vomit Similarly, and possibly even more rapidly, another stomach subject has powers of evacuation; yet vomiting has its intriguing characteristics. Herbivorous animals and rodents seldom or never vomit. When a horse does so, which is extremely rare, the vomit usually comes out of its nostrils. Carnivores and omnivores, except those which are rodents, vomit easily. The process of vomiting is not caused by the stomach, and is not a reversal of the customary procedure whereby food moves down the throat and through the stomach. Instead, co-ordinated by the brain, both abdominal wall and diaphragm contract together and squeeze the stomach lying between them. The cardiac sphincter at the oesphagus end of the stomach opens, the phyloric sphincter at the other end stays shut, and the stomach's contents have nowhere to go but up and out. Simultaneously the larynx is raised so that no vomit gets into the air passages. The stomach may have contributed marginally to the general effect by reversing its modest peristaltic flow; but, except for infants, this probably does not happen. In any case most babies have a vomiting ability of their very own: it is often not so much a mere outpouring with them as something akin to the firing of a projectile.

The act of vomiting can be triggered by any number of causes. There is an arresting book called *Understanding your Symptoms* (by Dr Joseph D. Wassersug) and vomiting is mentioned repeatedly in the index of such a book because it is associated with drugs, stomach diseases, many other abdominal complaints, pregnancy, smells, poisons, unpleasant movement, visual disturbances, brain tumours, cerebral abscesses, meningitis, smallpox, scarlet fever, typhus, cholera, too many green apples, or too much alcohol, or tobacco, or salt water, or with reasons that are never identified: people can just get up in the night, vomit, and return to bed feeling much better. Vomiting has been called God's gift to the hypochondriac. The frequent and customary event can be dramatically allied to a bewildering battalion of diseases, any one of which may be the cause. Or none.

Mere expulsion of air from the stomach is known as eructation. Such belches can have an electrifying effect in the wrong company; but they are not customarily explosive. Nevertheless some surgical operations have been rendered dangerous by a patient's sudden eructations of inflammatory gases. These mis-timed and chemically hyperactive belches have caused explosions. The true tale is also told of a parson who had every reason for emotional alarm: his breath caught fire each time he blew out the altar candles. No one cynically told him to pinch out the candles instead, but an operation was performed to enlarge the constricted opening at the far end of his stomach, the fault of an ulcer. 'Following which', according to the man who performed the surgery, 'he was able to carry out his duties in a more decorous fashion.'

Intestines Named after the Latin for 'internal' the intestines are a continuous tube leading from the stomach to the anus. An average figure for mammals is that the tube is eight times the

animal's length, but some herbivores have intestines proportionately much longer. Man has 28-30 ft (8·5-9m.) of intestine.

Various names have been somewhat arbitrarily attached to its
various sections. The first 10 or 12 inches (25-30 cms) are the
duodenum (or twelve finger widths), the next 8 or 9 feet (2·5m.)
are the jejunum, and the slightly longer subsequent length is
the ileum. The ileum joins up with the large intestine, but
apparently inaccurately: the actual join is more like a T-junction
at a main road, with the left-hand portion of the large intestine
being nothing more than a cul-de-sac called the caecum. The
right-hand turning is the true large intestine, or colon; it leads
after 6 ft (1·8m.) into the rectum and the anus. Little of nutritional
value, apart from water absorption, happens in the large intestine. It is to the caecum that the appendix is attached, a thin structure like a flexible pencil and, like pencils, from 2-6 ins (5-15cms.)
long.

Due to the convolutions and folds and twists in the intestinal
tube, the actual absorptive inner layer of the 28-ft. (8·5 m.) tube
has an area of over 100 sq. feet (9 sq. metres), or five times the skin
area of the body as a whole. Practically all human digestion is
carried out in the intestine, and an even greater proportion of the
actual absorption of food takes place through its walls. The
stomach has only played a small part in digestion, and an even
smaller role in absorption.

From the moment a piece of food is eaten it has a restless time;
yet the pace of its travels is by no means constant. William
Gladstone is alleged to have chewed every mouthful thirty-two
times, but most people are more hasty; they soon have the food
descending towards the stomach. Some four to six hours later all
of a large meal is out of the stomach and into the intestine. The
constant jostling by the waves of peristalsis hurry the food – now
called chyme – along the intestines at about an inch a minute.
Within five hours of leaving the stomach it is entering the far
wider tube of the large intestine, and about 12 ozs. (340 gms.) of
liquid chyme enter the large intestine every twenty-four hours.
The pace then slows down, as the chyme takes perhaps two dozen
hours to pass along the half dozen feet of the large intestine. It
takes a swallowed dye 15-25 hours to pass through the alimentary
canal from start to finish before it first begins to appear at the
rectum, but several days before the last of it has disappeared.

Nevertheless there is always movement. The peristaltic waves
are slower in the colon, and the movement of the rectum is slower
still as that is customarily only filled and emptied once every
twenty-four hours. The increasing pressure and the steady

accumulation of faeces cause the rectum to fill, thereby causing a desire to defaecate. When the faeces have been permitted to be expelled (the adult voluntarily opens the sphincter, but the child has to learn how to do this) the rectum is once again empty, and accumulation can begin anew.

The relatively slow rate of progress through the large intestine occurs primarily because that tube has a larger bore – 2½ ins. (6·3 cms.) diameter as against a maximum of 1½ ins. (3·cms.) for the small intestine; but the quantity actually flowing through the large intestine is also steadily decreasing. Throughout its 6 ft. (1·8 m.) length water is absorbed into the blood stream, making the flow less urgent and the chyme less bulky. Of the 12 ozs. (340 gms.) of chyme that flow daily into the large intestine only about 4 ozs. (113 gms.) of faeces pass daily out of it.

Like almost everything else in the body these figures are highly variable: a rich vegetable diet will produce 13 ozs. (368 gms.) a day, and a starvation diet will produce less than an ounce (28 gms.). Faeces are continually produced even throughout starvation. Children produce a relatively large amount of faeces, ranging from 2½-5½ ozs. (71-156 gms.) a day. Even though so much water has been extracted from the chyme to make faeces (after the Latin for dregs) water is still 65% to 80% of its bulk. Of the remainder (dry matter) a third to a half is bacteria, which are mainly dead. The rest consists of secretions from the intestine itself, of cellular remains from the alimentary canal and of very small amounts of food residue, all of which (except for the food residue) helps to explain how it is that faeces are still produced even during starvation.

Not much that is digestible, except in certain diseases, reaches the faeces. The food residue is mainly indigestible items, such as cellulose – hence the large faeces of a vegetable diet – and fruit skins and seeds. Faeces smell most on a meat diet, less so on a vegetable diet, and least of all on a milk diet.

Intestinal gases also vary in their potency. Essentially they are a mixture of swallowed air and of gases produced by the intestinal bacteria; the actual composition changes as diet, degree of constipation and bacterial activity are themselves changed. Flatus, the correct term for the more widely used expression – fart (from the Old English 'feortan') has been analysed, and an average finding is nitrogen 59%, hydrogen 21%, carbon dioxide 9%, methane 7% and oxygen 4%. Occasionally there is a little hydrogen sulphide, a gas responsible for the well-known odour of the rotten egg. Hydrogen and methane are always combustible but, given the right mixture with oxygen, they can be explosive. On rare occasions, notably when using diathermy or cautery, disastrous explosions have resulted when surgeons have opened up intestines containing dangerous proportions of air and the flammable gases.

Finally that appendix. It can still be dangerous: there were 606 appendicitis deaths in England and Wales in 1962, but there had been 1,405 in 1952. Today, for England and Wales, the total is nearer 150, with slightly more females than males dying from this complaint. Next to nothing is known about the causes of its inflammation, but the main cause is certainly not orange pips, grape seeds, apple pips, date stones, and other hard objects finding their way into it. Occasionally such relatively solid objects are found in the appendix, but only very occasionally. For the most part the complaining tube contains nothing but some hardened faeces or fragments of lime which are thought to be the effects of appendicitis and not its cause.

No one knows the purpose of the appendix, but many other primates have it and so do many rodents. Appendicitis commonly affects the young (half the cases are patients under twenty) and males are more frequently afflicted than females. It may be less dangerous these days, but 125,000 appendix patients in England and Wales are still admitted to hospital for treatment annually and, at any time, one in fifty of the general hospital beds is occupied by someone who has every reason to wonder what on earth the human appendix is all about, and why, even if it has a point, it is so painfully unreliable.

Liver   It is the largest gland in the body. In an infant it weighs 4% of the total body weight, occupies two-fifths of the abdominal cavity, and is largely responsible for the distended silhouette of the abdominally podgy child. In an adult it weighs 3 – 4 lbs. (about 1·5 kgs.), according to the adult's size, and represents about a fortieth of the total body weight. It lies on a person's right-hand side, close up against the diaphragm, and roughly matches the smaller stomach's position high up on the left-hand side. At rest a quarter of the body's blood is within the liver, but a pint or two ($\frac{1}{2}$ to 1 litre) of this blood will suddenly leave it when exercise is either taken or is imminent.

Frequently called the chemical factory, the liver performs innumerable roles – some 500 or more have already been counted – in the body's metabolism. It only plays a minor role in intestinal digestion even though it arises in embryology as a gland leading from the gut wall in the duodenal area. Each liver (the word is an Anglo-Saxon one, possibly linked with the verb to live) has a double blood supply: its oxygen comes primarily in the hepatic artery, but it also receives the portal vein. It is this vein that brings to the liver all the absorbed materials from the intestines. No one can live without a liver. The experimental removal of the organ from laboratory animals leaves them with apparent well-being for a little time, but thereafter deterioration is rapid.

No vertebrate is without a liver, but many are without gall-bladders, the small, often troublesome, containers of bile that man can dispense with and frequently has to. Lampreys, many birds, the horse, the deer, the rat, for example, never have one. Strangely the striped gopher has one, and the pocket gopher does not. Stranger still the giraffe sometimes has one and sometimes does not.

Whether or not there is a gall-bladder there is always bile. In the human being this contains no digestive enzymes. Its production is over a pint a day, but the gall-bladder can only store about a tenth of that amount. Bile does include some substances which assist digestion, but it also includes a lot of waste products. Gall-stones, observed only in man and domesticated animals, are usually made of cholesterol, the steroid so frequently maligned for its association with circulatory failure and disease. It is reckoned that a quarter of all women and a tenth of all men will develop gall-stones sometime before they reach sixty. 'Fair, fat and forty' is the old tag for possible victims, and the complaint is even older, being recorded in the earliest medical writings. There may be one gall-stone, even up to hen's egg size, or there may be several hundred minute grains. No one has learnt how to dissolve the stones away, and almost all operations for their extraction remove the gall-bladder as well.

In the past the liver was called the seat of life. A cult of liver investigation and interpretation seems to have started in Mesopotamia, spread westwards, reached Greece, and then the Etruscans and the Romans. When viscera were examined for propitious signs the liver was the most carefully inspected organ. Ezekiel xxi. 21 tells of the King of Babylon who, at a parting of the ways necessitating decision, 'looked in the liver'. Fortunately for hepatoscopy, as this form of divination was called, and for the diviners, the loose structure of the liver, plus its varying appearance, plus its susceptibility to disease, presented each investigator with sufficient changeable evidence on which to base his imagination and his predictions. The right half of the liver was allegedly relevant to the questioner, the left half to everyone and everything else. (Palmistry draws a similar parallel between the two hands.)

The belief that the liver was also the origin of the veins lent additional authority to the organ's importance, and the later view that it was the source of two of the four cardinal humours (choler and melancholy, leaving only blood and phlegm) added even greater prestige to the liver's position of power. As remnants of these previous beliefs there are many current words – lily-livered, choleric, bilious, melancholic, liverish, gall – that make unintentional obeisance to the liver's ancient authority and none to medical accuracy. The liver is certainly important but it is neither the origin of veins nor the creator of anger and sadness any more than it is soothsayer to all who consult its five-lobed oracle.

The multifarious functions of the liver make a formidable, wide-ranging and awe-inspiring list. They include the destruction of red blood cells, the manufacture of plasma proteins and of blood clotting agents, the storage of carbohydrates in the form of glycogen, some storage of fats and proteins, the conversion of fats and proteins to carbohydrate, the change of galactose – milk sugar – into glucose, the extraction of ammonia from amino-acids, the conversion of ammonia into urea, the production of bile salts for the digestion and absorption of fat, the storage of fat-soluble vitamins, the conversion of depot fat into ketone bodies (a more combustible form) and the modification of various drugs and poisons – the short-acting barbiturates, for example, are destroyed by the liver. The organ, therefore, is vital and emphatically the central organ of metabolism.

It is also more than capable of replacing its tissues. A dog with 90% of its liver cut away is still able to produce bile at the normal rate; and if three-quarters of its liver is removed the remaining quarter will undergo extremely active cell division until the entire organ reaches its original dimensions six to eight weeks later. The liver's powers of regeneration are so good, particularly when much of its substance has been destroyed by some hepatic poison, that it has been called the immortal organ. If any organ has to be called the seat of life, perhaps the liver does most deserve the title, even if its importance is now seen in biochemical terms rather than those of mood and prophecy.

Pancreas and spleen   These two organs are frequently spoken of in the same breath, presumably because they lie fairly near each other, they are about the same size, and are both in the abdominal cavity. However, they are as chalk and cheese. One of them is little understood, and not essential to life. The other has two distinct roles, both thoroughly investigated, and both essential to normal life: no surgeon removes the pancreas without there being an urgent reason to do so, and life without it needs careful supervision. Whereas both organs are customarily similar in weight (one is 3 ozs. or 85 gms. and the other is 6 ozs. or 170 gms.) the spleen can swell up to 20 lbs. (9 kgs.) during infectious and other diseases. One is a gland; the other is not. The pancreas and the spleen are therefore two most distinct neighbours.

Pancreas Named after the Greek for 'all-meat', due to its boneless and fatless substance, and known gastronomically as sweetbread, the pancreas is a dual organ, and the second largest gland in the body (after the liver). Not only does it produce crucial digestive enzymes (trypsin, lipase, amylase) and pour one to two pints (½ to 1 litre) of its pancreatic juice into the duodenum every twenty-four hours, but it also manufactures insulin in the minute and vital quantities that are necessary. Total removal of the pancreas was first carried out in man in 1944, and this is still a rare operation. The loss of the pancreatic juice to the digestive system means that much food, notably fat, cannot be digested; therefore large quantities of undigested matter are passed out in the faeces; therefore the faeces are some three times bulkier than normal. The loss of insulin, following removal of the pancreas, means the existence of yet another diabetic, although the disease of diabetes customarily arises naturally without any such brutal interference to the pancreas's function.

Normally the organ weighs 3 ozs. (85 gms.), it is soft, it is at the back of the abdomen behind the lower part of the stomach, and it looks pink or yellowish grey. Along the pancreatic duct, which connects the gland to the alimentary tract, flow the pancreatic juices; in every hour of flow these secretions weigh almost as much as the gland itself. Although vital for sound digestion the pancreas's fame in the eyes of the world is primarily associated with its importance for the diabetic. (When the term diabetes is used on its own it always means diabetes mellitus. Diabetes insipidus is quite a different disease, and has no relationship with it.)

The great year for diabetics was 1922. Those who suffered from the disease before then had little hope, but in Canada in that year Sir Frederick Banting, in conjunction with C. H. Best and J. J. R. Macleod, discovered insulin. The story really began over three decades before this momentous year. Two German physiologists, von Mehring and Minkowski, had wanted to know more about the pancreas. So they removed the organ from a dog, and later on observed the attraction that particular dog's urine had for flies and wasps. Contrary to Samson's riddle, where the buzzing bees had caused the sweetness, the buzzing flies were being attracted to the urine by its sweetness. (Others had noted the presence of sugar in a diabetic's urine, but they had not caused it to be there.) Later it was realised that sugar was excreted into the urine because it had reached high levels in the blood, and later still it was assumed that the pancreas was liberating something which somehow kept the sugar level in check. (I once asked a lady who had known H. G. Wells to explain for me his considerable attraction for women. 'He smelt of honey,' she replied, gazing fondly into the past. A sweetish odour on the breath is an occasional sign of diabetes, and Wells was a sufferer.)

It was Banting, Best and Macleod who managed to extract the insulin, an extraction made more complex by the fact that the pancreas produces large quantities of protein-splitting enzymes in its pancreatic juices as well as the insulin which is itself a protein. Fortunately calves secrete insulin before they secrete their enzymes, and from calf pancreas some insulin was obtained in sufficent quantity to quell the rising glucose levels in the blood of a diabetic dog. The Canadian team also tied off the pancreatic ducts of dogs, thereby causing atrophy of the pancreas's enzyme production and thereby enabling the insulin to be safely extracted. The calf and the dog were to save diabetic man. (The name insulin, from the Latin for island, was coined because it is produced in the pancreas's small and scattered endocrine glands called the islets of Langerhans. The name diabetes is after the Greek for 'a passer through', and was given to the disease by Aretaeus in the second century AD.)

Insulin is a small-sized protein whose production (or injection for the diabetic: it cannot be swallowed as the digestive enzymes pepsin and trypsin would destroy it) has to be correctly balanced. Too little, and there is too much blood sugar, fatigue and loss of weight. Too much and there is insufficient blood sugar, leading eventually to irritability, sweating, hunger and coma. There is a fairly narrow margin between too much and too little sugar in the blood, and the normal range is between 0·1% and 0·18%. Above 0·18% there is likely to be sugar in the urine, and Minkowski noted that the sugar level in his dogs without a pancreas rose to 0·3%. The tight-rope control of insulin, and thus of sugar-levels, cannot swerve very far from normal without disastrous consequences.

The disease is common. There are over 300,000 sufferers in Britain, and it is assumed there are tens of thousands more who are undetected – random surveys always uncover more. Surveys in the United States suggest that nine people in every 1,000 are undiagnosed diabetics. It is suggested that the figure of 2,000,000 known American diabetics is possibly matched by just as many who are still undiagnosed. Diabetes is increasing all over the world, and primarily in the well-fed communities. Blacks in Africa, who frequently have a low fat and low calorie diet, do not suffer from it, but the disease or lack of it is not a racial characteristic. Blacks in the United States suffer as much as European Americans, and Europeans in Europe only suffer less when there is a war on. This beneficial effect of food rationing was first noticed in the Franco-Prussian war, and was still being confirmed during World War Two.

There is also a hereditary risk. If one identical twin gets the disease the other twin will get it in seven cases out of ten, although possibly not for ten or more years. People with the disease are more likely to have diabetic relatives than those without it, and diabetics wishing both to marry diabetics and produce children should be aware of the greater likelihood of their children following suit. Overweight predisposes people to diabetes, and diabetic women have greater problems with their pregnancies. There is also considerable discrimination against the 0·5–1% of the population who are diabetics: many jobs are denied them, and there has been mounting pressure to eradicate the injustice.

Spleen This organ has no right whatsoever to appear in a section dedicated to digestion and its allied organs, save that the pancreas and spleen – strange bed-fellows that they are – are so frequently bracketed together. On the other hand it is possible to hold a slightly splenetic view that the spleen (sometimes called lien. Spleen is Greek, lien is Latin) has no right to appear anywhere, so vague and unassuming is the organ. It is 6 ins. (15 cms.) long, and it weighs about 6 ozs. (170 gms.). Its form is roughly fist-shaped. Its position is high up behind the stomach. Its appearance is a ruddy-red and its texture much smoother than the pancreas. Its function – well, that is the problem, for it is certainly not essential to life. Its total removal can be carried out with minor changes to the blood being the only detectable after-effects.

In some animals, such as the cat and the dog, it is an important reservoir of blood, a built-in transfusion system against times of stress or oxygen lack. In man, as it weighs just a few ounces, the addition of just a few ounces of blood – assuming it contained nothing but blood – to the 12 lbs. (5·4 kgs.) of blood in a normal body would be of negligible assistance. Therefore it is scarcely identifiable as a store-house any more than is a larder containing just one can of beans.

However, the spleen does act upon blood cells. In the foetus it is important in manufacturing both red and white blood cells, but after birth it loses its ability to make red cells although it perseveres with white cells. In adult life it reverses its foetal role and helpfully destroys old blood cells. It also prepares old red cells for their destruction elsewhere. Normally the spleen cannot be palpated, or felt from the outside, unless it is three or four times its ordinary weight, say over 1 lb. ($\frac{1}{2}$ kg.). The spleen usually enlarges when it is suffering from disease, and it also often enlarges when the body itself is afflicted with disease, notably with malaria. In extreme cases the spleen can grow from 6 ozs. (170 gms.) to 20 lbs. (9 kgs.), a fifty-fold increase. One distinct disadvantage of a large spleen is its vulnerability; engorged as it is with blood, disastrous internal bleeding can follow quite trivial bumps or knocks.

Although malaria is characterised by a lack of red blood cells, and although the spleen's traditional malarial enlargement can be linked with the need for more blood, the organ's expansion in most diseases is associated more precisely with its role as a producer of antibodies. It plays a part in the defensive system whereby, as soon as foreign protein or infection invades the body, some matched protein is manufactured to counteract the foreign material; antibody is produced to counter the antigen. However, even here the spleen's role can be carried out competently elsewhere in the body.

Is the spleen therefore totally valueless? No, but it happens that whatever it does seems to be better done elsewhere. The poor organ was not even granted a positive role in the early and more imaginative days of medicine. Andrew Boorde, the sixteenth-century medical monk, diplomatically said it 'doth make a man to be merry and to laugh, although melancholy resteth in the spleen if there be impediments in it'. Neither happy nor sad the organ is still viewed equivocally, and Boorde's casual definition is virtually valid today.

34. Excretion

Its evolution/The kidneys/Urine/The bladder

The excretory system has several dual aspects to it. Firstly, as a process it is inextricably involved with the reproductive system: ducts and organs are shared, and the amalgam of interests is customarily known as the urino-genital system. Secondly the kidney has two distinct jobs to do: it has to eliminate waste products via the urine *and* it has to regulate the salt and liquid content of the body. Thirdly, although the kidneys are the organs of excretion, and without them a man will quickly die, the kidneys do not do all the excreting; the sweat glands, the lungs and the intestines also excrete waste products. Finally, as if to emphasise such duality, the two kidneys are not situated evenly in the human body: the left kidney is a little longer and narrower, and slightly higher up the back than its right hand partner. When one kidney fails to develop, as does happen on rare occasions, the missing kidney is more likely to be the left one, and the condition affects males to females in the dual proportion 2:1.

It is with the removal of waste products that the kidney is principally renowned. Fats and carbohydrates are made up of carbon, hydrogen and oxygen. When these two types of food are broken down during their combustion, or oxidation, the end products are carbon dioxide and water. Carbon dioxide is removed from the lungs, and water – if this vital requirement can ever be considered a waste product – can also be removed by the lungs, and is so to the extent of about a pint ($\frac{1}{2}$ litre) a day.

Excretion thus far is no problem, but it becomes so when the proteins are broken down. Their main constituent, after the big three of carbon, hydrogen and oxygen, is nitrogen. Unfortunately, nitrogen is quite different. It cannot be removed as a gas like carbon dioxide for there are energy problems about nitrous oxide (NO_2) that only some bacteria have mastered and it cannot be turned into something simple, like ammonia (NH_3), because this is extremely poisonous to all tissue. A few marine animals excrete their nitrogen as ammonia, but hastily and into the big wide sea. The land-based compromise is to form urea, a far less poisonous substance whose molecule contains two atoms of nitrogen, four of hydrogen, and one each of carbon and oxygen.

Frogs skilfully blend their two existences; when free-swimming as tadpoles they excrete ammonia, and when adult they excrete urea.

Birds and reptiles encounter a third problem, one neither solved by the immediate removal of ammonia in water nor by the slightly more leisurely removal of urea, also in water. Their eggs, developing and experiencing metabolism, and therefore creating waste products for weeks at a time without any extra supply of water, have no means for excreting anything. The excellent compromise is to produce uric acid, a larger molecule (4 nitrogen, 4 hydrogen and 3 oxygen atoms) than either ammonia or urea, and one that is quite insoluble. Therefore crystals of it can be produced, and stored here and there within the egg. These are then incapable of doing any harm to the developing creature whose egg it is. A human embryo, when within its mother's uterus, and seemingly egg-like, is by no means comparable with the constricted, pre-packed prisoners of the reptilian and avian eggs. The human is bathed in ever-changing liquid and continually refreshed with blood that happily yields the waste products collected from the embryo to the mother's blood supply, and therefore to her kidneys.

The kidneys The word's two syllables are taken from two Anglo-Saxon words meaning uterus and, oddly, kidney. As extras the Latin for kidney is ren, plural renes, and thus responsible for renal; the Greek nephros is the stem for nephritis, etc. In human beings the kidneys are each about $4\frac{1}{2}$ ins. (11·4 cms.) long, $2\frac{1}{2}$ ins. (6·3 cms.) wide, $1\frac{1}{2}$ ins. (3·8 cms.) thick and weigh 5 ozs. (142 gms.). They are not in the small of the back, but higher up and in front of the twelfth ribs where they lie close to the spine.

Basically each kidney is nothing more than a collection of filter units, each one of which initially absorbs virtually everything small from the blood – broken-down food molecules, water, waste products – and then a similar collection of tubes whose dominant function is to put back into the blood everything still required. (One way of tidying a desk is to remove everything and then only replace objects still of value. Another is to pick out the valueless objects without disturbing the remainder. The kidney makes use of the former system.) Therefore, as actual urine production is only two to three pints a day, and as the initial extraction of fluids and solids is about a hundred times greater – perhaps as much as 50 gallons (227 litres) a day, it must be the fate of much liquid and solid matter to pass through the kidneys very many times every twenty-four hours. Moreover, some $2\frac{1}{4}$ pints (1·3 litres) of blood

473

are pumped through the kidneys every minute – or one-quarter of the heart's resting output, and therefore much of the blood is going through the kidneys a mere five minutes after its previous passage. Of the one-fifth of a pint (0·1 litre) of water filtered by the kidneys every minute, over 99% is passed back again to the blood and less than 1% goes into the urine.

The filter units responsible for all this activity are called nephrons. A human being has about 1,250,000 in each kidney. Consequently, despite the enormous quantities being filtered by the kidneys as a whole, the work done by each minute nephron is appropriately modest – roughly 2 cubic millimetres an hour. The first person to see a glomerulus, the bulb-like part of each nephron that does the initial extracting, was Marcello Malpighi, the seventeenth-century Italian who was accustomed to such small-scale work: he also saw the capillary network of blood vessels, thereby putting final proof on William Harvey's theory of blood circulation. It is through the tube connecting each glomerulus to the main ducts of the kidney that the wanted substances, like that 99% of the water, like glucose, like amino-acids, are returned again to the blood system via the renal vein. In a sense each glomerulus is a crude filter, just sorting according to size; the tubule is more refined, being highly specific about what it does and does not pass through its walls to the blood.

Study of the kidney is highly relevant to an understanding of evolution. Palaeontologists (in general) believe that the earliest fishes lived in rivers and lakes, and only later took to the seas. It was a study of kidneys that helped to clinch this point. Fresh-water fish, living in a fluid less salty than their own body fluid, require quite a different kidney system from the one necessary in the sea where the surrounding water is saltier than body fluids. Osmosis, that process encountered in school biology when water is shown to flow through a membrane to dilute the saltier side, is also encountered everywhere in living tissue. A marine fish is threatened with dehydration by osmosis; its fluids are trying to dilute the salt water all around it. Conversely, a fresh-water fish is saltier than the fresh water, and is therefore threatened with dilution. In both cases the kidney system reflects the particular threat, even with animals such as salmon – which can happily swim from one to the other, and the conclusion is that primitive fishes, our ancestors 500 million years ago, struggled into their vertebrate existence in the rivers and the lakes rather than in the sea. Their osmotic problem is thought to have been excessive water rather than a lack of it, and was therefore caused by life in fresh water rather than salt.

Urine Normal daily production is 1¾–3 pints (1–1·7 litres). Disease can cause this quantity to go up, as in diabetes insipidus and various kidney ailments, or down, as in conditions causing fever and diarrhoea. Of course heavy drinking causes a rise, and so do diuretics like tea, coffee and cocoa, as well as excitement or nervousness. Hot weather causes production to go down. Normally, whatever the actual liquid output, a day's urine contains 2 ozs. (57 gms.) of solids. These make it slightly heavier than water, for urine has a specific gravity of 1,010 to 1,025, but this will fall with a lot of drinking or rise with a lot of sweating.

The colour of urine is proportional to the specific gravity, but colour is no one thing because many pigments are involved. Also some diseases can bring about coloration of their own, ranging from the orange of many fevers, to the brown of blackwater fever – an early hazard of West Africa, to brownish-black as in alkaptonuria, and to greenish-blue with cholera and typhus. The dyes of some foods can produce garish and somewhat hair-raising effects. Beetroot, blackberries and quite a few drugs can change the colour, but even without the senna and methylene blue of drugs the urine can be kaleidoscopically intriguing, whether a different diet or disease is the cause. In mediaeval times the study of urine had a high status, and early woodcuts frequently show some phial receiving careful examination while the unfortunate victim of plague, childbirth, syphilis or whatever lies nearby and blatantly in need of help.

Nothing is known about the ordinary smell of urine, but this usual odour can change, notably after eating asparagus. That urine smells of ammonia is really only true of stale urine, when bacteria have had time to act upon the urea to produce far more ammonia than exists in fresh urine. Urine is virtually bacteria-free when liberated, and has even been used as a disinfectant in extreme conditions, such as the battlefield. Faeces, whose bulk is largely bacteria, and urine are therefore entirely distinct in this regard.

The composition of urine is a bewilderment of organic and inorganic chemistry. It also varies widely, as is reasonable, with all changes in diet. On an average day it contains an ounce of urea, and far smaller amounts of differing acids, bases, salts and organic compounds all adding up to another ounce. Some glucose is always present, but not in the 3%–5% proportion of the diabetic. Some protein is present, but if easily detectable may be a sign of one of a plethora of diseases which can put it there.

The mediaeval medicine men were right in a way, in that the composition of urine is a valuable guide to the condition of the patient, but holding a phial to the light is a poor substitute for today's microanalysis. This can, for instance, tell of a person's stage in starvation, for so long as fat is available fat will be consumed to produce energy, but when fat is exhausted it is the turn of tissue, of protein, to supply the energy. Twice as much protein as fat has to be used to produce a similar supply of energy. Therefore, as protein contains nitrogen and fat does not, the sudden extra metabolism of tissue means the sudden presence of more nitrogen in the urine. Known as the pre-mortal rise the urine is giving warning of possible death from starvation.

Kidneys have achieved considerable notoriety due to efforts being made to transplant them from one person to another, from a donor or a corpse to someone with renal disease. The surgery has been successfully worked out, but the problems of immunity have been and still are formidable. Up to June 1966 over 600 human kidney transplants were performed, 500 of them after March 1963. Most of the patients have since died, but survival time is improving. By the end of 1965 some 37% of all recipients had lived more than a year after the operation.

The first such transplant was performed in Chicago in 1950 (and the kidney lasted until 1951). The first transplant from one identical twin to another was performed in 1954, and this was successful. In the past three decades the surgery has, of course, improved, but so has tissue-typing (finding a suitable donor) and the suppression of the rejection process, first by X-rays (in 1959) and by drugs (since 1961). In the United Kingdom and by 1980 there were 3,145 functioning transplants, 80% of which had come from cadavers and the rest from living donors. It is still best if the donation can be from an identical twin. It is second best, in general, if it comes from a living person and third best if from a corpse. By 1980 the two-year survival rate was 45% from the dead and 80% from the living. Plainly most progress is being sought in the lower percentage, as the greatest and easiest supply of kidneys for transfer is likely to come from the dead.

Artificial kidneys were first developed in the early 1940's by W. J. Kolff, working in German-occupied Holland. After the war he gave some of his kidney machines to major hospitals in Europe and the United States. In 1961 the first patient able to dialyse himself was taught to do so in London, but the technique was cumbersome. Even by 1970 dialysis – the process of removing waste products from the blood – demanded about thirty hours a week of the patient's time. Today some 12–14 hours a week are necessary, and in Britain there are 1,200 people on hospital dialysis and over 2,000 doing it at home. The need for this treatment is not going to vanish. Between thirty and forty people develop permanent kidney failure each year, and without dialysis they would die.

Bladder   'Of Pissing in the Bedde', as Thomas Phaire called it in *The Boke of Chyldren* in 1553, there has already been discussion in the section on child growth. However, it is also a problem for the old. In one group of 500 elderly people, all waiting to go into a hospital for the aged sick, seventy of them were incontinent. It was not so much a lack of control with them as of possessing a bladder which emptied itself long before the stage at which conscious control was previously exercised. The sphincters were relaxing before any feeling of fullness was made manifest.

A normal bladder appears thick and stratified when empty, but it is only a layer or two of scale-shaped cells when full. Customarily no sensation is felt initially as the bladder fills and its pressure rises, but the pressure rise is small at first because the bladder expands to accommodate the extra urine. The pressure is about 4 ins. (10 cms.) water gauge, or some $2\frac{1}{2}$ ozs. per square inch of bladder, until the contents start nearing a pint. The bladder still expands to accommodate increasing urine, but reflex contractions give rise to a conscious desire to urinate (or micturate). This can be suppressed for a time, but suppression fails when the internal pressure has built up to around 40 ins. (100 cms.) water gauge (or $1\frac{1}{2}$ lbs. per sq. in. – about a tenth of normal atmospheric pressure).

Like any flexible container filled internally with a positive pressure, the bladder tries to assume a spherical shape. This means that half a pint ($\frac{1}{4}$ litre) of urine, the normal level for emptying, distends the baldder into a sphere 3·2 ins. (8 cms.) in diameter. Should there be a whole pint ($\frac{1}{2}$ litre) within, when the pressure is likely to be either rising or painfully seeming to, the diameter will be a fraction over 4 ins. (10 cms.) Therefore, because a sphere's volume goes up as the cube of the radius ($\frac{4}{3}\pi r^3$), the doubling of the contents only pushes up the diameter another 0·8 ins. (2 cms.), although the subjective impression is of a far greater increase. Were it possible to double even that accumulation of one pint up to two pints (1 litre) the diameter would only increase by one more inch (2·5 cm.). If relief is impossible, as may happen in unconsciousness or paralysis, or if relief is strongly suppressed, the swelling bladder may cause a visible fullness in the lower part of the abdomen. Normally no change in external shape is detectable.

The process of relief is straightforward, but involved. The full bladder causes contraction waves. These are regular, and can be timed subjectively with a stop-watch. They stimulate pressure receptors in the muscles of the bladder wall. From them nervous impulses travel to the spinal cord and the brain. The desire for relief is caused by these messages, and nothing more will happen – beyond the despatch of reminding impulses – until it is convenient to micturate. The brain then starts to cancel various automatic controls. Nervous impulses from it cause a relaxation of the bladder's internal sphincter plus a contraction of the bladder muscle. Similarly the bladder's external sphincter is relaxed. Then the breath is held and the diaphragm forced down while the abdominal wall is contracted. Both acts increase the pressure on the bladder, and help it to get rid of its accumulated urine.

Such an assortment of nervous control, inhibition and relaxation makes it entirely comprehensible why a child takes time to learn the act, and most children can walk before they can urinate with any degree of controlled competence. Even an adult, maturely adept at such control, can find it impossible initiating both micturition and defaecation simultaneously. The problem is likely to be resolved only by initiating one process and then the other, but the attempt should indicate the complexity latent in both systems.

An intriguing extra complexity, more a matter of hydro-dynamics than nervous physiology, is the twist imparted to the masculine urine stream. Obviously with hygienic advantages the system ensures a compactness of flow which must be preferable to the widespread spraying which might otherwise result.

Adult females are generally credited with the ability to micturate less frequently than the adult male. There is no greater bladder capacity in the female; instead it is assumed there is a greater suppression of the emptying reflexes. The urethra, the tube leading from the bladder to the exterior, is of course far longer in males than females, roughly 8 ins. (20 cms.) as against 1½ ins. (3·8 cms.) and the male uretha goes through the enlarging prostate gland, which can be a source of masculine suffering in later years. In fact the bladder then has a tendency to revert to a second childhood of its own; not only is there increasing incontinence with age, but the business of having to rise in the middle of the night is an additional chore for many over fifty-five. Sir George McRobert and Lord Brabazon both drew attention during the early 1960s to the 'current lack of a chamber pot' in hotel rooms: this 'defect in sanitation causes a dark shuffling along unknown corridors in the small hours, an infliction much deprecated by the elderly'. The chamber pot has vanished from hotel rooms since those mediaeval days *c.* 1960, and lavatories are frequently *en suite,* but the need for night-time rising by the elderly has not diminished. It merely requires a shorter journey for the necessary relief.

35. Respiration and Blood

Smaller solutions/The need for a system/The lungs/Breathing/Its control/Altitude/Drowning/Pollution/Choking, coughing, sneezing/Laughing, yawning, hiccups/Oxygen, excess and lack/ The heart/Harvey and Galen/One pump into two/Heartbeat/ Blood pressure/Its changes/Blood distribution/The pull of gravity/Blood; red, white and platelets/Plasma/Blood-letting/ Blood-giving

If man were a single-celled animal there would be no need for a respiratory system; oxygen would be able to diffuse through the cell wall in ample quantities. If man were a small multi-celled animal there would still be no need provided his dimensions were modest, and the distance from oxygen source to remotest cell was short, perhaps half a millimetre at most. The vital gas could still diffuse through the intermediate tissues in sufficient quantity. Even if man were an insect there would also be no need for lungs and pulmonary arteries because insects are small enough to rely upon air-carrying tubes which ramify throughout their bodies. No part of the insect is more than a small distance from a branch of one of these tubes, and therefore no part is without oxygen. (Such a system can only supply small animals. H. G. Wells's giant wasps, however god-like their nourishment, could never have existed. For which great thanks.)

Man is a large animal, too big by far for simple diffusion, too big for the insect's tracheal network. Consequently there has to be a better method of distributing oxygen around the body. In all the higher vertebrates this advanced system consists basically of one air pump and two liquid pumps. The air pump, a combination of the diaphragm and the chest, draws air into the passive lungs and then forces it out again. The two liquid pumps are the two ventricles of the heart. They have quite separate functions. The right ventricle pumps its blood into the capillary network of the lungs, and so brings blood short of oxygen into the proximity of air possessing oxygen. The left ventricle pumps its blood, now rich with oxygen, into the capillary network of the body, and so brings rich blood to the tissues needing oxygen. The system is only an extension of the simple diffusion mechanism acceptable to the smallest creatures. Blood has to be brought near to the surface so that the oxygen of air can diffuse through into it. The blood is then moved near all tissues so that its oxygen can diffuse through into them. The diffusion is similar; it is only the distances which are greater, and blood is the transport agency which eliminates the distances.

First, some measurable facts about the respiratory system. A relaxed adult breathes in and out some 10 to 14 times a minute, with each breath therefore lasting 4 to 6 seconds. In each minute a resting adult breathes in 9 to 12 pints (about 5-7 litres) of air. Should this person then leap into violent physical exercise his demand for air will increase perhaps fifteen-fold, perhaps twenty-fold, and he will pant in and out some 20 gallons (91 litres) of air every minute, with only a second or so for each breath. The urgency for more air is great because the relaxed body carries slender reserves of oxygen. Within 20 seconds of starting hard work all the reserve capacity will have been consumed, but long before that the panting will have begun. A normal day's breathing involves about 3,300 gallons of air, or 15,000 litres, or 530 cubic feet of it, or a cube of air roughly 8 ft. by 8 ft. by 8ft. In one night, therefore, a person's lungs will require a third of that amount – namely a block 6 ft. by 6 ft. by 5 ft. Although replenishment with fresh air is necessary, the traditional gale that blows through so many English bedrooms is more than adequate compensation for the quite moderate requirements of a night's respiration. In a lifetime of nights and days each human being will have breathed in a phenomenal quantity of air, say 13 million cubic feet, or 368,000 cubic metres, or twice the capacity of the airship *Hindenburg*.

A lot of people can use up the available oxygen in any unventilated hall or shelter in quite a short time, particularly if they are active or even walking about. For instance, assuming no removal of carbon dioxide and no replenishment with oxygen, 650 people could stay in a shelter of 42,000 cubic feet (say 45 ft. by 45 ft. by 21 ft.) or 1,200 cubic metres only for three hours even if they were sitting down. If moving about the time would be halved. A black hole of Calcutta is easy to create without ventilation. Oxygen replenishment is about 1 cubic foot (0·03 cubic metre) of oxygen per person per hour if the people are sitting down or lying, and 2 cubic feet (0·06 cubic metre) of oxygen if they are active. At the same time carbon dioxide and other noxious accumulations, like smoke, have to be removed.

The traditional English telephone booth, with a tight fitting door and no effective ventilation, contains about 37 cubic feet (1 cubic metre) of air. On the figures quoted above, and allowing a small displacement bulk for the person telephoning, his call ought to be no longer than 30 minutes, certainly no more than 45, otherwise he will faint and fall foul of the built-in curtailment system for excessive private use of a public service.

Lungs   A pair of human lungs weighs about $2\frac{1}{2}$ lbs. (just over 1 kg.). The right one is heavier than the left, and both extend upwards to an inch (2·5 cms.) or so above the collar bone. In various aspects the lungs are inefficient organs; they are a brilliant execution of a poor design. Much like most London buses the exit and the entrance are at the same place; the air both enters and leaves by the mouth. Consequently there is only a partial inter-change of gas with each breath, and about five-sixths of the air present in the lungs is still there when the next breath begins. The next breath again leaves five-sixths, and so on. In fact, due to molecular minuteness, it could even be argued that some mole-cules present in the lungs at the start of life are still there at the end of it; the relentless removal of one-sixth of their number at each breath is insufficient to ensure their total extraction in a life time of breaths. Not so with fishes. They use a flow system, and water is passing over their gills in a steady stream. The human mechanism, which had to make do with and adapt the air bladder developed as a buoyancy device by fishes generations beforehand, is far less attractive in theory, although some awkward valving would be necessary were we too to possess outlet vents on the sides of our chests.

The dominant reason why mankind, and any large animal, has to possess a respiratory system is because bulk becomes so much larger than skin area with any increase in size. However good the diffusion of oxygen through the skin the area for diffusion would be inadequate, but this area for diffusion has to exist somehow. It is in fact created by the architecture of the lungs, and human lungs have an area some forty to fifty times greater than the surface area of the body's skin. (Traditionally this huge lung area of 800 to 1,000 square feet or 90 square metres is referred to as being tennis-court-sized. I personally find it hard to equate the ramifying delicacy of the human lung with a flat slab, apart from the difficulties of courts being single, double, two sided and bordered or not.) It is the alveoli of the lungs, the round endings of each culmination of all the sub-divisions of the one original wind-pipe, that cause this huge area. There are said to be 300 million of them, and they are spots where gaseous diffusion most readily occurs. (The number of alveoli fluctuates wildly in the estimates, and a figure of 750 million is also often quoted. This must be a prime example of follow-my-leader, as any thought of personally checking the astronomical number is highly intimidating.)

Each alveolus is covered with a tracery of blood capillaries. Each capillary is just wide enough to let the red blood cells pass through one after the other, just as a narrow street only permits cars in single file. Each blood cell is being jostled along by pumping from the heart's right ventricle, and it only stays in the capillary for three-quarters of a second. In that time, aided by the fact that the walls of the alveolus are a twenty-five thousandth of an inch thick, the red blood cells have picked up their oxygen atoms, and the blood's content of carbon dioxide will have been yielded up to the lungs. Should the heart be pumping vigorously, as during exercise, the blood will only be in the pulmonary capillaries for a third of a second, but the time is still adequate for the exchange. The haemoglobin molecules of each red blood cell, it need hardly be said, have a great affinity for oxygen.

Pumping 17 pints (about 10 litres) of blood a minute through the capillary network of the lungs, and over those 300 million alveoli endings, is quite a task, but a tenth of the pressure is required for this act than for the greater problem of pumping oxygenated blood through every other capillary network in the body. The left ventricle's more powerful role demands a pressure of 4 ins. (10 cms.) of mercury, but both its pressure and the $\frac{1}{2}$ in. (1 cm.) of mercury needed for pulmonary circulation seem ridiculously small when the fact is borne in mind that hundreds of miles of capillaries are involved in each instance. Part of the explanation lies in the fact that all blood vessels are slightly elastic (an elasticity which diminishes with age) and there is only a small volume of blood actually in the capillaries at any one moment. The huge network of pulmonary capillaries, for example, only contains about 5 cubic ins. (80 cubic cms.) of blood, an amount roughly equal to the output of each single pump of the right ventricle.

Breathing in, with the inrush of air through the nostrils, or the mouth, or both, gives the impression that some system within the nose or the mouth sucks the air in. Not so. To expand the lungs, which are themselves elastic and prone to collapse upon themselves – as in operations, the diaphragm pulls itself downwards and the rib muscles enlarge the rib cage. In quiet breathing the diaphragm moves downwards about two-thirds of an inch (or almost 2 cms.), but in deep breathing it moves nearly 3 ins. (7·6 cms.). When the diaphragm is pulled down, and the chest wall expands, the lungs are pulled both downwards and outwards. Air then rushes in with nature's customary abhorrence of vacuums. The reverse action causes a slight increase of pressure within the lungs. Therefore, as nature has to balance pressures, air leaves the lungs to join the atmosphere again. Normally the pressure differentials are small, and breathing is quietly carried out without anything more than a gentle flowing to and fro.

Diaphragm and chest wall can each maintain breathing on their own should one fail, although the diaphragm is normally responsible for three-quarters of the tidal respiration. The diaphragm, in particular, is capable of creating quite considerable pressure, as in exercise or in blowing up balloons. A pressure of about 2 lbs. per square inch, or a seventh of an atmosphere, can be made to force air out of the lungs, but a smaller pressure differential is involved when air is breathed in. The forces associated with breathing seem quite powerful, as someone puffs and blows, but their actual levels can be demonstrated by attempting to emulate those film heroes who submerge themselves in the swamp with only a straw as an air link to the surface. As everyone knows, the pressure of water increases with depth, but at only 9 ins. (23 cms.) below the surface the pressure is sufficient to stop breathing. Some heroes might be able to go a fraction deeper, and the depth does depend upon the position adopted beneath the surface; but no hero, however strong the straw, can submerge himself to any great extent beyond 9 inches. The pressure of water around him is too much for his powers of respiration.

The iron lung kind of respirator also works upon the outside of the chest, but beneficially, by lowering the pressure surrounding the patient's chest. This has the effect of expanding the thorax, enlarging the lungs and of drawing air in through the mouth. A slight reversal of the system will cause the reverse procedure, and the patient breathes out. Naturally there has to be a seal to prevent air entering or leaving around the patient's neck.

The normal automatic control of the rate of breathing is dependent mainly upon the level of carbon dioxide in the blood, and therefore in the expired air; but there is some uncertainty. Mild exercise neither reduces the amount of oxygen in the arteries nor does it increase their content of carbon dioxide, and yet the breathing rate goes up. Severe exercise certainly depletes the oxygen levels and raises the carbon dioxide levels, and the breathing rate increases considerably. Therefore it does seem logical to presume that the changing levels are linked with the changed breathing.

Normal air possesses virtually no carbon dioxide – perhaps three parts in ten thousand. Expired air may be 4% carbon dioxide, and breathing may become twice as fast when this proportion goes up to 4·2%. Should the inspired air also be rich in carbon dioxide we start feeling faint, and will probably pass out if the proportion goes up to 5%. There may still be the normal 20% of oxygen in the inspired air, but the abnormal quantity of CO_2 getting into our blood is more than a match for us.

Although high levels of CO_2 can cause panting and an unpleasant swimming feeling, the reduced quantity of oxygen at altitude can cause unconsciousness without any consciousness of its approach. Pilots and others have been placed in pressure chambers to demonstrate the insidiousness of oxygen lack. At simulated altitude both unconsciousness and recovery can be effected by cutting off the oxygen supply without any awareness of either event by the individual pilot. Of course mountaineers are well aware of the thinness of the air through which they are climbing, as their lungs are made to gasp for more, but pilots have little physical work to do, and can quietly pass out. It was a World War Two regulation that they switched on their oxygen supply at 15,000 ft. (4,500 m.), when atmospheric pressure is nearly 40% lower than at sea level, but work at Farnborough suggests that a person's ability to cope with a novel flying situation is impaired if he is breathing ordinary air at any pressure level above 5,000 ft. (1,500 m.).

Much of humanity lives above 5,000 ft. (1,500 m.). It is the height of Nairobi and many other regions of the African continent. In the Andes there are settlements three times as high and the limit for permanent acclimatisation, according to the physiologist L. G. C. E. Pugh, is between 15,000 ft. (4,500 m.) and 17,500 ft. (5,300 m.) – roughly 3 miles up. Mountain climbers have spent many weeks at higher altitudes, and early Everest attempts reached 28,000 ft. (8,500 m.) or thereabouts. F. S. Smythe, for example, spent three nights at 27,400 ft. (8,350 m.) in 1933, an altitude where the pressure is less than one-third of that at sea level. All the early successful climbs of Mount Everest were with the aid of oxygen cylinders, but it is an odd coincidence that the earth's highest mountain just about represents the highest peak which man can achieve solely with the aid of his lungs.

Condors fly over the Andes and geese have been seen at 30,000 ft. (9,100 m.), but birds have a different lung system from mammals. Also, one assumes, flight is less of an effort than stumbling along some snow-encrusted wind-swept mountain ridge. Human beings already fly higher, of course, and with even less effort, in their pressurised airliners. Cabin pressure is generally equivalent to a height of 6,000–8,000 ft. (1,800–2,400 m.), according to the height of the aircraft. Therefore the altitude is already substantial, and can be disturbing for some people. Should an aircraft flying above 30,000 ft. (9,100 m.) suddenly lose its cabin pressure to become in equilibrium with the outside air, and should the passengers fail to breathe through the emergency oxygen masks carried on board, their lungs will reverse their traditional role; the pulmonary capillaries will then start giving up oxygen to the lungs rather than taking it from them. Naturally such a procedure will be short-lived, and so too will be the passengers unless their oxygen is boosted either artificially at altitude or by swift descent to denser regions of the atmosphere.

At a post-mortem the lungs can have much to say about life and death. For instance, the lungs of a new-born infant that never took a breath will sink in water. After a few breaths any portion of the lung will have been sufficiently inflated to float. Not all who drown have water in their lungs; some 10% to 20% of them are asphyxiated quite simply when air is prevented by water from getting to their lungs. There is a great difference whether someone drowns in fresh or salt water. Fresh water is less salty than blood; therefore, if it reaches the lungs, it rapidly invades the blood and, quite apart from blocking off the oxygen supply by its existence in the lungs, the presence of so much extra water in the blood can cause paralysis of the circulatory system. Salt water is more salty than blood; when it is in the lungs it extracts water from the blood causing what is known as haemoconcentration. This too produces heart failure but for quite a different reason and after a longer time.

Between the lungs of an Eskimo, who has lived his life in clean air, and of a coal miner, who has not, there are the lungs of the rest of us. An Eskimo's lungs stay clean, and can easily be cut with a knife. A miner's lungs, particularly in the days before coal being extracted was dowsed with water, could be black, rich in mineral matter, and hard to cut. An ordinary city dweller, according to Professor Julius Comroe of California University, breathes in 20,000 million particles of foreign matter on an average day. It does not matter that the air may be very cold or hot or dry because it will be warm and moist by the time it reaches the alveoli, but it does matter that it is polluted, although less than might be imagined.

Experimental animals have breathed in dirty air at 950°F (500°C) and at –148°F (–100°C), and in both cases the air was at

486

body temperature or nearly so before it had penetrated very deeply into the lungs. At the same time it was rendered virtually free of particles before reaching the delicate alveoli. Of course some particles are breathed out just as they were breathed in, and some are vehemently sneezed out if particularly irritating, but a large number do land on the trachea, the bronchi and the huge branching network of bronchioles. All these tubes are lined with cilia, minute hair-like structures which can row back and forth. Their rowing is all co-ordinated, much like the leg movement of a millipede, and it has the effect of wafting particles up and out of the lungs. The cilia all work within a thin layer of mucus, and it is this mucus sheath plus its trapped particles that is pushed out of the lungs at the astonishing speed of about two-thirds of an inch (about 17 cms.) a minute. This means that a particle which has penetrated 1 ft. (30 cms.) into the lungs will have been evicted about a quarter of an hour later. The moving staircase of mucus and particles eventually reaches the junction with the pharynx. It can then be quietly swallowed or accumulated and spat out.

The alveoli have neither mucus glands nor cilia. Particles which reach this far can still be removed either by special phagocytes or by being forced into contact with the mucus sheath of the bronchioles; but some stay there for ever. It is these which can stridently testify at a post-mortem whether the lungs in question belonged to Eskimos, coal-miners, or city-dwellers on the wrong side of town.

Large particles of noxious substances result in immediate coughing and choking. Any attempt by food to go down the wrong way, into the lungs rather than the stomach, is met with explosive protests from the lungs. Customarily the procedure of swallowing blocks off the glottis, stops respiration for a moment or two, and ensures correct division of air and food. An unconscious person generally has no such reflex. The forcing of a drink between his clenched teeth, however well intentioned, may lead to the pouring of this drink straight into his lungs. Even during consciousness the system is not always perfect. Fruit stones, nails, bits of toys, parts of false teeth, and of course, food have all found their way into the bronchi. Respiratory distress is generally instantaneous, but often the objects cannot be choked out, and they remain within the lungs for decades, or for ever.

One of the most famous chokings was suffered by Isambard Kingdom Brunel, better known for engineering the Great Western Railway. He swallowed a half sovereign in 1843 and, which was far worse, it chose to go the way of his lungs rather than his stomach. For two days he suffered, and was then summarily strapped on to a plank. This was raised nearly to the vertical, with Brunel's head at the lower end; his back was then hit. This rough approach failed entirely. It caused much more choking to arise, but no half sovereign. Further treatment in the same sledge-hammer manner three and a half weeks later did in fact budge the coin from his right bronchus, but not far enough. His trachea was then cut open, but unavailingly. Two more weeks passed, and he again submitted to the upside-down shock treatment – it sounds entirely his own devising – and the coin dropped from his mouth. Richer then by ten shillings and an unpleasant six-week experience Brunel was a lot luckier than some who never lose the impediment in their lungs.

The force of a cough can be exceedingly powerful. It involves a slight breathing in, a closing of the glottis, a forcible respiratory effort to build up pressure, and then a sudden release of the trapped air. Speeds up to 500 ft. (150 m.) a second have been quoted. A sneeze can be even more explosive, and supersonic speeds have been alleged for the expelled gust. Unlike a cough the glottis stays open before and during the sneeze. There are different varieties of sneeze, some modified by a desire not to douche the immediate environment, but basically a sneeze is always initially accompanied by a raising of the tongue to block the mouth. This causes the nose to receive, and pass on, the first blast, with only the second part coming from the mouth. According to a report from Dr David Mezz, of New York, polite sneezes, or internal attempts to muffle the approaching storm, can lead to nose bleed, a ringing in the ears, sinus trouble and much else. He recommends an open mouth, a natural sneeze, a temporary disregard of popularity, and a handkerchief if there is time. Not only the obvious and inhaled irritants and approaching attacks of influenza, measles, colds and hayfever can initiate sneezing, but more bizarre phenomena like changes in external temperature or exposure to bright lights. They too can trigger off paroxysms of supersonic sneezing. (Or even, as I have just noted, on re-reading this paragraph.)

There are other oddities of respiration. Laughter is a deep breathing in followed by a succession of short and spasmodic breathings out. Crying, right at the other end of the pleasure scale, or sometimes right next to it, is caused by similar respiratory movements. Yawning, a deep breathing in with the mouth opened exceptionally wide, seems curiously pointless, as has already been described (in the chapter on sleep). Sighing is the reasons. Hiccups, or hiccoughs, are spasmodic inspirations which opposite, for it is a breathing out, but it is equally vague in its end with a click due to the sudden closing of the vocal cords. Here it is the diaphragm which is at fault, or probably the nerves controlling it. There are innumerable home remedies, such as drinking out of the other side of a glass (which concentrates attention) or breathing into a paper bag (which must concentrate carbon dioxide) or eating sugar lumps (which is nice – for some) or drinking gin.

Every now and then someone gets the hiccups and keeps them. Lucy McDonald, an Atlanta waitress, got them in 1963 and still had them in 1965. Naturally she tried all the old remedies, some two thousand of them, and she saw over one hundred doctors. Finally, forty pounds weaker and millions of hiccups after the attack had begun, still dosed with tranquillisers and unable to get work to feed her three children, she underwent surgery. Her right phrenic nerve was deadened with a drug, and the hiccups immediately stopped. The nerve was then crushed, and the hiccups did not come back even when the drug was no longer active. The result – a cure. The cost – a loss of a quarter of her breathing capacity due to the loss of that nerve. The conclusion – entirely satisfactory, as a normal person's breathing capacity is more than enough in normal circumstances. In fact men have lived long lives with only a fifth of their lungs.

Breath-holding is possible to a certain extent, say for 45 seconds (or longer for some), but thereafter becomes impossible. Breathing is resumed before the lack of oxygen has been damaging. Should someone take either a few puffs of pure oxygen first, or gasp deeply solely with ordinary air, the time of breath-holding can be extended by perhaps two minutes. Should he first breathe pure oxygen for one minute he may thereafter be able to hold his breath for five minutes. In all these cases the voluntary restriction does not cause any damage due to oxygen lack.

Although oxygen is vital to almost every living thing on this planet (there are exceptions, such as the anaerobic bacteria) it can also be a poison. Animals kept in pure oxygen for several days have suffered lung damage as a result. The American astronauts who have been living off pure oxygen when in orbit have been receiving it at much less than atmospheric pressure. This is known to be safe – at least physiologically, although prone to disastrous fires. Should men be given pure oxygen at four atmospheres pressure they will generally suffer severe symptoms, such as convulsions or faintness, in less than an hour. Four atmospheres is equivalent to a water depth of 130 ft. (40 m.). As men have already lived quite extensively in sea-houses at that pressure, and even deeper, the problems lie jointly in seeing that they get enough to breathe yet not too much oxygen.

Asphyxia was probably the cause of Christ's death, and of all who were crucified. Dr Jacques Bréhant, of Algiers, who has examined the subject, and reviewed reports of crucifixions from World War Two, believes suffocation resulted when the victim's arms carried the whole weight of the body. This made breathing difficult and later, when exhaustion set in, impossible. Temporary relief could be gained by using the nail through the feet to take some of the body's weight; but again exhaustion would intervene or, as in the case of the two thieves, the legs would be broken making such relief impossible. Therefore the correct anatomical picture of the dead Christ on the cross should show, according to Dr Bréhant, the head tipped forward – not upwards or to the side, and the nails through the wrists – not through the palms through which they would tear. Had the Renaissance painters been as familar with this method of execution ('the most cruel and horrible of all' said Cicero) as were the Romans and their subjects, they would probably have been more strict in their interpretations.

The Heart Weighing less than 1 lb. ($\frac{1}{2}$ kg.), the size of a fist, having scant resemblance to the simple arrow-pierced emblem carved on trees, pumping the body's entire blood content through its chambers every minute, beating throughout life, growing from less than an ounce (28 gms.) at birth, starting work months before and continuing to beat thereafter, the heart is a formidable pump. In fact it is two pumps, each having similar output. One sends the blood through the pulmonary network; the other sends it through the body, the systemic network. Each pump produces its 2,000 gallons (9,100 litres) a day, or 50 million gallons (227 million litres) in a lifetime. Bearing in mind its continuous labour and responsibility it does seem entirely reasonable that failure of the heart and its blood vessels is the major cause of death in many countries.

Willam Harvey, the seventeenth-century physician whose *De Motu Cordis* so shook the world, had much of Charles Darwin about him. He knew he was right, and knew many would think him wrong; therefore he polished and verified his ideas for year after year. In 1616 he wrote: 'The movement of the blood is constantly in a circle, and is brought about by the beat of the heart.' It was this – to us – innocuous thought which had to be proved beyond disbelief, and it was not for another 12 years before he published this thought, then expanded to 72 pages. To us it seems incredible that Harvey's work should have encountered opposition. Many men other than anatomists had seen arms lopped off and had seen rich arterial blood gush forth in great jets, but the belief was that these portrayed the ebbing and flowing of the vascular system. Galen, the second-century know-all of medicine, had laid down the law about blood flow, and this mentioned nothing about circulation. Even fourteen hundred years later Galen still commanded unswerving respect in most quarters. Jean Riolan, professor of anatomy in Paris, countered Harvey by saying that if dissections no longer agreed with Galen it was because nature had changed and not because Galen had been wrong. Guy Patin, professor of medicine in Paris, said Harvey's theory was paradoxical, useless, false, impossible, absurd, harmful.

Harvey was of course quite right. Blood does go 'in a circle'. He was right even though no one had then seen blood capillaries, the minute links between the outgoing arterial system and the incoming venous system. He was right even though he knew nothing of oxygen, and not too much about the reason for circulation – 'whether for the sake of nourishment or for the communication of heat, is not certain'. Marcello Malpighi, of Bologna, was to see capillaries later in the seventeenth century, but Joseph Priestley, of Leeds and America, did not isolate oxygen, and Antoine Lavoisier, of Paris, did not name the gas or discover its nature until the second half of the succeeding century. However, Harvey's attackers had been vanquished long before either capillaries or oxygen had set their seals upon the argument.

The heart's two pumps consist of a total of four chambers. Unhelpfully for clarity the primitive vertebrate heart is also of four chambers, but for quite a different reason. The fishes, having no lungs and making use of a more primitive pumping system, have an in-line arrangement for their chambers. Blood flows through all four chambers consecutively, and then leaves the heart to make its circuit through the gills and round the body. With the development of lungs, in those vertebrates that struggled on land to make use of air, a complexity arose. Venous blood returning to the heart first had to be pumped through the lungs, and the subsequent arterial blood had to be pumped round the body. It was plainly unsatisfactory if the two streams were to be mixed in the heart; therefore – and this is loose teleological speaking of a high order – the heart's single pump had to become two pumps, the pulmonary pump and the systemic pump. The wall between the two main pumping ventricles had to be a total barrier. (Galen was so wide of the mark in his theory of blood's tidal flow that he had to assume this wall, the ventricular septum, was somehow porous even though its porosity was impossible to demonstrate.)

The complexity of making one pump into two was not solved overnight during evolution. Both in the lung fishes and in the later amphibia there is still mixing of the blood. In the reptiles there is less mixing, as a wall exists between the two parts of the pumping ventricles, but the wall is not quite complete, and so mixing of the two types of blood still occurs. Only in the birds and mammals is the wall complete. Only in these two classes of the animal kingdom is the venous blood pumped entirely separately to be oxygenated, and the oxygenated blood is then pumped just as separately to the body. The dual role of a single heart is thus satisfactorily accomplished.

Venous blood returning to the heart first enters the right auricle (named after its alleged resemblance to a little ear). It is then pumped by this auricle into the right ventricle (the name meaning a little stomach). This ventricle pumps the blood around the lungs, and it flows back from them into the left auricle. This third chamber pumps it into the most powerful chamber of all, the left ventricle. It leaves this ventricle via the aorta, a vessel one inch in diameter which receives the heart's output of a fifth of a pint in every beat. In mechanical terms this means the relentless exertion of between 35 and 50 ft-lbs. every minute, or the exceptional exertion of 500 ft-lbs. or more during strenuous exercise.

The human heart beats roughly 70 times a minute, or four times for every breath. It will therefore have beaten 2,500,000,000 times before it finally calls it a day at the end of an average life. In general the bigger the animal the slower the beat. An elephant's 48 lbs. (22 kgs.) heart beats only 25 times a minute. The mouse's heart trips along at 600 to 700 times a minute. Birds tend to be faster for their size; a chicken's beat is between 200 and 400, and a canary's is 1,000 times a minute. The dog species varies according to the size of the dog; big dogs are 80, small dogs are 120. Each individual slows down his heart as he or she grows; the human infant has a pulse of about 130 which will slow down to the adult 70. Before birth it was even faster. These infantile speeds are only again matched during strenuous exercise, or anxiety, or both, when the pulse rate may triple its customary pace.

When increased demands are being made upon the heart, and when its beat quickens, it also pumps more with each beat. Many machines become less efficient as they go faster, but the heart's output – its stroke volume – can be nearly doubled from the

492

normal resting output. Consequently, with both more beats and more blood per beat, the heart can pump five times the quantity of blood that it normally pumps when at rest. If a large output is maintained, as in a long distance race, the heart shrinks during the race by about 15%; the larger the heart the greater the shrinkage. Often athletes have large hearts, as might be expected both by virtue of their proven abilities and of their training, but their hearts are not abnormally large and are well within the ordinary range of human variation. Most mammals, including humans, have hearts which weigh from less than 0·5% up to 1% of the total body weight. Ordinary dogs have 1% hearts; sedentary dogs have 0·5% hearts. The human proportion of heart to body is nearer that of the sedentary kind of dog. An ordinary horse has a 9 lbs. (4 kgs.) heart; the famous thoroughbred Eclipse had a 14 lbs. (6 kgs.) heart.

The human heart does not lie entirely on the left side, as is frequently alleged. It lies fairly near the midline with about one-third of its bulk on the right side and two-thirds on the left. An anatomical awkwardness is that the flatter base of the heart faces backwards, and the sharper apex of the heart faces outwards and downwards. It is this apex which reaches out to a person's left side, almost to the nipple, and because the apex pulses with every beat the heart has a reputation for being at that spot, rather than stretching to it with its pointed end. This apex beat can both be seen and felt. Its location in a standing person is 3½ ins. (9 cms.) to the left of the midline, and between the fifth and sixth ribs. (As a helpful nearby reference point, although not for everyone, the nipples are between the fourth and fifth ribs.)

Whereas this apex is seen to beat at the normal 70 or so times a minute, and whereas the pressure waves of the pulse can be felt to possess an identical rhythm, the heart itself makes a two-fold sound for every beat. Listen to it with a stethoscope, or just by placing your ear in the vicinity of someone else's heart (personal auscultation demanding excessive contortion), and the noise is distinctly double. Spelt as lip-dup or lub-dub or even lub-dup the first sound is louder, lower and longer than the second. That first lip or lub is itself a mixture of sounds; it is the ventricles contracting and then the valves between them and the auricles slamming shut. The dup or dub, the second sound, is caused when the valves in the big blood vessels leading to the lungs and the body each slam shut after the pulse of blood has gushed past them.

These four valves are all passive one-way valves. They are of varied shapes. Between the auricles and the ventricles lie the mitral and the tricuspid valves, so called because they have two

493

flaps (like a mitre) or three, and they permit the blood to flow in the direction of the ventricles but prevent it flowing back again into the auricles. The two valves in the aorta and the pulmonary artery are called semilunar, and they shut to prevent the blood flowing back again into the ventricles. Lock gates of the traditional river type also work in this one-way fashion; they slam shut should the water attempt to flow back. It is this steady shutting, of necessity twice for every beat and of necessity with a time lag between them, that contributes the remorseless lip-dup of the active heart. Should a valve weaken or leak there will be an equally steady rhythm of gurgling or back-wash as the blood flows backwards past the faulty valve.

Blood pressure is contrary to heart beat. Instead of falling from infancy to adulthood it rises throughout childhood, and then continues to rise in later life. The first man to demonstrate and measure arterial blood pressure was a clergyman from Teddington, near London, named Stephen Hales. He also found time to devise a windmill for ventilating Newgate Prison to reduce gaol-fever, but his principal claim to renown came after using a horse, a goose and a tall glass to examine blood pressure. In 1733 he tied the horse to a post, removed a goose's wind-pipe in lieu of flexible tubing, and inserted a small glass tube into one of the horse's leg arteries. By means of the goose's windpipe he connected this small tube to a very tall and vertical tube, also of glass. The blood then shot up it to a height of 9 ft. (2·8 m.). The pressure of arterial blood caused this high level and it then oscillated steadily with every heart beat. Soon blood clotting curtailed the experiment.

Nowadays the more ordinary human practice is to use a sphygmomanometer (pulse pressure measurer) in conjunction with a stethoscope. By listening to an artery's beat, and applying variable pressure to the arm, it is possible to estimate both the pressure of each beat and the residual pressure between each beat. The two are known as systolic and diastolic respectively, with the former always being greater than the latter and the two always written down in that order, such as 180:110. The figures are a measurement of the pressure in millimetres of mercury, as is the 760 of normal atmospheric pressure.

Systolic pressure of a new-born baby is about 40. At ten days it may be 70, and at the end of the first month it may be around 80. By ten years it is still under 100. An average young man at rest has a systolic pressure of 120, and a diastolic pressure of 80. Nearly all mammals, large or small, have adult systolic pressures between

100 and 200, and man is not one of the few exceptions. The human combination of 120:80 stays reasonably constant for a time, but after the twenty-fifth birthday may creep up by 0·5 a year. By the age of sixty it will be near 140, and by eighty it will on average be about 160. No one asserts that it ought to rise like this, any more than middle age fatness ought to happen. It is just that blood pressure does rise with age, and a modest rise is better than a steep one.

The insurance companies are often enthusiastic to discover the blood pressure of potential customers, and insurance statistics are cold-blooded in showing how the odds lie against the hypertensive population, although not necessarily against the hypertensive individual. In general the higher the pressure the greater the risk of death. There are always exceptions, but a high blood pressure in a middle-aged man is more dangerous than the same blood pressure in an elderly one; for instance, 160:90 men in their forties are nearly three times more likely to die than average men of the same age, while 160:90 men in their sixties are only twice as likely to die as average sexagenarians.

The lack of any blood pressure is of course more rapidly fatal than the mere excess of it. For those whose hearts stop beating, and whose pressure slumps in consequence, irreversible brain damage follows within a few minutes. Until 1960 the treatment for such an emphatic condition was to cut open the chest, expose the heart and massage it manually; but in that year a method of external massage was first published. W. B. Kouwenhoven, of the United States, and his colleagues excited the world by describing how an internal circulation could be maintained by rhythmic pressure from outside. Just by leaning heavily on the sternum sixty times a minute they achieved blood circulation and considerable blood pressure despite the heart's failure. Simultaneous efforts were made to start the heart again, and first reports indicated considerable success. Other attempts have not always been so satisfactory, and the firm pummelling has frequently led to broken ribs and sternums; but the 1960 report, coupled with the resurgence of interest in mouth to mouth artificial respiration (which has a venerable, even biblical history: 2 Kings iv. 32), has helped to open up the subject of starting stopped hearts without the hazardous extra of opening up the chest.

Artificial pacemakers, which deliver regular electric shocks to hearts no longer capable of beating correctly on their own, have also kept hearts going in recent years. Between 1958 and 1964 over 3,000 patients in the United States had these devices implanted in their bodies, and since then the operation has become even more frequent. In fact modified pacemakers, beating much more frequently than the conventional 70 times a minute and implanted in the neck rather than near the heart, have even been capable of lowering the blood pressure. William Harvey was undoubtedly the first man to prove the blood flows 'in a circle'. Now, almost four centuries later, mankind is making great efforts to make it continue to go round and round instead of coming prematurely to a stop.

Blood distribution, so far as the heart is concerned, is elementary. It pumps it all into the aorta, and that is that. At a customary speed of 15 ins. (38 cms.) a second the aorta receives the blood, and almost immediately the branches begin which distribute the blood around the body. With all this ramification the speed slackens, mainly because the cross-sectional area of all the arterioles and then of the capillaries is so much greater than that of the one inch fat aorta. Within a capillary the blood flow is only about a fiftieth of an inch a second.

The heart, as has already been said, can boost its delivery of blood almost five-fold if need be. From over a gallon (4·5 litres) a minute it can rise to $5\frac{1}{2}$ gallons (25 litres) a minute at times of maximum output. Such a delivery means pumping the body's entire blood supply several times round the body in every 60 seconds. However, even this flow is inadequate. The muscles need more. They are bearing the brunt of the strenuous exercise, and need increasing quantities.

In fact they do get more, but at the expense of certain other organs. Blood is diverted away from these other organs, and the muscles receive eighteen times as much blood in times of extreme exertion even though the heart's output is only five times as great. Organs to suffer are: the kidneys, which receive about a quarter of their normal supply; the skin, which gets more initially with moderate exertion, although this is severely cut back with extreme exertion; and the digestive system, which only gets a fifth of normal. The heart, being muscle itself and undeniably working harder, receives about four times as much during the exertion. The brain is apparently totally unmoved by all exertion or diversion. At rest it receives $1\frac{1}{3}$ pints (three-quarters of a litre) of blood a minute; at peak demand it still receives $1\frac{1}{3}$ pints of blood per strenuous minute. In the meantime blood flow to the muscles, excluding the heart, has been boosted from some 2 pints (just over 1 litre) a minute up to almost 40 pints (22·7 litres). At this time of top exertion the muscles are receiving 88% of the heart's furious output; at rest they receive only 20%. Conversely the abdomen normally gets 24% of the supply; during exertion its share of the greater output is only 1%.

An extra work load for the heart, never encountered in nature – at least not to any marked degree – but developed by man, is caused by artificially increasing the normal gravitational pull upon the system. When a man leaps out of bed in the morning, moving from the horizontal to the vertical, his heart suddenly has to pump blood upwards to his head rather than along to it. This represents a greater effort than pumping it horizontally. Sometimes the leaping man finds himself sitting dizzily down on the bed when the load has proved too much or too sudden. Aircraft pilots become extremely familiar with increased work load when steep turns cause them to black out. The heart can no longer adequately pump blood to the brain against the extra pull involved in the turn, a pull equivalent to a few times that of normal gravity. An astronaut, subjected to the tremendous accelerations of rocket travel, suffers still greater pulls upon his system – and upon his heart. To help to counter them he sits so he is lying facing the direction of travel; therefore his heart is pumping mainly horizontally. A report from the Soviet Union suggested that a man could withstand 26·5 times the pull of gravity, and would not black out, if he was inclined at eighty degrees to the direction of acceleration. A 200 lbs. (90 kgs.) man at that acceleration can be said to weigh two and a quarter tons.

Blood   The transport medium of the body, the principal fluid of the circulatory system, is blood. Slightly heavier than water, three times as viscous, half plasma and half corpuscles, human blood is pushed around the body once every minute even under quiet conditions. It is squirted out of the heart, it falls back to the valves, it is squeezed through the capillaries, and then sucked and forced along the veins to start the whole cycle once more. During its tempestuous circulation it carries water, vital to every cell, it takes oxygen from the lungs and carbon dioxide to them, it carries nutrients to the cells and waste products from them, it transports heat from the hotter to the cooler regions, it distributes hormones, it is a circulator of antibodies – the anti-infective agents – and of its own white cells, and it carries its own self-sealing mechanism for whenever its essential fluidity has to be congealed to block an open wound.

The average human body, weighing 156 lbs., 11 stones, or 70 kgs. contains $10\frac{1}{2}$ pints (6 litres) of blood. This means, owing to the different measures, that the average American possesses 13 pints of blood. The Briton with his 10½ pints therefore has about one imperial pint of blood for every stone of his body; the American has one American pint for every 12 lbs. In both communities blood weight is about one-twelfth of total body weight. Whereas 1 pint (about ½ litre) can be given at a transfusion centre with equanimity (the amount given is usually nearer three-quarters of a pint), the loss of blood from, say, a wound can never be regarded casually, even though the body is capable of losing at least a quarter, and possibly a third of its vital transport fluid without any apparent severe physiological consequences. However, although a quarter can be lost with equanimity, and a third generally without serious effect, the loss of half is likely to be fatal. Transfusion must intervene to prevent death in such cases. Experiments on dogs have shown that the occasional dog can lose two-thirds of its original blood volume without dying and without transfusion.

At any ordinary moment three-fifths of a man's blood is in his veins and returning to his heart. A further fifth is in his lungs, and the final fifth – say 2 pints or 1 litre for the average man – is in his heart, arteries, arterioles and systemic capillaries. The universal colour of blood, bright red in the arteries and dark blue in the veins, is belied if some blood is taken and allowed to stand a while. Provided something is added to the blood to stop it clotting all the colour will sink to the bottom. This part contains the red corpuscles. On top, forming 55% of the whole, will be the blood plasma, a straw-coloured liquid. In between the two, as a very thin boundary layer, will be the white corpuscles, not so heavy as the red ones but heavier than the plasma.

The white corpuscles are not only lighter than the red, but far less numerous. In each pinprick of blood measuring a cubic millimetre there are only between 4,000 and 10,000 white corpuscles. In the same pinprick there will be about a thousand times as many red corpuscles, perhaps five million for men and 4·5 million for women. (By no means are distinctions between the sexes solely confined to chapters on reproduction.) Thirdly there are about 250,000 platelets in each pinprick; these play a part in stopping bleeding after injury.

For those for whom large numbers are comprehensible, 10 pints (5 litres) of human blood therefore contain a total of between 20,000,000,000 and 50,000,000,000 white corpuscles, 1,250,000,000,000 platelets, and 25,000,000,000,000 red corpuscles: they all fit into the 45% of blood which is not plasma. Furthermore, to remain a little longer at this level, each red cell has a short life. Some say a month, some say four months. A month will have involved 43,000 journeys round the body; four months will have involved 172,000 journeys. Even allowing the longer period this means a replacement every day of 200,000,000,000 red cells by the bone marrow. In fact only nine ounces of bone marrow is sufficient to carry out this task. Although probably unhelpful in themselves, these large numbers ought, by their very scale, to give some meaning to the statement that a capillary is only the diameter of one red cell, and therefore help to stress the minute size of the transport network at the opposite end to the heart. Every minute the heart's pumping squeezes all the 26,300,000,000,000 corpuscles and platelets through this ubiquitous filigree of blood vessels.

It is the red corpuscles that contain haemoglobin, and haemoglobin both picks up oxygen from the lungs and delivers it to the tissues. When carrying oxygen it is, for some reason, bright red. When without oxygen it is dark blue, almost black. 'Blue babies' are suffering from oxygen lack, and whenever pulmonary function is unsatisfactory a person will appear blue. Haemoglobin has been the subject of much research in recent years leading both to Nobel prizes and an understanding of the molecule. There are about 280 million molecules of haemoglobin in each red corpuscle. Each molecule is very big, having 10,000 atoms and a molecular weight of 64,500, but it can readily pick up and give up four molecules (eight atoms) of oxygen. Therefore this figure of eight multiplied by the number of molecules in a corpuscle multiplied by the number of corpuscles flowing through the lungs will indicate the body's ability to transport oxygen – or 56 thousand million million million atoms of oxygen per average minute. Were such a carrier as haemoglobin not available, and were the oxygen to be dissolved in blood as it is dissolved in water, the amount of blood would have to be seventy times greater than it is. (But that is quite enough of such astronomical numbering.)

White cells, or leucocytes, are of different kinds. None of them is plate-shaped, like the red corpuscles, and they are all fairly transparent. They have the power of movement, unlike red cells, and they progress in an amoeboid manner pushing out part of themselves and then advancing into it. Customarily they move along the sides of blood vessels, rather than being bowled along in the middle like the pulsing tide of red cells, and they can even intrude their way through the walls of the capillaries. They can advance upon solid particles or upon bacteria, and they can ingest them by flowing around them. For a time the twenty or so bacteria imprisoned within a white corpuscle remain alive, but this is not a one-sided battle with the outcome inevitable. The corpuscles can die from the effects of the bacterial toxins, and the subsequent pus is accumulation of dead leucocytes. It is also not one-sided in the sense that bacteria can win, however vigilant and active the leucocytes; but most bacterial invasions – and they are constant – come to nothing partly because of the omnivorous leucocytes. These cells can also absorb the polluted particles which stick in the lung, they can slowly ingest splinters, and they do attack practically anything foreign to the system. Too many white cells can be as disastrous as too few. In leukaemia, which is an over-production, there may be sixty times as many white cells in the blood as normal. By no means is everything understood about the different kinds of leucocyte, why they are different, why their numerical proportions are so dissimilar, and how many roles each type can play.

Blood platelets, or thrombocytes, smaller than the red cells but even smaller and much more numerous than the leucocytes, are also elusive about their many tasks. They are concerned with the clotting of blood, and with the immediate needs at the site of an injury – such as constriction of the blood vessels. Like the red blood cells, but unlike the white, platelets have no nucleus. They are just broken off fragments of the large marrow cells that made them.

Finally, blood plasma. This 55% of the whole blood, the part that is not cells, is itself 91% water. The rest of it is proteins, salts, and most of the blood cargo, such as those nutrients, hormones, waste products and antibodies that have already been mentioned. The proteins are crucial to the maintenance of blood pressure, to the regulation of blood volume, to the clotting of blood, and to the manufacture of antibodies. A dramatic benefit of plasma has been its use as a substance for transfusion. Quite often a patient needs the plasma part of blood more than he needs all the corpuscles; death from blood loss is more a matter of loss of blood bulk than of loss of cells capable of carrying oxygen, vital as oxygen is. Unlike whole blood, which has to be of the right blood group, plasma can be mixed from all donors and is then useful for all those in need. It can also be dried, unlike whole blood which cannot and which has a short shelf-life, and dried plasma only requires the addition of sterile water to make it fit for a needy recipient.

Quite the most remarkable aspect to the history of blood, of this revered and vital fluid, was mankind's willingness and enthusiasm to rid himself of it voluntarily and often disastrously.

Every major civilisation has practised blood-letting. Every region seems to have had a different rationalisation for opening the veins, whether it was a form of sacrifice, a necessary outlet for the fourth bodily juice, the calming down of an over-excited metabolism, a driving out of the sickness and fever, an evacuation of malevolence, or a precaution against malaise. In Europe the practice of derivatio flourished, both with ancient support from Hippocrates and with mediaeval argument from the doctors. It entailed the letting of large amounts of blood from the centre of disease. A rival system had been the modest extraction of blood from a spot opposite to the diseased area, but derivatio won the day in the fifteenth and sixteenth centuries. Two pints (1 litre) was the customary letting. Louis XIII of France once suffered 47 bleedings in 6 months. Louis XIV was bled 38 times. Charles II of England was a victim of countless lettings and purges, and no less so when on his deathbed. Frederick the Great even had his veins opened during battles to calm his nerves.

Then, suddenly, the old idea of leeches became popular. Instead of cutting blood vessels, and often encouraging septicaemia, millions of leeches were dragged from ponds, and made to do the work in their own quiet way – half an ounce of blood per leech. The leech trade reached its zenith, not in some Aztec period of ancient history, but in Paris in the first few decades of the last century. The fiercest enthusiast of this practice, France's Dr Broussais, was alleged to be shedding more blood than all the bloody wars going on at that time. The allegation is not unfair for, according to the medical historian H. S. Glasscheib, Broussais instigated the letting of some 35–50 million pints (20 to 30 million litres) of blood in France alone.

Suddenly, in the middle of the last century, blood-letting fell out of favour. It only failed to overlap by a few decades the medical revolution of blood transfusion. From pouring it out to pouring it in, the total reversal of policy was extremely abrupt, and extremely welcome.

36. Skeleton and Muscle

The bone total/Development/Growth/Sex differences/Dual
function of bone/Framework and minerals/The hand/The
muscle total/All or none/The working of muscle/A.T.P. and
A.D.P./Cramp, stiffness and stitch

The traditional complement of bones in each human being is 206,
but this is a general rule, not a law. About one person in twenty,
for example, has a thirteenth pair of ribs, and mongols frequently
have only eleven pairs. A baby is born with about 350 bones,
some of which fuse in later life, some of which retain their
individual identity throughout life. All bone fusion is over by the
end of the growth period, perhaps in the twenty-fifth year for
men, younger for women. An old person will, accidents apart, retain his or her mature complement of bones, but many – notably
women – will have lost half of their bone content by the time
seventy is reached. Bones not only provide skeletal support for
muscles, but perform other roles. In fact there is even argument
that the structural and protective tasks of the skeleton should be
considered secondary to the more important bone duty of providing a mineral reservoir. It is hard contemplating a blancmange of
a human being, totally without a rigid framework, but it is harder
still imagining the proper function of metabolism without extra
supplies of calcium and phosphorus being always readily
available.

   The total of 206 is achieved more from the limbs than from the
main axial skeleton of trunk and head. Each arm has thirty-two
bones namely: one collar bone, one shoulder blade, one humerus,
one radius and one ulna, eight wrist bones in two rows, five
metacarpals in the palm, and fourteen phalanges, three to each
finger and two to the thumb. Each leg has only thirty-one bones;
being one hip bone, one femur, one kneecap, one tibia and one
fibula, seven tarsals in the instep and heel, five metatarsals in the
foot, and fourteen phalanges, three to each ordinary toe and two
to the biggest toe, the hallux. With two legs and two arms this
means a limb total of 126 bones. The axial skeleton has eighty.
There are twenty-nine in the skull, of which eight are in the

cranium, fourteen are in the face, six are in both ears as the ear ossicles, and one – the hyoid bone – is in the throat between the lower jaw and the upper larynx. The spine has twenty-six bones, of which – in descending order – seven are cervical vertebrae, twelve are thoracic vertebrae, five are lumbar vertebrae, one is the sacrum, and one is the coccyx or tail. The chest has twenty-five bones, of which one is the breast-bone and twenty-four are ribs.

Unlike all other apes (as was mentioned in the chapter on the male) the human species has no bone in its penis; it has to make do without such an aid. Also man can acquire bone during life according to his mode of existence. Cavalrymen have been known to acquire bones in their buttocks and thighs, quite distinct and separate objects from the traditional hip and femur bones. As such extra bones have arisen entirely as a result of the cavalryman's continual horse-work this makes one wonder again about humanity's lack of an os penis.

The mammals, despite unevenness in size, ranging as they do from shrews to whales, are remarkably consistent in their bony skeleton. The quaintest example is the system of cervical vertebrae. These bones are quite distinct from the thoracic vertebrae which follow them, and yet the long-necked giraffe only has seven and the no-necked whale also has seven. Exceptions to this rule of seven are the manatees and the tree-sloths. The horse, able to transport a man with ease, has 205 bones as against 206 for the man. The horse has eighteen pairs of ribs (against the human twelve), and fifty-four bones in the backbone (against twenty-six); but, because the manner of its evolution forced the modern horse to be prancing around on its middle fingers and middle toes, there has been considerable bone loss; horse limbs have only twenty bones each, against thirty-two and thirty-one for human limbs.

As everyone knows, but tends to disregard, the fact that two children both happen to be the same chronological age is relevant but frequently misleading. Neither intellectual nor sexual maturation pay such diligent regard to the calendar as do human beings. Skeletal maturity is equally casual about birthdays but its degree of development can be of terrifying importance to some children. Does this girl's above average size mean that she will end up taller than average and therefore useless for ballet? Does this girl's current height indicate that she may be over 6 ft. (1.8 m.) when eighteen, and ought she to try to slow down her growth while there is still time? An X-ray of the wrist will probably give the answer. Bone development follows a definite routine, and the stage reached indicates the proximity of the final stage, the end of growth.

On average, growth in height has virtually ceased at $17\frac{3}{4}$ for boys and $16\frac{1}{4}$ for girls. However it has not entirely ceased. The vertebral column increases by 3–5 mms. between twenty and thirty years, a fraction of an inch which will worry neither ballet dancer nor too-tall girl. After fifty stature diminishes.

Sex differences in adult size and shape come about in different ways. The longer male legs result from the greater time spent by boys growing before puberty, a time when legs are growing faster than the trunk. Conversely the longer male forearm is established at birth, and it continues to be longer than the female forearm throughout life. Similarly the female tendency to have a longer second finger (index) than the fourth finger (ring) exists from birth. Yet another cause of sex differential is direct intervention by male and female hormones. This happens both in the shoulders (broader for men) and in the hips (broader for women) when androgens and oestrogens respectively stimulate cartilage growth.

Bone growth and form is undoubtedly genetic, but it is influenced by the demands being made upon it. Unlike some structural girder it is altered and improved, provided the load is insufficient to cause damage, by any increased demands being made upon it. Both volume and density of bone can be made to increase by work. For example, two groups of growing rats were fed differently. One group received hard food which needed chewing, and the other group received soft food which did not. Not only did the soft-fed rats acquire slightly smaller heads and faces – by 1–2%, but the bones of their head and jaw were lighter – by 12%. One wonders in consequence whether the steaks of America help to produce the thick-set jawline characteristic of the well-fed American citizen, or if it is just the good feeding?

Mice have been kept in centrifuges at four times the force of gravity, and have thicker and differently shaped bones as a result. Human beings kept in bed promptly lose bone, this is not just decalcification but bone itself which is lost when the skeleton is abnormally idle. For the same sort of reason the sudden change and call-up from a sedentary life to an Army battle course can lead to certain types of fracture. Demands are being made of a skeleton unprepared for such relentless effort. It is amazing that astronauts, particularly the Soviet variety who have spent many months continuously in orbit, have been able to walk within such a short time (apparently less than twenty-four hours) of returning to earth's gravitational pull after so long in the weightlessness of space.

Bone grows to a large extent where it ought to grow, where the stresses demand that it should be grown. Should a child break a femur, and should that bone fracture be wrongly aligned, the healing will look initially unsightly as a large callus is formed, but the site of the fracture ought to be invisible a couple of years after the break. Bone construction cells will have combined with bone destruction cells not only to join the break but to erase the broken outline of the faulty alignment. Unfortunately the repair system grows increasingly inefficient with increase in age, and faulty alignment is more likely then to lead to permanent thickening.

The word skeleton is derived from a Greek word meaning dried up. Bones are most frequently seen when dried up, but their living strength depends upon their not being dry. Chemically, bone tissue is 70% inorganic matter and 30% organic matter. Dissolve away much of the inorganic component with an acid, and the result is like a dog's rubber bone; in fact a long bone like the human femur can be tied in a knot. Remove the organic component, either by burning or by letting decay seek it out, and the result will be a dry, brittle, hard object with more of the properties of cast iron than the flexible and stronger steel of the living tissue.

Although the body's 206 bones are long (like the femur), short (as in the wrist), flat (like the shoulder blade) or irregular (like the vertebrae), all bones have an outer, denser layer of compact bone and an inner meshwork of porous material. The outer layer benefits from the principles which cause a tube to be almost as strong as a solid rod of the same diameter. The inner layer, often called spongy because it looks like that, is phenomenally strong. More like hard coral than sponge its strength also follows from the engineering principles which explain why many meshworks, such as honeycomb, can be so strong and yet apparently so delicate.

Both inner and outer strength are necessary. When a parachutist touches down, or when someone jumps from a wall, the traditional loads are exaggerated many times. The leg bones are capable of withstanding compression of a ton or more. Later, particularly in post-menopausal women, bones can be unfortunately vulnerable to far smaller loads, and the neck of the femur is then a frequent victim of quite modest accidental burdens imposed upon it.

The function of bone is diverse. Its rigidity provides a shape and support. It protects. It acts as an anchor point for muscles. Its internal marrow manufactures all red blood corpuscles. It also

505

produces other constituents of blood, as well as destroying old red cells. It contains minerals, chiefly calcium and phosphorus but also some magnesium, fluorine, chlorine, iron. More important than their mere presence, the bone's minerals are frequently being removed and replaced. For all its structural solidity, bone is remarkably changeable.

In the earliest fishes the skeleton was external, much like the armour of a mediaeval knight. Not only is such external plating a restriction to movement but it can have little to do with the body's metabolism. Like teeth it was probably deposited in a one-way process; and, once deposited, the minerals were as good as lost. The calcium carbonate and calcium phosphate of bone can instead be deposited, then removed, then replaced, and so on. This ability to have a supply of phosphorus readily to hand is vital considering the huge number of chemical processes in which this element takes a crucial part. The ability to have a supply of calcium equally available is also necessary for many other chemical reactions and for the correct balance of the body's fluids. Animals living in salt or fresh water, or a mixture of the two, or moving from one to the other, have to be able to cope with the changing osmotic situation. Adjusting the level of calcium in bodily fluids is an essential part of keeping the internal environment stable whatever is happening to the external environment. Bone acts helpfully both as a storehouse and as a dumping ground. At different times the two aspects have equal merits, but the ability to control these vital elements, both so crucial to metabolism, has created the argument that bone's metabolic functions are primary to its functions of structure and support. Admittedly each is vital, and pride of place is therefore entirely hypothetical, but at least the argument makes the point that bone is not the static framework of structure and support which a skeleton might seem to be. It is not comparable with the permanent frame of an office block.

Moreover, unlike most structures, the skeleton not only has to provide firm rigidity but extreme flexibility. The muscles have to be strongly anchored, and yet there has to be articulation. As with so much else dictated by all the opposing forces of natural selection, the result is compromise. Man is a mixture of stiffness and relaxation, restriction and freedom, joints without movement and joints with great versatility. The wrist is free, the ankle far less so. The thumb is opposable, the big toe is not. Just as an architect indicates on his drawings how a door will open, so have attempts been made to indicate all possible human movement. The result is always a muddle. The hand can touch everywhere else on the body, even the middle of the back for most people, but it is hard to show this. It is harder still to show how a child's ability to squat, with an apparent total disregard of any knee or ankle joint limitations, gradually loses this rubbery simplicity and is transformed into the more awkward posture of the typical adult. A child's elasticity is particularly important at birth when even its cranial bones can be moulded, and the narrow passage of the birth canal is not so restrictive as would otherwise be the case.

Considering that the hand of man has played an integral part in man's development, and that both its structure and nervous connections are more highly advanced than in any other creature, it still has great limitations. Nothing much can be done with any of its fingers except flexion and extension. The thumb has slightly greater freedom, and the customary explanation of the human hand's dexterity is the capacity for that thumb to oppose itself to every finger. With a little co-operation from each finger both thumb tip and finger tip can be made to touch. However, the fact that the hand possesses two distinct grips is believed to be even more remarkable. Two billiard balls, for example, can be picked up, and each held independently by two parts of the same hand. One is held between the last two fingers and the palm; the other is gripped between thumb and forefinger. Being able to hold two balls in this manner may seem a poor asset, but it is certainly not unimportant that these two grips exist. Try taking the top off a pen with the hand that is holding it, and imagine doing it without two grips. Try doing anything with only the single, claw-like grip of a baby's hand, and imagine such restriction.

Incidentally, with regard to the hand, the professions have been confused by nomenclature for the five digits. Elsewhere in the body the anatomists never seem to have been at a loss, but the medical and legal professions, plus the ordinary citizen, do not agree about nomenclature. Courts of law, insurance documents, doctors and mere people should at least be referring to the same digit, but such unanimity is frequently lacking. Is the thumb the first finger, or is the fore-finger? Do we have four fingers or five? Is the order thumb, index, middle, ring and little? If so, where is the fore-finger, and where the first finger, and do all people wear rings, and on the left or the right hand? And how many people in England know that the pinky is an American little finger?

By the end of 1965 every single one of the fifty States of the Union except for California had adopted standard systems of naming the five (or four) fingers, but in their collective attempts at standardisation they had made use of ten different systems. *The Lancet* said every system had its drawbacks, and wondered about a man 'who has lost one digit from a hand showing a six-fingered polydactyly but having no properly differentiated thumb'. The dictionaries, in attempting to clarify, outline the problem. In general they say a finger is 'one of the five terminal members of the hand, or one of the four other than the thumb'.

Finally, some of the earliest human-type skeletons give strong indications of human-type behaviour. According to *Bones, Bodies and Disease* (by Richard Fiennes), skulls of Australopithecines have been found showing a form of fracture consisting of two depressions close together. The humerus bone of antelopes has been discovered nearby. The end of such a bone fits well into the skull depressions. In other words, a million years ago people were already being hit on the head by people. Burial grounds and skeletal remains of more recent date give permanent evidence of mankind's perpetuation of this custom.

Muscle. All animal movement is the work of muscles. Muscles can only work by pulling, never by pushing and however much a man may be pushing down a wall, every single muscle doing work is doing it by pulling; the body's engineering sees to it that the pulling becomes pushing. Human muscle tissue is laid down before birth or shortly afterwards. In this aspect it is similar to nervous tissue. An infant's complement of muscle fibres is its complement for life; it will probably acquire no more. Strength comes after the expansion of each fibre and after it has been made to do work. For the same reason the limbs of a blacksmith and of a girl contain a similar number of fibres, although he may have several times her strength. Big people are customarily stronger than small people but, per pound of body weight, bantam class champions can lift about half a pound more than heavyweight champions. In terms of maximum energy output it has been calculated that the theoretical limit for man is about six horse-power, that the highest recorded is 4·5 horsepower, that 0·6 horsepower can be sustained for five minutes, and that 0·5 horsepower can be sustained indefinitely. (Which makes one wonder about the sustained horsepower of one horse.)

Muscle fibres can be long – up to $1\frac{1}{2}$ ins. (4 cms.), and they can be minute – a millimetre or less, but their diameter is always far smaller. Most fibres are from a tenth of a millimetre to a hundredth of a millimetre across. Many individual fibres make up each distinct anatomical muscle, and there are about 656 muscles in the body, or over three times as many muscles as there are bones. Some 42% of male weight is muscle, some 36% of female weight. Most muscles attached to the skeleton are linked to a bone either at one end by a tendon or, less frequently, at both ends. In man these tendons (or sinews) may be very small, or more than a foot long. Ligaments are also composed of strong fibrous tissue, but their role is to bind bones together.

Each muscle fibre obeys the same all-or-none principle as the nerve fibre; it either does contract or it does not. However, movement is not a series of robot-like jerks because each muscle contains innumerable fibres, and each separate fibre has to be triggered to effect its pull. Should the stimulus be strong, as when a hand touches a hot surface, every fibre will pull and the result will be a jerk. Should the nervous stimuli to contract continue to arrive very rapidly at the muscle it will not contract and relax again, as generally happens, but will remain contracted in a state of tetanus.

The impulses which can cause a fibre to twitch are electrical, mechanical, thermal, and chemical. The time between the arrival of a stimulus at the fibre and the start of that fibre's contraction is between two and four thousands of a second. A muscle works by the conversion of chemical energy into mechanical energy; only about 25% of the potential energy is correctly converted, and the remaining 75% is assumed to be lost as heat, assisting the rising temperature of someone suddenly doing strenuous work. A maximum efficiency of 25% is similar to the efficiency of an internal combustion engine; that too loses most of its energy as heat, and systems are necessary to dissipate this heat.

However much a muscle may either contract or fail to contract on an all-or-none basis, and however much this rule may be generally applicable, there is no such uniformity about the total time taken either for a contraction or between regular contractions. On heart beat alone there is great diversity: an elephant's will contract every two and a half seconds, a canary's will contract seventeen times every second. The human heart, as everyone knows, can change from a solemn beat of forty-five a minute to the furious pace three or four times faster during extreme exercise. Even the fast heart is not the fastest muscular rhythm achievable by the body. Finger tapping can be far faster – for a time, and the tongue and the teeth and the eyelids can all be made to move with a far swifter tremor. Such human speeds are completely outclassed by the wing beats of many insects. Butterflies are ponderous as they flap their casual course, but wings of beetles can oscillate up to 175 times a *second*, of bees up to 247 cycles a second, of mosquitoes up to 587 cycles, and one midge has been recorded with a wing beat faster than 1,000 cycles a second.

Such vibrations are only possible because the wing muscles do not conform to the conventional relationship between nerve and muscle. Almost always a voluntary muscle twitches once in response to a single nerve impulse. The remarkably high speed insect wing is twitching many times in response to each stimulus. It is unique in this respect for the muscle is not going through the normal routine of relaxing after contracting. This phenomenon means that, weight for weight, the wing muscle of some insects generates more energy than any other animal tissue, certainly much more than human muscle tissue.

When a human finger tip, for example, is tapped rapidly on a table in a feeble attempt to emulate the insects it will soon tire. Eventually, suffering a paralysis of its own, the finger becomes immobile. The muscle, it seems, can do no more. In fact, if that muscle is then stimulated electrically and externally, as with those massage machines used by therapists, the finger will start tapping again. It is therefore not the muscle which is tired, for that it still responsive; nor is it the nerve which is tired, for that is still capable of transmitting the customary stimuli; rather it is the gap between the two, the synapse, which fails to conduct the stimulus any more from nerve to muscle. Of course, such finger-tapping paralysis is short-lived. Within a few moments the synapse will have had its capabilities restored, and tapping can begin again. Muscular fatigue itself exists quite independently of any synapse fatigue, and is believed to be due to an accumulation of lactic acid which has arisen as one result of the temporary lack of oxygen.

How does muscle work? The subject was opened up notably after World War Two, and is still in the state of exciting development. Basically the decade of the fifties ended with a generality that muscle contracts because it possesses two kinds of filament. One is thick and one is thin, and they slide past each other to produce the shortening. It was known that this contraction is extremely rapid – each fibre shortens at a speed equal to several times its length in a second – and it was known that the power of its tension is 40 lbs. per sq. inch of muscle cross-section. It is now known that the thin filaments (composed of the protein actin) are drawn further and further in between the thick filaments (composed of the protein myosin), and it was felt valid in the 1950's to compare the system to a rachet action. Those doing this work still like the comparison. They like it because it appears that the sliding action progresses in a series of distinct notches.

The biochemistry is extremely complex, but it is understood that the energy for pushing each filament on a notch comes from A.T.P. (or adenosine triphosphate) which itself gets energy from glucose. A.T.P. uses up its energy in the notch-advancement process and becomes A.D.P. (or adenosine diphosphate) when it does so. Probably myosin has been the protein which acted as an enzyme to split off the phosphate group, the act which turned the triphosphate into the diphosphate. Such a remark is easy to make, but how proteins act as enzymes – which they are doing all the time in every bodily activity – is another matter. Anyhow, glucose, the simplest sugar, then supplies the energy to reverse the process, and assists in putting that phosphate group back again – another easy remark of hideous complexity. Each molecule of glucose can recharge many molecules of A.D.P. to form A.T.P.

Even if this over-simplified version is comprehensible it is by no means the whole story. The full story is really the full story of how proteins work, how they catalyse biochemical reactions, how they are affected yet unchanged. Such questions are not just unanswered problems of muscular contraction. They are among the fundamental questions of the chemistry of living tissue. A.T.P. is not just another set of initials; they stand for the molecule that is the universal carrier of energy in the living cell.

Work on muscular contraction has taken such a leap forward since its critical strides in the 1950's (which led also to Nobel prizes) that it may solve not only many of its own problems but others equally fundamental to all forms of tissue. The biochemical examination of contraction, and of the two proteins myosin and actin, has already shown itself to be a favourable platform for the examination of protein as a whole. As with some tortuous knot the careful unravelling of one small part of it can suddenly expose the simplicity of the remainder. However it has not done so yet.

Cramp, stiffness and stitch have been painfully encountered by most people. Although common they are certainly not fully understood. Cramp, for example, is certainly a muscular contraction, and may well be caused by the nervous system, notably without recourse to the higher control centres; but there is cramp and cramp. It frequently comes during sleep, or when there has been discomfort, or when a body is starved of salt, or when cold and exertion – as in swimming – combine to induce it. There must be many possible causes, each leading either to different varieties of cramp or even to the same punitive and spasmodic kind of contraction.

Stiffness is equally emphatic in its symptoms and equally vague in its causes. It may affect joints, ligaments, tendons or muscles. It usually occurs after exceptional exercise, it may be either an accumulation of products of contraction, whatever they may be, or merely the manifestation of slight injury. The tissue between muscle fibres may have been torn or pulled causing the modest pain and modest impairment.

A stitch can give the sharpest pain of the lot. Generally it is felt below the ribs, and generally on the left. It is frequently the accompaniment of jolting exercise, as in running with a loaded stomach. Cross-country running, with all its irregularities, is more of a jolt. Sir Adolphe Abrahams, in *Fitness For the Ordinary Man,* says that exposure to cold can help to bring it on, that the wearing of mittens has sometimes proved beneficial, that it helps if the stomach is empty, that disappointment usually follows attempts at prevention, and some people just do seem to have a constitutional tendency for stitch.

37. Radiation

Ionising radiation/Its effects/Death and injury/Safe levels/
Fall-out/Sensing radiation/A glossary

When this book was first written – between 1965 and 1967 – there was so much current talk about the effects of radiation upon a body that it seemed entirely correct to include a chapter on the subject. Today that inclusion seems odd but still worthy, despite or because of the profusion of nuclear activity in the past two decades. There are now more nuclear power stations than ever before, and there are more nuclear weapons in existence. This radiation chapter has not been removed in this revised edition, partly because the subject is just as interesting but partly for the suspicion that today's citizens know less about ionising radiation than those of twenty years ago. We are now deeper into the nuclear age but do not have fall-out falling quite so liberally upon our heads. The fear has remained but, in my opinion, much of the knowledge has gone. At all events this final chapter, coming so strangely after Skeleton and Muscle, is here retained and amended very slightly.

Background Ionising radiation is no new thing. It has been bombarding this planet ever since the world began, partly from the sun, partly from space. And earth itself has been bombarding itself from its own naturally radioactive materials. What is new is man-made radiation. Man has been indulging in this type of bombardment for almost a century. In 1895 Röntgen demonstrated his newly discovered X-rays. In 1898 the Curies isolated radium. By 1900, in one way or another, 170 cases of radiation injury had been reported, but no one was alarmed. X-ray machines were used almost as toys, and certainly without restriction.

By the 1920's, owing to the mounting toll of radiation injuries, with fingers and toes eaten into and turned cancerous, there was increasing alarm, but little understanding. Lord Rutherford was gathering a brilliant team of men around him at Cambridge, but even he could see little future for the atomic energy they were starting to unleash. Between the wars atoms were split and transmuted, but only experimentally in the laboratories. Scarcely anyone had any idea of the nuclear holocaust to come, or of the sudden explosive flowering of man-made radiation. People were being urged to drink radioactive waters. Watches and clocks

513

glowed brightly from their radioactive luminous paint. The girls who painted it on used to lick their brushes to make the job easier. Some trichologists used to get rid of surplus hair by subjecting it to X-rays. The hair fell out all right, but subsequent malignancies were common. (The last such depilatory machine in the United States was working until 1949.) X-ray machines in shoe shops used to keep the children happy. Pregnant women were routinely X-rayed. No one worried very much about 'maximum permissible doses'. Those days were still to come.

In August 1939 Albert Einstein wrote his famous letter to President Roosevelt warning him of the possibilities of nuclear energy. Project Manhattan, dedicated to the exploitation of nuclear energy, was put in hand and, at the end of 1942, a disused Chicago squash court was the site of man's first nuclear chain reaction. Enrico Fermi, born in Italy, watched over the 'pile' of uranium as it went critical, as atom after atom of uranium split more and more atoms of uranium, and left the radioactive products of all this fission in their wake. That first chain reaction was carefully monitored and controlled; but later, once over new Mexico in July 1945, and twice over Japan in August that same year, the chain reactions were permitted to do their damnedest. This time the fission products were violently scattered over the land and into the atmosphere. Eighty per cent of the Hiroshima and Nagasaki victims died from blast and heat, much as millions have died from blast and heat since high explosives were first unleashed, but the remaining 20% died a quite new death from radiation.

The subsequent testing of nuclear weapons began modestly enough, then accelerated as first the United States, then Russia, then Britain, then France, and then China loosed off their devices. The fission products rained down as fall-out, and increasingly as the bombs increased in numbers. In 1961 the northern hemisphere fall-out rate of strontium-90, the long lived radioactive element which settles in bone, was 1·2 microcuries per square kilometre. By 1962 it was 5·2. By 1963 it was 9·1. Fortunately the escalation was stopped by the Test Ban Treaty of December 1962, signed by all nuclear powers except France and China. By then the United States had set off some 250 'devices' of one kind or another, Russia about 140, and Britain 23. The fission products of all this atomic splitting, the famous isotopes like strontium-90 and caesium-137, were then part of every one of us, rich, poor, old or newly-born.

Despite the huge destructiveness of these weapons the amount of radioactive fission products released by them was very small.

The Hiroshima bomb, although it exploded with the power of 20,000 tons of T.N.T., was all the work of 2–3 lbs. (0·9–1·4 kgs.) of uranium. More uranium is actually necessary to set the thing off, but most of this is scattered in the explosion. A mere 2·2 lbs. (about 1 kg.) is equivalent, in explosive energy, to those 20,000 T.N.T. tons.

When uranium splits, many of its resulting products are harmless to man. They either lose all their radioactivity too quickly, or too slowly, for it to be harmful. Some, like strontium-90, and caesium-137, are definitely harmful, or have the power to be. But these two products comprise only about 6% each of the weight of the split uranium, of those 2·2 lbs. In other words, Hiroshima's huge bomb left, as a residue of its splitting, a couple of ounces (or about 56 gms.) of harmful strontium and another couple of caesium. The scale of the atomic world is incomprehensibly minute.

No one knows quite how much fission product has been produced by the nuclear total of over 400 'devices', for they were all secret and some were dirtier than others in producing radioactive residue. This number of bombs could have gone off leaving only 900 lbs. (408 kgs.) of fission products, or 45 lbs. (20 kgs.) each of strontium and caesium. Even if those 45 lbs. are multiplied by ten or a hundred the total is still seemingly small when compared with the 196 million square miles of the earth's surface. Yet, despite the huge dilution, despite the great oceans and the atmosphere, despite the size of the continents, those fission products were more than sufficient to be present in every one of us. Take any fragment of bone, take a sliver of that fragment, whether from Eskimo or Londoner, and it will instantly record on a suitable instrument the amount of man-made radioactivity that it contains. We all possess fall-out. We all drink it and eat it. No one can escape.

But then neither can we escape from all the natural radiation. It is just that man has added to it since the nuclear age began, gently with Röntgen's X-rays, explosively with the bombs of 1945 and since that nuclear dawn.

The effects of radiation   It was after America's Pacific Island atomic tests of the early 1950s that the world suddenly woke up to the dangers of fall-out. Beforehand there had been sympathy entirely for the Japanese. After the Bikini test, and after the huge clouds of radioactivity had encircled the earth to deposit active iodine in the milk and strontium and caesium (among other elements) in the soil, the sympathy became more introspective.

What will this radiation do to us? Since then pronouncements, official and otherwise, have either soothed or scared. Usually the pronouncers have had more than one axe to grind. The director of Britain's Atomic Energy Authority's Health and Safety Division once gave a skilful summing up of the conflicting axes when he wrote: 'Firstly, the public should be protected from unnecessary and harmful exposure to ionising radiation; secondly, from unnecessary and harmful propaganda about it; thirdly, from unnecessary and harmful interference with the ability to enjoy the many benefits that radiation can bring now and in the future.'

Ionising radiation exists. It can do great good. Britain takes millions of X-rays every year. Radioactive chemicals are regularly used in diagnosis. Many kinds of radiation source (e.g. for killing cancer) are manufactured. Ionising radiation can also do great harm, and the extent of this harm is the subject of continuing debate. To try and clarify this wrangle I have split the harm story into three. There is obvious harm when it kills; there is conspicuous harm when it suddenly causes a particular disease; and there is suspected harm if sufficient people receive a dose appreciably higher than normal.

Radiation deaths, apart from those thousands in the two Japanese cities, have been extremely rare. The first accidental death by radiation occurred in America a few days after Hiroshima and Nagasaki. On August 21, 1945, Harry Dagnian accidentally gave his right hand a huge dose. Apart from a slight tingling sensation in his fingers, he felt well. Three and half weeks later he was dead. On May 21, 1946, in another American laboratory, the screwdriver of Louis Slotin slipped. After the blinding flash, caused by two masses of uranium coming temporarily too close together, Slotin coolly calculated the dosage that everyone in the room had received. He sketched their positions on the blackboard, realised that his own dosage had been fatal, and not that of anyone else. The whole group went to hospital, and nine days later Slotin was dead. Since then a small handful of other such nuclear fatalities have occurred, notably in Yugoslavia and the United States.

Such a death hit a worker at the Rhode Island uranium-recovery plant in July 1964. This was the world's first fatal industrial accident from radiation (which is not too bad a record considering that the British coal-mining industry alone was killing two to three hundred men a year at the time). For some reason the nuclear worker poured enriched uranium into a small tank. Once again there was a sudden flash of neutrons and gamma rays. The man received 8,800 rads, or ten times the lethal dose (more about rads and dosages later in the glossary). Two hours later he was in hospital, with a slight headache, a normal temperature, and normal blood pressure.

The clinical report in the *New England Journal of Medicine* skates over the man's personal feelings at this time, but suffering massive radiation must be emotionally akin to swallowing poison irretrievably. The body feels well temporarily, but the mind knows a little of what is in store for it. In the case of this Rhode Island man his blood pressure suddenly dropped after four hours, and steps were taken to keep it up. Eight hours later it was still falling, and his temperature was up to 102°F. (38·9°C). Parts of his body, those which had been nearest to the tank, were red and swollen; but he still felt physically well. Thirty-six hours later the swellings were worse, his blood pressure was becoming more difficult to maintain, and his blood was increasingly full of the products of decay. After forty-four hours his blood pressure was falling too low for it to be maintained, and after forty-nine hours he died.

Death had been inevitable throughout those two days. Its immediate cause was circulation failure, but the destruction of tissues could equally well have caused death in plenty of other ways. A man just cannot take more than a certain amount of radiation, just as he cannot take more than a certain amount of poison, or injury, or blood loss, or fever, or cold. There are limits, and somewhere between 500 and 1,500 rads is the limit for nearly all of us. A Yugoslav, who suffered 1,350 rads in a reactor accident near Belgrade, holds the authenticated world record in this respect because he survived this dose. Bone-marrow grafting in Paris was vital to his survival. Current thinking is that no treatment would be effective for doses higher than 1,500 rads. The Mol (Belgium) reactor accident at the end of 1965 helped to confirm this view. One man received up to 5,000 rads on his left foot, but only 200 rads on his trunk. Six months later he was discharged from hospital, but with a mid-thigh amputation of his left leg.

Radiation injury   Apart from killing, radiation can also maim or cause definite injuries. The injuries may prove fatal in time, or hinder life in the meantime, but they come into a different category to the inevitably fatal class. Mrs Mijjua Job of Rongelap came into that category following the events that began on a sunny day in March 1954. Rongelap is one of the Marshall Islands of the Pacific; so is Bikini. She saw the flash and then the huge growth of cloud following the detonation of one of America's hydrogen bombs. The 120 mile gap between her atoll and the one pulverised in the explosion had been considered sufficient, particularly as the forecast wind would take all radioactive debris to the west and away from Rongelap. The wind, unfortunately, did nothing of the kind. It blew to the east, and then subjected Mrs Job and eighty-one Rongelapese to two days of intensive fall-out. Radioactive iodine, strontium, caesium and other isotopes all came their way, and landed silently on the inhabited atoll.

Consequently nearly everybody suffered radiation burns. These are much like ordinary burns, but harder to heal. Ninety per cent of the children and 40% of the adults suffered hair loss. Practically everybody's white blood cell count slumped to half its former level. Something like 175 rads of radiation dose had been suffered by these people from the direct consequences of fall-out. They had also suffered further rads when they drank water and ate food contaminated by the Bikini debris, perhaps 160 rads extra for the adults and even more for the children. The wind miscalculation had turned the island, as one American bluntly and accurately put it, into a unique human laboratory colony.

In the years after 1954 the colony's thyroid glands were of particular concern. By 1965, eighteen of the eighty-two, including Mrs Job, were found to have thyroid abnormalities. Six of the islanders, again including Mrs Job, were flown to Boston for surgery. Fortunately most of the thyroid nodules were thought to be benign, but one turned out to be a carcinoma of the gland. Seventy-five of the Rongelap children were on another island at the time, and significantly not one of them has developed a thyroid nodule. Other deleterious changes to the eighty-two have been that the bone marrow status of some of them was still below normal even ten years after the mishap, that more miscarriages happened in the first four years, that some benign lumps appeared near some of the burnt skin areas, and that the children who were less than five at the time are slightly smaller or less developed than normal. Many investigators of the Rongelap disaster are already beginning to sigh with relief that such huge doses have led to consequences no worse than those already recorded.

Safe radiation levels   There is no such thing as a 'safe level' of radiation, a threshold below which no harm can be done. Radiation is harmful. It does do damage. It comes in various forms, all with variously damaging potentialities, but all do harm. Whether as alpha or beta particles, as gamma rays, as neutrons, or as X-rays, all do harm, but differently. Energetic gamma rays pass, bullet-like, straight through the human body. Alpha particles are stopped in the skin. None of them is detectable by the five human senses (although some animals seem to be aware of them). A lethal dose is painless – until its effects start showing up as painful symptoms some hours later. Sub-lethal doses are naturally equally painless, but even quite high doses, such as 100 rads, produce no symptoms at all. Between 100 and 200 rads the symptoms are mild, if obvious at all, and they really only become apparent after doses of more than 200 rads have been received. However, symptoms are quite different from both long-term effects and from genetical effects upon subsequent generations. It is these effects that cause so much reasonable concern. High doses, of the sort received at Rhode Island or Rongelap, will – one hopes – be rare. Low doses are with us all the time; and so, presumably, are their effects.

In Britain the maximum permitted dose to the whole body for an ordinary member of the public is half a rem a year. (In this connection rems and rads can be considered equivalent.) For ordinary workers at atomic energy establishments in Britain the maximum dose is one and a half rems a year. For workers actually dealing with radioactive materials the maximum is five rems a year, with a subsidiary maximum of three rems in any period of thirteen weeks. Such doses are minute when compared with the 175 rads received by Mrs Job as the direct consequence of the fall-out, and the 160 rads extra that she received indirectly.

This total of 335 rads in a few weeks is about half as much again as an atomic worker would be permitted to receive in a working lifetime of fifty years. It is almost 700 times as much as the average member of the public, subjected to his or her share of natural radioactivity, X-rays and fall-out, is permitted to receive as a maximum in any one year. And this maximum dose was still not reached by a long way even during the worst fall-out year – 1963, or during the year for greatest contamination of food by strontium – 1964. The Rhode Island worker received 8,800 rads, and was dead forty-nine hours later. Mrs Job received a total of about one twenty-fifth of that amount, and had to have her thyroid gland removed ten years later. Each ordinary man in the street receives much less than one seven-hundredth of her dose from fall-out in any year, let alone in a few weeks, and yet there is great concern.

So there should be. There is a tremendous lack of knowledge about the effects of radiation both genetically – upon the future generations – and somatically – upon the current generation. There are men who stress the dangers; there are those who belittle them. *Time* magazine has reported that in 'almost all arguments involving fall-out and its potential hazards, equally reputable scientists can be found on both sides'. Despite this lack of knowledge, or perhaps because of it, there is more concern about the genetic than the somatic possibilities. From the genetic point of view there is no such thing as a safe level because at any level radiation is likely to increase the number of mutations, the irreversible changes between one generation and the next, and almost all of them will be harmful. J.B.S. Haldane once likened this harmful inevitability to a crude attempt at watch-repairing. If you take an erratic watch, and jab a finger in its mechanism, there is just a chance that this jab will improve matters. There is a far greater chance that it will not. So too with mutations. They are almost always harmful. Radiation causes more of them. Therefore radiation is harmful.

Long-term experiments are being carried out to discover just how harmful it is. Some mice at Los Alamos, for example, were subjected to 200 rads of gamma radiation (roughly Mrs Job's direct dose and some 5,000 times the amount customarily received from natural sources) and allowed to breed. By 1967 this mouse colony was nine years old, and had produced forty-four generations. There were by then two visible effects; one good, one bad. There were fewer mice per litter, but more litters per female lifetime. The result was just as many mice as in the control group, and a growing conviction that nothing short of sufficient radiation to bring about complete sterility will cause the genetic death of a mammalian population. Admittedly mice are not men, but one past fear had been that in an irradiated world so many genetic monstrosities would result that breeding would be impossible. Although only mice, the Los Alamos colony helped to dispel this fear.

Future generations are also in jeopardy when young embryos and foetuses are X-rayed. When X-ray machines were new and obstetricians were revelling in their new-found ability to check on foetal development, on twins, and on the correctness of foetal posture, the wombs of pregnant women were regularly pierced with ionising radiation. Now this is less true. The maximum permissible dose to be given to any foetus in its nine-month gestation used to be one rem (again virtually equivalent to a rad), with the mother's abdomen not permitted to receive more than 1.3 rems.

Since those rulings of the 1960's the belief has grown that neither foetus nor maternal abdomen should be subjected to *any* radiation unless absolutely necessary. With the arrival of ultra-sound, providing allegedly harmless moving pictures of events within the uterus, the reasons for using X-rays have been reduced virtually to zero. (The technique of ultra-sound was given a further clean bill of health when two major articles on the subject were published in *The Lancet* during November 1984. The suspicion that a link existed between the use of ultra-sound and cancers among young children was considered to be unfounded. Doctors have asserted that, if there are risks, they are outweighed by the benefits of knowing more about events within the uterus – but then they also said that about the use of X-rays.)

In May 1966, almost twenty-one years after the first atomic bombs, twelve years after the large Bikini tests, and three and a half years after the test ban treaty, the Medical Research Council's Committee on Protection against Ionising Radiations summed up the situation. It concluded that 'as a result of the atmospheric tests that have taken place up to the end of 1965 it is estimated that by the year AD 2000 three additional cases of leukaemia, two of all other types of fatal cancer and two of thyroid cancer per million of the population exposed may, at worst, have been caused in the United Kingdom'.

Only France and China have conducted atmospheric nuclear tests since the treaty and the number and size of their detonations have been modest compared with the peak years of testing by the major powers during the early 1960's. Even so it is reprehensible that any atmospheric tests should be conducted by any nation at any time.

Radioactive fall-out, which first became a major issue after the American nuclear tests of 1954 (although the word was actually coined back in 1945) is still falling and will fall in significant amounts for many years to come. Far more forcibly than the early Röntgen rays, and even the holocaust over Japan, it was fall-out that alerted the world to the dangers inherent in increased levels of radioactivity. Yet even at its worse time the amount of radiation suffered, on average, by anyone as the consequence of fall-out has always been less than the amount suffered as the result of medical radiology. And the amount of radiation from such X-ray equipment has always been less on average than the inescapable amount to which we are all subjected from natural sources, from space, from the rocks of the earth. Fall-out sounded the shrill alarm, but proved to be less alarming than had originally been feared.

A major fall-out scare was concentration. It is all very well to talk in general terms about whole body doses, but radioactive materials can be concentrated in very small areas. There is 30,000 times as much radiation in some foods as in others. Brazil nuts are outstandingly high. So too are some cereals. Various plants and animals collect it in amazing amounts. Certain fish swimming in the waters near the Bikini atoll staggered the scientists by their ability to concentrate radioactivity within them. One fish collected radioactive zinc which could only have come from the casing of the exploded bomb. Many seaweeds collect radioactive iodine, a major product of nuclear fission.

Mrs Job's thyroid trouble was undoubtedly due to radioactive iodine, for iodine will always collect in the thyroid, whether radioactive or not. If radioactive, it can do harm primarily in that area. Other common radioactive substances have other propensities. Radium, soluble plutonium, strontium and calcium settle in the bone; insoluble plutonium in the lungs. The kidneys accumulate gold, uranium and other heavy metals. The bone marrow collects phosphorus, tritium, sodium and chlorine. Radioactive caesium, after strontium the most notorious product of nuclear fission, is much more general, and gets carried around the blood. Therefore it can come more into contact with the precious gonads than, say, strontium, which settles in the bone. Strontium is worrying enough, being superbly placed to induce bone cancers and leukaemia, but caesium presents the extra genetical worry of the future.

Professor Willhelm Röntgen first demonstrated the power of his discovery to himself on November 7, 1895. He told the world about it in his article 'A new kind of ray' that Christmas. The world found it difficult to comprehend, this something which could not be felt or seen. Several firms made much money by exploiting this bewilderment, and selling X-ray proof underclothes for women. To a large measure the incomprehension still exists. Neither sight, sound, touch, taste, nor smell can detect ionising radiation under ordinary circumstances. So the Japanese fishermen of the *Lucky Dragon,* liberally dusted with fall-out in the Pacific, swept up the stuff to put it under their beds and in their lockers, thereby storing up inevitable disease and sickness for themselves. The Mexican family, who died in 1966 when one member brought back a lump of cobalt, felt and saw nothing strange; but it was strongly radioactive cobalt 60, a powerful emitter of gamma rays, and it killed them. One measuring instrument could have saved their lives. Geiger counters stammer away audibly as they count the nuclear disintegrations hitting them, and give warning of levels of radioactivity. They are extraordinarily efficient, detecting in each gram of ordinary rain water some fall-out from a quite modest explosion on the other side of the globe. Human beings are quite incapable of doing this.

Like aircraft flying in cloud they are totally dependent upon their instruments.

Nevertheless certain creatures have smelt out X-rays and the like. No one knows how they do it. Rats have proved their ability to detect minute doses, such as 1/50 of a rad. Cats can detect X-rays, and it has been proved that their eye is not involved; they seem literally to smell them out. If small beams of radiation are aimed at the delicate and receptive horns of snails these horns will retract. Barnacles withdraw their tentacles in the presence of undue radiation, worms change course, fleas hop away. Even plants, like the sensitive *Mimosa pudica,* which collapses when touched or irritated in any way, will react under the invisible, inaudible effect of a modest beam of radiation. How? It is not known, but it is suspected that the cats and rats may be smelling the extra ozone in air caused by the irradiation.

In this nuclear age, when a built-in geiger counter of a sense organ would be an invaluable asset, it would be nice to know. In the meantime, like miners carrying canaries to detect firedamp, we might well carry snails. Their delicate horns could then be induced to give warning of high levels of radiation.

Finally the nuclear age has caused a nuclear language of its own. Words likely to be encountered are:

Curie	The principal unit of radioactivity. One gram of radium has the activity of one curie. It emits 37,000,000,000 disintegrations a second.
Millicurie	With one thousandth of the activity of a curie.
Microcurie	With one millionth of the activity of a curie.
Picocurie	With one thousandth of one millionth of the activity of a curie (i.e. thirty-seven disintegrations a second). Levels of strontium-90 in human bone, for example, are measured in picocuries per gram of calcium.
Rad	The unit of absorbed ionising radiation. It is defined as the energy absorption of 100 ergs per gram of tissue. Between 500 and 1,500 rads will kill most people.
Röntgen	The unit of radiological dose. Defined as the dose which corresponds to the release of 83·3 ergs of energy in a gram of air.
Rem	The unit of biological dose of ionising radiation, being an abbreviation of Röntgen Equivalent Man.

(The differences between Rads, Röntgens and Rems are real, and not just the hair-splitting of physics. Röntgens are always measured in air. Rads measure the radiation which has been absorbed by something. Rems measure the radiation absorbed by man. For practical purposes rems and rads are the same for all of us who receive our radiation predominantly from X-rays, fall-out and natural background activity. Rads are likely to be less than rems, perhaps only one-tenth as much, for those people involved in reactor accidents, who survive atomic bomb attacks and who either eat or breathe the very heavy radioactive elements, such as radium and plutonium. To use another language, rads are similar to rems for beta particles, gamma rays and X-rays. They are much less than rems for alpha particles and neutrons, particularly the energetic kind.)

Radiation An all-embracing word taking in all ways in which energy can be given off by an atom. It thus includes X-rays, gamma rays, all charged particles and all neutrons.
(Radiation is also used with other meanings, such as radiators and radiating heat, that are distinct from the word's use in atomic energy.)

Radioactivity Most atoms are stable, and non-radioactive. Those which are unstable and radioactive will give off either particles or gamma radiation. Although a piece of radioactive material is continually active and emitting radiation, each atom in it is either about to emit energy or has already done so.

Half-life All radioactive substances form stable substances in time. Their half-life occurs when they have lost half their activity. Half-lives can vary from less than a millionth of a second to millions of years. The substances in fall-out whose half-lives pose the greatest harm biologically are neither the very short nor the very long. The very short have lost all their activity before falling out. The very long take so long to lose their radioactivity that their gentle smouldering is virtually inactive. Iodine 131 has a half-life of eight days, cobalt 60 of 5·2 years, strontium-90 of twenty-eight years, caesium-137 of thirty years, and carbon-14 of 5,600 years.

Decay
An atom decays when it disintegrates, when it changes from instability to stability. Half of the atoms have decayed in a half-life.

Isotope
Two atoms are isotopes if they are of the same element but have different masses. Less accurately, but more frequently, an isotope means a radioactive element.

Disintegration
This occurs when an atom's nucleus emits a particle.

Fission
The splitting of an atom's nucleus into fragments. These fragments are called fission products. Fission is always accompanied by the release of energy.

Gamma rays
Electromagnetic radiation similar to X-rays.

Ionising radiation
Radiation which knocks electrons from atoms, thereby leaving ions behind. Conversely ions are charged atoms which have lost or gained an electron.

Nucleus
The heart of the atom. About a millionth of a millionth of a centimetre wide.

38. Postscript

What do you think of Western civilisation?
'I think it would be a good idea.'

MAHATMA GANDHI

'If we could first know where we are and whither we are tending, we could better judge what to do and how to do it.' Abraham Lincoln's remark, already used in this book, should be used time and time again. But where on earth are we?

In the past we led our hazardous lives with a mixture of fear, common sense and superstition. We watched our children die, and we ourselves died – probably – before our time. In the recent past the hazards have diminished (save for those of our own making), much of the superstition has faded away, and we can expect to live to a decent age; the triumphs of hygiene, sanitation, medicine and science have caused such expectation wherever a country's economy enables these triumphs to be applied. In a sense they have merely put the balance right. Man's body has a life of 70 years: hygiene and the rest are permitting it to achieve those 70 years. And that is where we are.

But whither are we tending? With every chapter of this book there will be quite a different tale to tell in the future. I do not mean solely that fewer babies will be dying, that ulcers may be a thing of the past or that athletes will run a 3 mins. 30 secs. mile: I believe the differences will be revolutionary, Also, whereas the ethics of today impel society to prolong life wherever feasible, the ethics of tomorrow will be vastly more complex.

Take abortion. Today we are destroying two-, three- and four-month-old foetuses as a form of contraception. Today we can keep some babies alive who are born after only six months in the uterus. Skills in the prematurity wards will grow more skilful, and even younger babies will survive. An abortion can then no longer be called contraception but will be the killing of a baby that could be kept alive, like denying a transfusion to a bleeding man. (There is current belief that the legal limit to abortion should be reduced from 28 to 24 weeks for this very reason. Babies born at 26 weeks without handicaps now stand a chance of survival greater than 50%.) Perhaps transfusions should be denied more to some bleeding men. Perhaps euthanasia should have wider applications. Perhaps the plug should be pulled out more readily, as is discussed in the chapter on death. Perhaps grossly malformed babies should be deliberately killed at birth, rather than permitted to die with assistance being withheld. And should steps be taken to prevent their production? Should certain marriages and matings be forbidden?

Conversely, should positive eugenics, which aims to create better stock, be actively encouraged? Should man breed men as he breeds his animals? Should woman, who has her imperfections as a brood-chamber for foetuses because she smokes, takes drugs, falls and succumbs to illness, be supplanted by the test-tube? Should children be entrusted to parents who, according to George Bernard Shaw, are the last people who should have children? Should growth be adjusted? Should mental development? Should senility, or the age of death? All such questions could well be the chapter headings of a book such as this in a few years' time.

On a different level, and bearing in mind that food production may frequently be inadequate, should man permit himself the inefficient luxury of nourishment like beef? And what will he do in a world of leisure? In a cosseted community will his virtues decay? La Bruyère wrote that 'Liberty is not idleness; it is the free use of time, the choosing of work and exercise. To be free, in short, is not to do nothing: it is to be the sole arbiter of what one does or does not do.' Just as retirement today seems frequently to hasten death, will the automated life of the future be disastrous when man is free. Will we merely prolong senility, and remember Somerset Maugham's address on his eightieth birthday: 'There are many values in growing old. ... For the moment, I cannot think of one of them.'

Drugs have already become today's problem. If a newspaper headlines 'Youth Takes Drugs' as against 'Drugs Sent To Marooned Youth', our ambivalence at least is clear. 'The desire to take medicine is perhaps the greatest feature which distinguishes man from animals' wrote Sir William Osler. Drugs have enabled us to survive so well; they are wonder drugs, miracle drugs. Will mankind's propensity for self-medication continue to be mainly beneficial?

Spare-part surgery did not come within the compass of this book, but it will surely grow more and more crucial. Not only will artificial replacement of valves and joints become more widespread, but one accident victim can provide the organs for a number of ailing patients. Will people be allowed to bury their dead? Will many of the current dignities of life be assaulted when they interfere with the new ethical wave?

'Life', said Blaise Pascal 300 years ago, 'is a maze in which we take the wrong turning before we have learnt to walk.' Few would disagree, and still fewer that the maze ahead of us is infinitely more complex than anything encountered thus far.

Index

529

Index